ELECTRICAL CIRCUIT PRINCIPLES

W. Bolton

Longman Scientific & Technical
Longman Group UK Limited
Longman House, Burnt Mill, Harlow,
Essex CM20 2JE, England
and Associated Companies throughout the world.

First published 1992

ISBN 0 582 08802 X

British Library Cataloguing in Publication Data
A catalogue record for this book is available from the British Library

Set by 8 in 10/12 pt Linotron Times

Printed in Malaysia by PA

Contents

Preface

The overall aim of this book is to introduce and develop the principles of electrical circuit analysis such that:

1 circuit theory can be applied to the solution of d.c. circuits involving series and parallel combinations of resistors;
2 the transient behaviour of simple *RL*, *RC* and *RLC* circuits can be explained;
3 phasors can be used to explain the behaviour of, and solve problems involving, single phase a.c. circuits;
4 complex notation can be applied to the solution of single-phase a.c. circuits involving series and parallel combinations of resistors, inductors and capacitors;
5 circuit theory can be applied to magnetically coupled circuits;
6 circuit theory can be applied to tuned circuits;
7 circuit theory can be developed and applied to two-port networks;
8 the behaviour of signals transmitted along transmission lines can be determined;
9 the basic theory of three-phase circuits can be developed and applied to both balanced and unbalanced circuits;
10 the Laplace transform can be used in circuit analysis;
11 circuit problems involving complex waves can be solved.

While it is assumed that the reader has some basic knowledge of electrical circuit principles, the opportunity has been taken in Chapters 1 and 2, and where appropriate in later chapters, to give a quick consideration of the basic principles. A basic knowledge of algebra, the handling of angles and calculus has been assumed. While it is recognized that circuit analysis of any complexity is generally carried out using a computer, it was felt that within a book devoted to the establishment of principles the exercise of working out things 'by hand' was more appropriate and would then enable the reader to use computer programs intelligently and be more able to interpret the result effectively. The book includes 169 worked examples and 264 problems, all with answers supplied.

The book aims to give a comprehensive coverage of all electrical circuit principles. It is seen as being particularly relevant to students taking Business and Technician Education Council Higher National Certificate and Diploma courses in Electrical and Electronic Engineering and students in the early years of degree courses in Electrical and Electronic Engineering. The book covers all the electric circuit principles of the Business and Technician Education Council Bank of Objectives for Electrical and Electronic principles (13683B) for Higher National Courses, revising all those circuit principles covered in the Bank of Objectives for Electrical and Electronic principles (U86/329) for National Courses. The following is an identification of chapters in the book with the sections of those banks.

BTEC Higher National Bank	Chapter
Revision of National Bank	Mainly 1, 2
A	3
B	4, 5
C	6
D	7, 8
E	12
F	11
G	9, 10
H	10

Sections K, L, M, N and O are covered in the companion book *Electrical and Magnetic Properties of Material*, sections J and P in *Control Engineering*.

W. Bolton

1 Circuit components

Introduction

This chapter is a brief overview of the characteristics of some basic circuit components, namely resistors, inductors and capacitors. It is assumed that this material will not be completely unfamiliar to the reader so the treatment is fairly brief.

The convention is adopted in this book of using capital letters for quantities which are not varying with time, e.g. I, and lower case letters for those which are the values at some instant of time of a varying quantity, e.g. i.

Current, voltage and power

Current is the rate of movement of charge. Thus

$$i = \frac{dq}{dt} \qquad [1]$$

where dq/dt is the rate of movement of charge q. The unit of current is the ampere or amp (A), the unit of charge the coulomb (C). By convention, the *direction of a current* is taken to be in the direction of the flow of positive charge, i.e. in the opposite direction to the flow of electrons. If the current direction through a circuit element is constant then it is said to be a *direct current* (d.c.). If the current direction varies with time, e.g. in a sinusoidal manner, then it is said to be *alternating current* (a.c.).

In order to make charge move through a circuit element, energy must be supplied. The energy required to move a positive charge of one coulomb through a circuit element is called the *potential difference* or *voltage* across the element. Thus the potential difference v between two points at some instant of time when energy w is required to move the charge q between the points is

$$v = \frac{w}{q} \qquad [2]$$

A potential difference of one volt (V) is said to exist between two points in an electrical circuit if one joule (J) of energy is required to move one coulomb (C) of charge between the points.

For a source of energy, e.g. a battery, charge is made to move internally between the terminals of the source so that it builds up positive charge at the positive terminal; the positive terminal is thus at a higher potential than the negative terminal. The arrow used on circuit diagrams indicates the direction in which the potential increases and is thus from the terminal labelled negative to the one labelled positive (Fig. 1.1(*a*)). When an external circuit is connected across the source terminals positive charge flows from the positive terminal to the negative terminal, hence the current direction is from the positive terminal to the negative terminal. If an external source is being used to supply the energy to move the charge through a circuit element then the terminal of the element where the current enters is defined as being positive and where it leaves as negative (Fig. 1.1(*b*)). The positive terminal indicates that the voltage at that terminal is higher than that at the negative terminal. The arrow used on circuit diagrams to show for a circuit element the direction of the potential difference points from the low-potential terminal to the high-potential terminal.

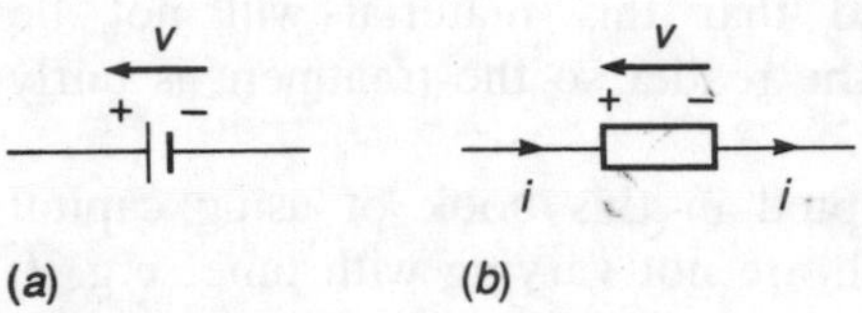

Fig. 1.1 Direction of potential difference

When charge is continuously moved through a circuit, i.e. there is a current in the circuit, then energy has to be continually expended to move the charge through the circuit. The rate at which energy is expended is called the *power*. This can be expressed as

$$p = \frac{\mathrm{d}w}{\mathrm{d}t} \qquad [3]$$

with $\mathrm{d}w/\mathrm{d}t$ being the rate of expenditure of energy w. The power p is in watts (W) when the rate is in joules/second.

When there is a current i between two points in a circuit and the potential difference between the points is v then if a charge δq is moved between the points in a time δt, equation [1] gives $i = \delta q/\delta t$ and equation [2] gives $v = \delta w/\delta q$, where δw is the energy expended in moving the charge δq between the points. Thus

$$iv = \frac{\delta q \delta w}{\delta t \delta q} = \frac{\delta w}{\delta t}$$

Thus the rate at which energy is expended $\delta w/\delta t$ is iv. Hence, using equation [3], the power p is

$$p = iv \qquad [4]$$

Kirchoff's laws

If charge is not to accumulate at a circuit junction the current entering a circuit junction must equal the current leaving it. This is known as *Kirchoff's current law*. The law can be formally stated as: at any node in an electrical circuit the algebraic sum of currents is zero. The term *node* is used for a point at which two or more conductors are joined to form a junction. The term 'algebraic sum' means that we have to take account of the directions of the currents. A simple convention which can be used is to take the current entering the node as being positive and that leaving as being negative. The actual sign we attach to the currents is, however, quite arbitrary. Reversing the signs leads to exactly the same sum. Thus for the node at point a in Fig. 1.2

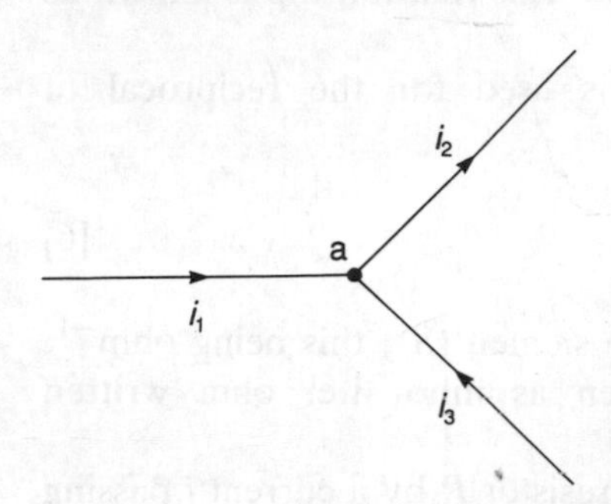

Fig. 1.2 Currents at a node

$$i_1 - i_2 + i_3 = 0$$

Thus for components in series, since the current entering one component must equal the current leaving it, the same current flows through each component. For components in parallel, the current entering the parallel arrangement must be equal to the sum of the currents through each component.

If we have components connected in series then the energy required to drive a current through them is the sum of the energies required to drive it through each component. Thus the potential difference across the arrangement is the sum of the potential differences across each component. *Kirchoff's voltage law* states that around any closed path in a circuit the algebraic sum of the voltages across all the components in the closed path is zero. A *closed path* can be defined as a traversal through a series of nodes which ends up at the starting node without encountering a node more than once. A closed path is often called a *loop*. Figure 1.3 shows a closed path or loop. For that closed path, if we traverse the path in a clockwise direction from node a to b to c to d and so back to a, taking voltages in our direction of traverse as positive and those in the opposite direction as negative, then

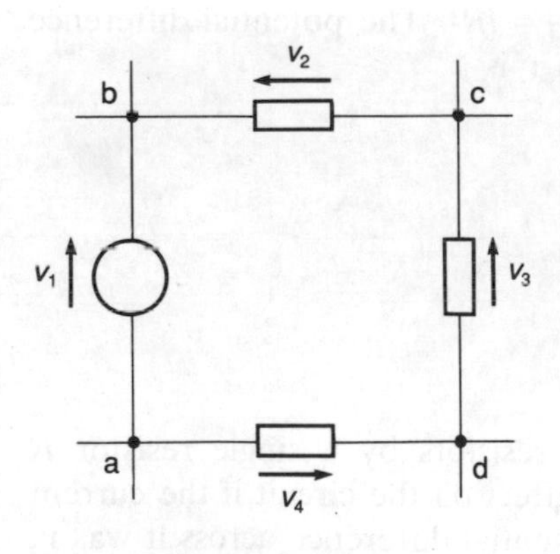

Fig. 1.3 A closed path

$$v_1 - v_2 - v_3 - v_4 = 0$$

For this closed path the movement round the path was in a clockwise direction; the direction in which we traverse the path is, however, quite arbitrary. Reversing the path and the voltage signs leads to exactly the same equation.

Resistors

The *resistance R* of a circuit element is the physical property which describes the extent to which the element impedes the flow of current, and is defined as being

$$R = \frac{v}{i} \qquad [5]$$

where v is the potential difference across the conductor and i the current through it. The unit of resistance is the ohm (Ω) when the current is in amps and the potential difference in volts.

For many materials, if the temperature does not change, the current is directly proportional to the potential difference and thus the resistance is a constant. This relationship is known as *Ohm's law*.

The term *conductance* G is used for the reciprocal of resistance, i.e.

$$G = \frac{1}{R} = \frac{i}{v} \qquad [6]$$

The unit of conductance is the siemen (S), this being ohm^{-1}. Sometimes the unit is written as mho, i.e. ohm written backwards.

The power p dissipated in a resistor R by a current i passing through it is given by equation [4] as $p = iv$, where v is the potential difference across the resistor. This equation can be expressed in a number of alternative ways by substitutions using $R = v/i$.

$$p = iv = i^2R = \frac{v^2}{R} \qquad [7]$$

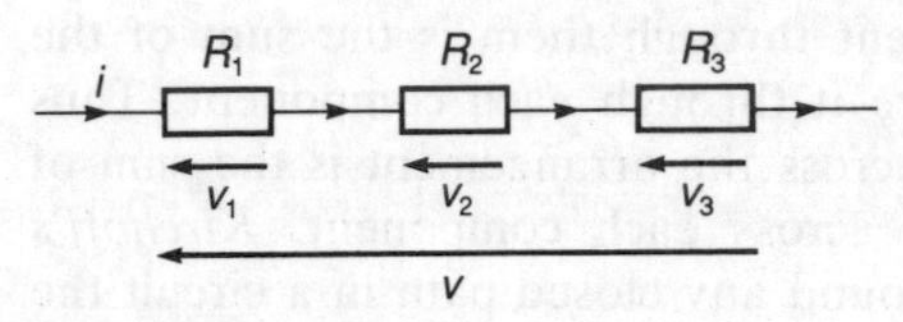

Fig. 1.4 Resistors in series

For resistors in series (Fig. 1.4) the current i will be the same through each of them. Thus the potential difference across the resistor R_1 is given by $v_1 = iR_1$, for resistor R_2 as $v_2 = iR_2$ and resistor R_3 as $v_3 = iR_3$. The potential difference across the series combination v is

$$v = v_1 + v_2 + v_3$$

Hence

$$v = iR_1 + iR_2 + iR_3$$

$$\frac{v}{i} = R_1 + R_2 + R_3$$

We could replace the three resistors by a single resistor R which would have the same effect on the circuit if the current through it was i when the potential difference across it was v, i.e. $R = v/i$. Thus

$$R = R_1 + R_2 + R_3 \qquad [8]$$

A series resistor circuit can be considered to be a *voltage divider* circuit since the potential difference across any one resistor is a fraction of the total potential difference applied across the series arrangement, the fraction being determined

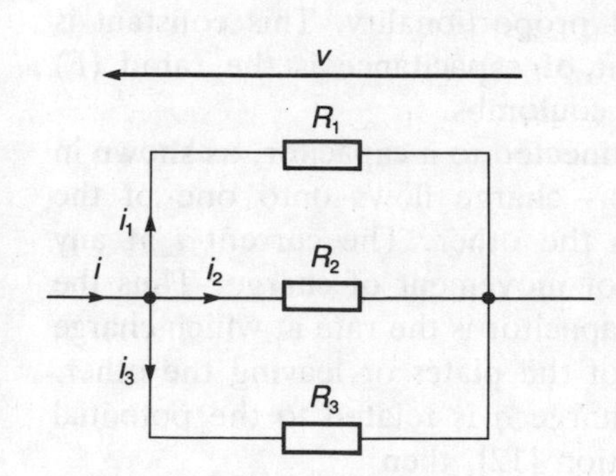

Fig. 1.5 Resistors in parallel

by the values of the resistances. Thus the potential difference across R_1 is $v_1 = iR_1$ and since the potential difference across the three series resistors is $v = i(R_1 + R_2 + R_3)$ then

$$\frac{v_1}{v} = \frac{R_1}{R_1 + R_2 + R_3} \qquad [9]$$

For resistors in parallel (Fig. 1.5) the potential difference v across each resistor is the same. The current i entering the parallel arrangement is equal to the sum of the currents through each resistor, i.e.

$$i = i_1 + i_2 + i_3$$

Hence

$$i = \frac{v}{R_1} + \frac{v}{R_2} + \frac{v}{R_3}$$

The three resistors can be replaced by a single resistance R if the current through it is i when the potential difference across it is v, i.e. $i = v/R$. Thus

$$\frac{1}{R} = \frac{1}{R_1} + \frac{1}{R_2} + \frac{1}{R_3} \qquad [10]$$

A parallel resistor circuit can be considered to be a *current divider* circuit in that the current through one resistor is a fraction of the total current, the fraction depending on the values of the resistors. Thus for R_1 we have $i_1 = v/R_1$ and for the total current $i = v/R$, where R is the total resistance of the parallel arrangement and is given by equation [10]. Thus

$$\frac{i_1}{i} = \frac{(v/R_1)}{v(1/R_1 + 1/R_2 + 1/R_3)} = \frac{R_2\,R_3}{R_1 + R_2 + R_3} \qquad [11]$$

Capacitors

A capacitor is essentially just a pair of parallel conducting plates separated by an insulator (Fig. 1.6). When the plates are connected to a d.c. supply one of the plates becomes positively charged and the other negatively charged. This is because the applied voltage causes current to flow in the circuit such that electrons move away from one of the plates and into the circuit while at the other plate electrons move from the circuit onto the plate. The amount of charge q on a plate at any instant depends on the potential difference v between the plates, q being directly proportional to v. Hence, for the charge and potential difference values at some instant of time

$$q = Cv \qquad [12]$$

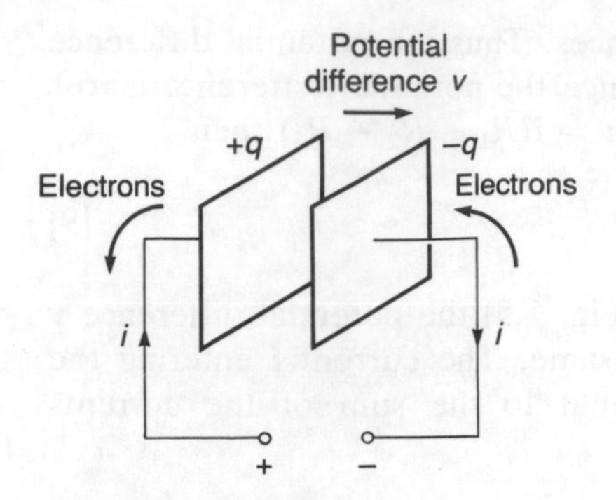

Fig. 1.6 Capacitor

where C is the constant of proportionality. This constant is called *capacitance*. The unit of capacitance is the farad (F) when v is in volts and q in coulombs.

When a battery is first connected to a capacitor, as shown in Fig. 1.6, a current flows as charge flows onto one of the capacitor plates and leaves the other. The current i at any instant of time is the rate of movement of charge. Thus the current in the leads to the capacitor is the rate at which charge is either flowing onto one of the plates or leaving the other. Since for a capacitor the charge q is related to the potential difference v across it, equation [12], then

$$i = \frac{dq}{dt} = \frac{d(Cv)}{dt}$$

Since the capacitance does not vary with time then

$$i = C\frac{dv}{dt} \quad [13]$$

Equation [13] can be rearranged to give

$$dv = \frac{1}{C}i\,dt$$

Hence

$$\int_0^v dv = \frac{1}{C}\int_0^t i\,dt$$

if the potential difference is 0 at $t = 0$ and v at time t. Thus

$$v = \frac{1}{C}\int_0^t i\,dt \quad [14]$$

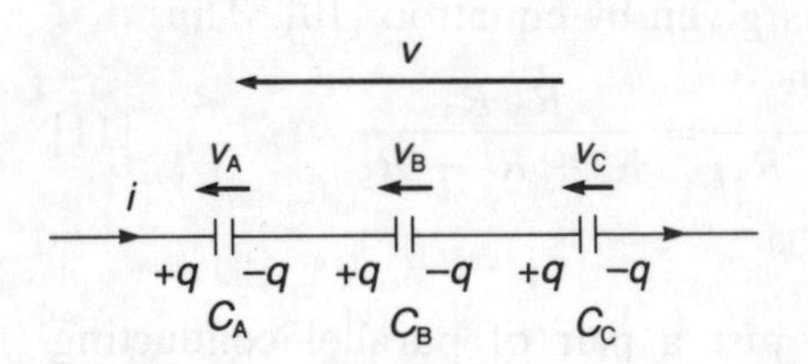

Fig. 1.7 Capacitors in series

Consider three capacitors in series and connected to a voltage supply so that there is a potential difference across the series combination (Fig. 1.7). The potential difference v across the combination is the sum of the potential differences across each capacitor.

$$v = v_A + v_B + v_C$$

The currents in the circuit result in the end plate of capacitor A losing electrons and so becoming positively charged, and the end plate of capacitor C gaining electrons and becoming negatively charged. These charged plates cause currents to flow between the inner connected plates so that the charge on the end plate A of $+q$ causes electrons to move onto its facing plate and so a charge of $-q$ occurs on the opposing plate and a charge of $+q$ on the plate of the second capacitor which is connected to it and which has supplied the electrons which have moved. The result of such movements of electrons is that each capacitor will have plates with charges of $+q$ and $-q$. Dividing the above equation throughout by q gives

$$\frac{v}{q} = \frac{v_A}{q} + \frac{v_B}{q} + \frac{v_C}{q}$$

But the capacitance of capacitor A, C_A, is q/v_A, that of capacitor B, C_B, is q/v_B, and capacitor C, C_C, is q/v_C. Hence, since the overall capacitance C of the system is q/v,

$$\frac{1}{C} = \frac{1}{C_A} + \frac{1}{C_B} + \frac{1}{C_C} \qquad [15]$$

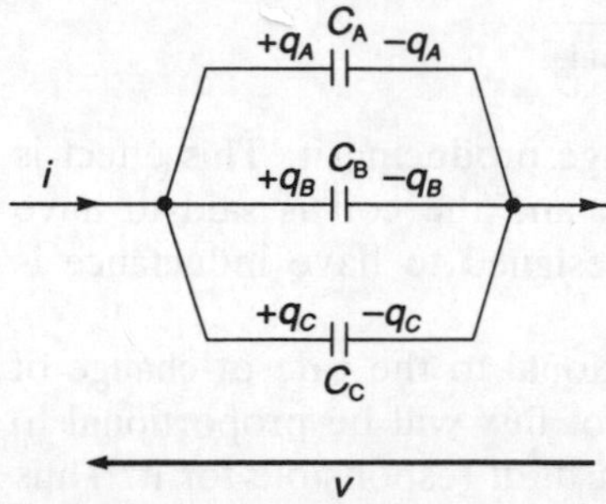

Fig. 1.8 Capacitors in parallel

For three capacitors in parallel (Fig. 1.8) the potential difference v across each capacitor will be the same. The charges on each capacitor will depend on their capacitances. The total charge q is the sum of the charges on the plates connected together, i.e.

$$q = q_A + q_B + q_C$$

Dividing each term by v gives

$$\frac{q}{v} = \frac{q_A}{v} + \frac{q_B}{v} + \frac{q_C}{v}$$

But for capacitor A $C_A = q_A/v$, for capacitor B $C_B = q_B/v$ and for capacitor C $C_C = q_C/v$. Since the capacitance C of the system can be written as $C = q/v$, then

$$C = C_A + C_B + C_C \qquad [16]$$

Consider a capacitor being charged. When the potential difference between the plates is v then adding a small increment of charge δq means that work of $v\delta q$ has to be done (equation [2]). Thus the total work done in charging a capacitor to have charge Q is

$$\text{energy} = \int_0^Q v\,dq = \int_0^Q \frac{q}{C}\,dq = \frac{Q^2}{2C}$$

Since $Q = CV$ this can also be written as

$$\text{energy} = \tfrac{1}{2}Q^2/C = \tfrac{1}{2}QV = \tfrac{1}{2}CV^2 \qquad [17]$$

This energy is stored in the electric field that is produced between the plates of the capacitor and can be released when the capacitor is discharged.

Inductors

When the current in a coil changes, the magnetic flux originating from that current must change (Fig. 1.9). This changes the flux linked by the coil and hence an e.m.f. is

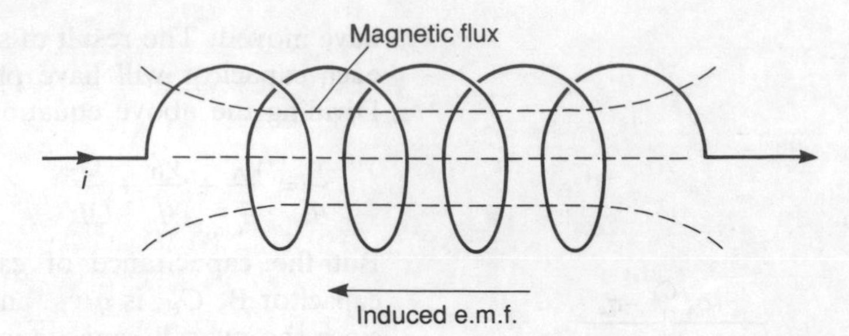

Fig. 1.9 Inductance

induced which opposes the change producing it. This effect is called electromagnetic induction and the coil is said to have *inductance*. A circuit element designed to have inductance is called an inductor.

The induced e.m.f. is proportional to the rate of change of linked flux. The rate of change of flux will be proportional to the rate of change of the current $\mathrm{d}i/\mathrm{d}t$ responsible for it. Thus

$$\text{e.m.f. is proportional to } -\frac{\mathrm{d}i}{\mathrm{d}t}$$

The constant of proportionality L is called the *inductance* of the circuit. Thus

$$\text{e.m.f.} = -L\frac{\mathrm{d}i}{\mathrm{d}t} \tag{18}$$

If we are concerned with a pure inductance, i.e. a circuit element which has only inductance and no resistance, then there will be no potential difference across the inductor due to resistance or capacitance. To maintain the current though the inductor the source must just supply a potential difference v to cancel out the induced e.m.f. Thus the potential difference v is

$$v = L\frac{\mathrm{d}i}{\mathrm{d}t} \tag{19}$$

The inductance of a circuit is said to be 1 henry (H) when the e.m.f. induced is 1 V as a result of the current changing by 1 A/s.

Equation [19] can be rearranged to give the current i after some time t. If the current is 0 at $t = 0$,

$$\int_0^i \mathrm{d}i = \frac{1}{L}\int_0^t v\,\mathrm{d}t$$

$$i = \frac{1}{L}\int_0^t v\,\mathrm{d}t \tag{20}$$

Figure 1.10 shows three inductors in series. Provided the magnetic fluxes from each inductor do not interact then

$$v = v_1 + v_2 + v_3$$

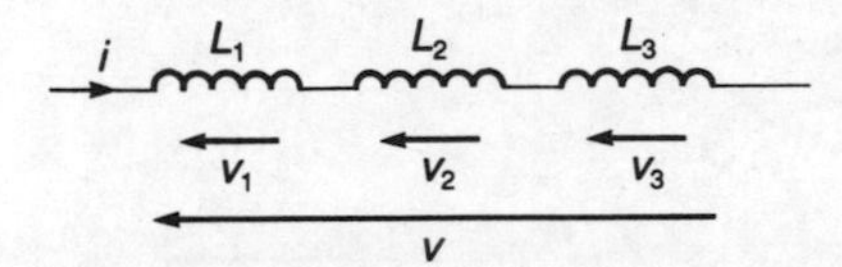

Fig. 1.10 Inductors in series

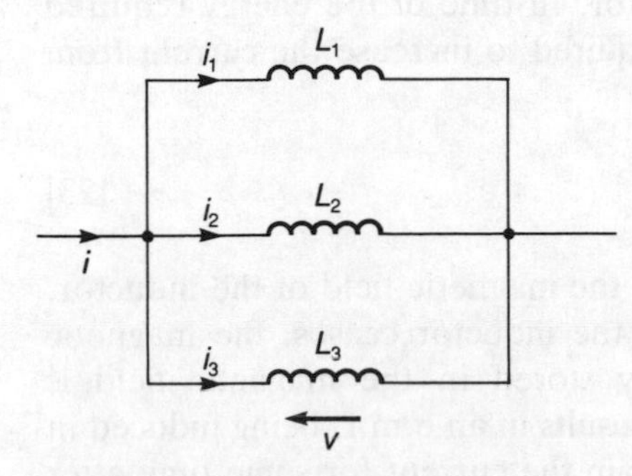

Fig. 1.11 Inductors in parallel

But, using equation [19], since there will be the same current through each inductor and the same rate of change of current with time,

$$v = L_1\frac{di}{dt} + L_2\frac{di}{dt} + L_3\frac{di}{dt}$$

$$v = (L_1 + L_2 + L_3)\frac{di}{dt}$$

The three inductors could be replaced by a single inductor of inductance L where $v = L(di/dt)$. Thus

$$L = L_1 + L_2 + L_3 \qquad [21]$$

Figure 1.11 shows three inductors in parallel. Provided the magnetic fluxes from each inductor do not interact we have

$$i = i_1 + i_2 + i_3$$

and thus, using equation [20],

$$i = \frac{1}{L_1}\int v\,dt + \frac{1}{L_2}\int v\,dt + \frac{1}{L_3}\int v\,dt$$

Because the inductors are in parallel there is the same potential difference across each of them and hence the integral of $v\,dt$ is the same for each. Thus

$$i = \left(\frac{1}{L_1} + \frac{1}{L_2} + \frac{1}{L_3}\right)\int v\,dt$$

The three can be replaced by a single inductor of inductance L where

$$i = \frac{1}{L}\int v\,dt$$

Thus

$$\frac{1}{L} = \frac{1}{L_1} + \frac{1}{L_2} + \frac{1}{L_3} \qquad [22]$$

When the current through an inductor changes, an e.m.f. is induced which opposes the change producing it. If at some instant of time the current is changing at the rate di/dt, then the e.m.f. induced is $-L(di/dt)$. The value of the supply voltage v at this instant of time needed to overcome this and maintain the current through the inductor is thus

$$v = L\frac{di}{dt}$$

Energy is thus required to maintain the current and the power required is the product of the voltage that has to be applied

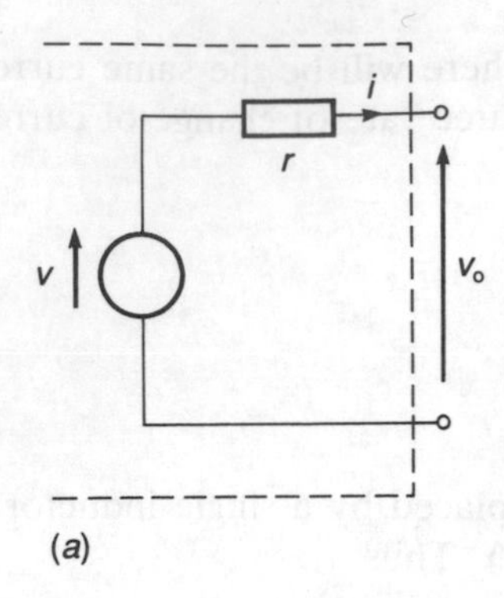

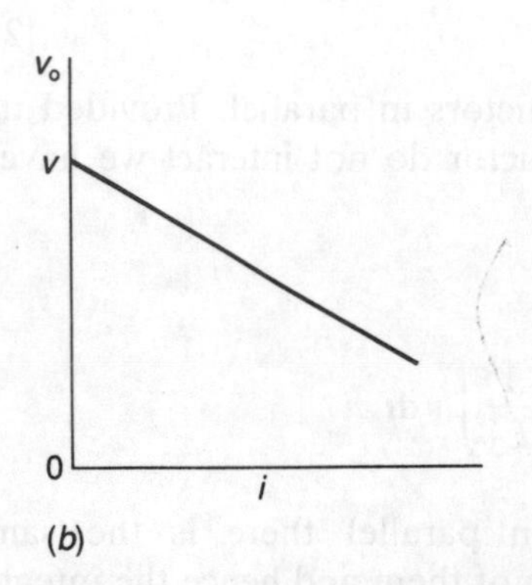

Fig. 1.12 Non-ideal voltage source, (*a*) as an ideal voltage source in series with internal resistance, (*b*) the output characteristic

and the current. Hence

$$\text{power } p = iv = iL\frac{\mathrm{d}i}{\mathrm{d}t}$$

When the current is switched through an inductor it grows to reach a final steady value. At the steady value $\mathrm{d}i/\mathrm{d}t$ is zero and the power input becomes zero. The energy that has been supplied to the inductor has been used to establish the magnetic field of the inductor. In time $\mathrm{d}t$ the energy required is $p\,\mathrm{d}t$, hence the energy required to increase the current from zero to I is

$$\text{energy} = \int_0^I Li\,\mathrm{d}i = \tfrac{1}{2}LI^2 \qquad [23]$$

This is the energy stored in the magnetic field of the inductor. When the current through the inductor ceases, the magnetic field collapses. The energy stored in the magnetic field is returned to the circuit and results in an e.m.f. being induced in it which continues to maintain the current for some time after the current supply to the inductor has been cut off.

Voltage and current sources

The term *voltage source* is used to describe a source of energy which establishes a potential difference across its terminals. If that potential difference is completely unaffected by the current taken from it by a circuit, then the source is said to be *ideal*. The term *current source* is used to describe a source of energy which provides a current, the term *ideal* being used when it maintains a current which is completely unaffected by the circuit into which the current is flowing.

A non-ideal voltage source may be considered to be an ideal voltage source connected in series with a resistance, this resistance being referred to as the *internal resistance* of the non-ideal source (Fig. 1.12(*a*)). The output potential difference v_o between the terminals of the non-ideal source is then given by

$$v_o = v - ir \qquad [24]$$

where v is the potential difference of the ideal source, r the internal resistance and i the current being taken by the external load at that time. The potential difference v is the

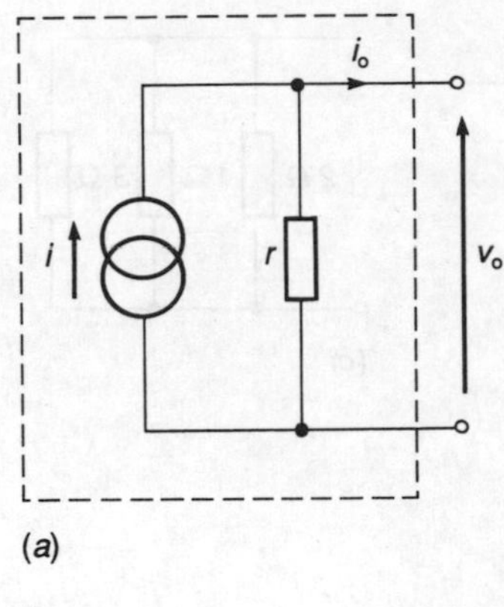

(a)

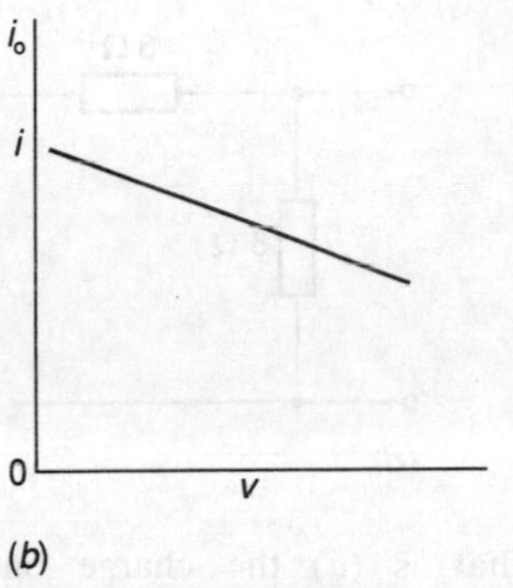

(b)

Fig. 1.13 Non-ideal current source, (a) as an ideal current source in parallel with an internal resistance, (b) the output characteristic

potential difference between the terminals of the source when no current is drawn. It is sometimes referred to as the electromotive force (e.m.f.). Figure 1.12(*b*) shows the graph described by the above equation of output potential difference v_o against load current i.

A non-ideal current source may be considered to be an ideal current source connected in parallel with a resistance (Fig. 1.13(*a*)). The output current i_o is then equal to the current from the ideal source i minus that part of the current that passes through the parallel resistance, i.e.

$$i_o = i - (v/r) \qquad [25]$$

where v is the potential difference between the output terminals and r the internal resistance. Figure 1.13(*b*) shows the graph described by the above equation of output current i_o against load potential difference.

Sources for which the voltage or current between the output terminals does not depend on any other factor are said to be *independent* sources. Thus an independent voltage source provides a voltage independent of any other voltage or current. A *dependent* source provides a voltage or current between its output terminals which depends on another variable, such as voltage or current. Thus, for example, a voltage amplifier can be considered to be a dependent voltage source since it can provide an output voltage which is dependent on another voltage, that of the input to the amplifier.

Problems

1 A 12 V car battery is used to supply a constant current of 1 A to a car's headlights, when the engine is not running. What is (*a*) the power delivered to the headlights, (*b*) the energy expended over a period of 2 minutes?

2 A car battery is charged with a constant current of 2 A for 4 hours. If the potential difference v between the terminals of the battery varies with time according to

$$v = 10 + 0.5t$$

where the time t is in hours, what will be the total amount of energy delivered to the battery during the charging time?

3 What is the current through a circuit element which dissipates energy at the rate of 20 W when the potential difference across it is 5 V?

4 Determine the resistance between the terminals of each of the circuits shown in Fig. 1.14.

5 What is the capacitance when capacitors of 4 μF and 8 μF are connected (*a*) in series, (*b*) in parallel?

6 Two capacitors of capacitances 1 μF and 2 μF are connected in series and charged by a potential difference of 12 V being applied

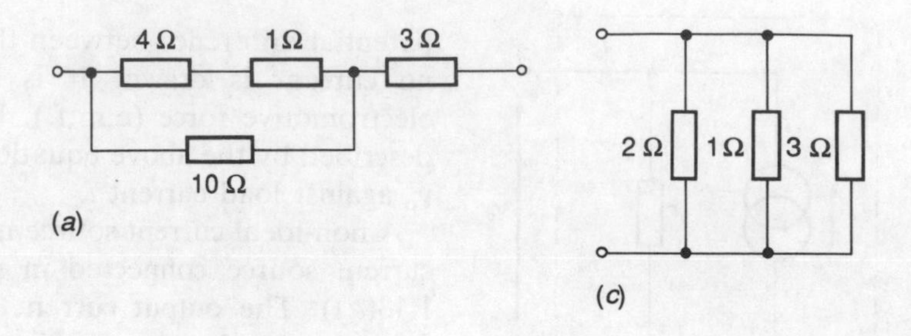

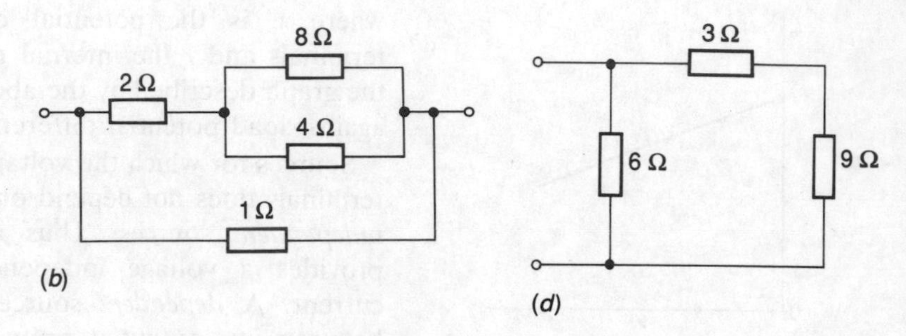

Fig. 1.14 Problem 4

across the combination. What is (*a*) the charge on each capacitor, (*b*) the potential difference across each capacitor and (*c*) the energy stored by each capacitor?

7 What is the current through an inductor of inductance 0.5 H when the potential difference v across it varies with time according to $v = 2t^2$ if $i = 0$ at $t = 0$?

8 What is the potential difference across an inductor of inductance 10 mH if the current is changing at the rate of 5 A/s?

9 What is the total inductance when inductors of 2 mH and 10 mH are connected in (*a*) series, (*b*) parallel?

10 What is the energy stored in an inductor of inductance 0.1 H when the current through it is 0.5 A?

11 A voltage source has an e.m.f. of 10 V and an internal resistance of 0.5 Ω. What will be the potential difference between the terminals of the source when the circuit connected to them draws a current of 2 A?

12 A current source gives an output of 2.0 A when the potential difference between its terminals is 8 V and 1.8 A when it is 10 V. What is the internal resistance of the source?

2 Transients

Introduction

This chapter is a brief consideration of the transient currents and voltages that occur when circuits involving resistors, capacitors and inductors have a voltage suddenly applied, a so-called step input. It leads to differential equations and a technique for solving them which considers the response to an input as the sum of a forced response and a natural response. This topic is considered in more depth in Chapter 11.

Charging a capacitor

If an initially uncharged capacitor C is connected in series with a pure resistor R, as in the circuit shown in Fig. 2.1, then when the switch is closed the supply voltage V is connected across the series arrangement of C and R. Such a sudden jump in voltage to the circuit is called a *step input* (Fig. 2.2). At any instant of time the sum of the potential differences across C and R will be V, i.e.

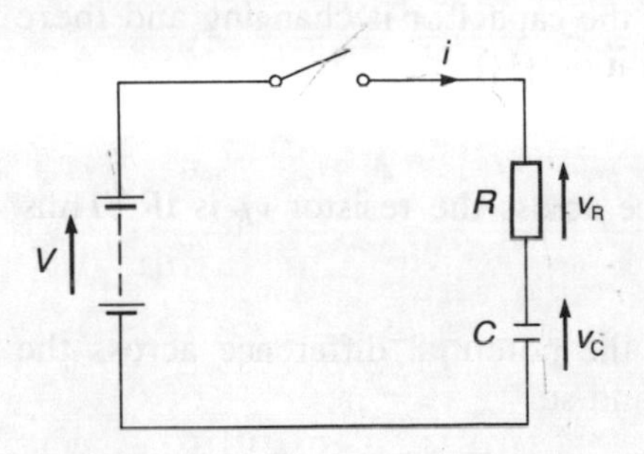

Fig. 2.1 Capacitor and resistor in series

$$V = v_R + v_C \qquad [1]$$

where v_C is the potential difference across the capacitor at that instant and v_R is the potential difference across the resistor at the same instant of time. But $v_R = iR$ and $v_C = q/C$, where i is the current in the circuit and q the charge on the capacitor at the time concerned. Thus

$$V = iR + \frac{q}{C} \qquad [2]$$

To start with the capacitor is uncharged; thus at the instant the switch is closed $q = 0$ and so there is no potential difference across the capacitor, i.e. $v_C = 0$, and all the supply voltage is across the resistor. Hence, initially, equation [2] is

$$V = iR + 0$$

The initial current is thus V/R and so initially the capacitor acts as a short circuit.

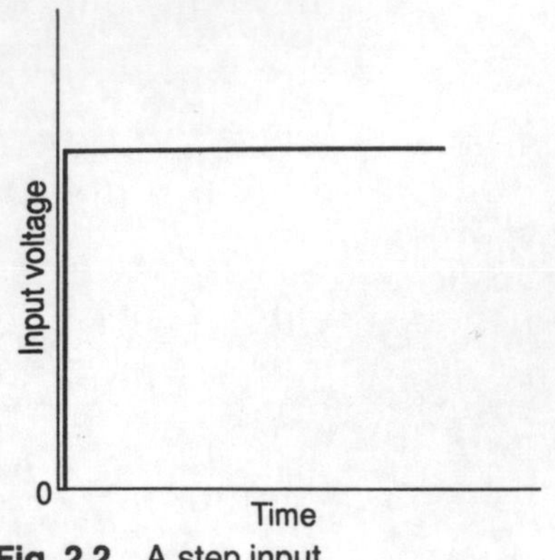

Fig. 2.2 A step input

The current shows that charge is being moved onto one of the capacitor plates and leaves the other. As time elapses so charge builds up on the capacitor plates. The potential difference across the capacitor, v_C, thus increases. Hence the potential difference across the resistor, v_R, decreases, since the sum of v_R and v_C remains constant. This can only happen if the current i decreases. Thus as the capacitor charges up so the current decreases. We thus have, for equation [2]

$$V = \underbrace{iR}_{\text{decreasing}} + \underbrace{\frac{q}{C}}_{\text{increasing}}$$

The potential difference across the capacitor continues to increase until it becomes equal to V. When this happens the potential difference across the resistor has become zero and so the current in the circuit is zero. Thus when the potential difference across the terminals of the capacitor is not changing it acts as an open-circuit. Then we have, for equation [2],

$$V = 0 + \frac{q}{C}$$

So $V = q/C$ and, since there is no further change, the capacitor is said to be fully charged.

For a step voltage input of size V, at some time when the potential difference across the capacitor is changing and there is a current, we have (equation [1])

$$V = v_R + v_C$$

But the potential difference across the resistor v_R is iR. Thus

$$V = iR + v_C$$

The current is related to the potential difference across the capacitor by $i = C\mathrm{d}v_C/\mathrm{d}t$ and so

$$V = RC\frac{\mathrm{d}v_C}{\mathrm{d}t} + v_C \qquad [3]$$

This equation is called a *first order differential equation* because for the variable quantity v_C there is the term $\mathrm{d}v_C/\mathrm{d}t$, i.e. the rate of change of v_C with time. (Note that a second order equation would have $\mathrm{d}v_C^2/\mathrm{d}t^2$, i.e. the rate of change of $\mathrm{d}v_C/\mathrm{d}t$ with time.)

There are a number of ways of solving the differential equation and thus writing an equation involving just v_C and time t. The method that follows is a general method which can be applied to other types of differential equations and other forms of input (there are shorter methods of solution which apply to this particular form of differential equation but cannot be applied to other types of differential equation and other inputs). The method involves rearranging the differential equation into a form to which we can try a solution. We first of all consider v_C to be made up of two parts and write it as

$$v_C = v_n + v_f \quad [4]$$

where v_n is what is called the *natural* or free or transient part and v_f is called the *forcing* part. The reasons for this will become clear shortly. Since we now have

$$\frac{dv_C}{dt} = \frac{dv_n}{dt} + \frac{dv_f}{dt}$$

then the differential equation becomes, when v_C is replaced,

$$V = RC\frac{dv_n}{dt} + RC\frac{dv_f}{dt} + v_n + v_f$$

This gives, on rearrangement,

$$V = \left(RC\frac{dv_n}{dt} + v_n\right) + \left(RC\frac{dv_f}{dt} + v_f\right)$$

We consider the natural part of v_C to be the response that would occur if there was no input to the circuit and the forcing part that part of the response which is due to there being an input. Hence if we take

$$RC\frac{dv_f}{dt} + v_f = V \quad [5]$$

then we must have

$$RC\frac{dv_n}{dt} + v_n = 0$$

For this differential equation for the natural part of the response we try a solution of the form

$$v_n = A\,e^{st}$$

With this solution we have

$$\frac{dv_n}{dt} = As\,e^{st}$$

and thus the natural differential equation can be written as

$$RCAs\,e^{st} + A\,e^{st} = 0$$

and so $s = -1/RC$. The reason for choosing this form of solution was to achieve such a cancellation of $A\,e^{st}$. Thus

$$v_n = A\,e^{st} = A\,e^{-t/RC}$$

The forced differential equation [5] can be solved by

considering a particular solution for it. The form of solution chosen will depend on the type of input. For a step input, since it is not varying with time, we choose $v_f = k$, where k is a constant. For this solution we have $dv_f/dt = 0$, since the differentiation of a constant is zero. Thus equation [5] can be written as

$$RC(0) + k = V$$

Thus $k = V$ and so $v_f = V$.

The full solution for v_C is the sum of the natural and the forced responses and so

$$v_C = A\,e^{-t/RC} + V$$

The constant A can be determined by using the condition that at time $t = 0$ we have $v_C = 0$. The equation then becomes

$$0 = A\,e^0 + V$$

Since $e^0 = 1$, then $V = -A$. Hence the complete solution is

$$v_C = -V\,e^{-t/RC} + V$$

The exponential part of the solution is the natural or transient element. As time t increases so this term will die away, leaving only the forced element of V. The equation is usually written as

$$v_C = V(1 - e^{-t/RC}) \qquad [6]$$

This describes an exponential relationship for the variation of v_C with time, Fig. 2.3(a) showing the form of the graph.

Since $V = v_R + v_C$, then

$$v_R = V - v_C = V - V(1 - e^{-t/RC})$$

and so

$$v_R = V\,e^{-t/RC} \qquad [7]$$

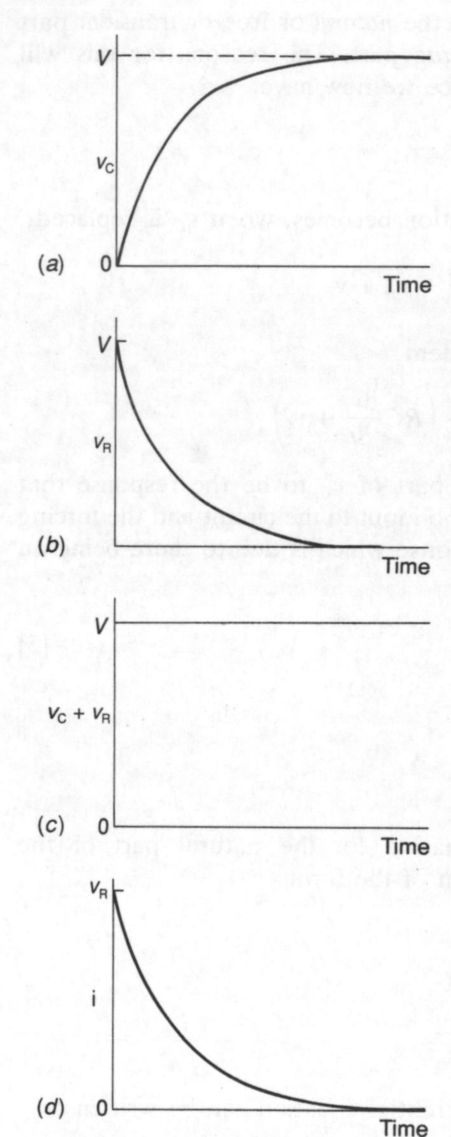

Fig. 2.3 Response of an RC circuit to a step input

Figure 2.3(b) shows the graph of this equation, i.e. how v_R varies with time. At all times the sum of v_R and v_C is V (Fig. 2.3(c), so adding together the graphs in Fig. 2.3(a) and (b), or equations [6] and [7], will always give V. The current i at any time in the circuit is v_R/R, hence

$$i = \frac{V}{R}e^{-t/RC}$$

V/R is the initial charging current I. Thus

$$i = I\,e^{-t/RC} \qquad [8]$$

Figure 2.3(d) shows the graph.

The time taken for the current to decrease to half its value, i.e. to $\frac{1}{2}I$, is given by equation [8] as

$$\tfrac{1}{2}I = I\mathrm{e}^{-t/RC}$$

$$\ln(\tfrac{1}{2}) = -\frac{t}{RC}$$

and so $t = 0.69RC$. The term RC is called the *time constant* τ and has the units of time, being seconds when C is in farads and R in ohms. In a time of 0.69τ the current decreases by half. In each succeeding time of 0.69τ the current drops by a half. If we had considered the voltages we would have found a similar type of relationship. In a time of 0.69τ v_C and v_R both drop by a half. Table 2.1 summarizes some of the key data points for the circuit with a capacitor and resistor in series, when being charged.

Table 2.1 Series RC circuit when being charged

Time	q	v_C	v_R	i
Initially	0	0	V	$I = V/R$
After 0.69τ	$Q/2$	$V/2$	$V/2$	$I/2$
After $2 \times 0.69\tau$	$3Q/4$	$3V/4$	$V/4$	$I/4$
After $3 \times 0.69\tau$	$7Q/8$	$7V/8$	$V/8$	$I/8$
Steady state	$Q = CV$	V	0	0

Example 1

An initially uncharged 8 μF capacitor is connected through a series 1.0 MΩ resistor to a voltage supply of 20 V. What is (*a*) the initial current, (*b*) the initial potential difference across the capacitor, (*c*) the current after 4.0 s, (*d*) the potential difference across the resistor after 4.0 s, (*e*) the potential difference across the capacitor after 4.0 s, (*f*) the potential difference across the fully charged capacitor?

Answer

(*a*) Initially the potential difference across the resistor is that of the voltage supply, thus

$$i = \frac{V}{R} = \frac{20}{1.0 \times 10^6} = 20\,\mu\mathrm{A}$$

(*b*) Because it is uncharged the initial potential difference is 0 V.
(*c*) Using equation [8] above

$$i = \frac{V}{R}\mathrm{e}^{-t/RC} = \frac{20\,\mathrm{e}^{-4/8}}{1.0 \times 10^6} = 12\,\mu\mathrm{A}$$

(*d*) Since $v_R = iR$, then

$$v_R = 12 \times 10^{-6} \times 1.0 \times 10^6 = 12\,\mathrm{V}$$

(*e*) Since $V = v_R + v_C$, then

$$v_C = 20 - 12 = 8\,\text{V}$$

(*f*) The capacitor is fully charged when the current is zero and the potential difference across the resistor is zero. Then the potential difference across the capacitor equals that of the voltage supply, i.e. 20 V.

Example 2

A 10 μF uncharged capacitor is in series with a 0.5 MΩ resistor when a voltage of 12 V is connected across them. What will be the time taken for (*a*) the potential difference across the resistance to decrease to 6 V, (*b*) the potential difference across the capacitor to increase to 9 V?

Answer

The time constant τ is *RC*, thus

$$\tau = 10 \times 10^{-6} \times 0.5 \times 10^{6} = 5.0\,\text{s}$$

(*a*) The time taken for the potential difference across the resistor to decrease to 6 V, by half, is 0.69τ and so 3.45 s.

(*b*) The time taken for the potential difference to increase to 6 V is 0.69τ. The time taken for it to then increase by a further half, i.e. 3 V, is 0.69τ. Thus the total time to increase to 9 V is 2 × 0.69τ, i.e. 6.9 s.

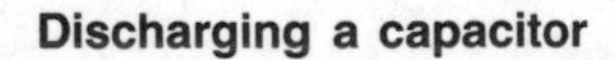

Discharging a capacitor

Assume a capacitor has been charged to some potential difference *V* and is now connected in series with a resistor *R*, as in Fig. 2.4. When the switch is closed the capacitor begins to discharge. Since the e.m.f. in the closed circuit loop is zero the sum of the potential difference across the capacitor v_C and potential difference across the resistor v_R must at all times be zero (Kirchoff's voltage law).

$$v_C + v_R = 0 \qquad [9]$$

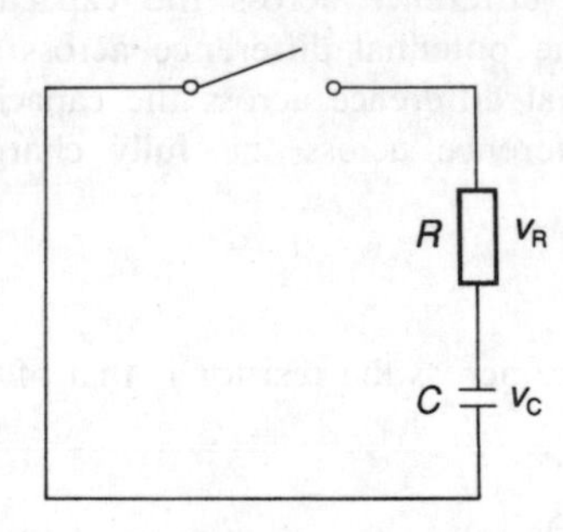

Fig. 2.4 Discharging a capacitor

But $v_R = iR$, where *i* is the current in the circuit. Thus

$$v_C + iR = 0$$

But $i = C\text{d}v_C/\text{d}t$ (equation [13] in Chapter 1) and so we can write the equation as

$$v_C + RC\frac{\text{d}v_C}{\text{d}t} = 0 \qquad [10]$$

This is a first order differential equation describing how the potential difference across the capacitor v_C varies with time.

It can be solved in the same way as the differential equation for the charging of a capacitor (see earlier this chapter). This

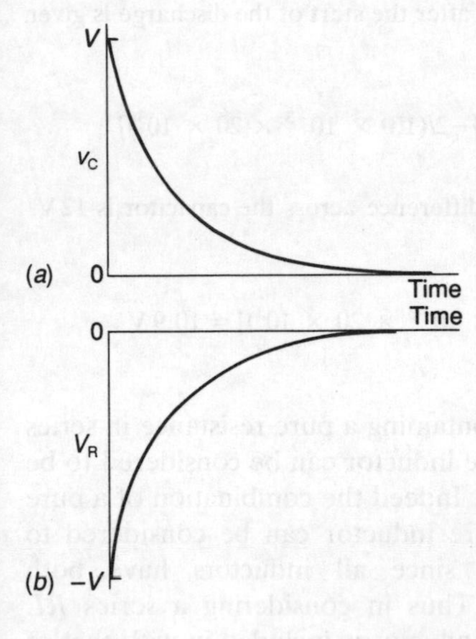

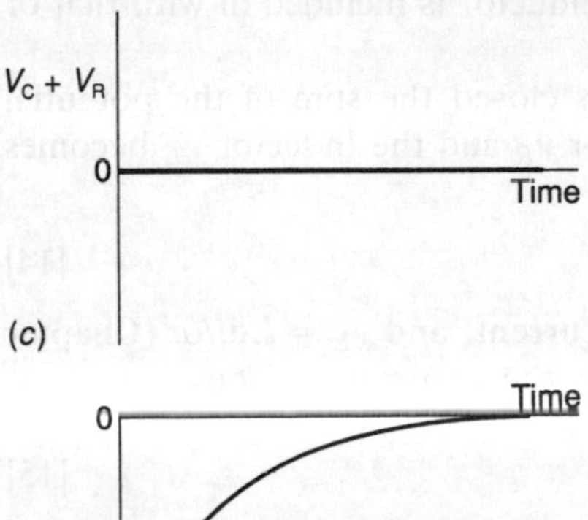

Fig. 2.5 Transients during discharging in an *RC* circuit

means considering v_C as having a natural part and a forcing part. But in this case there is no forcing input in the circuit, no step voltage or other form of input. The differential equation is just in the form of the natural differential equation considered earlier. Thus we only have the natural component and hence the solution is

$$v_C = V\mathrm{e}^{-t/RC} \qquad [11]$$

This describes an exponential relationship for the variation of v_C with time, Fig. 2.5(*a*) showing the form of the graph.

Since we have $v_C + v_R = 0$, equation [9], then

$$v_R = -v_C = -V\mathrm{e}^{-t/RC} \qquad [12]$$

Figure 2.5(*b*) shows a graph of this equation. Since $v_R = iR$, then

$$i = -\frac{V}{R}\mathrm{e}^{-t/RC}$$

But V/R is the initial value of the current I, hence

$$i = -I\mathrm{e}^{-t/RC} \qquad [13]$$

Figure 2.5(*d*) shows a graph of this equation.

The term RC is called the *time constant* τ. Table 2.2 shows the values of the charge, potential difference across the capacitor, potential difference across the resistor and current at various points during the capacitor discharge.

Table 2.2 Discharge of a capacitor through a resistor

Time	q	v_C	v_R	i
Initially	$Q = CV$	V	$-V$	$I = -V/R$
After 0.69τ	$Q/2$	$V/2$	$-V/2$	$-I/2$
After 2 × 0.69τ	$Q/4$	$V/4$	$-V/4$	$-I/4$
After 3 × 0.69τ	$Q/8$	$V/8$	$-V/8$	$-I/8$
Steady state	0	0	0	0

Example 3

A 1.0 mF capacitor which was initially charged to a potential difference of 12 V is discharged through a 20 kΩ resistor. What are (*a*) the current and (*b*) the potential difference across the capacitor (i) initially and (ii) 2 s after the start of the discharge?

Answer

(*a*) (i) The initial value of the current is the result of a potential difference of 12 V being applied across the resistor. Hence $I = 12/(20 \times 10^3) = 0.60$ mA.

(ii) The current at a time t after the start of the discharge is given by equation [13],

$$\begin{aligned} i &= -I\mathrm{e}^{-t/RC} \\ &= -0.60 \times 10^{-3} \exp[-2/(1.0 \times 10^{-3} \times 20 \times 10^{3})] \\ &= -0.54\,\mathrm{mA} \end{aligned}$$

(*b*) (i) Initially the potential difference across the capacitor is 12 V.
(ii) Equation [11] gives

$$\begin{aligned} v_C &= V\mathrm{e}^{-t/RC} \\ &= 12 \exp[-2/(1.0 \times 10^{-3} \times 20 \times 10^{3})] = 10.9\,\mathrm{V} \end{aligned}$$

Series *RL* circuit

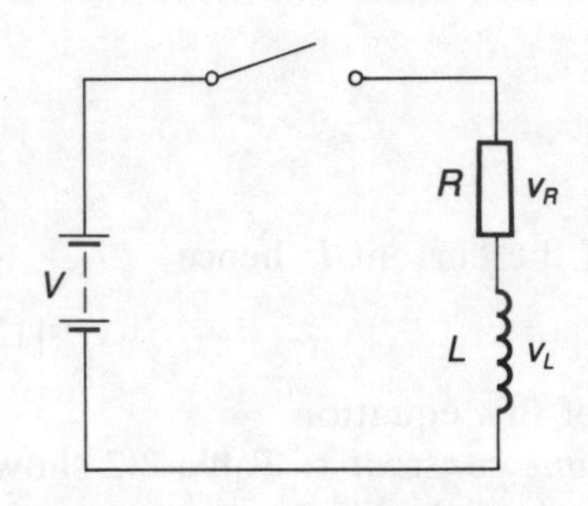

Fig. 2.6 Series *RL* circuit

Figure 2.6 shows a circuit containing a pure resistance in series with a pure inductor. A pure inductor can be considered to be one which has no resistance. Indeed the combination of a pure resistor in series with a pure inductor can be considered to represent a real inductor since all inductors have both inductance and resistance. Thus in considering a series RL circuit the resistance of the inductor is included in with that of the resistor.

When the circuit switch is closed the sum of the potential differences across the resistor v_R and the inductor v_L becomes equal to V.

$$V = v_R + v_L \tag{14}$$

But $v_R = iR$, where i is the current, and $v_L = L\mathrm{d}i/\mathrm{d}t$ (Chapter 1, equation [19]), where L is the inductance. Thus

$$V = iR + L\frac{\mathrm{d}i}{\mathrm{d}t} \tag{15}$$

This is a first order differential equation describing how the current changes with time. If there is a steady, non-changing, current in the circuit then $\mathrm{d}i/\mathrm{d}t = 0$ and so the steady state current I is given by $V = IR$. The equation can thus be written as

$$IR = iR + L\frac{\mathrm{d}i}{\mathrm{d}t}$$

$$I = i + \frac{L}{R}\frac{\mathrm{d}i}{\mathrm{d}t} \tag{16}$$

There are a number of ways of solving the differential equation and thus writing an equation involving just i and time t. Using the method given for the charging of a capacitor we first of all consider i to be made up of two parts and write it as

$$i = i_{\mathrm{n}} + i_{\mathrm{f}} \tag{17}$$

where i_{n} is what is called the natural or free or transient part and i_f is called the forcing part. Since we now have

$$\frac{di}{dt} = \frac{di_n}{dt} + \frac{di_f}{dt}$$

then the differential equation becomes, when i is replaced,

$$I = \frac{L}{R}\frac{di_n}{dt} + \frac{L}{R}\frac{di_f}{dt} + i_n + i_f$$

This gives, on rearrangement,

$$I = \left(\frac{L}{R}\frac{di_n}{dt} + i_n\right) + \left(\frac{L}{R}\frac{di_f}{dt} + i_f\right)$$

We consider the natural part of i to be the response that would occur if there was no input to the circuit and the forcing part that part of the response which is due to there being an input. Hence if we take

$$\frac{L}{R}\frac{di_f}{dt} + i_f = I \qquad [18]$$

then we must have

$$\frac{L}{R}\frac{di_n}{dt} + i_n = 0$$

For this differential equation for the natural part of the response we try a solution of the form

$$i_n = A\,e^{st}$$

With this solution we have

$$\frac{di_n}{dt} = As\,e^{st}$$

and thus the natural differential equation can be written as

$$\frac{L}{R}As\,e^{st} + A\,e^{st} = 0$$

and so $s = -(R/L)$. The reason for choosing this form of solution was to achieve such a cancellation of $A\,e^{st}$. Thus

$$i_n = A\,e^{st} = A\,e^{-Rt/L}$$

The forced differential equation [18] can be solved by considering a particular solution for it. The form of solution chosen will depend on the type of input. For a step input, since it is not varying with time, we choose $i_f = k$, where k is a constant. For this solution we have $di_f/dt = 0$, since the differentiation of a constant is zero. Thus equation [18] can be written as

$$\frac{L}{R}(0) + k = I$$

Thus $k = I$ and so $i_f = I$.

The full solution for i is the sum of the natural and the forced responses and so

$$i = A\,e^{-Rt/L} + I$$

The constant A can be determined by using the condition that at time $t = 0$ we have $i = 0$. The equation then becomes

$$0 = A\,e^0 + I$$

Since $e^0 = 1$, then $I = -A$. Hence the complete solution is

$$i = -I\,e^{-(R/L)t} + I$$

The exponential term is the natural or transient element and as the time t increases will eventually die away, leaving just the forced element I. In other words, given enough time the transient element will die away and the current will just be I. The equation is usually written as

$$i = I(1 - e^{-(R/L)t}) \qquad [19]$$

Figure 2.7(d) shows the graph of the above equation. Since $v_R = iR$, then equation [19] gives

$$v_R = V(1 - e^{-Rt/L}) \qquad [20]$$

Figure 2.7(a) shows the graph of the above equation. Since $v_R + v_L = V$ then

$$v_L = V - v_R = V - V(1 - e^{-Rt/L})$$

$$v_L = V\,e^{-Rt/L} \qquad [21]$$

Figure 2.7(b) shows the graph of the above equation.

The time taken for the potential difference across the inductor to decrease to half its value is

$$v_L = \tfrac{1}{2}V = V\,e^{-Rt/L}$$

$$\ln(\tfrac{1}{2}) = -Rt/L$$

and so $t = 0.69(L/R)$. The quantity (L/R) is known as the *time constant* τ and has the units of seconds when L is in henries and R in ohms. The potential difference v_R across the resistor is given by equation [20] which on rearrangement is

$$V - v_R = V\,e^{-Rt/L}$$

The time taken for $(V - v_R)$ to decrease to half its value, e.g. from V at $t = 0$ to $\tfrac{1}{2}V$, is

$$\tfrac{1}{2}V = V\,e^{-Rt/L}$$

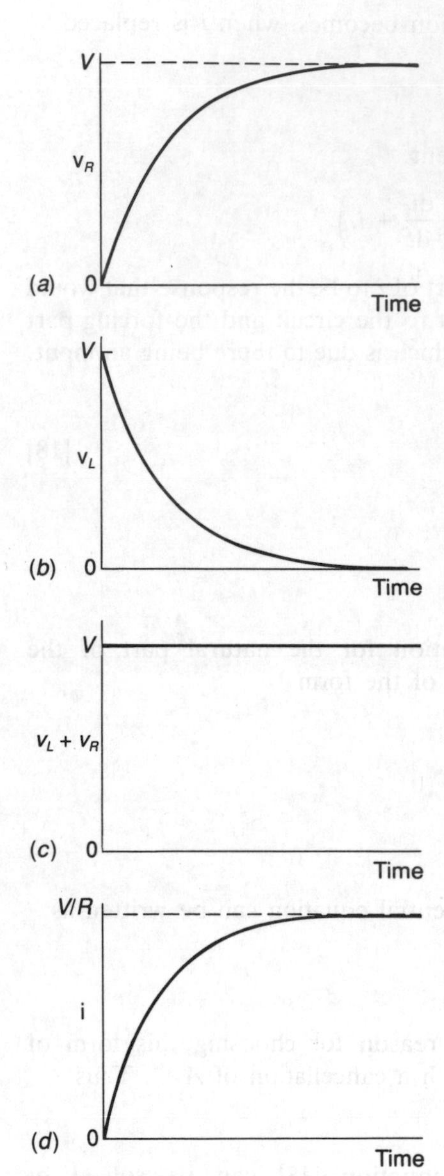

Fig. 2.7 Transients during rise of current in *RL* circuit

$$\ln(\tfrac{1}{2}) = -Rt/L$$

and so $t = 0.69(L/R) = 0.69\tau$. Thus the time taken for $(V - v_R)$ to decrease by half is 0.69τ. Similarly, the time taken for the $(I - i)$ to decrease to half its value is 0.69τ. Table 2.3 shows values of the current, potential difference across the inductor and potential difference across the resistor at various times during the growth of the current.

Table 2.3 Series *RL* circuit when the current is rising

Time	i	v_L	v_R
Initially	0	V	0
After 0.69τ	$I/2$	$V/2$	$V/2$
After $2 \times 0.69\tau$	$3I/4$	$V/4$	$3V/4$
After $3 \times 0.69\tau$	$5I/8$	$V/8$	$5V/8$
Steady state	$I = V/R$	0	V

If, after a steady current has been flowing through a series *RL* circuit, the applied voltage is removed and the circuit short-circuited, then the current does not immediately fall to zero. This is because the collapsing magnetic field of the coil induces an e.m.f. in it which opposes the collapse and so prolongs the current. For this circuit

$$v_R + v_L = 0$$

and since $v_R = iR$ and $v_L = L\,\mathrm{d}i/\mathrm{d}t$, then

$$iR + L\frac{\mathrm{d}i}{\mathrm{d}t} = 0$$

This is just the natural part of the previous solution. Thus

$$i = I\mathrm{e}^{-Rt/L} \quad [22]$$

$$v_R = iR = RI\mathrm{e}^{-Rt/L} = V\mathrm{e}^{-Rt/L} \quad [23]$$

$$v_L = -v_R = -V\mathrm{e}^{-Rt/L} \quad [24]$$

The quantity (L/R) is called the time constant τ of the circuit. Table 2.4 shows values of the current, potential difference across the inductor and potential difference across the resistor at various times during the decay of the current.

RLC series circuit

Consider a circuit having a resistor, inductor and capacitor in series, as in Fig. 2.8. When the switch is closed the circuit is subject to a step input, i.e. an abrupt increase in voltage from

Table 2.4 Series *RL* circuit when the current is decaying

Time	i	v_L	v_R
Initially	$I = V/R$	$-V$	V
After 0.69τ	$I/2$	$-V/2$	$V/2$
After 2 × 0.69τ	$I/4$	$-V/4$	$V/4$
After 3 × 0.69τ	$I/8$	$-V/8$	$V/8$
Steady state	0	0	0

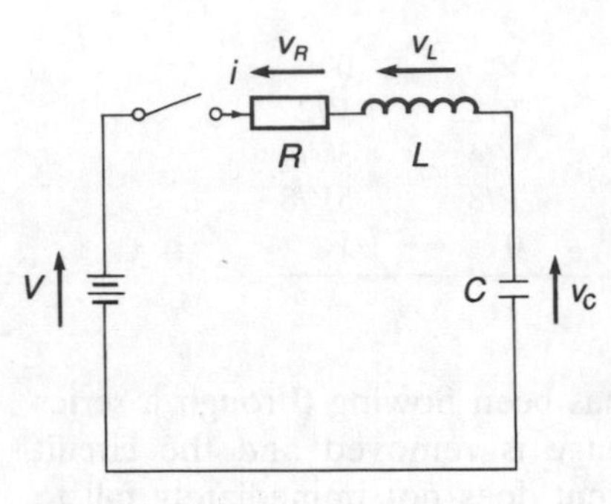

Fig. 2.8 *RLC* series circuit

0 to V. At any instant we must have, as a result of applying Kirchoff's voltage law,

$$V = v_R + v_L + v_C$$

where v_R is the potential difference across the resistor, v_L that across the inductor and v_C that across the capacitor. Now consider the situation at some instant of time when there is a current i. The potential difference across the resistor will be iR, across the inductor $L\mathrm{d}i/\mathrm{d}t$ and across the capacitor v_C. Thus

$$V = L\frac{\mathrm{d}i}{\mathrm{d}t} + iR + v_C$$

But i is given by $i = C\mathrm{d}v_C/\mathrm{d}t$. If this equation is differentiated then

$$\frac{\mathrm{d}i}{\mathrm{d}t} = C\frac{\mathrm{d}^2 v_C}{\mathrm{d}t^2}$$

Thus substituting for i and $\mathrm{d}i/\mathrm{d}t$ gives

$$V = LC\frac{\mathrm{d}^2 v_C}{\mathrm{d}t^2} + RC\frac{\mathrm{d}v_C}{\mathrm{d}t} + v_C \qquad [25]$$

This is a second order differential equation which describes how v_C varies with time.

This differential equation can be solved by the same method as described earlier for the first order differential equation. We can consider v_C to be made up of two elements, a natural or transient part and a forced part, i.e. a part responsible for the situation that prevails when all the transients have died away. Then

$$v_C = v_n + v_f$$

where v_n is the natural part of the solution and v_f the forced part. Differentiating this equation gives

$$\frac{\mathrm{d}v_C}{\mathrm{d}t} = \frac{\mathrm{d}v_n}{\mathrm{d}t} + \frac{\mathrm{d}v_f}{\mathrm{d}t}$$

and differentiating yet again,

$$\frac{d^2v_C}{dt^2} = \frac{d^2v_n}{dt^2} + \frac{d^2v_f}{dt^2}$$

Thus the differential equation can be written as

$$V = LC\left(\frac{d^2v_n}{dt^2} + \frac{d^2v_f}{dt^2}\right) + RC\left(\frac{dv_n}{dt} + \frac{dv_f}{dt}\right) + v_n + v_f$$

$$V = \left(LC\frac{d^2v_n}{dt^2} + RC\frac{dv_n}{dt} + v_n\right) + \left(LC\frac{d^2v_f}{dt^2} + RC\frac{dv_f}{dt} + v_f\right)$$

Thus, for the forcing differential equation we can write

$$V = LC\frac{d^2v_f}{dt^2} + RC\frac{dv_f}{dt} + v_f \qquad [26]$$

and for the natural differential equation we must thus have

$$0 = LC\frac{d^2v_n}{dt^2} + RC\frac{dv_n}{dt} + v_n \qquad [27]$$

To solve the natural equation we can try a solution of the form

$$v_n = A\,e^{st}$$

With such a solution

$$\frac{dv_n}{dt} = As\,e^{st}$$

$$\frac{d^2v_n}{dt^2} = As^2\,e^{st}$$

Thus equation [27] becomes

$$LCAs^2\,e^{st} + RCAs\,e^{st} + A\,e^{st} = 0$$

$$LCs^2 + RCs + 1 = 0 \qquad [28]$$

Thus $v_n = A\,e^{st}$ can only be a solution provided the above equation is true. This equation is called the *characteristic equation* or *auxiliary equation*. The roots of the equation can be obtained by factorizing or using the formula

$$s = \frac{-b \pm \sqrt{(b^2 - 4ac)}}{2a}$$

where a, b and c are the constants in an equation of the form $ax + bx + c = 0$. Thus

$$s = \frac{-RC \pm \sqrt{(R^2C^2 - 4LC)}}{2LC}$$

$$s = -\frac{R}{2L} \pm \sqrt{\left[\left(\frac{R}{2L}\right)^2 - \frac{1}{LC}\right]} \qquad [29]$$

There are thus two values of s which will satisfy equation [28].

$$s_1 = -\frac{R}{2L} + \sqrt{\left[\left(\frac{R}{2L}\right)^2 - \frac{1}{LC}\right]}$$

and

$$s_2 = -\frac{R}{2L} + \sqrt{\left[\left(\frac{R}{2L}\right)^2 - \frac{1}{LC}\right]}$$

So the solution to the natural differential equation can be written as

$$v_{\mathrm{n}} = A \exp s_1 t + B \exp s_2 t \qquad [30]$$

To solve the forcing equation [26], we consider a particular form of input signal and then try a solution. Thus for the step input, since we expect there to be a steady state constant value of v_C, we can try a solution $v_{\mathrm{f}} = k$, where k is a constant. Thus

$$\frac{\mathrm{d}v_{\mathrm{f}}}{\mathrm{d}t} = 0$$

and

$$\frac{\mathrm{d}^2 v_{\mathrm{f}}}{\mathrm{d}t^2} = 0$$

and so equation [26] becomes

$$0 + 0 + k = V$$

Thus $v_{\mathrm{f}} = V$.

The solution to the differential equation for v_C, i.e. equation [25], is thus

$$v_C = v_{\mathrm{n}} + v_{\mathrm{f}} = A \exp s_1 t + B \exp s_2 t + V \qquad [31]$$

The values of A and B depend on the initial conditions in the circuit. Thus if the capacitor is initially uncharged then $v_C = 0$ at $t = 0$, and so equation [31] becomes

$$0 = A\,\mathrm{e}^0 + B\,\mathrm{e}^0 + V$$

$$0 = A + B + V \qquad [32]$$

There is also zero current at $t = 0$. But zero current means, since $i = C\mathrm{d}v_C/\mathrm{d}t$, that $\mathrm{d}v_C/\mathrm{d}t = 0$ at $t = 0$. Thus differentiating equation [31] gives

$$\frac{\mathrm{d}v_C}{\mathrm{d}t} = As_1 \exp s_1 t + Bs_2 \exp s_2 t$$

and so

$$0 = As_1 + Bs_2 \qquad [33]$$

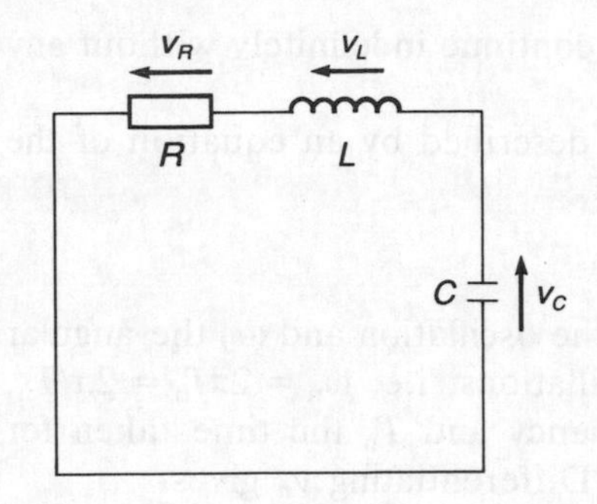

Fig. 2.9 *RLC* series circuit

Equations [32] and [33] can be solved to give the values of A and B.

If the circuit had been just a charged capacitor discharging through a resistor and inductor (Fig. 2.9) then we just have equation [25] with zero voltage input, i.e.

$$v_C + LC\frac{d^2v_C}{dt^2} + RC\frac{dv_C}{dt} = 0$$

This is just the same as the natural element of v_C (equation [27]) The solution to this equation is thus

$$v_C = A\exp s_1t + B\exp s_2t \qquad [34]$$

where s_1 and s_2 are the roots and given by equation [29].

The roots

The way in which v_C varies with time for the series *RLC* circuit depends on the values of the roots of the characteristic equation, i.e. s_1 and s_2 in equation [31]. The roots are given by equation [29] as

$$s = -\frac{R}{2L} \pm \sqrt{\left[\left(\frac{R}{2L}\right)^2 - \frac{1}{LC}\right]}$$

The values of R, L and C can be such that the roots fall into three categories:

1. $(R/L)^2 > (1/LC)$
For this condition the square root is the square root of a positive number and yields two real roots. The circuit is said to be *over damped* because the natural, i.e. transient, part of the voltage just very slowly decays with time.

2. $(R/L)^2 = (1/LC)$
For this condition the square root term is zero and so there are two real equal roots. The circuit is said to be *critically damped*, this representing the dividing line between the over damped and under damped condition. The natural, i.e. transient, part of the voltage decays to leave just the final steady value in the minimum amount of time without oscillations occurring.

3. $(R/L)^2 < (1/LC)$
For this condition the square root term is the square root of a negative number and yields two complex roots. The circuit is said to be *under damped* because the natural, i.e. transient, part of the voltage oscillates about the final steady value with the oscillations eventually dying away to give the steady value.

The resistance in the series *RLC* circuit is responsible for damping down the change in v_C with time. If there was no resistance then the motion would be definitely under damped, indeed the oscillations would continue indefinitely without any reduction in amplitude.

Such an oscillation can be described by an equation of the form (Fig. 2.10)

$$v_n = A \sin \omega_n t$$

where A is the amplitude of the oscillation and ω_n the angular frequency of the natural oscillations, i.e. $\omega_n = 2\pi f_n = 2\pi/T_n$, where f_n is the natural frequency and T_n the time taken for one cycle of that oscillation. Differentiating v_n gives

$$\frac{dv_n}{dt} = \omega_n A \cos \omega_n t$$

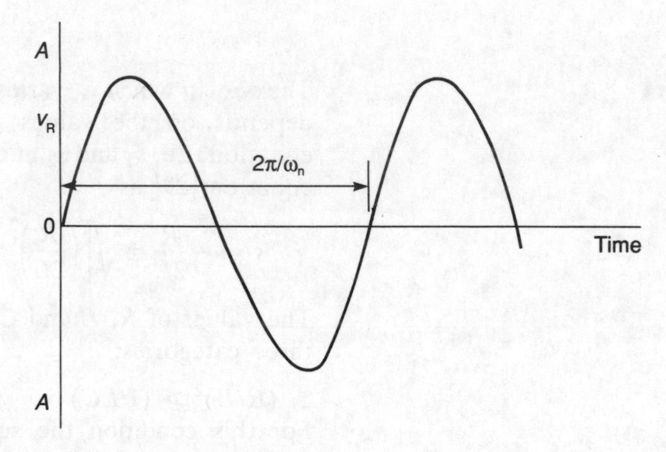

Fig. 2.10 $v_n = A \sin \omega_n t$

Differentiating this again gives

$$\frac{d^2 v_n}{dt^2} = -\omega_n^2 A \sin \omega_n t = -\omega_n^2 v_n$$

and so

$$\frac{d^2 v_n}{dt^2} + \omega_n^2 v_n = 0$$

But this is just the *RLC* differential equation for the natural, i.e. transient, variation of v_C with time, with the R term missing (equation [27]) and slightly rearranged:

$$LC\frac{d^2 v_n}{dt^2} + RC\frac{dv_n}{dt} + v_n = 0$$

$$\frac{d^2 v_n}{dt^2} + \frac{1}{LC} v_n = 0$$

The *natural angular frequency* ω_n is thus given by

$$\omega_n = \frac{1}{\sqrt{LC}} \quad [35]$$

This is thus the frequency with which the potential difference in the series *RLC* circuit will oscillate in the absence of any resistance.

If we define a term ζ as being given by

$$\zeta\omega_n = (R/2L) \quad [36]$$

and use equation [35] then we can write the equation for the roots as

$$s = -\zeta\omega_n \pm \sqrt{(\zeta^2\omega_n^2 - \omega_n^2)}$$

$$s = -\zeta\omega_n \pm \omega_n\sqrt{(\zeta^2 - 1)}$$

The value of ζ thus determines whether the square root term will be positive, zero or negative and so determines the damping. For this reason ζ is called the damping ratio. The circuit is thus over damped when ζ is greater than 1, critically damped when ζ is 1 and under damped when ζ is less than 1. Figure 2.11 shows how the voltage varies with time for different values of ζ.

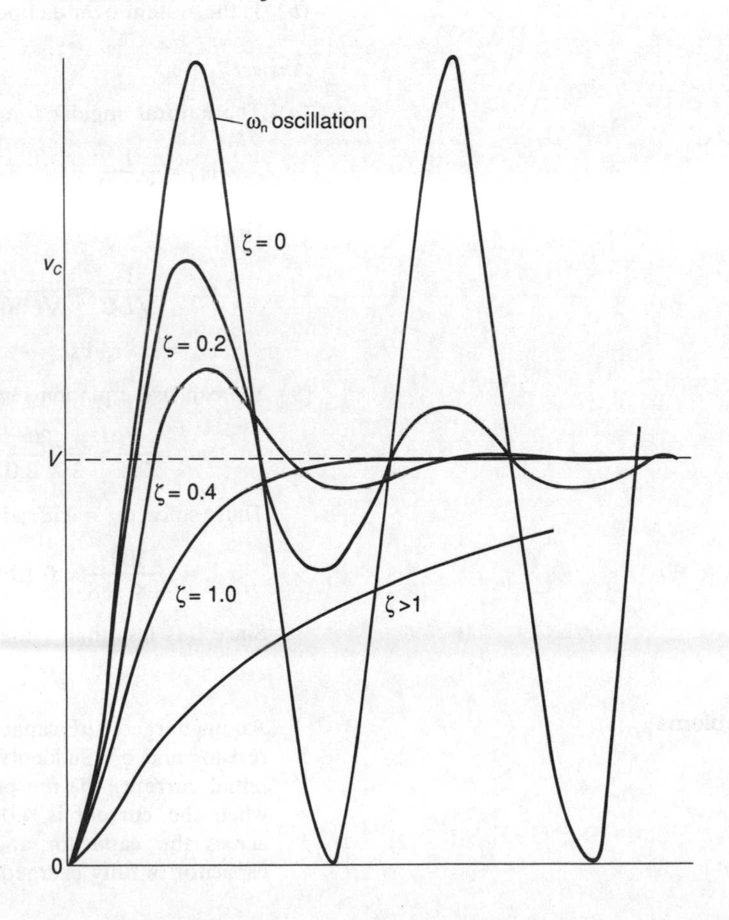

Fig. 2.11 Effect of damping ratio on v_C

Example 4

A series *RLC* circuit has a resistance of 2 kΩ, an inductance of 5 mH and a capacitance of 2 μF. Is the circuit over, critically or under damped?

Answer

The determining factor is whether $(R/L)^2$ is greater than, equal to, or less than $(1/LC)$.

$$\left(\frac{R}{L}\right)^2 = \left(\frac{2 \times 10^3}{5 \times 10^{-3}}\right)^2 = 0.16 \times 10^{12}$$

$$\frac{1}{LC} = \frac{1}{5 \times 10^{-3} \times 2 \times 10^{-6}} = 1 \times 10^8$$

The circuit is thus over damped.

Example 5

A series *RLC* circuit has $R = 100\,\Omega$, $L = 2.0\,\text{H}$ and $C = 20\,\mu\text{F}$ and there is a step input of V at $t = 0$, the capacitor prior to this being uncharged.

(*a*) What is the natural frequency of the circuit?

(*b*) Is the system over damped, critically damped, or under damped?

Answer

(*a*) The natural angular frequency ω_n is given by equation [35] as

$$\omega_n = \frac{1}{\sqrt{LC}}$$

Hence

$$\omega_n = \frac{1}{\sqrt{LC}} = \frac{1}{\sqrt{(2.0 \times 20 \times 10^{-6})}}$$

$$\omega_n = 158\,\text{rad/s}$$

(*b*) We can use equation [36] to determine the damping ratio.

$$\zeta\omega_n = \frac{R}{2L} = \frac{100}{2 \times 2.0}$$

Thus, since $\omega_n = 158\,\text{rad/s}$,

$$\zeta = \frac{100}{4 \times 158} = 0.16$$

Since ζ is less than 1 the system is under damped.

Problems

1 An uncharged 8 μF capacitor is connected in series with a 1.0 kΩ resistor and 6 V suddenly switched across them. What is (*a*) the initial current, (*b*) the potential difference across the capacitor when the current is 4.0 mA, and (*c*) the potential difference across the capacitor and the charge on its plates when the capacitor is fully charged?

2 A series RC circuit contains a $0.10\,\mu F$ capacitor, initially uncharged, and a $1.0\,M\Omega$ resistor. If a $20\,V$ supply is suddenly connected to the circuit what will be (*a*) the initial current, (*b*) the initial potential difference across the capacitor, (*c*) the current after $0.01\,s$, (*d*) the potential difference across the capacitor after $0.01\,s$, (*e*) the current after a time equal to the time constant, (*f*) the potential difference across the capacitor after a time equal to the time constant, (*g*) the steady state current, (*h*) the steady state potential difference across the capacitor?

3 A $10\,\mu F$ capacitor in series with a $1.0\,M\Omega$ resistor is fully charged by a $20\,V$ supply. (*a*) What will be the charge on the fully charged capacitor and the potential difference across it? (*b*) The supply voltage is then disconnected and the capacitor is then discharged through the resistor. What will be the charge on the capacitor and the potential difference across it $5.0\,s$ after the start of the discharge?

4 A $5\,\mu F$ capacitor is charged to a potential difference of $10\,V$ and then allowed to discharge through a $20\,k\Omega$ resistor. What will be (*a*) the initial potential difference across the resistor, (*b*) the initial current, (*c*) the initial potential difference across the capacitor, (*d*) the current after $0.1\,s$ from the start of the discharge, (*e*) the potential difference across the capacitor after $0.1\,s$, (*f*) the potential difference across the resistor after $0.1\,s$?

5 A coil having a resistance of $40\,\Omega$ and an inductance of $0.10\,H$ is connected across a d.c. supply of $4\,V$. What will be (*a*) the current after $1\,ms$, (*b*) the current after $2\,ms$ and (*c*) the steady state current?

6 An inductor has an inductance of $2\,H$ and a resistance of $2\,\Omega$. It is connected in series with a resistance of $4\,\Omega$. What will be the equation describing how the current in the circuit varies with time if a step voltage of $6\,V$ is applied to the series RL arrangement?

7 A coil has an inductance of $0.1\,H$ and a resistance of $4\,\Omega$ and is in series with a $16\,\Omega$ resistor. What is (*a*) the time constant τ of the circuit, (*b*) the current after a time equal to 0.69τ and (*c*) the current after a time equal to $2 \times 0.69\tau$, if a d.c. supply of $10\,V$ is switched across the series RL arrangement?

8 A relay coil has an inductance of $0.10\,H$ and a resistance of $100\,\Omega$. If the relay contacts close when the current is $30\,mA$, determine the time delay before the contacts close when a voltage of $4.0\,V$ is applied across the relay.

9 Write down the differential equations for the potential difference v_C across the capacitor for the following circuits:

(*a*) a series circuit of $10\,k\Omega$, $2\,H$ and an initially uncharged $8\,\mu F$ capacitor when subject to a step voltage of size $10\,V$,

(*b*) a series circuit of $50\,\Omega$, $50\,mH$ and an initially uncharged $2\,\mu F$ capacitor when subject to a step voltage of $4\,V$,

(*c*) a series circuit of $200\,\Omega$, $3\,H$ and a $4\,\mu F$ capacitor which is initially charged before being discharged through the circuit, there being no voltage source in the circuit.

10 Are the circuits in problem 9 over, critically or under damped?

11 What resistance should be used in a series *RLC* circuit with $L = 1\,\text{H}$ and $C = 10\,\mu\text{F}$ if the circuit is to be critically damped?

12 A series *RLC* circuit has $R = 50\,\Omega$, $L = 1.0\,\text{H}$ and $C = 4\,\mu\text{F}$ and there is a step input of 5 V at $t = 0$, the capacitor prior to this being uncharged.

(*a*) What is the natural angular frequency of the circuit?

(*b*) Is the system over damped, critically damped, or under damped?

3 a.c. circuits

Introduction

This chapter is concerned with an introduction to the analysis of circuits containing resistors, inductors and capacitors, when the input signals vary sinusoidally with time. Such signals are particularly important in electrical engineering because practically all the electrical power generation in the entire world is of low frequency sinusoidal voltages. Sinusoidal signals are also widely used in telecommunications. When sinusoidal signals are first switched on or off in a circuit there will be signals which die away with time, so-called transients, before steady state conditions are realised. The concern of this chapter is however only with the steady state response and involves a consideration of phasors and the use of complex notation to represent them.

Alternating signals

An alternating signal is one that oscillates from positive to negative values in a regular, periodic manner. The time T taken for one complete cycle is called the *periodic time*. The number of cycles occurring per second is called the *frequency* f, this having the unit hertz (Hz). Thus $f = 1/T$. The maximum value of the alternating signal, i.e. the amplitude of a sinusoidal signal, is referred to as the *peak or maximum value*, I_m in the case of a current and V_m for a voltage.

Meters used for the measurement of alternating current or voltage are often calibrated in terms of the direct current or voltage that would give the same power dissipation in a resistor. With a sinusoidal waveform the average power is $\frac{1}{2}I_m V_m = \frac{1}{2}I_m^2 R$ (see p. 54 for proof of this). Thus for a steady current I, which is known as the *root-mean-square current*, to give the same power as the alternating current we must have $I^2R = \frac{1}{2}I_m^2 R$. Hence

$$\text{r.m.s. value of current} = \frac{I_m}{\sqrt{2}} = 0.707 I_m \qquad [1]$$

Similarly for the root-mean-square voltage,

$$\text{r.m.s. value of voltage} = \frac{V_m}{\sqrt{2}} = 0.707V_m \quad [2]$$

In the case of alternating currents or voltages with other waveforms the relationships between the root-mean-square values and the maximum values are different.

Graphical representation of sinusoidal quantities

Consider a line OA rotating in an anticlockwise direction about O with a constant angular velocity ω, as in Fig. 3.1. If it starts from the horizontal position then the angle θ rotated in a time t is ωt. But

$$AB = OA \sin \omega t \quad [3]$$

AB is the perpendicular height of the line at some instant of time, OA being its length. The way in which AB varies with time is a sinusoidal waveform. Since OA is the maximum value of the waveform and AB the value at some instant, then in the case of current we can write equation [3] as

$$i = I_m \sin \omega t \quad [4]$$

with i being the current at some instant and I_m the maximum value, and for a voltage

$$v = V_m \sin \omega t \quad [5]$$

with v being the voltage at some instant and V_m the maximum value.

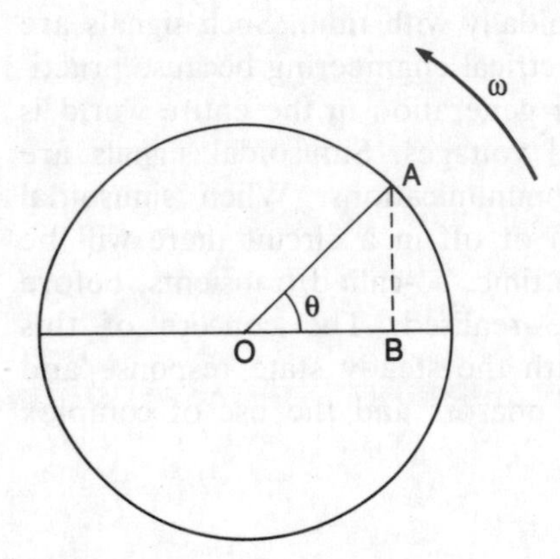

Fig. 3.1 Generating a sinusoidal waveform

The line OA is rotating at constant angular velocity ω and thus if ω is in radians/second then since there are 2π radians in one complete revolution the time taken to complete one revolution T is $2\pi/\omega$. But T is the time taken for one complete cycle of the sinusoidal waveform. The frequency f is $1/T$ and so

$$\omega = 2\pi f \quad [6]$$

Because ω is proportional to the frequency it is often referred to as the *angular frequency*.

Figure 3.2(a) shows how two rotating lines can be used to generate two sinusoidal signals which differ only in maximum values. The lengths of the rotating lines equal the size of the maximum values and they both rotate with the same angular velocity because they have the same frequency, starting together at the same time and keeping in step as they rotate. Figure 3.2(b) shows how two rotating lines can be used to generate two sinusoidal signals which differ only in that they do not start at the same value. The two rotating lines always have an angle ϕ between them. This angle is called the *phase angle difference*. For the sinusoidal signals represented in Fig.

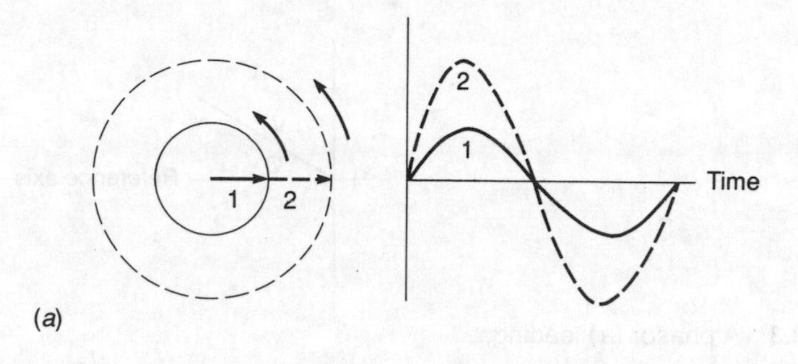

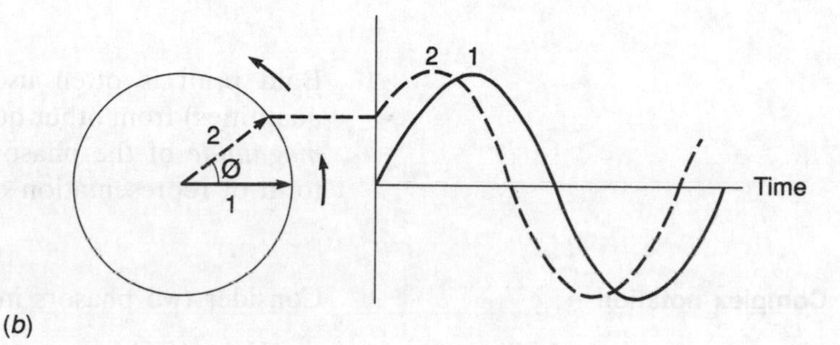

Fig. 3.2 Two sinusoidal signals differing only in a (*a*) maximum values, (*b*) phase

3.2(*b*) we talk of signal 2 *leading* signal 1 by ϕ. We could likewise talk of signal 1 *lagging* signal 2 by ϕ.

For signal 1 we have

$$i = I_m \sin \omega t$$

while for signal 2 we have

$$i = I_m \sin(\omega t + \phi)$$

This is because this signal is always ϕ ahead of the angle ωt.

Phasors

As the previous discussion illustrates, it is possible to describe a sinusoidal signal in terms of a rotating line provided we know its length and its phase angle. The term *phasor*, an abbreviation of the term 'phase vector', is used for such lines. A phasor is a line with a length that represents the maximum value, or r.m.s. value since this is purely 0.707 times the maximum value, of a sinusoidal signal. They are drawn with an arrow at one end, indicating the end that rotates and are taken to rotate anticlockwise. The angle between two phasors is the phase angle between the two signals. The horizontal line stretching out to the right from the point of rotation is taken as the reference axis and phase angles are measured relative to it, a positive angle being referred to as leading and a negative angle as lagging (Fig. 3.3). A phasor can be written as

$$\mathrm{V} = V_m \angle \phi \qquad [7]$$

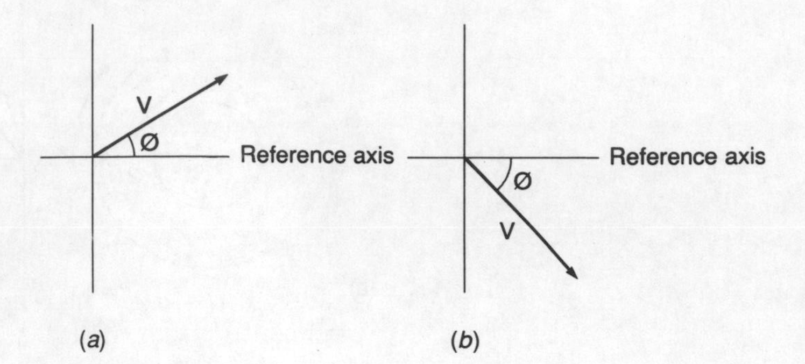

Fig. 3.3 A phasor (*a*) leading, (*b*) lagging

Bold print is often used to distinguish phasors (or vector quantities) from other quantities. V_m is often referred to as the *magnitude* of the phasor and sometimes written as $|V|$. This form of representation of a phasor is called the *polar form*.

Complex notation

Consider two phasors in polar form,

$$\mathbf{V_1} = V \angle \phi$$

$$\mathbf{V_2} = V \angle \phi + 90°$$

The two phasors have the same magnitude but differ in phase by 90°. $\mathbf{V_2}$ is just $\mathbf{V_1}$ rotated by 90°. Suppose we write this relationship as

$$\mathbf{V_2} = \mathrm{j}\mathbf{V_1}$$

where j indicates that $\mathbf{V_2}$ is obtained by rotating $\mathbf{V_1}$ by 90° in an anticlockwise direction (Fig. 3.4(*a*)).

Now suppose we rotate the $\mathbf{V_2}$ phasor by a further 90° to give

$$\mathbf{V_2} = V \angle \phi + 180°$$

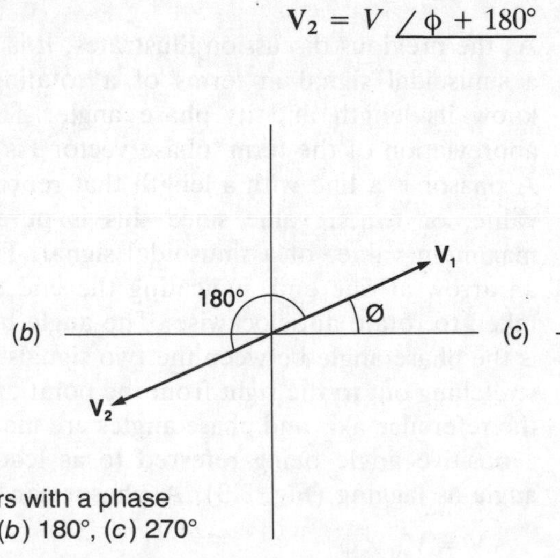
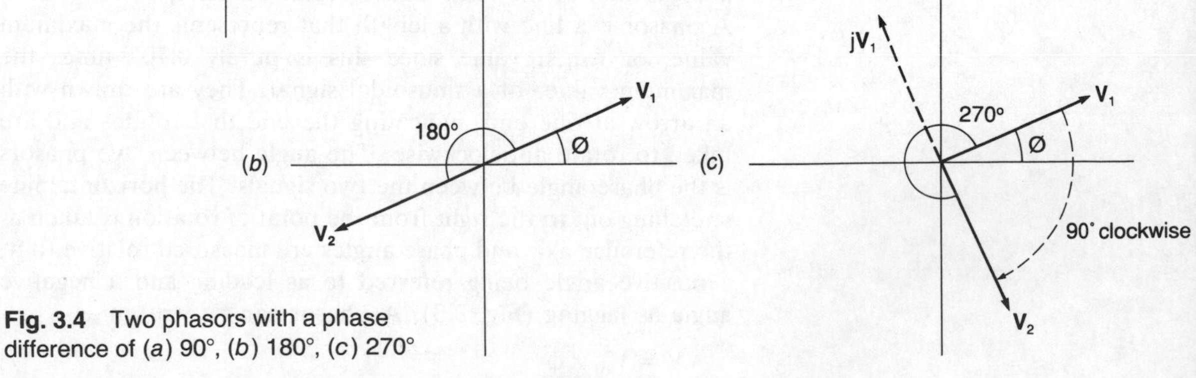

Fig. 3.4 Two phasors with a phase difference of (*a*) 90°, (*b*) 180°, (*c*) 270°

Then we can write

$$V_2 = jjV_1 = j^2V_1$$

Figure 3.4(*b*) shows the phasor. But rotation through 180° means V_2 is just $-V_1$. Thus we must have with this notation that

$$j^2V_1 = -V_1$$

and this leads to the idea that $j^2 = -1$.

Similarly if we rotate V by a further 90° to give

$$V_2 = V \angle \phi + 270°$$

then

$$V = jjjV_1 = j^3V_1$$

Figure 3.4(*c*) shows the phasor. But rotation through 270° means that V_2 is just $-jV_1$. The $-j$ indicates that this phasor is just V_1 rotated clockwise by 90°. This thus leads to the idea that $j^3 = -j$.

Thus this j notation implies

j means a rotation anticlockwise by 90°
j^2 means a rotation anticlockwise by 180°
j^3 means a rotation anticlockwise by 270°
$-j$ means a rotation clockwise by 90°

There is a ready association of j with $\sqrt{-1}$. This then gives $j^2 = -1$ and $j^3 = j^2j = -j$. The square root of -1 is called an *imaginary number*, since such an entity does not exist as a real number.

Rectangular form of notation

Consider the phasor V shown in Fig. 3.5. The phasor can be considered to be composed of two components, one of V_a along the reference axis and one of V_b at right angles to it. In order to indicate that V_b is at right angles to the reference axis we can write it as jV_b. Thus we can write

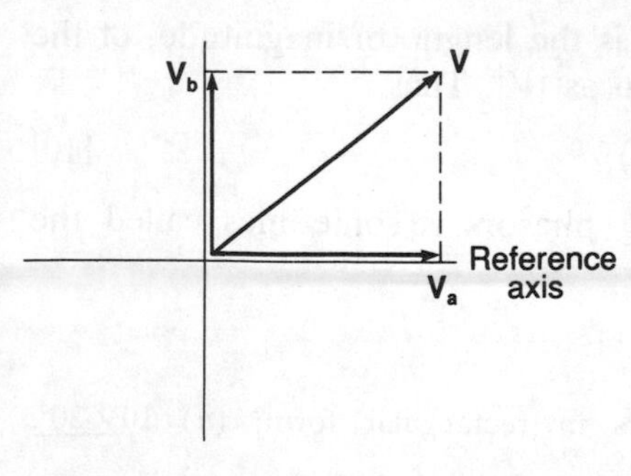

Fig. 3.5 Rectangular components of a phasor

$$V = V_a + jV_b$$

We can consider that V_a is a unit length phasor multiplied by a and V_b is a unit length phasor multiplied by b. Thus

$$V = a1 + jb1$$

This is usually written as

$$V = a + jb \qquad [8]$$

with it being understood that the quantity $(a + jb)$ is being multiplied by a phasor of unit length. Such a representation of a phasor is called its *rectangular form*.

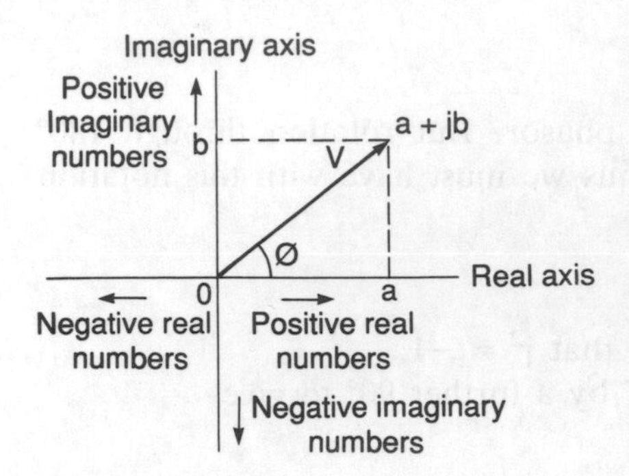

Fig. 3.6 The Argand diagram

Equation [8] enables us to represent a phasor by the lengths of its components along the reference axis and at right angles to it. Because the right angle axis is always a number multiplied by j, it is called the imaginary axis, the reference axis being called the real axis. Such a system of axes is referred to as an *Argand diagram*. Figure 3.6 shows these axes. Thus, for example, we can represent the phasor V by (1 + j2). This tells us that the resolved component of the phasor along the real axis is of length +1 and along the imaginary axis of length +2.

We can relate the a and jb values to the length V and phase angle ϕ of a phasor (Fig. 3.6) since, using Pythagoras,

$$V^2 = a^2 + b^2$$

$$V = \sqrt{(a^2 + b^2)}$$

and using trigonometry

$$\frac{b}{a} = \tan\phi$$

Thus

$$\mathrm{V} = V\angle\phi = \sqrt{(a^2 + b^2)}\angle\tan^{-1}(b/a) \qquad [9]$$

Alternatively, if we want to convert from polar form to rectangular form, since we have

$$\frac{a}{V} = \cos\phi$$

and

$$\frac{b}{V} = \sin\phi$$

then

$$\mathrm{V} = a + \mathrm{j}b = V\cos\phi + \mathrm{j}V\sin\phi$$

To indicate clearly that V is the length, or magnitude, of the phasor V it is often written as $|V|$. Thus

$$\mathrm{V} = |V|(\cos\phi + \mathrm{j}\sin\phi) \qquad [10]$$

This form of representing phasors is sometimes called the trigonometrical form.

Example 1

Write the following phasors in rectangular form: (*a*) $10\angle 20°$, (*b*) $10\angle 140°$.

Answer

(*a*) Using equation [10] we have

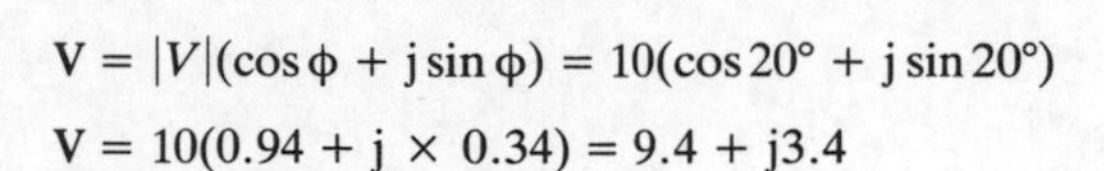

$$V = |V|(\cos\phi + j\sin\phi) = 10(\cos 20° + j\sin 20°)$$

$$V = 10(0.94 + j \times 0.34) = 9.4 + j3.4$$

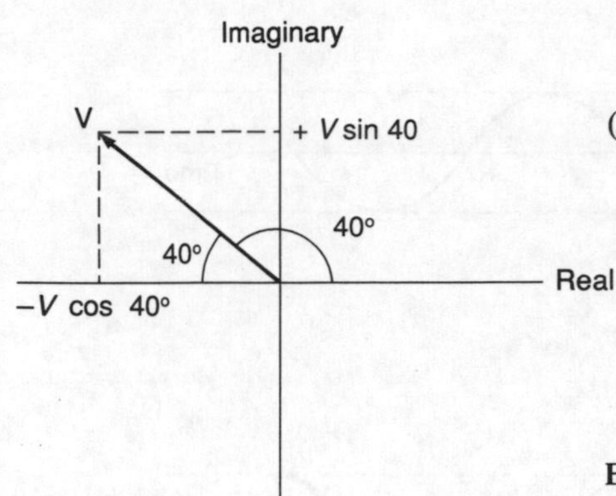

Fig. 3.7 Example 1

(*b*) Using equation [10] we have

$$V = |V|(\cos\phi + j\sin\phi) = 10(\cos 140° + j\sin 140°)$$

If we look at the situation on an Argand diagram (Fig. 3.7) then it is obvious that $10\cos 140°$ is equal to $-10\cos(180° - 140°)$ and that $10\sin 140°$ is equal to $+10\sin(180° - 140°)$. Thus

$$V = -7.7 + j6.4$$

Example 2

Write the following phasors in polar form: (*a*) 2 + j3, (*b*) 2 − j3.

Answer

(*a*) Using equation [9]

$$V = V\angle\phi = \sqrt{(a^2 + b^2)}\angle\tan^{-1}(b/a)$$

$$V = \sqrt{(2^2 + 3^2)}\angle\tan^{-1}(3/2) = 3.6\angle 56.3°$$

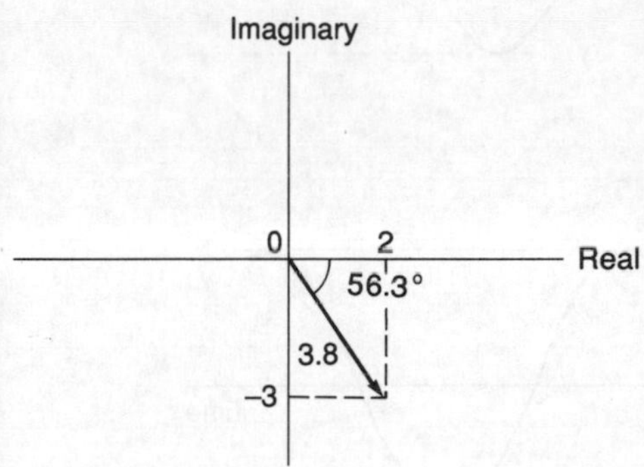

Fig. 3.8 Example 2

(*b*) Using equation [9]

$$V = V\angle\phi = \sqrt{(a^2 + b^2)}\angle\tan^{-1}(b/a)$$

$$V = \sqrt{[2^2 + (-3)^2]}\angle\tan^{-1}(-3/2)$$

Fig. 3.8 shows the Argand diagram for this phasor. From this it is obvious that $\tan^{-1}(-3/2)$ is equal to $-\tan^{-1}(3/2)$. Hence $V = 3.6\angle -56.3°$.

Addition and subtraction of phasors

Suppose we have two phasors $V_1 = V_1\angle\phi_1$ and $V_2 = V_2\angle\phi_2$. We can obtain the sum of the two sinusoidal voltages that the phasors represent by adding together the two sinusoidal graphs, so obtaining the result shown in Fig. 3.9(*c*). However, exactly the same result is obtained by adding together as vector quantities the two phasors which generated the graphs (*a*) and (*b*) and then using the resultant to generate graph (*c*). The resultant line has a length V_R given, since V_1 and V_2 are at 90°, by the use of Pythagoras' theorem as

$$V_R^2 = V_1^2 + V_2^2 \qquad [11]$$

and it is at a phase angle ϕ, where

$$\tan\phi = \frac{V_2}{V_1} \qquad [12]$$

Thus phasor V_1 plus phasor V_2 gives phasor V, where

$$V = [\sqrt{V_1^2 + V_2^2}]\angle\phi$$

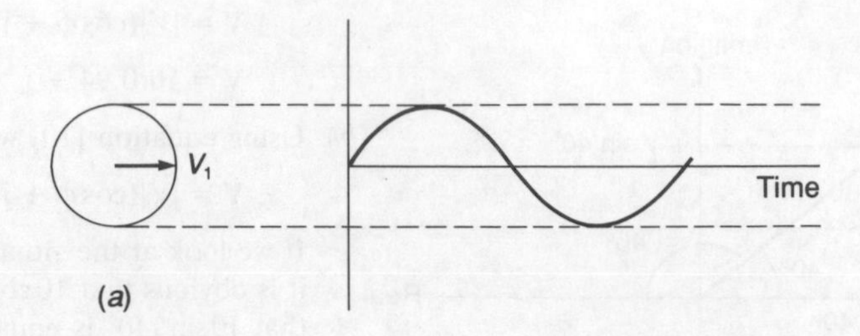

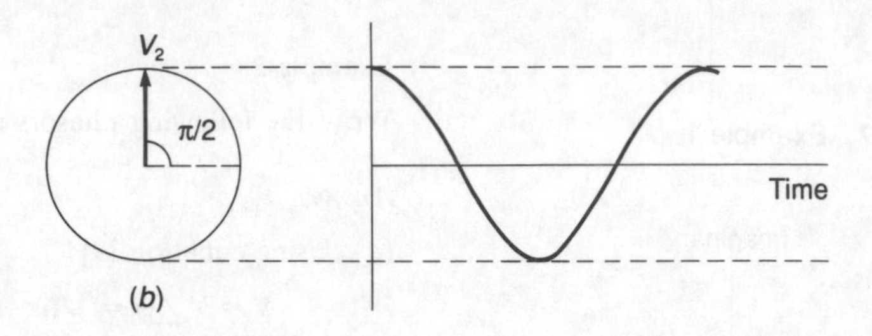

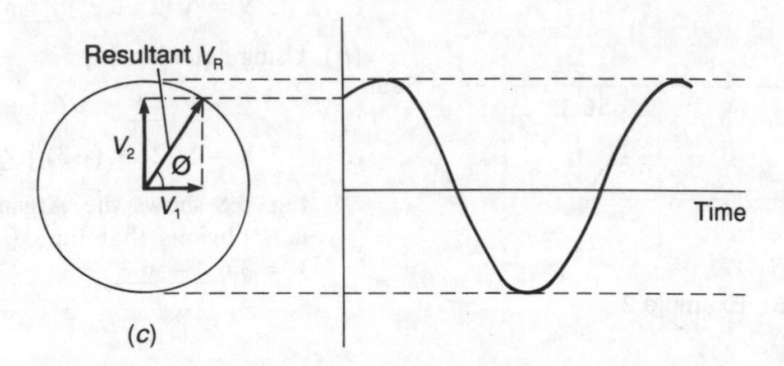

Fig. 3.9 Adding two sinusoidal signals of the same frequency

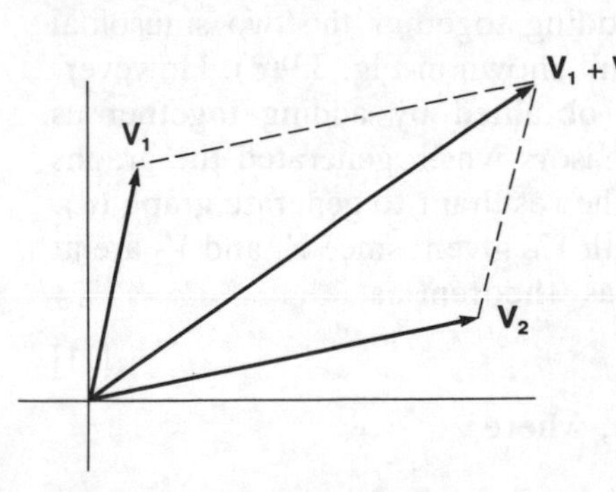

Fig. 3.10 Addition of phasors

Phasors are added by considering them as representing the adjacent sides of a parallelogram (Fig. 3.10). Their sum is then the diagonal of the parallelogram.

If one sinusoidal signal is to be subtracted from another, the technique to be adopted is to add by vector methods the negative equivalent of the signal to be subtracted. The negative equivalent is that signal which would be generated by a phasor which is phase shifted by 180°. Phasors are thus subtracted by first adding a phase change of 180° to the phasor being subtracted. Thus, for example, $\mathbf{V_1}$ becomes $-\mathbf{V_1}$. This is then added to $\mathbf{V_2}$, by means of the parallelogram method described above, to give the difference $\mathbf{V_2} - \mathbf{V_1}$ (Fig. 3.11).

The addition and subtraction of phasors is, however, simpler if they are in rectangular form. Thus with

$$\mathbf{V_1} = a_1 + \mathrm{j}b_1$$
$$\mathbf{V_2} = a_2 + \mathrm{j}b_2$$

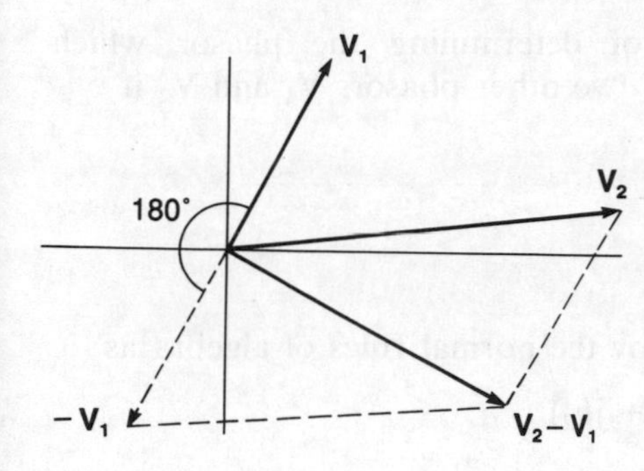

Fig. 3.11 Subtraction of phasors

inspection of Fig. 3.12 shows that the resultant phasor, **V**, obtained by addition is

$$\mathbf{V} = (a_1 + a_2) + \mathrm{j}(b_1 + b_1) \qquad [13]$$

The subtraction of the two phasors involves adding $-\mathbf{V}_2$ to $\mathbf{V}_1$. Using rectangular form (Fig. 3.13) this is

$$\mathbf{V} = (a_2 - a_1) + \mathrm{j}(b_2 - b_1) \qquad [14]$$

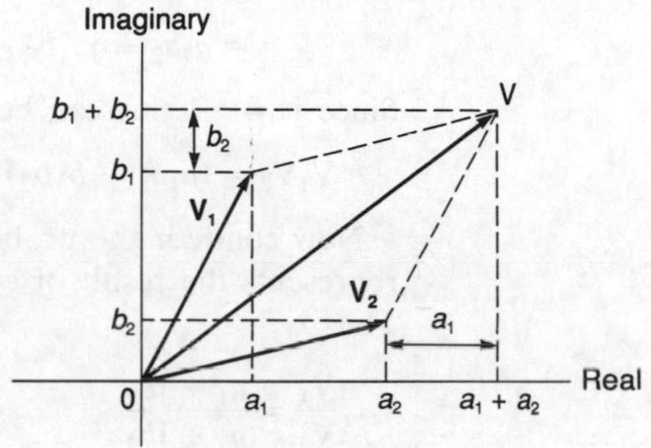

Fig. 3.12 Addition of phasors

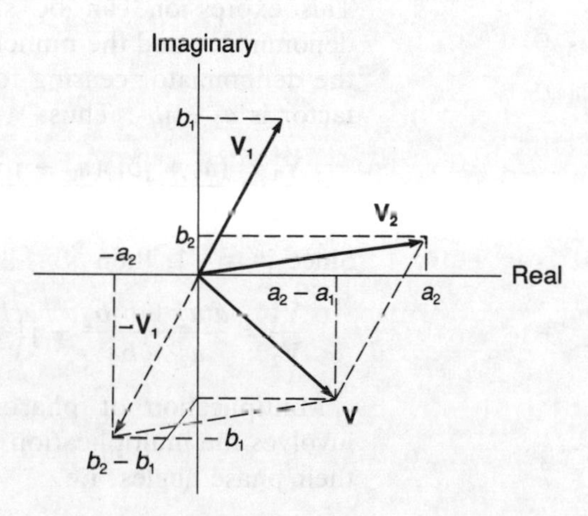

Fig. 3.13 Subtraction of phasors

Example 3

Determine the rectangular form of the phasor for the current which is (*a*) the sum, (*b*) the difference of the two currents $\mathbf{I}_1$ and $\mathbf{I}_2$, where $\mathbf{I}_1 = 2 + \mathrm{j}3$ and $\mathbf{I}_2 = 4 - \mathrm{j}5$.

Answer

(*a*) The phasor sum is, using expression [13] above,

$$\mathbf{I} = (2 + 4) + \mathrm{j}[3 + (-5)] = 6 - \mathrm{j}2$$

(*b*) The phasor difference is, using equation [14],

$$\mathbf{I} = (2 - 4) + \mathrm{j}[3 - (-5)] = -2 + \mathrm{j}8$$

Multiplication and division of phasors

Consider the problem of determining the phasor which represents the product of two other phasors $\mathbf{V}_1$ and $\mathbf{V}_2$ if

$$\mathbf{V}_1 = a_1 + \mathrm{j}b_1$$

and

$$\mathbf{V}_2 = a_2 + \mathrm{j}b$$

The product is obtained by the normal rules of algebra as

$$\begin{aligned}\mathbf{V}_1\mathbf{V}_2 &= (a_1 + \mathrm{j}b_1)(a_2 + \mathrm{j}b_2)\\ &= a_1a_2 + \mathrm{j}a_1b_2 + \mathrm{j}b_1a_2 + \mathrm{j}^2b_1b_2\end{aligned}$$

Since $\mathrm{j}^2 = -1$ this can be simplified to give

$$\mathbf{V}_1\mathbf{V}_2 = (a_1a_2 - b_1b_2) + \mathrm{j}(a_1b_2 + b_1a_2) \qquad [15]$$

Now consider the problem of determining the phasor which represents the result of one phasor being divided by another. Thus

$$\frac{\mathbf{V}_1}{\mathbf{V}_2} = \frac{a_1 + \mathrm{j}b_1}{a_2 + \mathrm{j}b_2}$$

This expression can be simplified by multiplying both the denominator and the numerator by a factor which will result in the denominator ceasing to have any imaginary term. Such a factor is $a_2 - \mathrm{j}b_2$. Thus

$$\frac{\mathbf{V}_1}{\mathbf{V}_2} = \frac{(a_1 + \mathrm{j}b_1)(a_2 - \mathrm{j}b_2)}{(a_2 + \mathrm{j}b_2)(a_2 - \mathrm{j}b_2)} = \frac{a_1a_2 - \mathrm{j}^2b_1b_2 + \mathrm{j}b_1a_2 - \mathrm{j}a_1b_2}{a_2^2 - \mathrm{j}^2b_2^2 + \mathrm{j}b_2a_2 - \mathrm{j}b_2a_2}$$

Since $\mathrm{j}^2 = -1$, then

$$\frac{\mathbf{V}_1}{\mathbf{V}_2} = \frac{a_1a_2 + b_1b_2}{a_2^2 + b_2^2} + \mathrm{j}\left(\frac{b_1a_2 - a_1b_2}{a_2^2 + b_1^2}\right) \qquad [16]$$

Multiplication of phasors when they are in polar form involves the multiplication of their lengths and the addition of their phase angles, i.e.

$$\mathbf{V}_1\mathbf{V}_2 = V_1V_2\angle\phi_1 + \phi_2 \qquad [17]$$

A simple proof of the above can be obtained using the trigonometrical representation of phasors. Thus if

$$\mathbf{V}_1 = V_1(\cos\phi_1 + \mathrm{j}\sin\phi_1)$$

and

$$\mathbf{V}_2 = V(\cos\phi_2 + \mathrm{j}\sin\phi_2)$$

then

$$\begin{aligned}\mathbf{V}_1\mathbf{V}_2 &= V_1V_2(\cos\phi_1 + \mathrm{j}\sin\phi_1)(\cos\phi_2 + \mathrm{j}\sin\phi_2)\\ &= V_1V_2[(\cos\phi_1\cos\phi_2 - \sin\phi_1\sin\phi_2)\end{aligned}$$

$$+ \mathrm{j}(\sin\phi_1\cos\phi_2 + \cos\phi_1\sin\phi_2)]$$

$$= V_1V_2[\cos(\phi_1 + \phi_2) + \mathrm{j}\sin(\phi_1 + \phi_2)]$$

The product is thus a phasor of magnitude V_1V_2 and phase angle $(\phi_1 + \phi_2)$.

Division of phasors in polar form is

$$\frac{\mathbf{V_1}}{\mathbf{V_2}} = \frac{V_1\angle\phi_1}{V_2\angle\phi_2} = \frac{V_1\angle\phi_1}{V_2\angle\phi_2} \times \frac{V_2\angle-\phi_2}{V_2\angle-\phi_2} = \frac{V_1V_2\angle\phi_1-\phi_2}{V_2^2\angle\phi_2-\phi_2}$$

$$\frac{\mathbf{V_1}}{\mathbf{V_2}} = \frac{V_1}{V_2}\angle\phi_1-\phi_2 \qquad [18]$$

Example 4

Determine the value of the phasor, in polar form, that represents $\mathbf{V_1}/\mathbf{V_2}$, when (*a*) $\mathbf{V_1} = 10\angle 20°$ and $\mathbf{V_2} = 5\angle 40°$, (*b*) $\mathbf{V_1} = 4\angle -10°$ and $\mathbf{V_2} = 8\angle 30°$.

Answer

(*a*) Using equation [18]

$$\frac{\mathbf{V_1}}{\mathbf{V_2}} = \frac{10}{2}\angle 20° - 40° = 5\angle -20°$$

(*b*) Using equation [18]

$$\frac{\mathbf{V_1}}{\mathbf{V_2}} = \frac{4}{8}\angle -10° - 30° = 0.5\angle -40°$$

Example 5

Determine the phasor representing the product of the two phasors 2 + j3 and 3 − j4.

Answer

From first principles the product is

$$(2 + \mathrm{j}3)(3 - \mathrm{j}4) = 6 - \mathrm{j}^2 12 + \mathrm{j}9 - \mathrm{j}8 = 18 + \mathrm{j}1$$

Example 6

If two voltages are represented by the phasors $\mathbf{V_1} = 5 + \mathrm{j}2$ and $\mathbf{V_2} = 4 + \mathrm{j}3$, determine the value of $\mathbf{V_1}/\mathbf{V_2}$.

Answer

From first principles

$$\frac{\mathbf{V_1}}{\mathbf{V_2}} = \frac{5 + \mathrm{j}2}{4 + \mathrm{j}3}$$

Multiplying the numerator and the denominator by 4 − j3 gives

$$\frac{\mathbf{V_1}}{\mathbf{V_2}} = \frac{(5 + \mathrm{j}2)(4 - \mathrm{j}3)}{(4 + \mathrm{j}3)(4 - \mathrm{j}3)} = \frac{20 - \mathrm{j}^2 6 + \mathrm{j}8 - \mathrm{j}15}{16 - \mathrm{j}^2 9 + \mathrm{j}12 - \mathrm{j}12}$$

$$= \frac{26}{16+9} - \mathrm{j}\frac{7}{16+9}$$
$$= 1.04 - \mathrm{j}0.28$$

Current and voltage phasors

For a pure *resistor* the current through it and the potential difference across it are in phase. Thus the phasors are as shown in Fig. 3.14 and can be represented as $\mathbf{V} = V\angle 0°$ and $\mathbf{I} = I\angle 0°$.

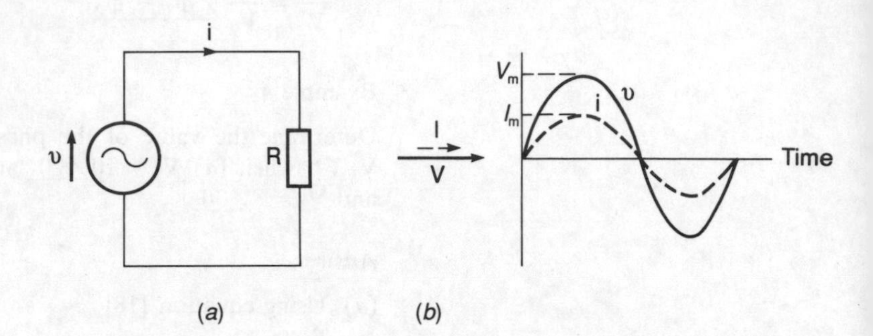

Fig. 3.14 An a.c. circuit containing only resistance

For a circuit containing only *inductance* (Fig. 3.15) a changing current produces a back e.m.f. of $L\mathrm{d}i/\mathrm{d}t$, where L is the inductance and $\mathrm{d}i/\mathrm{d}t$ the rate of change of the current. Thus applying Kirchoff's voltage law to the circuit,

$$v - \frac{\mathrm{d}i}{\mathrm{d}t} = 0$$

and so

$$\frac{\mathrm{d}i}{\mathrm{d}t} = \frac{v}{L} = \frac{V_\mathrm{m}}{L}\sin\omega t$$

$$\int_0^i \mathrm{d}i = \frac{V_\mathrm{m}}{L}\int_0^t \sin\omega t$$

$$i = -\frac{V_\mathrm{m}}{\omega L}\cos\omega t$$

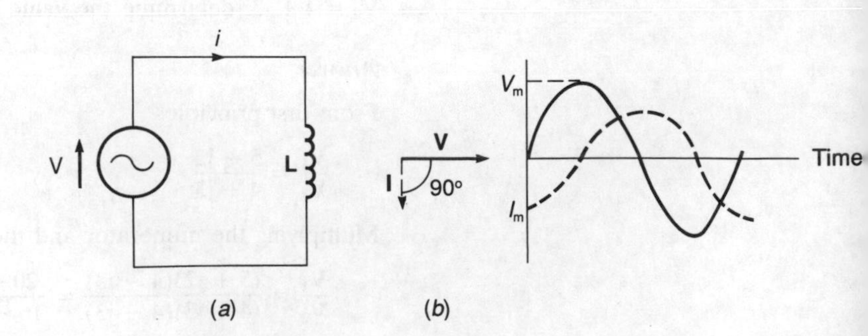

Fig. 3.15 An a.c. circuit containing only inductance

But $-\cos\omega t = \sin(\omega t - 90°)$. Hence

$$i = \frac{V_m}{\omega L}\sin(\omega t - 90°) \qquad [19]$$

The current thus lags the voltage by 90° (Fig. 3.15(*b*)). This means that the maximum value of the current does not occur at the same time as the maximum voltage. The maximum value of the current is $I_m = V_m/\omega L$. The term V_m/I_m is called the *inductive reactance* X_L. Thus

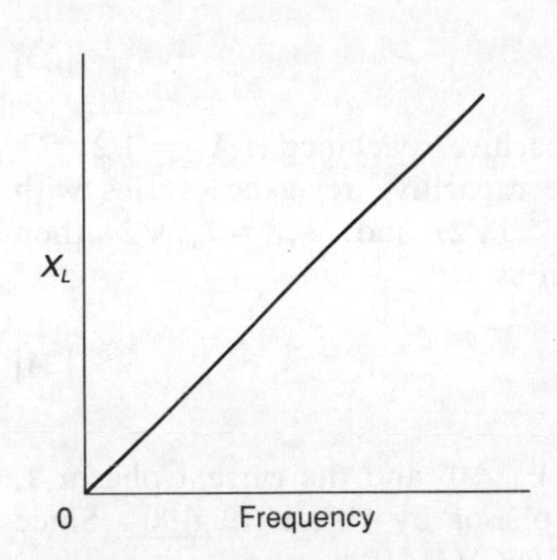

Fig. 3.16 Variation of X_L with frequency

$$X_L = \frac{V_m}{I_m} = \omega L \qquad [20]$$

Since reactance is volts/amps it has the unit of ohms. Since $\omega = 2\pi f$, equation [20] can also be written as $X_L = 2\pi fL$. The inductive reactance is thus directly proportional to the frequency (Fig. 3.16). Since $V_{rms} = V_m/\sqrt{2}$ and $I_{rms} = I_m/\sqrt{2}$ then equation [19] can be written as

$$\frac{V_{rms}}{I_{rms}} = \frac{V_m}{I_m} = \omega L = X_L \qquad [21]$$

The voltage phasor V is $V_m\angle 0°$ and the current phasor **I**, since it lags the voltage phasor by 90°, is $I_m\angle -90°$. Since $V_m = \omega L I_m$ then V is $\omega L I_m\angle 0°$.

Consider an a.c. circuit in which a sinusoidal voltage is applied to just *capacitance* (Fig. 3.17(*a*)). The current in the circuit is given by $i = C\mathrm{d}v/\mathrm{d}t$. Thus if

$$v = V_m \sin\omega t$$

then

$$i = CV_m\frac{\mathrm{d}(\sin\omega t)}{\mathrm{d}t} = \omega CV_m\cos\omega t$$

and thus, since $\cos\omega t = \sin(\omega t + 90°)$,

$$i = \omega CV_m\sin(\omega t + 90°) \qquad [22]$$

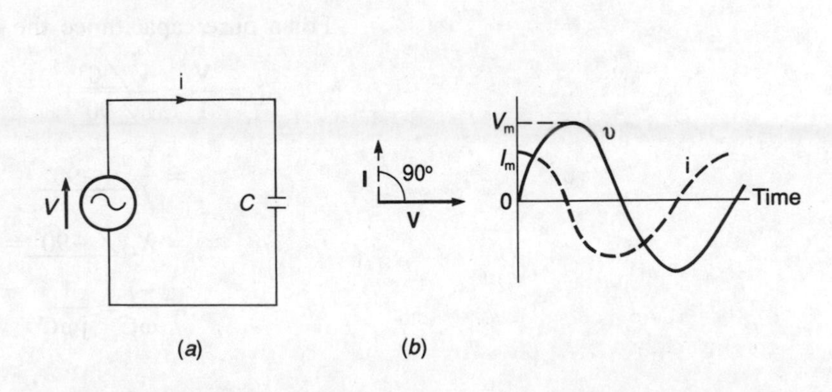

Fig. 3.17 An a.c. circuit containing only capacitance

The current in the circuit thus leads the voltage by 90° (Fig. 3.17(*b*)). The maximum value of the current I_m occurs at a different time to the maximum voltage, the maximum current being given by $I_m = \omega C V_m$. The term *capacitive reactance* X_C is used for V_m/I_m, having the units of ohms, and thus

$$X_C = \frac{V_m}{I_m} = \frac{1}{\omega C} \qquad [23]$$

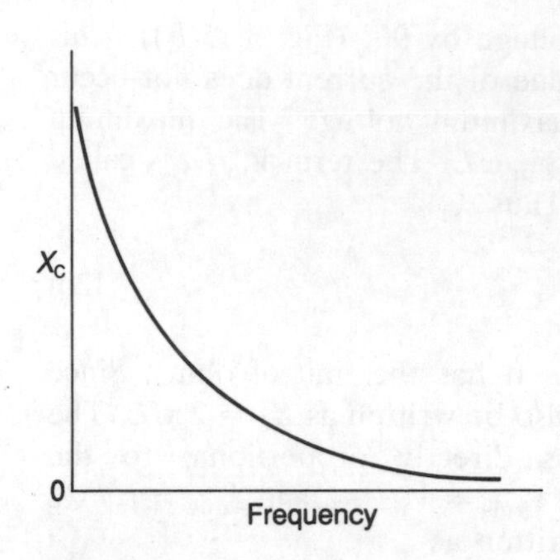

Fig. 3.18 Variation of X_C with frequency

Since $\omega = 2\pi f$ then the capacitive reactance is $X_C = 1/2\pi f C$. Figure 3.18 shows how the capacitive reactance varies with frequency. Since $V_{rms} = V_m/\sqrt{2}$ and $I_{rms} = I_m/\sqrt{2}$ then equation [23] can be written as

$$\frac{V_{rms}}{I_{rms}} = \frac{V_m}{I_m} = \frac{1}{\omega C} = X_C \qquad [24]$$

The voltage phasor **V** is $V_m\angle 0°$ and the current phasor **I**, since it leads the voltage phasor by 90°, is $I_m\angle 90°$. Since $V_m = (1/\omega C)I_m$ then **V** is $(1/\omega C)I_m\angle 0°$.

Impedance and admittance

The term *impedance* Z is used to describe the relationship between the voltage phasor **V** and the current phasor **I**, this including both their magnitude and phase relationships.

$$Z = \frac{\mathbf{V}}{\mathbf{I}} \qquad [25]$$

The units of impedance are ohms (Ω). For a pure resistor

$$Z = \frac{V\angle 0°}{I\angle 0°} = \frac{V}{I}\angle 0° = R\angle 0° \qquad [26]$$

For a pure inductance the circuit impedance Z_L is

$$Z_L = \frac{\mathbf{V}}{\mathbf{I}} = \frac{V\angle 0°}{I\angle -90°} = \frac{V}{I}\angle 90°$$

$$= X_L\angle 90° = jX_L = j\omega L \qquad [27]$$

For a pure capacitance the circuit impedance Z_C is

$$Z_C = \frac{\mathbf{V}}{\mathbf{I}} = \frac{V\angle 0°}{I\angle 90°}$$

$$= \frac{V}{I}\angle -90°$$

$$= X_C\angle -90° = -jX_C$$

$$= \frac{-j}{\omega C} = \frac{1}{j\omega C} \qquad [28]$$

Dividing one phasor by another leads to a result which has a magnitude and a phase angle. Since we can represent phasors by the sum of a real number and an imaginary number, i.e. a number involving j, then we can also represent Z as the sum of a real number and an imaginary number.

$$Z = R + jX \qquad [29]$$

where the real term is resistance R and the j, or so-called imaginary, term is *reactance* X, both having units of ohms (Ω). This can be represented on the Argand diagram in the way shown in Fig. 3.19 and forms what is often termed the *impedance triangle*. In polar form the impedance can be represented as $|Z|\angle\phi$, where $|Z|$ is the magnitude of the impedance. Though impedance is the ratio of two phasors it is not itself a phasor. There is no sinusoidal waveform to be generated by the impedance. It is just a complex number which relates both the magnitudes and the phases of the V and I.

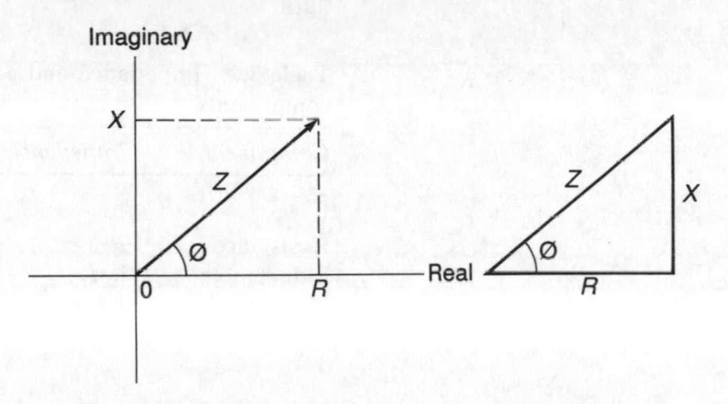

Fig. 3.19 The impedance triangle

For the impedance triangle we have for relationships between the magnitudes of Z, X and R

$$\tan\phi = \frac{X}{R} \qquad [30]$$

$$\cos\phi = \frac{R}{Z} \qquad [31]$$

$$\sin\phi = \frac{X}{Z} \qquad [32]$$

For a pure resistance, where $Z = R$, there is only a real term and no imaginary term, i.e. no reactance. For a pure inductance, where $Z = j\omega L$, there is no real term but only an imaginary term, i.e. only reactance. The reactance is thus ωL. For a pure capacitance, where $Z = 1/j\omega C = -j/\omega C$, there is no real term but only an imaginary term, i.e. only reactance.

Admittance Y is defined as being the reciprocal of impedance, i.e.

$$Y = \frac{1}{Z} \qquad [33]$$

Hence Y = I/V and has units of ohm^{-1} or siemens (S). While impedance in a.c. circuits is analogous to resistance in d.c. circuits, admittance in a.c. circuits is analogous to conductance in d.c. circuits. Like impedance, admittance can be written in terms of a real part and an imaginary part,

$$Y = G + jB \qquad [34]$$

where the real part G is called the *conductance* and the imaginary part B is the *susceptance*. Both conductance and susceptance have units of siemens (S).

For a pure resistance, where $Z = R$, then $Y = 1/R$ and thus since this is a real term $G = 1/R$. For a pure inductance, where $Z = j\omega L$, then $Y = 1/j\omega L = -j/\omega L$. Thus the susceptance is $-1/\omega L$. For a pure capacitance, where $Z = 1/j\omega C$, then $Y = j\omega C$. Thus the susceptance is ωC. Table 3.1 tabulates this data.

Table 3.1 Impedance and admittance of basic components

Component	*Impedance*	*Admittance*
Resistance	R	$1/R$
Inductance	$j\omega L$	$1/j\omega L - j\omega L$
Capacitance	$1/j\omega C = -j/\omega C$	$j\omega C$

Example 7

Determine the impedance and the admittance for an inductor with purely inductance of 0.5 H when the angular frequency is 1000 rad/s.

Answer

The impedance Z is $j\omega L$ and thus j500 Ω. The admittance is $1/j\omega L$ or $-j/\omega L$, hence $-$j0.002 S. The impedance can also be written as $500\angle 90°$ Ω and the admittance as $0.002\angle -90°$ S.

Example 8

Determine the voltage, of angular frequency 1000 rad/s, which when applied to a 2 μF capacitor results in a current of $2.5\angle 90°$ A.

Answer

The circuit impedance Z is $R + X_C$. Since $R = 0$ then $Z = X_C$. Since

$$X_C = -\frac{j}{\omega C} = -\frac{j}{1000 \times 2 \times 10^{-6}} = -j500\ \Omega$$

then $Z = -j500 = 500\angle{-90°}\ \Omega$. Hence, since V/I = Z, then

$$\mathbf{V} = (2.5\angle 90°)(500\angle -90°) = 1250\angle 0°\ \text{V}$$

Kirchoff's laws with complex notation

Consider Kirchoff's laws applied to varying voltages and currents at an instant of time. Thus the voltage law can be written, for a closed circuit path, as

$$v_1 + v_2 + v_3 + \ldots = 0$$

where v_1, v_2, v_3, etc. are the values of the voltages across the components in that circuit path at the same instant of time. For sinusoidally varying voltages with

$$v_1 = V_{m1} \sin(\omega t + \phi_1)$$

$$v_2 = V_{m2} \sin(\omega t + \phi_2)$$

$$v_3 = V_{m3} \sin(\omega t + \phi_3), \text{ etc.}$$

then at steady state we have

$$V_{m1} \sin(\omega t + \phi_1) + V_{m2} \sin(\omega t + \phi_2) + V_{m3} \sin(\omega t + \phi_3) + \ldots = 0$$

But each of these alternating voltages can be represented by a phasor. Thus we have

$$\mathbf{V}_1 + \mathbf{V}_2 + \mathbf{V}_3 + \ldots = 0 \qquad [35]$$

Thus the sum of the phasor voltages round a closed circuit path is zero.

In a similar way Kirchoff's current law can be shown to give

$$\mathbf{I}_1 + \mathbf{I}_2 + \mathbf{I}_3 + \ldots = 0 \qquad [36]$$

The sum of the phasor currents at a node is zero.

Impedances in series and parallel

Consider three impedances connected in series (Fig. 3.20). Kirchoff's voltage law gives

$$\mathbf{V}_1 + \mathbf{V}_2 + \mathbf{V}_3 = \mathbf{V}$$

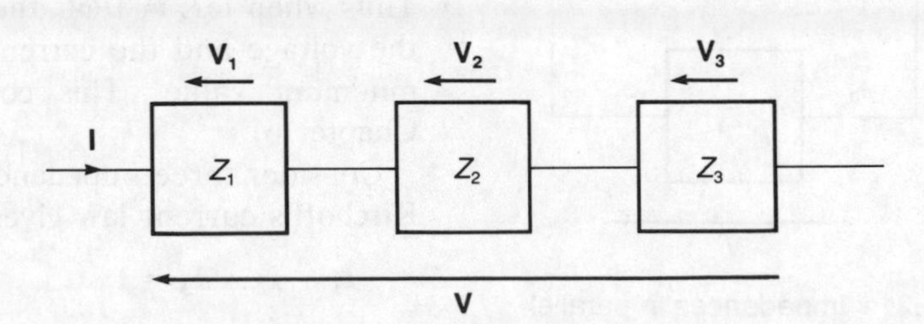

Fig. 3.20 Impedances in series

The same phasor current **I** will flow through each component. Thus

$$\frac{\mathbf{V}_1}{\mathbf{I}} + \frac{\mathbf{V}_2}{\mathbf{I}} + \frac{\mathbf{V}_3}{\mathbf{I}} = \frac{\mathbf{V}}{\mathbf{I}}$$

Thus, since $\mathbf{V}/\mathbf{I} = Z$,

$$Z_1 + Z_2 + Z_3 = Z \qquad [37]$$

The equivalent impedance of a number of series connected impedances is thus the sum of their impedances.

Thus for a resistance R in series with an inductance L, the total impedance Z is the sum of the impedances of the two components. Since the impedance for the resistor is R and that for the inductor is $\mathrm{j}\omega L$, then

$$Z = R + \mathrm{j}\omega L$$

In polar notation, the impedance has a magnitude of $\sqrt{[R^2 + (\omega L)^2]}$ and a phase angle of $\tan^{-1}(\omega L/R)$.

For a capacitor C in series with a resistor R, the total impedance Z is the sum of the impedances of the two components. Since the impedance of the resistor is R and that of the capacitor $-\mathrm{j}/\omega C$, then

$$Z = R - \mathrm{j}/\omega C$$

This can also be written as

$$Z = R + 1/\mathrm{j}\omega C$$

In polar notation, the impedance has a magnitude of $\sqrt{[R^2 + (1/\omega C)^2]}$ and a phase angle of $\tan^{-1}(1/R\omega C)$.

For a resistance R, capacitance C and inductance L in series, the total impedance Z is the sum of the impedances of the three components. Since the impedance of the resistor is R, that of the inductor $\mathrm{j}\omega L$ and that of the capacitor $-\mathrm{j}/\omega C$, then

$$Z = R + \mathrm{j}\omega L - \mathrm{j}/\omega C$$

$$Z = R + \mathrm{j}(\omega L - 1/\omega C)$$

In polar notation, the impedance has a magnitude of $\sqrt{[R^2 + (\omega L - 1/\omega C)^2]}$ and a phase angle of $\tan^{-1}[(\omega L - 1/\omega C)/R]$. Thus when $\omega L = 1/\omega C$ there is no phase difference between the voltage and the current and the impedance is just R, the minimum value. This condition is called resonance (see Chapter 6).

Consider three impedances in parallel (Fig. 3.21). Applying Kirchoff's current law gives

$$\mathbf{I}_1 + \mathbf{I}_2 + \mathbf{I}_3 = \mathbf{I}$$

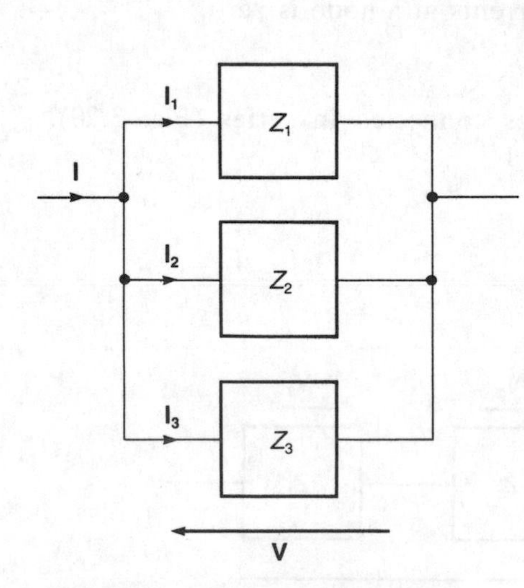

Fig. 3.21 Impedances in parallel

But there is the same voltage phasor across each impedance. Thus

$$\frac{\mathbf{I}_1}{\mathbf{V}} + \frac{\mathbf{I}_2}{\mathbf{V}} + \frac{\mathbf{I}_3}{\mathbf{V}} = \frac{\mathbf{I}}{\mathbf{V}}$$

Since $\mathbf{I}/\mathbf{V} = 1/Z$, then

$$\frac{1}{Z_1} + \frac{1}{Z_2} + \frac{1}{Z_3} = \frac{1}{Z} \qquad [38]$$

Often a more convenient way of considering parallel circuits is in terms of the admittances. Since $Y = \mathbf{I}/\mathbf{V}$, then

$$Y_1 + Y_2 + Y_3 = Y \qquad [39]$$

Thus the equivalent admittances of a number of parallel admittances is the sum of their admittances.

For a resistance and capacitance in parallel the total admittance Y is the sum of the admittances of the two components. Since the admittance for the resistance is $1/R$ and that for the capacitor $\mathrm{j}\omega C$, then

$$Y = (1/R) + \mathrm{j}\omega C$$

In polar form, the admittance has a magnitude of $\sqrt{[(1/R)^2 + (\omega C)^2]}$ and a phase angle of $\tan^{-1}(\omega CR)$.

For a resistance and inductance in parallel the total admittance Y is the sum of the admittances of the two components. Since the admittance for the resistance is $1/R$ and that for the inductor is $-\mathrm{j}/\omega L$, then

$$Y = (1/R) - \mathrm{j}/\omega L$$

In polar form, the admittance has a magnitude of $\sqrt{[(1/R)^2 + (1/\omega L)^2]}$ and a phase angle of $\tan^{-1}(1/\omega LR)$.

For a resistance, inductance and capacitance all in parallel then the total admittance is

$$Y = (1/R) - (\mathrm{j}/\omega L) + \mathrm{j}\omega C$$

$$Y = (1/R) + \mathrm{j}(\omega C - 1/\omega L)$$

In polar form, the admittance has a magnitude of $\sqrt{[(1/R)^2 + (\omega C - 1/\omega L)^2]}$ and a phase angle of $\tan^{-1}[R(\omega C - 1/\omega L)]$. When $\omega C = 1/\omega L$ there is no phase difference between the voltage and the current and the admittance is just $1/R$, the minimum value (maximum impedance). This condition is called resonance (see Chapter 6).

Example 9

Determine the voltage, of angular frequency 100 rad/s, across an inductor of inductance 50 mH and resistance 10 Ω when a current of $0.50\angle 0°$ flows through it.

Answer

The inductor can be considered to be a pure inductor in series with a pure resistor. The impedance of the inductor Z_L is just jX_L, the impedance of the resistor Z_R being just R. Hence the circuit impedance Z, which is the sum of the impedances of the series components, is

$$Z = Z_R + Z_L = R + jX_L = R + j\omega L$$
$$= 10 + j(100 \times 0.050)$$
$$= 10 + j5$$

Therefore, since $0.50\angle 0° = 0.50 + j0$,

$$\mathbf{V} = Z\mathbf{I} = (10 + j5)(0.50 + j0) = 5 + j2.5\,\text{V}$$

This can be put into polar form, using equation [9]:

$$\mathbf{V} = V\angle\phi = \sqrt{(a^2 + b^2)}\angle \tan^{-1}(b/a)$$
$$\mathbf{V} = \sqrt{(5^2 + 2.5^2)}\angle \tan^{-1}(2.5/5) = 5.59\angle 26.6°\,\text{V}.$$

Example 10

What is the potential difference across a circuit consisting of a 1 kΩ resistor, a 0.2 H inductor and a 0.1 μ*F* capacitor in series if the current through them, at a frequency of 1 kHz, is $2.0\angle 0°$ mA?

Answer

The impedance of the resistor will be 1000 Ω, that of the inductor $j\omega L = j(2\pi \times 1000 \times 0.2) = j1257\,\Omega$, and that of the capacitor $-j/\omega C = -j/(2\pi \times 1000 \times 0.1 \times 10^{-6}) = -j1592\,\Omega$. Since the three components are in series then the total impedance is the sum of the individual impedances. Thus

$$Z = 1000 + j1257 - j1592 = 1000 - j335\,\Omega$$

Hence

$$\mathbf{V} = Z\mathbf{I} = (1000 - j335)(0.002 + j0) = 2.0 - j0.67\,\text{V}$$

This can be represented in polar form, using [9], as

$$\mathbf{V} = V\angle\phi = \sqrt{(a^2 + b^2)}\angle \tan^{-1}(b/a)$$
$$\mathbf{V} = \sqrt{[2.0^2 + (-0.67)^2]}\angle \tan^{-1}(-0.67/2.0) = 2.11\angle -18.5°$$

Example 11

For the circuit shown in Fig. 3.22, determine the circuit current when the voltage input is 250 sin 1000*t* V.

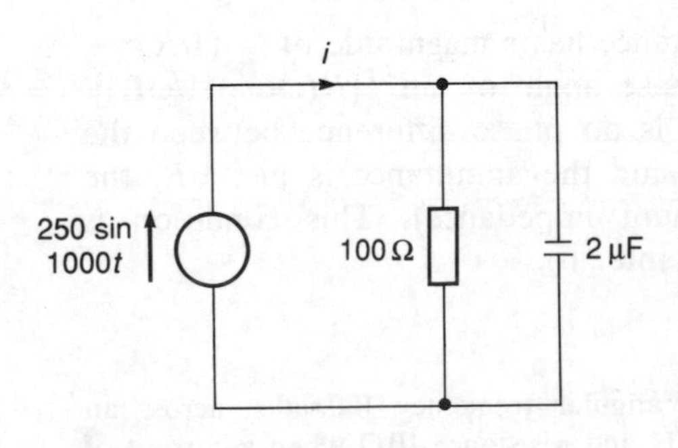

Fig. 3.22 Example 11

Answer

The admittance of the resistor is 1/100 = 0.01 S. The admittance of the capacitor is $j/X_C = j\omega C = j(1000 \times 2 \times 10^{-6}) = j0.002\,\text{S}$. Because the two components are in parallel, the total admittance is

$$Y = 0.01 + j0.002\,\text{S}$$

Since $Y = \mathbf{I}/\mathbf{V}$, and $\mathbf{V}$ can be written as $250 + j0$, then

$$\mathbf{I} = (0.01 + j0.002)(250 + j0) = 2.5 + j0.50$$

This can be represented in polar form, using [9], as

$$\mathbf{I} = I\angle\phi = \sqrt{(a^2 + b^2)}\angle\tan^{-1}(b/a)$$

$$\mathbf{I} = \sqrt{(2.5^2 + 0.50^2)}\angle\tan^{-1}0.50/2.5 = 2.55\angle 11.3°$$

Thus

$$i = 2.55 \sin(1000t + 11.3°)$$

Example 12

Determine the potential difference across the capacitor in Fig. 3.23 when there is a current input of $10 \sin(1000t + 30°)$.

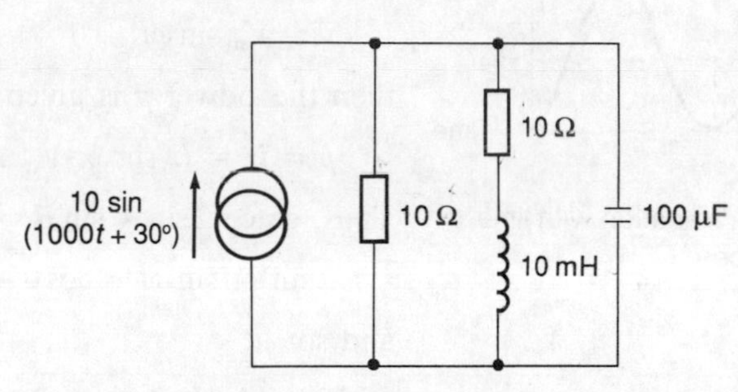

Fig. 3.23 Example 12

Answer

The admittance of the single resistor is $1/10 = 0.1$ S. The impedance of the resistor and inductor in series is $R + j\omega L$ and hence the admittance is $1/(10 + j10)$ S. The admittance of the capacitor is $j\omega C = j0.1$ S. The total admittance of the parallel arrangement is thus

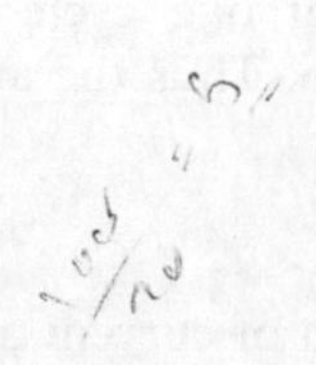

$$Y = 0.1 + \frac{1}{10 + j10} + j0.1$$

$$= 0.1 + \frac{(10 - j10)}{(10 + j10)(10 - j10)} + j0.1$$

$$= 0.1 + \left(\frac{10 - j10}{200}\right) + j0.1$$

$$= 0.15 + j0.05$$

This can be represented in polar form, using [9], as

$$Y = \sqrt{(a^2 + b^2)}\angle\tan^{-1}(b/a)$$

$$= \sqrt{(0.15^2 + 0.05^2)}\angle\tan^{-1}(0.05/0.15)$$

$$Y = 0.158\angle 18.4° \text{S}$$

Since $Y = \mathbf{I}/\mathbf{V}$, then

$$\mathbf{V} = \frac{10\angle 30°}{0.158\angle 18.4°} = 63.3\angle 11.6°$$

or

$$v = 63.3 \sin(1000t + 11.6°)$$

Power consumption

Figure 3.24 shows how the current i through and the potential difference v across a pure *resistance* vary with time, the current and the potential difference being in phase. Since when the current is positive the potential difference is positive and when the current is negative the potential difference is negative, the product iv is always positive. The power p at an instant of time is the product iv. The power variation with time is thus of the form shown in Fig. 3.24, being zero when the current and potential difference are zero and a maximum when the current and potential difference are a maximum.

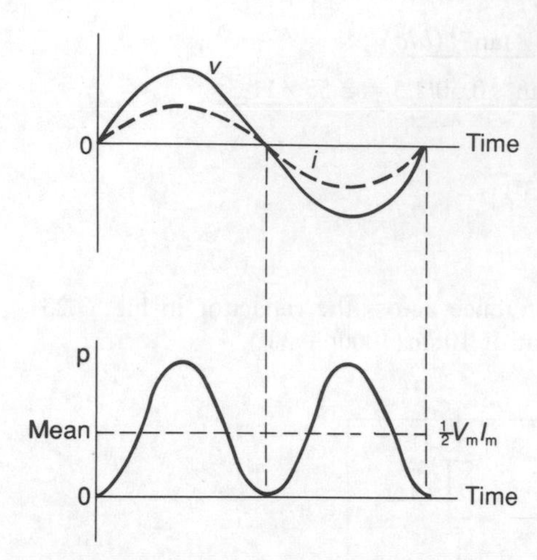

Fig. 3.24 Power variation with time for pure resistance

With the current i and the voltage v being given by

$$i = I_m \sin \omega t$$

and

$$v = V_m \sin \omega t$$

then the power p is given by

$$p = iv = I_m \sin \omega t \, V_m \sin \omega t$$

Thus, since $2 \sin A \sin B = \cos(A - B) - \cos(A + B)$,

$$2 \sin \omega t \sin \omega t = \cos 0 - \cos 2\omega t$$

and so

$$p = \tfrac{1}{2} I_m V_m (1 - \cos 2\omega t) \quad [40]$$

This is the equation describing how the power varies with time for a pure resistance. It is a cosine variation with time about $\frac{1}{2}I_m Vm$ with a frequency which is twice that of the alternating current or potential difference. Over one cycle the average value of the cos $2\omega t$ term will be zero. Thus the average value of the power is

$$\text{average power} = \tfrac{1}{2} I_m V_m \quad [41]$$

For a pure *inductance* the variation of current and potential difference with time is as shown in Fig. 3.25. The power is iv and thus is zero whenever the current i or the potential difference v are zero. The variation of power with time is thus of the form shown in Fig. 3.25.

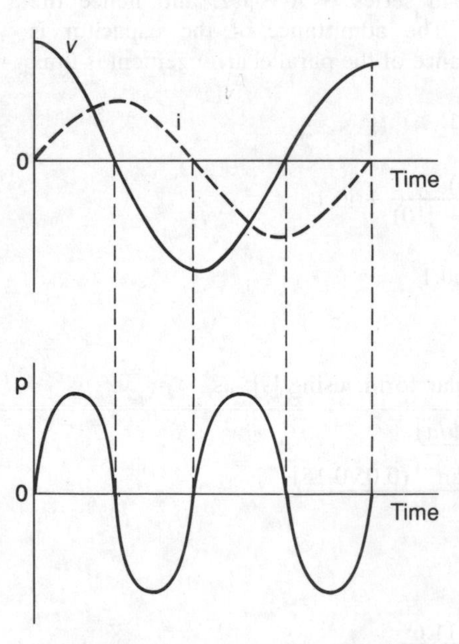

Fig. 3.25 Power variation with time for pure inductance

$$p = iv = I_m \sin \omega t \, V_m \cos \omega t$$

Thus, since

$$2 \sin A \cos B = \sin(A + B) + \sin(A - B)$$

then

$$2 \sin \omega t \cos \omega t = \sin 2\omega t + 0$$

and so

$$p = \tfrac{1}{2} I_m V_m \sin 2\omega t \quad [42]$$

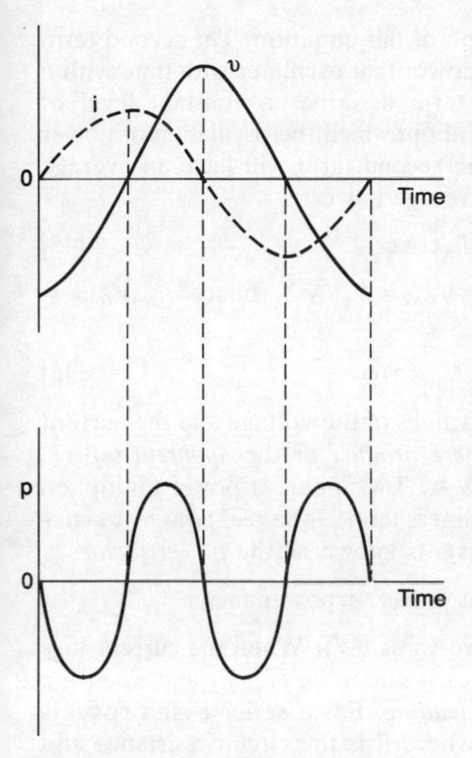

Fig. 3.26 Power variation with time for pure capacitance

Figure 3.25 shows the graph of this equation. The power alternates about the zero axis with a frequency which is twice that of the current or potential difference. Over one cycle the average value of the power is zero. This is because in that part of the cycle where the current i is positive energy is being stored in the magnetic field of the inductor, while in that part of the cycle where the current is negative the magnetic field is releasing its energy into the circuit.

For a pure *capacitance* the variation of current and potential difference with time is as shown in Fig. 3.26. The power p is zero whenever the current i or the potential difference v is zero. The variation of the power with time is thus of the form shown in Fig. 3.26.

$$p = iv = I_m \sin \omega t \, V_m \sin(\omega t - 90°)$$

Thus, since

$$2 \sin A \sin B = \cos(A - B) - \cos(A + B)$$

then

$$2 \sin \omega t \sin(\omega t - 90°) = \cos 90° - \cos(2\omega t - 90°)$$
$$= 0 - \sin 2\omega t$$

and so

$$p = \tfrac{1}{2} I_m V_m \sin 2\omega t \qquad [43]$$

Figure 3.26 shows the graph of this equation. The power alternates about the zero axis with a frequency which is twice that of the current or potential difference. Over one cycle the average value of the power is zero. This is because while the voltage is positive the capacitor is storing energy in its electric field, while during that part of the cycle when the voltage is negative the capacitor is releasing energy into the circuit.

In general, when there is a phase difference ϕ between the voltage and the current (Fig. 3.27),

$$v = V_m \sin \omega t$$
$$i = I_m \sin(\omega t - \phi)$$
$$p = iv = V_m I_m \sin \omega t \sin(\omega t - \phi)$$

Thus, since

$$2 \sin A \sin B = \cos(A - B) - \cos(A + B)$$
$$2 \sin \omega t \sin(\omega t - \phi) = \cos \phi - \cos(2\omega t - \phi)$$

and so

$$p = \tfrac{1}{2} V_m I_m \cos \phi - \tfrac{1}{2} V_m I_m \cos(2\omega t - \phi) \qquad [44]$$

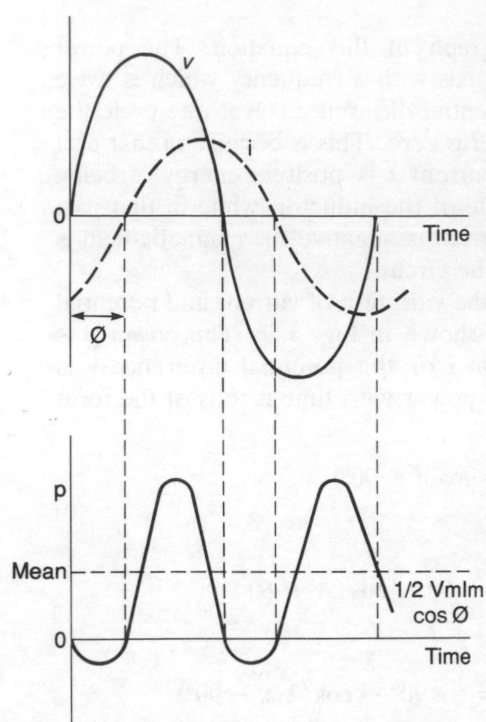

Fig. 3.27 Power variation with time when there is a phase difference ϕ between the voltage and the current

Figure 3.27 shows the graph of this equation. The second term in the equation describes power that oscillates with time with a frequency 2ω. The first term describes a constant level of power and is the base line about which the oscillation of power occurs. Over one cycle the second term will have an average value of zero. Thus the average power is

$$\text{average power} = \tfrac{1}{2}V_m I_m \cos\phi \qquad [45]$$

But $I_{rms} = I_m/\sqrt{2}$ and $V_{rms} = V_m/\sqrt{2}$, hence $I_{rms}V_{rms} = \tfrac{1}{2}I_m V_m$ and so

$$\text{average power} = V_{rms}I_{rms}\cos\phi \qquad [46]$$

The product of the r.m.s. values of the voltage and the current is known as the *volt–ampere product* or the *apparent power*. This has units written as V A. The apparent power multiplied by the factor $\cos\phi$ gives what is termed the *real power* or *active power*. For this reason $\cos\phi$ is known as the *power factor*.

$$\text{real power} = \text{apparent power} \times \text{power factor}$$

The units of real power are watts (W). When the current lags the voltage the power factor is said to be *lagging* and when the current leads the voltage *leading*. For a series circuit $\cos\phi$ is R/Z (see equation [31]), where R is the circuit resistance and Z the circuit impedance, and thus the power factor is R/Z.

An alternator may be specified as having an apparent power of 100 kVA. The phase difference between the current and the voltage will depend on the load connected to the generator. Thus if the power factor of $\cos\phi$ is unity, i.e. $\phi = 0°$, then the delivered power is equal to the apparent power of 100 kVA. If, however, the power factor is say 0.5 then the delivered power is only 50 kVA. The higher the power factor, i.e. the lower ϕ, the greater the power that can be delivered by a given alternator.

Example 13

A circuit has a sinusoidal voltage input of 100 V r.m.s., a current of 5 A r.m.s., and the phase angle between the current and the voltage is 30°. What is the average power consumed by the circuit?

Answer

Using equation [46]

$$\text{average power} = V_{rms}I_{rms}\cos\phi$$

$$= 100 \times 5 \times \cos 30° = 433\ \text{W}$$

Power triangle

Consider a circuit for which we have

$$v = V_m \sin\omega t$$

and

$$i = I_m \sin(\omega t - \phi)$$

The phasors V and I are thus as shown in Fig. 3.28(*a*). The current phasor can be replaced by two phasors, at right angles to each other, of lengths $I_m \cos\phi$ and $I_m \sin\phi$ (Fig. 3.28(*b*)). If we multiply all the sides of the phasor diagram in (*b*) by $\frac{1}{2}V_m$, then we have for one side $\frac{1}{2}V_m I_m \cos\phi$, i.e. the real power, and for the diagonal $\frac{1}{2}V_m I_m$, i.e. the apparent power. The other side is $\frac{1}{2}V_m I_m \sin\phi$ and this is called the *reactive power* and has units which are written as VAR or VAr. The apparent power, reactive power and real power are thus represented by the sides of a triangle, known as the *power triangle*, as shown in Fig. 3.28(*d*). Hence

$$\frac{\text{reactive power}}{\text{apparent power}} = \sin\phi$$

Thus

$$\text{apparent power} = V_{rms} I_{rms}$$

$$\text{real power } P = V_{rms} I_{rms} \cos\phi$$

$$\text{reactive power } Q = V_{rms} I_{rms} \sin\phi$$

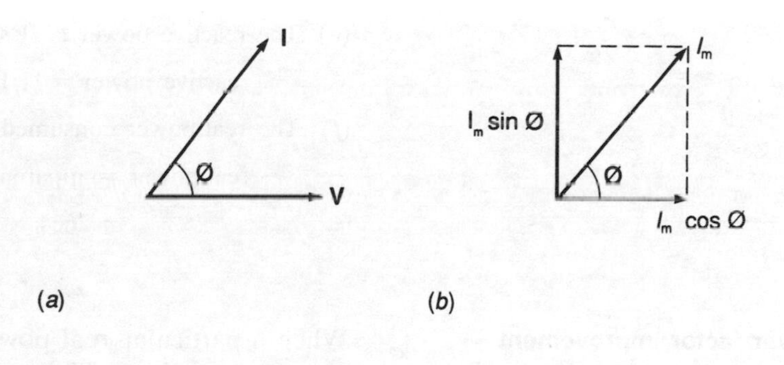

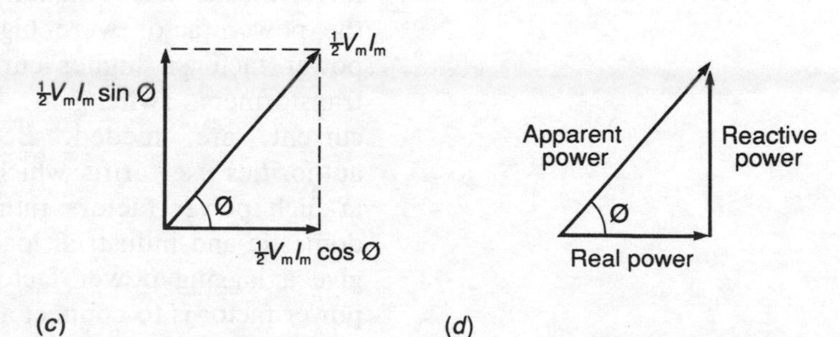

Fig. 3.28 (*a*) The phasors, (*b*) phasor diagram for the current, (*c*) the power diagram, (*d*) the power triangle

Example 14

A coil has an inductance of 50 mH and a resistance of 15 Ω and is connected to a 240 V r.m.s., 50 Hz, voltage supply. What is (*a*) the circuit impedance, (*b*) the current, (*c*) the apparent power, (*d*) the power factor, (*e*) the reactive power, (*f*) the real power?

Answer

(*a*) For a series arrangement of resistance and inductance, the impedance Z is given by

$$Z = \sqrt{(X_L^2 + R^2)}$$

Since $X_L = 2\pi fL$, then

$$Z = \sqrt{[(2\pi \times 50 \times 0.050)^2 + 15^2]} = 21.7\,\Omega$$

(*b*) Using $V = ZI$, then the r.m.s. current is

$$I = \frac{240}{21.7} = 11.1\,\text{A}$$

(*c*) The apparent power is $I_{rms}V_{rms}$ and thus

$$\text{apparent power} = 11.1 \times 240 = 2664\,\text{V A}$$

(*d*) For an inductance in series with a resistance, the power factor $\cos\phi$ is

$$\cos\phi = \frac{R}{Z} = \frac{15}{21.7} = 0.69$$

Since with such an arrangement the voltage leads the current then the power factor is said to be lagging.

(*e*) The reactive power is $IV\sin\phi$ and, since ϕ is 46.4°, is thus

$$\text{reactive power} = 11.1 \times 240 \times \sin 46.4° = 1929\,\text{V Ar}$$

(*f*) The real power consumed is thus

$$\text{real power} = \text{apparent power} \times \text{power factor}$$
$$= 2664 \times 0.69 = 1838\,\text{W}$$

Power factor improvement

When a particular real power is to be delivered, a low power factor means that a higher apparent power is required than if the power factor were high. This means that with the low power factor a higher current is required and thus cables, transformers, switchgear, etc., which can take the higher current, are needed. Because of this, electrical supply authorities use tariffs which encourage consumers to operate at high power factors rather than low power factors. Most domestic and industrial loads tend to be inductive and hence give a lagging power factor. One method of improving the power factor is to connect a capacitor in parallel with the load.

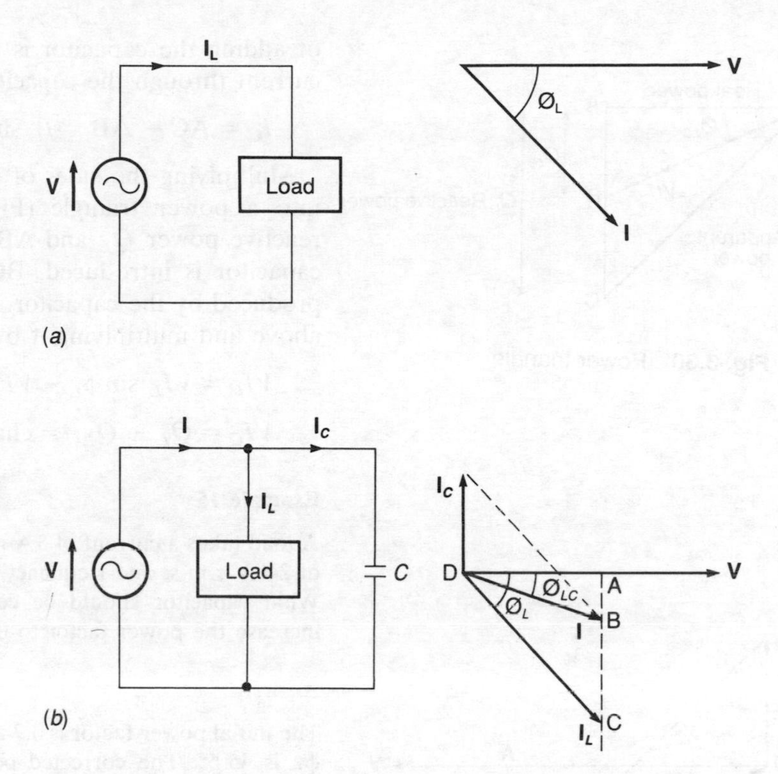

Fig. 3.29 Improving the power factor

This is equivalent to moving the load impedance closer to resonance at the supply frequency. Indeed a power factor of 1 means $\phi = 0°$ and so resonance occurs.

Figure 3.29(a) shows an inductive load, Fig. 3.29(b) the same load but with a capacitor in parallel with it to improve the power factor. With no capacitor we have a lagging phase angle ϕ_L. When the capacitor is connected in parallel, a current phasor $\mathbf{I_C}$ is added which leads the voltage by 90°. Thus the current **I** taken from the supply is now at a lower phase angle ϕ_{LC}.

Initially we have

$$\frac{\text{AC}}{I_L} = \sin \phi_L$$

$$\text{AC} = I_L \sin \phi_L$$

Reactive power is $IV \sin \phi$. Thus the initial reactive power is proportional to AC. After connecting the capacitor we have

$$\frac{\text{AB}}{I} = \sin \phi_{LC}$$

$$\text{AB} = I \sin \phi_{LC}$$

Thus the new reactive power is proportional to AB. The effect of adding the capacitor is to reduce the reactive power. The current through the capacitor I_C is

$$I_C = AC - AB = I_L \sin\phi_L - I \sin\phi_{LC} \qquad [47]$$

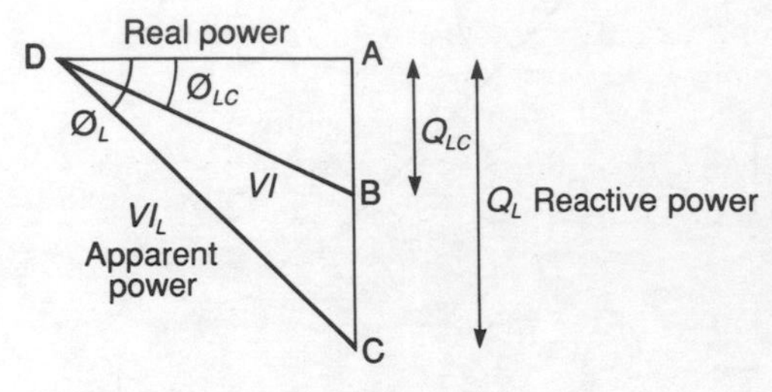

Fig. 3.30 Power triangle

Multiplying the sides of the triangle ACD by V converts it into a power triangle (Fig. 3.30). Then AC is the initial reactive power Q_L and AB the reactive power Q_{LC} after the capacitor is introduced. BC is the change in reactive power produced by the capacitor. But, using the equation developed above and multiplying it by V,

$$VI_C = VI_L \sin\phi_L - VI \sin\phi_{LC}$$

$$VI_C = Q_L - Q_{LC} = \text{change in reactive power} \qquad [48]$$

Example 15

A load takes a current of 5 A r.m.s. when connected to an a.c. supply of 240 V r.m.s. and frequency 50 Hz at a lagging power factor of 0.7. What capacitor should be connected in parallel with the load to increase the power factor to 0.9?

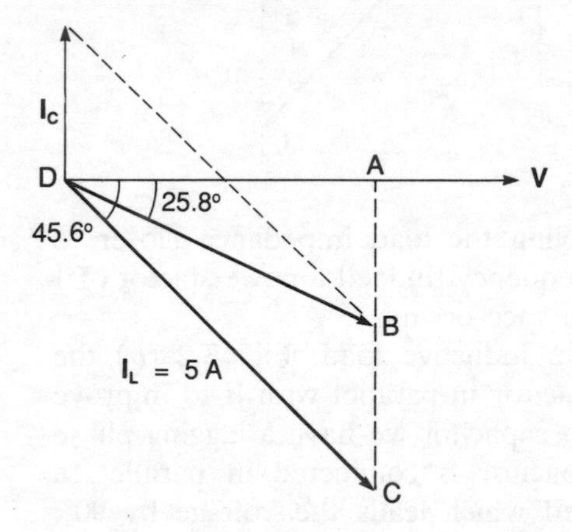

Fig. 3.31 Example 15

Answer

The initial power factor is 0.7 and, since this is $\cos\phi_L$, the phase angle ϕ_L is 45.6°. The corrected power factor $\cos\phi_{LC}$ is 0.9 and so the corrected phase angle ϕ_{LC} is 25.8°. The phasor diagram for the currents is thus as shown in Fig. 3.31. For the triangle ACD we have

$$\frac{AC}{5} = \sin 45.6°$$

$$AC = 3.57\,A$$

For the triangle ABD we have

$$\frac{AB}{AD} = \tan 25.8°$$

But AD = 5 cos 45.6°, hence AB = 1.69 A. Thus I_C, which is equal to BC, is 3.57 − 1.69 = 1.88 A.

For the capacitor we have $V = X_C I_C$, hence

$$X_C = \frac{240}{1.88} = 128\,\Omega$$

But $X_C = 1/2\pi fC$, hence

$$C = \frac{1}{2\pi f X_C} = \frac{1}{2\pi \times 50 \times 128} = 24.9\,\mu F$$

Example 16

Two loads are connected in parallel to an a.c. supply. One load has an apparent power of 5 kVA at a power factor of 0.8 leading and the other 8 kVA at a power factor of 0.6 lagging. What is the total load and power factor?

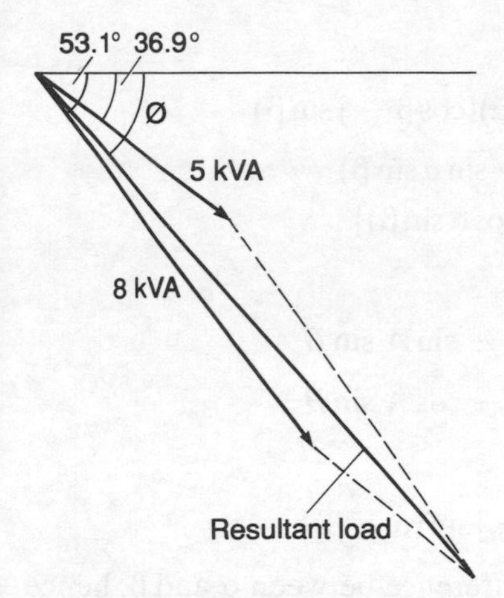

Fig. 3.32 Example 16

Answer

If the first load takes a current with a phasor of length I_1 then it will lag the voltage by ϕ_1, where $\cos\phi_1 = 0.8$. If the second load takes a current with a phasor of length I_2 then it will lag the voltage by ϕ_2, where $\cos\phi_2 = 0.6$. Since the potential difference V across each load is the same, then the lengths of the current phasors will be proportional to their apparent powers. Thus the phasor diagram is as shown in Fig. 3.32.

To simplify the calculations we can convert the phasor diagram in Fig. 3.32 into the form shown in Fig. 3.33. Then the resultant load is

$$\text{load} = \surd[(5\cos 36.9° + 8\cos 53.1°)^2 + (5\sin 36.9° + 8\sin 53.1°)^2]$$
$$= 12.9\,\text{kVA}$$

The power factor is lagging at

$$\cos\phi = \frac{5\cos 36.9° + 8\cos 53.1°}{12.9} = 0.68$$

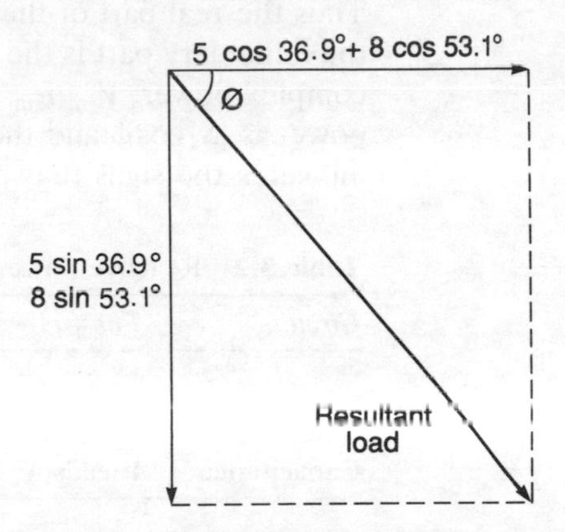

Fig. 3.33 Example 16

Complex power

Suppose we have a circuit supplied with a sinusoidal voltage of $V_{rms}\angle\alpha$ and the current taken by the circuit is $I_{rms}\angle\beta$ (Fig. 3.34). Now consider the significance of the product of $V_{rms}\angle\alpha$ and $I_{rms}\angle-\beta$. Note the minus sign with the phase angle for the current. The minus sign means that we are considering not the current phasor but what is termed the *conjugate*. If $\mathbf{I} = c + \mathrm{j}d$, then $I\angle-\beta$ is $c - \mathrm{j}d$ (Fig. 3.35). The complex conjugate phasor is denoted by $\mathbf{I}^*$. The product $\mathbf{V}_{rms}\mathbf{I}^*_{rms}$ is called the *complex power* (S).

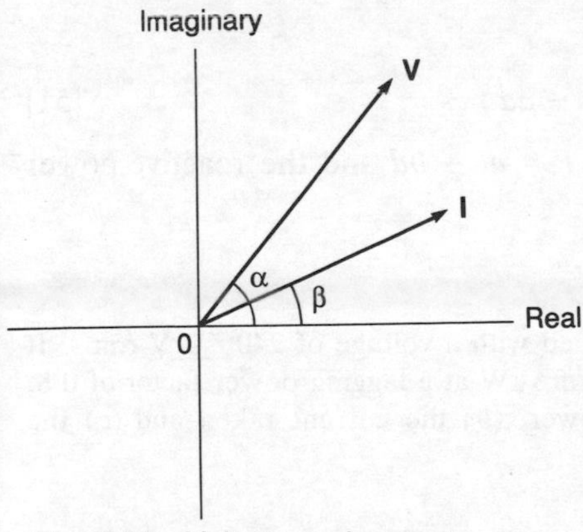

Fig. 3.34 Voltage and current phasors

$$S = \mathbf{V}_{rms}\mathbf{I}^*_{rms} \qquad [49]$$

The complex power has the unit VA. Using the trigonometrical way of expressing phasors, then $V_{rms}\angle\alpha$ is

$$\mathbf{V}_{rms} = V_{rms}(\cos\alpha + \mathrm{j}\sin\alpha)$$

and $I_{rms}\angle-\beta$ is

$$\mathbf{I}^*_{rms} = I_{rms}(\cos\beta - \mathrm{j}\sin\beta)$$

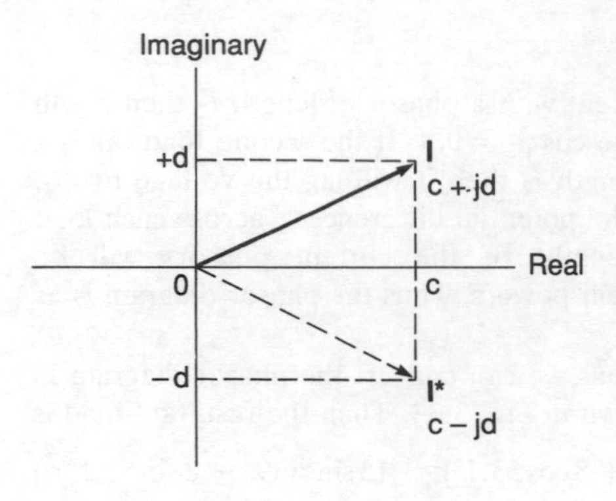

Fig. 3.35 The complex conjugate

Hence

$$S = V_{rms}I_{rms}(\cos\alpha + \mathrm{j}\sin\alpha)(\cos\beta - \mathrm{j}\sin\beta)$$
$$= V_{rms}I_{rms}[(\cos\alpha\cos\beta + \sin\alpha\sin\beta) + \mathrm{j}(\sin\alpha\cos\beta - \mathrm{j}\cos\alpha\sin\beta)]$$

But

$$\cos(A - B) = \cos A\cos B + \sin A\sin B$$
$$\sin(A - B) = \sin A\cos B - \cos A\sin B$$

Hence

$$S = V_{rms}I_{rms}[\cos(\alpha - \beta) + \mathrm{j}\sin(\alpha - \beta)]$$

But the phase angle ϕ is the difference between α and β, hence

$$S = V_{rms}I_{rms}\cos\phi + \mathrm{j}V_{rms}I_{rms}\sin\phi \qquad [50]$$

Thus the real part of the complex power is the real power and the imaginary part is the reactive power. The magnitude of the complex power, $V_{rms}I_{rms}$, or $|S|$ is the apparent power, the real power is $|S|\cos\phi$ and the reactive power is $|S|\sin\phi$. Table 3.2 indicates the signs that occur for reactive power.

Table 3.2 Reactive power

Circuit	*Phasors*	*Reactive power Q*	*Complex power S*
Inductive	**I** lags **V**	positive	$P + \mathrm{j}Q$
Capacitance	**I** leads **V**	negative	$P - \mathrm{j}Q$

If we represent $\mathbf{V}_{rms}$ as $(a + \mathrm{j}b)$ and $\mathbf{I}_{rms}$ as $(c + \mathrm{j}d)$, then $\mathbf{I}^*_{rms}$ is $(c - \mathrm{j}d)$ and so

$$S = (a + \mathrm{j}b)(c - \mathrm{j}d)$$
$$= (ac + bd) + \mathrm{j}(bc - ad) \qquad [51]$$

with then the real power $P = ac + bd$ and the reactive power $Q = bc - ad$.

Example 17

An electrical circuit is supplied with a voltage of $240\angle 0°$ V r.m.s. If the circuit absorbs a power of 5 kW at a lagging power factor of 0.8, what is (*a*) the complex power, (*b*) the current taken and (*c*) the circuit impedance?

Answer

(*a*) The real power P is 5 kW and so, since $P = |S|\cos\phi$, and $\cos\phi = 0.8$,

$$|S| = \frac{5000}{0.8} = 6250\,\text{V A}$$

Since $\cos\phi = 0.8$, then $\phi = 36.9°$ and $\sin\phi = 0.6$. Thus the reactive power Q is

$$Q = |S|\sin\phi = 6250 \times 0.6 = 3750\,\text{VAr}$$

Since **I** lags **V**, the reactive power is positive. Hence

$$S = P + jQ = 5000 + j3750\,\text{V A}$$

(*b*) Since $S = \mathbf{VI}^*$ then

$$\mathbf{I}^* = \frac{5000 + j3750}{240} = 20.8 + j15.6$$

Hence

$$\mathbf{I} = 20.8 - j15.6\,\text{A}$$

or, in polar form using equation [9],

$$\mathbf{I} = 26\angle -36.9°\,\text{A}$$

(*c*) The impedance Z is **V**/**I**, and so

$$Z = \frac{240}{26\angle -36.9°} = 9.2\angle 36.9°\,\Omega$$

Example 18

A load has an impedance of $(3 + j4)\,\Omega$ and is supplied with a voltage of $50\sqrt{2}\sin(10\,000t + 30°)$ V. What is (*a*) the current taken by the load, (*b*) the complex power, (*c*) the value of the capacitor that is needed in parallel with the load to improve the power factor to 0.97?

Answer

(*a*) The voltage is $50\angle 30°$ V r.m.s. and the impedance in polar form is $5\angle 53.1°$. Thus the current **I** is

$$\mathbf{I} = \frac{\mathbf{V}}{Z} = \frac{50\angle 30°}{5\angle 53.1°} = 10\angle -23.1°\,\text{A r.m.s.}$$

This can be written as $i = 10\sqrt{2}\sin(10\,000t - 23.1°)$ A.

(*b*) $\mathbf{I}^* = 10\angle 23.1°$ A r.m.s. and so the complex power is

$$S = \mathbf{VI}^* = (50\angle 30°)(10\angle 23.1°) = 500\angle 53.1°$$

This can be written in rectangular form as

$$S = 500\cos 53.1° + j500\sin 53.1° = 300 + j400\,\text{V A}$$

(*c*) A power factor of 0.97 means $\cos\phi = 0.97$ and so $\phi = 14.1°$ instead of 53.1°. This means we must have, for the same real power of 300 W,

$$P = 300 = VI\cos\phi = VI \times 0.97$$

Thus $VI = 309$ V A. Hence the new reactive power must be

$$Q = VI\sin\phi = 309\sin 14.1° = 75.3\,\text{VAr}$$

This means that the reactive power has been improved by adding $(400 - 75.3) = 324.7$ VAr leading. For the capacitor to do this then the current through the capacitor must be $I_C = Q/V = 324.7/50 = 6.5$ A r.m.s. Hence the reactance of the capacitor X_C must be $V/I = 50/6.5 = 7.7\,\Omega$. Hence, since $X_C = 1/\omega C$ and $\omega = 10\,000$ rad/s we have $C = 1/(10\,000 \times 7.7) = 13\,\mu$F.

Example 19

Three loads are connected in parallel across a $250\angle 0°$ V r.m.s. supply. Load 1 takes a power of 8 kW at a lagging power factor of 0.8, load 2 a power of 12 kW at a leading power factor of 0.6 and load 3 a power of 5 kW at a lagging power factor of 0.5. What are the currents through each load?

Answer

The complex power S for any load is $S = P + jQ$, where P is the real power and Q the reactive power. For load 1 we can write for the complex power

$$250\,\mathbf{I}_1^* = 8000 + jQ$$

But since $P = 8000 = IV \times 0.8$ then $IV = 10\,000$ and $\cos\phi = 0.8$, with the result that $\phi = 36.9°$. So Q, which is $IV \sin\phi$, is $10\,000 \times 0.6$. Hence

$$250\,\mathbf{I}_1^* = 8000 + j6000$$

$$\mathbf{I}_1^* = 32 + j24$$

Hence $\mathbf{I}_1 = 32 - j24$ A r.m.s. In polar form this becomes

$$\mathbf{I}_1 = 40\angle -36.9° \text{ A r.m.s.}$$

Similarly, for load 2 we can write

$$250\,\mathbf{I}_2^* = 12\,000 + jQ$$

Since $P = 12\,000 = IV \times 0.6$ then $IV = 20\,000$ and $\cos\phi = 0.6$, giving $\phi = 53.1°$. Then Q is $20\,000 \times \sin 53.1° = 16\,000$. Hence

$$250\,\mathbf{I}_2^* = 12\,000 - j16\,000$$

The minus sign is because the power factor is leading. Thus

$$\mathbf{I}_2^* = 48 - j64$$

$$\mathbf{I}_2 = 48 + j64 \text{ A r.m.s.}$$

In polar form this becomes

$$\mathbf{I}_2 = 80\angle 53.1° \text{ A r.m.s.}$$

Similarly, for load 3 we can write

$$250\,\mathbf{I}_3^* = 5000 + jQ$$

Since $P = 5000 = IV \times 0.5$ then $IV = 10\,000$ and $\cos\phi = 0.5$, giving $\phi = 60°$. Then Q is $10\,000 \times \sin 60° = 8660$. Hence

$$250\,\mathbf{I}_3^* = 5000 + j8660$$

$$\mathbf{I}_3^* = 20 + j35$$
$$\mathbf{I}_3 = 20 - j35 \text{ A r.m.s.}$$

In polar form this becomes

$$\mathbf{I}_3 = 40\angle 60.3° \text{ A r.m.s.}$$

Problems

1 What are the phasors, in polar form, that represent the following products and quotients?
(*a*) $10\angle 50°$ multiplied by $2\angle 10°$
(*b*) $2\angle 60°$ multiplied by $5\angle -10°$
(*c*) $10\angle 50°$ divided by $2\angle 20°$
(*d*) $4\angle -20°$ divided by $8\angle 40°$

2 Write the following phasors in rectangular form:
(*a*) $2\angle 30°$, (*b*) $10\angle 50°$, (*c*) $3\angle -30°$, (*d*) $5\angle 120°$

3 Write the following phasors in polar form:
(*a*) $2 + j5$, (*b*) $5 + j1$, (*c*) $3 - j2$, (*d*) $-2 + j3$

4 Determine the phasors which are the sums of the following pairs of phasors:
(*a*) $3 + j2$ and $2 + j5$
(*b*) $3 + j3$ and $5 - j2$
(*c*) $2 - j2$ and $-5 + j3$
(*d*) $-1 + j5$ and $-6 + j2$

5 Determine the phasors which represent the differences between the following pairs of phasors:
(*a*) $4 + j3$ and $2 + j2$
(*b*) $5 + j2$ and $3 + j4$
(*c*) $2 + j4$ and $5 + j1$
(*d*) $-2 - j3$ and $5 + j2$

6 Determine the products of the following pairs of phasors:
(*a*) $2 + j2$ and $3 + j1$
(*b*) $5 + j2$ and $2 + j5$
(*c*) $-2 + j3$ and $3 + j4$
(*d*) $1 + j3$ and $2 - j4$

7 Determine the quotients of the following pairs of phasors:
(*a*) $4 + j3$ and $2 + j4$
(*b*) $2 + j5$ and $6 + j2$
(*c*) $3 - j4$ and $5 + j2$
(*d*) $-2 + j4$ and $5 + j2$

8 Determine in complex and polar forms the impedance and admittance for the following pure components:
(*a*) Resistor of $20\,\Omega$
(*b*) Inductor of 10 mH at $\omega = 1000$ rad/s
(*c*) Capacitor of $20\,\mu$F at $\omega = 1000$ rad/s

9 Determine the voltage, at angular frequency 1000 rad/s, which when applied to a $4\,\mu$F capacitor results in a current of $0.5\angle 90°$ A.

10 A voltage supply of $20 \sin 100t$ is applied to a series combination of a $9\,\Omega$ resistor, a 10 mH inductor and a 1 mF capacitor. What will be the current?

11 A voltage supply of $100 \sin 5000t$ is applied to a series combination of a $100\,\Omega$ resistor, a 30 mH inductor and a $5\,\mu$F capacitor. What will be the current?

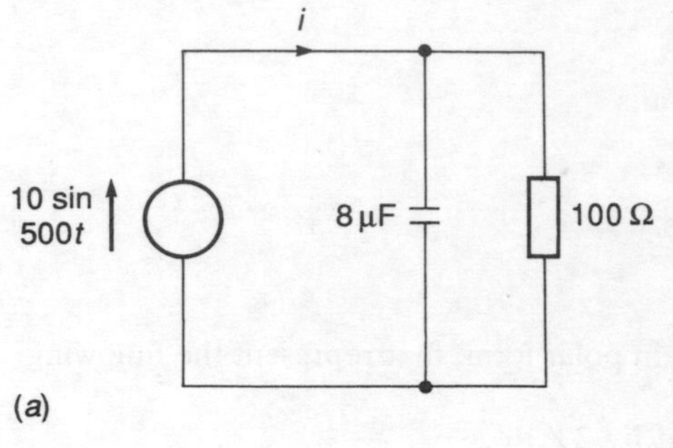

12 A circuit consists of a resistor in series with a capacitor. If the circuit has a total impedance of (50 – j400) Ω, what are the values of the resistance and capacitive reactance?

13 A 240 V, 50 Hz, sinusoidal supply is connected across an inductor of inductance 0.2 H and negligible resistance in series with a 50 Ω resistor. What is (*a*) the current through the circuit, (*b*) the potential difference across the resistor, and (*c*) the potential difference across the inductor?

14 The impedance of a coil is (20 + j5) Ω. What is its resistance and its inductive reactance?

15 Determine the current *i* in each of the circuits given in Fig. 3.36.

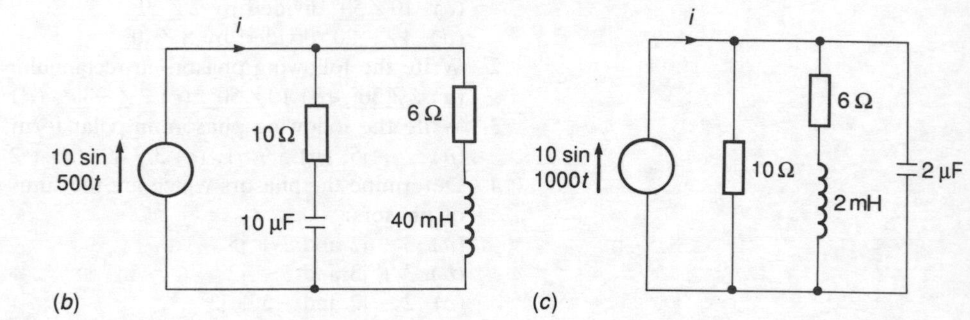

Fig. 3.36 Problem 15

16 A circuit consists of a 100 Ω resistance in parallel with a 10 μF capacitor. What will be (*a*) the current taken by the circuit, (*b*) the current through the resistor, and (*c*) the current through the capacitor, when a voltage supply of $20\angle 0°$ V and angular frequency 1000 rad/s is connected to it?

17 For the circuit shown in Fig. 3.37, determine (*a*) the current taken by the circuit, (*b*) the potential differences V_1 and V_2, (*c*) the currents I_1 and I_2.

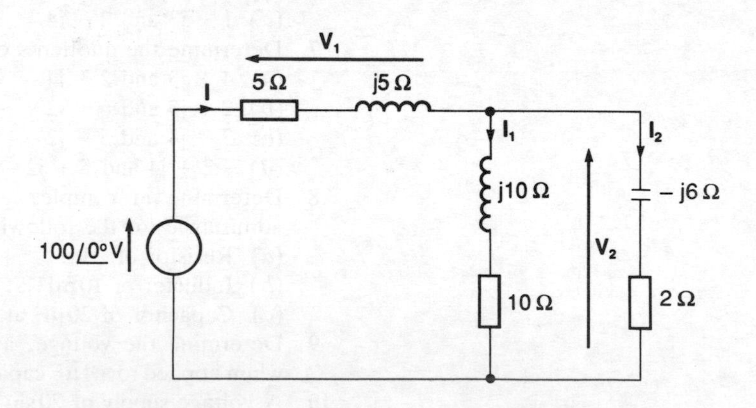

Fig. 3.37 Problem 17

18 For the circuit shown in Fig. 3.38 determine the relationship between *R* and *L* and *C* that is necessary if the circuit is to have a total impedance of $R\angle 0°$.

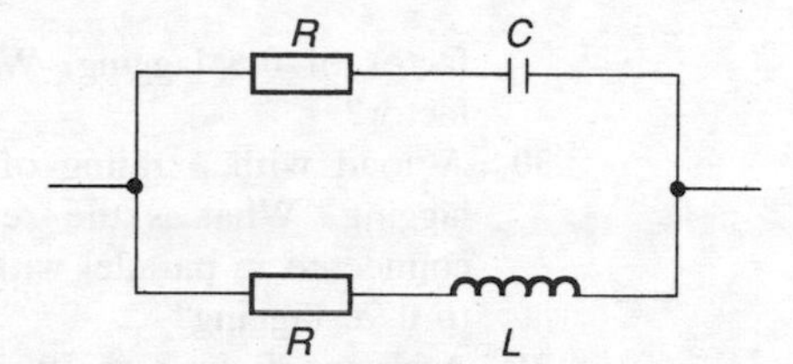

Fig. 3.38 Problem 18

19 Impedances of (10 + j6) and (12 − j5) are connected in parallel. What is the total admittance of the circuit and the phase angle between the current input to the circuit and the applied voltage?
20 What is (*a*) the complex power, (*b*) the real power and (*c*) the reactive power developed by a circuit of impedance (5 + j12) Ω when it is connected across a supply voltage of $10\angle 0°$ V r.m.s.?
21 Determine the real power and the reactive power developed when a voltage of $250\angle 0°$ V r.m.s. is applied to a circuit consisting of a resistance 40 Ω and an inductance of impedance j30 Ω.
22 What is (*a*) the real power and (*b*) the reactive power developed by a circuit of impedance 5 + j10 Ω when it is connected across a supply voltage of $50\angle 30°$ V r.m.s.?
23 A voltage of $v = 120\sqrt{2}\sin(400t + 30°)$ is applied to a circuit consisting of a 0.1 H inductance in series with a 15 Ω resistance. What is (*a*) the circuit impedance, (*b*) the circuit current, (*c*) the complex power, (*d*) the real power, (*e*) the reactive power?
24 An inductor with an inductance of 0.1 H and resistance 20 Ω is in parallel with a 50 μF capacitor and a sinusoidal supply of 240 V r.m.s. and frequency 50 Hz. What is (*a*) the current through the capacitor, (*b*) the current through the inductor, (*c*) the total current, (*d*) the power factor?
25 A load is supplied with an a.c. supply of 200 V r.m.s., at a frequency of 50 Hz and takes a current of 40 A at a power factor of 0.6 lagging. What is the value of the capacitor that should be connected in parallel with the load to increase the power factor to 0.9 lagging?
26 A motor takes a current of 6 A at a power factor of 0.65 lagging when connected to a 240 V r.m.s., 50 Hz, a.c. supply. What is (*a*) the value of the capacitor that has to be connected across the load to raise the power factor to 0.90 lagging and (*b*) the current then taken by the motor?
27 Loads of 5 kVA at power factor 0.75 lagging, 7 kVA at power factor 0.60 lagging and 9 kVA at unity power factor are connected in parallel across an a.c. voltage supply. What is the total load and power factor?
28 Two motors are connected in parallel across a 240 V r.m.s., 50 Hz, supply. If one motor takes a current of 8 A at power factor 0.8 lagging and the other 5 A at power factor 0.6 lagging, what is the total current taken from the supply and the power factor?
29 A plant has a purely resistive load of 50 kW for heating in parallel with an inductive load of motors of 100 kVA at power

factor of 0.8 lagging. What is the total power and the power factor?

30 A load with a rating of 300 kVA has a power factor of 0.70 lagging. What is the reactive power required of a capacitor connected in parallel with the load to improve the power factor to 0.90 lagging?

31 A circuit draws a real power of 200 W at a power factor of 0.20 lagging when connected to a voltage supply of $100\sqrt{2}\sin 1000t$. What is the value of the capacitor that needs to be connected in parallel with the circuit if the power factor is to be improved to 0.98 lagging?

32 A load when connected to a 240 V r.m.s. supply takes an average power of 8 kW at a power factor of 0.8 lagging. What is (*a*) the complex power and (*b*) the impedance of the load?

33 A load has an impedance of $(100 + j100)\ \Omega$. What is the value of the capacitor that needs to be connected in parallel with this load to give a power factor of 0.95 lagging? The angular frequency of the supply is 500 rad/s.

4 Circuit analysis

Introduction

This chapter shows the application to circuits of Kirchoff's laws, the simplifications that can be made to impedances in circuits by the star–delta and delta–star transformations, the simplifications that can be made to sources in circuits by voltage to current, or vice versa, source conversions, Thévenin's theorem, Norton's theorem and the superposition theorem, and node voltage and mesh current analysis. The condition for the maximum power transfer from a source network to the load is also considered.

Kirchoff's laws

Kirchoff's laws were considered in Chapter 1 for d.c. currents and voltages and in Chapter 2 for sinusoidal currents and voltages when represented by phasors. *Kirchoff's current law* can be stated as: at any node in an electrical circuit the algebraic sum of currents/phasor currents is zero. *Kirchoff's voltage law* can be stated as: around any closed path in a circuit the algebraic sum of the voltages/phasor voltages across all the components in the closed path is zero.

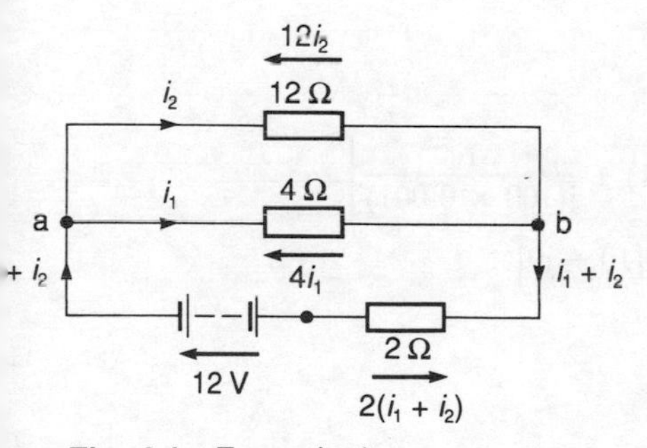

Fig. 4.1 Example 1

Example 1

Use Kirchoff's laws to establish the currents in each of the branches of the circuit shown in Fig. 4.1.

Answer

For node a we must have, as a consequence of Kirchoff's current law, the algebraic sum of the currents to be zero. Thus if we take the current through the 4 Ω resistor to be i_1 and that through the 10 Ω resistor i_2, then the current through the battery must be $i_1 + i_2$. Only then will we have

$$(i_1 + i_2) - i_1 - i_2 = 0$$

The potential difference across each resistor will be in the opposite direction to the current and given by the product of the current and resistance ($v = iR$). These potential differences have been indicated on Fig. 4.1. The direction of the potential difference across the battery is from negative to positive and this has also been indicated. Now applying Kirchoff's voltage law to the lower loop abc, by starting at node a and moving clockwise round the loop,

$$-4i_1 - 2(i_1 + i_2) + 12 = 0$$

Thus

$$6i_1 + 2i_2 = 12 \qquad [1]$$

Now applying Kirchoff's voltage law to the upper loop ab, starting at node a and moving clockwise round the loop gives

$$-12i_2 + 4i_1 = 0$$

$$i_1 = 3i_2 \qquad [2]$$

Hence substituting this value of i_1 into the earlier equation [1] gives

$$6(3i_2) + 2i_2 = 12$$

$$i_2 = 0.6\,\text{A}$$

Hence $i_1 = 3i_2 = 1.8\,\text{A}$ and the current through the battery branch of the circuit is $(i_1 + i_2) = 2.4\,\text{A}$.

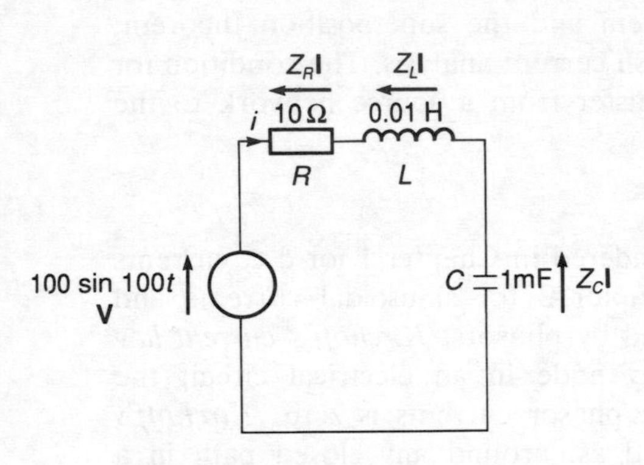

Fig. 4.2 Example 2

Example 2

Determine the steady state current in the steady state circuit shown in Fig. 4.2.

Answer

Applying Kirchoff's voltage law to the circuit gives

$$\mathbf{V} - Z_R\mathbf{I} - Z_L\mathbf{I} - Z_C\mathbf{I} = 0$$

$$\mathbf{V} = R\mathbf{I} + \text{j}\omega L\mathbf{I} + \frac{\mathbf{I}}{\text{j}\omega C}$$

$$\mathbf{V} = \mathbf{I}\left[10 + \text{j}(100 \times 0.01) + \frac{1}{\text{j}(100 \times 0.001)}\right]$$

$$\mathbf{V} = \mathbf{I}[10 + \text{j}1 - \text{j}10] = \mathbf{I}[10 - \text{j}9]$$

Thus

$$\mathbf{I} = \frac{\mathbf{V}}{10 - \text{j}9}$$

In polar form $10 - \text{j}9$ is $13.5\angle -42.0°\,\Omega$ and since $\mathbf{V} = 100\angle 0°$, then

$$\mathbf{I} = \frac{100\angle 0°}{13.5\angle -42.0°} = 7.4\angle 42.0°\,\text{A}$$

Therefore the current i is

$$i = 7.4\sin(100t + 42°)\,\text{A}$$

Star–delta transformation

Simplification of some circuits can be achieved by what is called the *star–delta transformation*. Figure 4.3(*a*) shows the form of a star, or T, grouping of impedances. This can be replaced by an equivalent grouping of impedances in the delta, or π, circuit shown in Fig. 4.3(*b*). These groupings are identical when there is the same impedance between any two of the nodes.

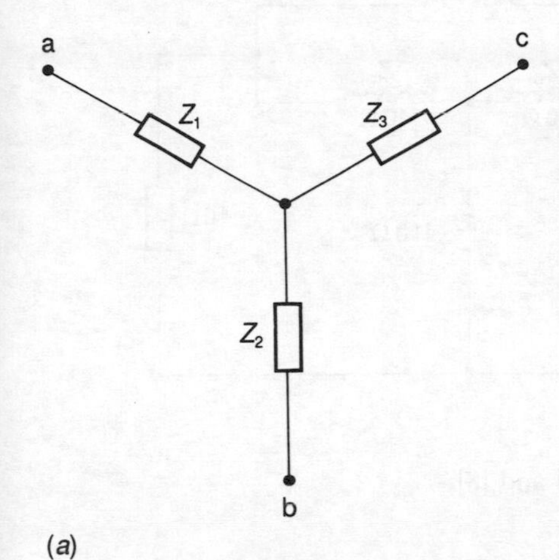

For the star circuit, the impedance between a and b is $Z_1 + Z_2$. For the delta circuit, the impedance between a and b is that of $(Z_A + Z_C)$ in parallel with Z_B, i.e.

$$Z_1 + Z_2 = \frac{Z_B(Z_A + Z_C)}{Z_A + Z_B + Z_C} \quad [3]$$

In a similar way we can write for the impedance between nodes a and c,

$$Z_1 + Z_3 = \frac{Z_A(Z_B + Z_C)}{Z_A + Z_B + Z_C} \quad [4]$$

and between nodes b and c,

$$Z_2 + Z_3 = \frac{Z_C(Z_A + Z_B)}{Z_A + Z_B + Z_C} \quad [5]$$

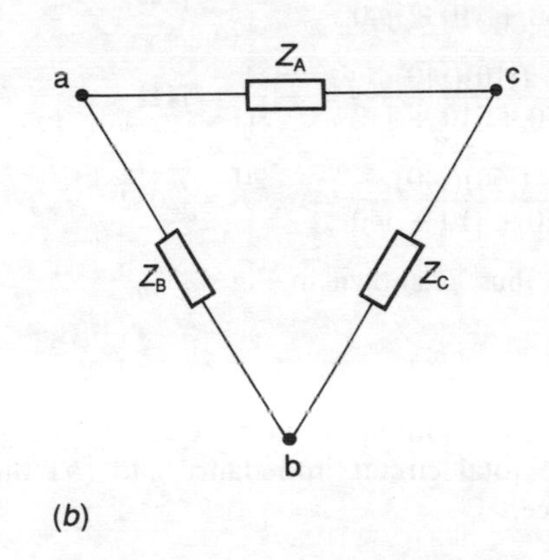

These three simultaneous equations [3], [4] and [5] can be solved to give Z_1, Z_2 and Z_3 in terms of Z_A, Z_B and Z_C, i.e.

$$Z_1 = \frac{Z_A Z_B}{Z_A + Z_B + Z_C} \quad [6]$$

$$Z_2 = \frac{Z_B Z_C}{Z_A + Z_B + Z_C} \quad [7]$$

$$Z_3 = \frac{Z_A Z_C}{Z_A + Z_B + Z_C} \quad [8]$$

or conversely Z_A, Z_B and Z_C in terms of Z_1, Z_2 and Z_3, i.e.

$$Z_A = \frac{Z_1 Z_2 + Z_1 Z_3 + Z_2 Z_3}{Z_2} \quad [9]$$

$$Z_B = \frac{Z_1 Z_2 + Z_1 Z_3 + Z_2 Z_3}{Z_3} \quad [10]$$

$$Z_C = \frac{Z_1 Z_2 + Z_1 Z_3 + Z_2 Z_3}{Z_1} \quad [11]$$

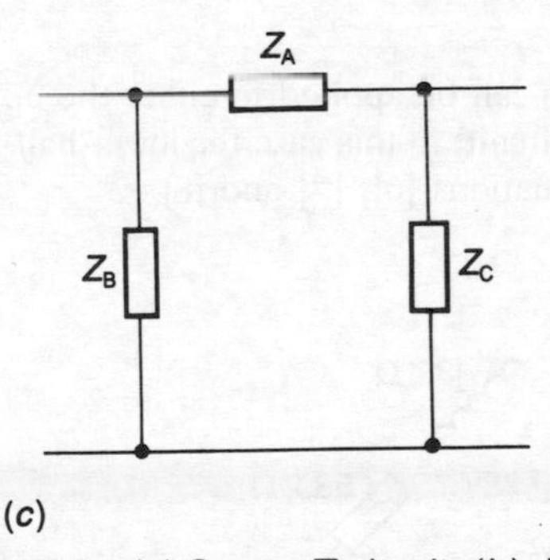

Fig. 4.3 (*a*) Star or T circuit, (*b*) delta circuit, (*c*) π version of delta circuit

Example 3

Simplify the circuit in Fig. 4.4 by using a delta–star transformation.

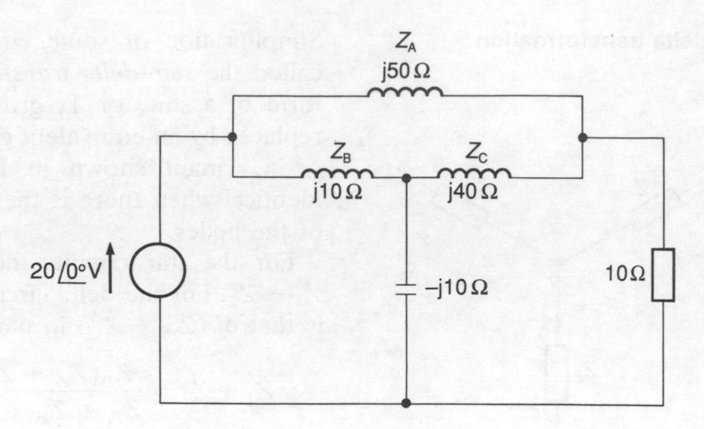

Fig. 4.4 Example 3

Answer

Using the equations [6], [7] and [8],

$$Z_1 = \frac{Z_A Z_B}{Z_A + Z_B + Z_C} = \frac{(j50)(j10)}{j50 + j10 + j40} = -\frac{5}{j} = j5\,\Omega$$

$$Z_2 = \frac{Z_B Z_C}{Z_A + Z_B + Z_C} = \frac{(j10)(j40)}{j50 + j10 + j40} = -\frac{4}{j} = j4\,\Omega$$

$$Z_3 = \frac{Z_A Z_C}{Z_A + Z_B + Z_C} = \frac{(j50)(j40)}{j50 + j10 + j40} = -\frac{20}{j} = j20\,\Omega$$

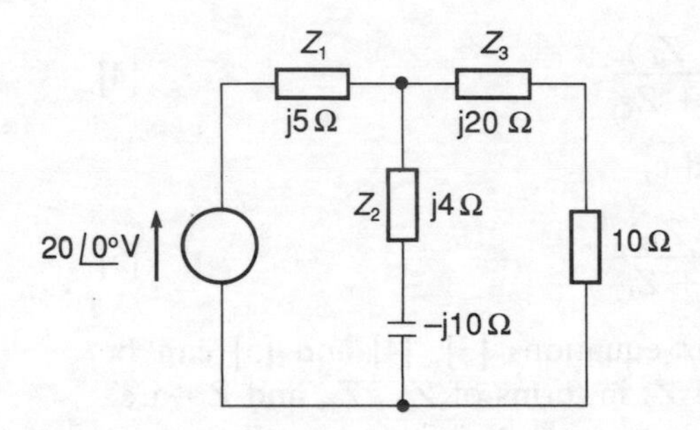

The star form of the circuit is thus as shown in Fig. 4.5.

Fig. 4.5 Example 3

Example 4

Use the delta–star transformation to simplify the bridge circuit in Fig. 4.6 and so determine (*a*) the total circuit impedance and (*b*) the current **I** taken from the source.

Answer

(*a*) The delta–star transformation can be applied to either the upper half or the lower half of the circuit; in this case the lower half has been chosen. Thus, using equations [6], [7] and [8]

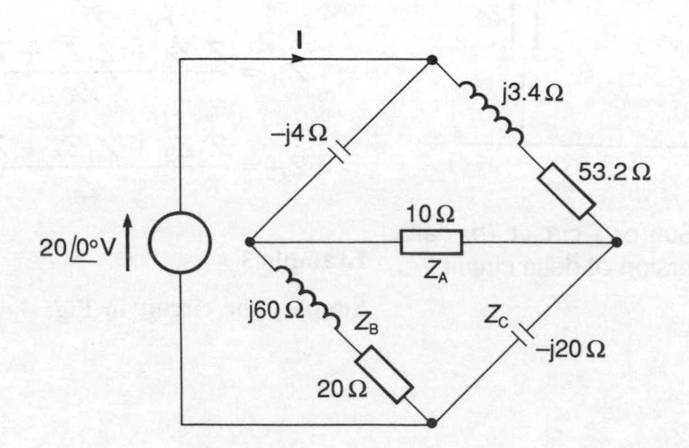

Fig. 4.6 Example 4

$$Z_1 = \frac{Z_A Z_B}{Z_A + Z_B + Z_C} = \frac{10(20 + j60)}{10 + 20 + j60 - j20} = \frac{200 + j600}{30 + j40}$$

$$= \frac{(200 + j600)(30 - j40)}{(30 + j40)(30 - j40)} = \frac{30\,000 + j10\,000}{2500}$$

$$= 12 + j4\,\Omega$$

$$Z_2 = \frac{Z_B Z_C}{Z_A + Z_B + Z_C} = \frac{(20 + j60)(-j20)}{30 + j40} = \frac{1200 - j400}{30 + j40}$$

$$= \frac{(1200 - j400)(30 - j40)}{(30 + j40)(30 - j40)} = \frac{20\,000 - j60\,000}{2500}$$

$$= 8 - j24\Omega$$

$$Z_3 = \frac{Z_A Z_C}{Z_A + Z_B + Z_C} = \frac{10(-j20)}{30 + j40} = \frac{-j200(30 - j40)}{(30 + j40)(30 - j40)}$$

$$= \frac{-8000 - j6000}{2500} = -3.2 - j2.4\,\Omega$$

The simplified circuit is thus as in Fig. 4.7.

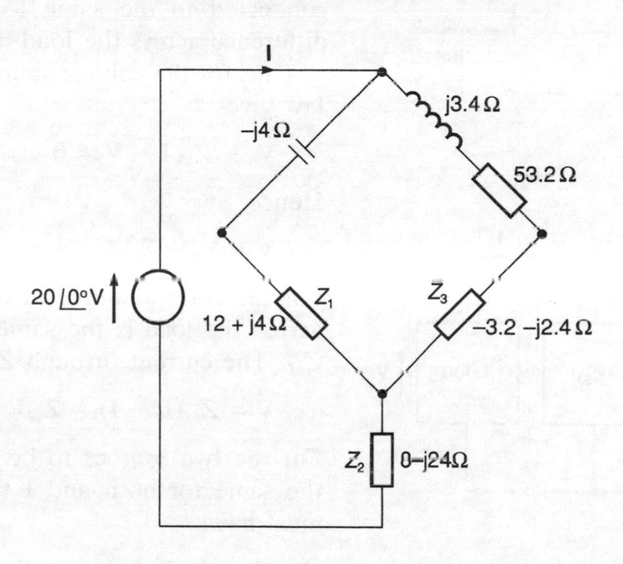

Fig. 4.7 Example 4

The circuit can now be further simplified by considering the series and parallel impedances. The parallel part of the circuit thus consists of (53.2 + j3.4 − 3.2 − j2.4) in parallel with (−j4 + 12 + j4). The parallel part has thus an impedance Z such that

$$\frac{1}{Z} = \frac{1}{50 + j1} + \frac{1}{12} = \frac{12 + 50 + j1}{12(50 + j1)}$$

$$Z = \frac{12(50 + j1)(62 - j1)}{(62 + j1)(62 - j1)} = \frac{37\,212 + j144}{3845} = 9.7 + j0.04$$

This impedance is in series with (8 − j24) and so the total circuit impedance is (17.7 − j23.06) Ω.

(*b*) The current **I** is the source voltage divided by the total circuit impedance. Thus

$$\mathbf{I} = \frac{20}{17.7 - 23.06} = \frac{20(17.7 + j23.06)}{(17.7 - j23.06)(17.7 + j23.06)}$$

$$= \frac{354 + j461.2}{325} = 0.42 + j0.55\,\Omega$$

In polar form this is $0.69\angle 52.6°$ A.

Source conversions

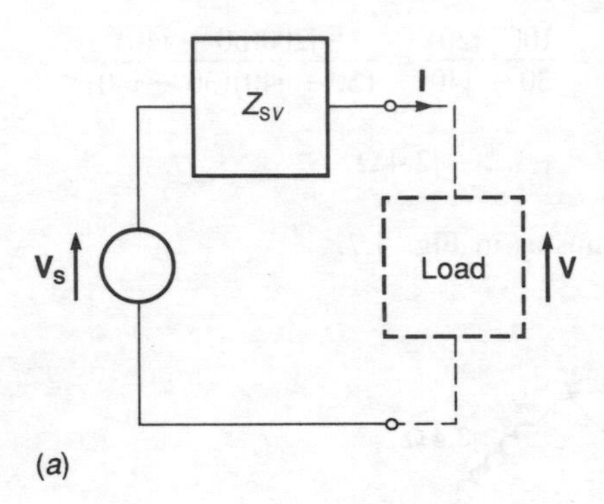

(*a*)

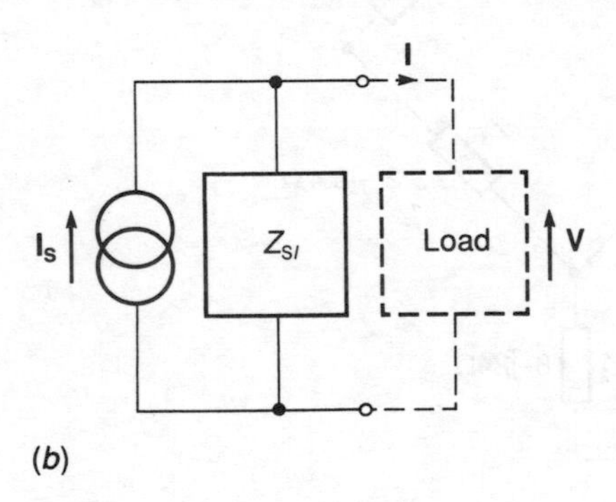

(*b*)

Fig. 4.8 (*a*) Converted practical voltage source, (*b*) converted practical current source

A practical voltage source consists of an ideal voltage source in series with an impedance, a practical current source consists of an ideal current source in parallel with an impedance or admittance (Fig. 4.8). This is just a statement of what was discussed in Chapter 1 for d.c. sources. The term 'source conversion' is used for the technique of converting a practical voltage source to an equivalent practical current source, or vice versa. Two sources are *equivalent* if, when they are connected to the same load, they give the same potential difference across the load and the same current through it.

Thus for the voltage source in Fig. 4.8(*a*), Kirchoff's voltage law gives

$$\mathbf{V_s} - Z_{sV}\mathbf{I} - \mathbf{V} = 0$$

Hence

$$\mathbf{V} = \mathbf{V_s} - Z_{sV}\mathbf{I}$$

For the current source in Fig. 4.8(*b*), the potential difference across the load is the same as the potential difference across Z_{sI}. The current through Z_{sI} is $(\mathbf{I_s} - \mathbf{I})$. Thus

$$\mathbf{V} = Z_{sI}(\mathbf{I_s} - \mathbf{I}) = \mathbf{Z}_{sI}\mathbf{I_s} - Z_{sI}\mathbf{I}$$

For the two sources to be equivalent, then, we must have **V** the same for both and **I** the same for both. This means we must have

$$\mathbf{V_s} = Z_{sI}\mathbf{I_s} \qquad [12]$$

and

$$Z_{SV} = Z_{sI} \qquad [13]$$

Example 5

Convert the circuit in Fig. 4.9 into one with a current source.

Answer

Using equation [12]

$$\mathbf{V_s} = Z_s\mathbf{I_s}$$

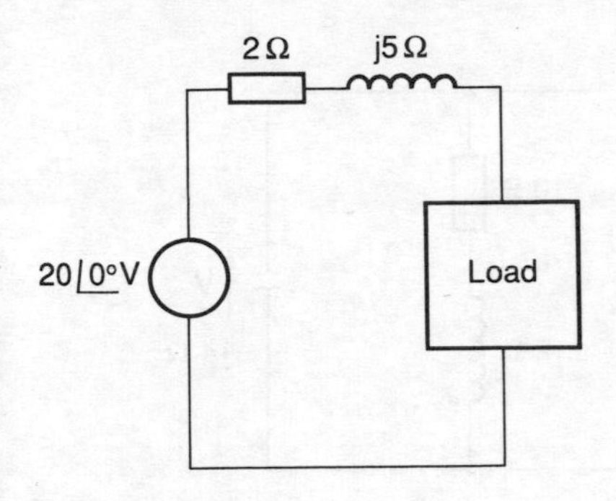

Fig. 4.9 Example 5

and, since Z_s is $(2 + j5)\ \Omega$, then

$$\mathbf{I}_s = \frac{20}{2 + j5} = \frac{20(2 - j5)}{(2 + j5)(2 - j5)} = \frac{40 - j100}{29} = (1.4 - j3.4)\ \text{A}$$

In polar notation this is $3.7\angle -68°$ A. The parallel impedance for the current source has the same value as the series impedance for the voltage source. The circuit is thus as shown in Fig. 4.10.

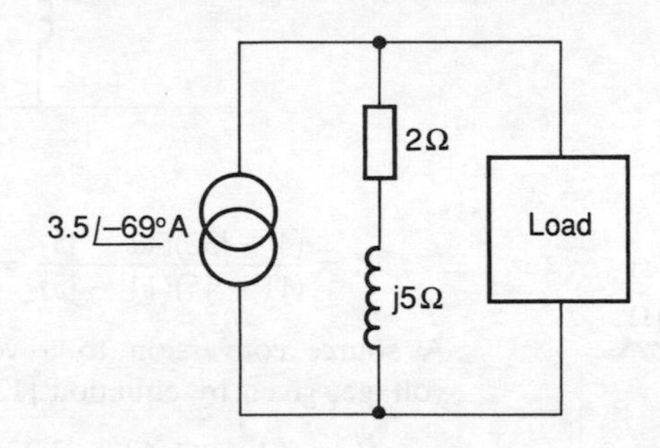

Fig. 4.10 Example 5

Example 6

Determine the voltage **V** for the circuit shown in Fig. 4.11 by converting the voltage source into a current source and then simplifying the circuit.

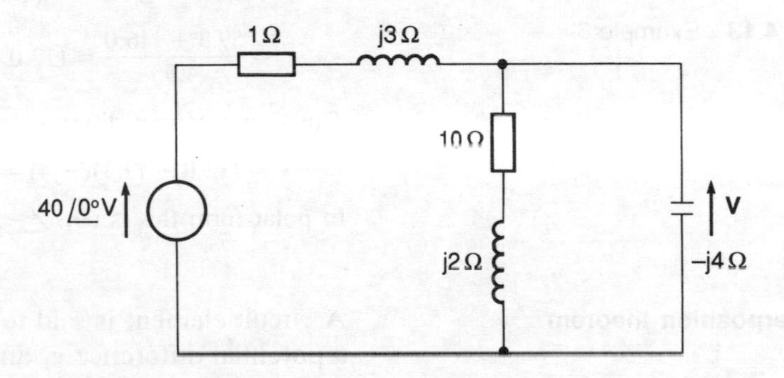

Fig. 4.11 Example 6

Answer

For the voltage source there is an impedance of $(1 + j3)\ \Omega$. Thus the current source must have an impedance of $(1 + j3)\ \Omega$ in parallel with it. The current source will have a current given by equation [12]:

$$\mathbf{V}_s = Z_s\mathbf{I}_s$$

$$\mathbf{I}_s = \frac{40}{1 + j3} = \frac{40(1 - j3)}{(1 + j3)(1 - j3)} = (4 - j12)\ \text{A}$$

The circuit thus becomes that shown in Fig. 4.12.

The two parallel branches of the circuit of $(1 + j3)\ \Omega$ and $(10 + j2)\ \Omega$ can be combined into a single parallel branch.

$$\frac{1}{Z} = \frac{1}{1 + j3} + \frac{1}{10 + j2} = \frac{10 + j2 + 1 + j3}{(1 + j3)(10 + j2)} = \frac{11 + j5}{4 + j32}$$

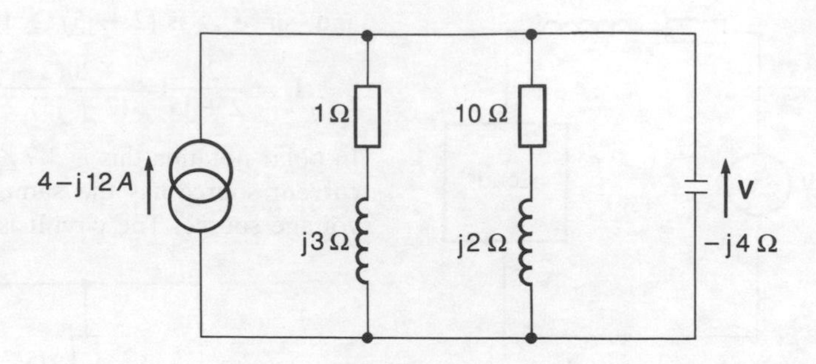

Fig. 4.12 Example 6

$$Z = \frac{(4 + j32)(11 - j5)}{(11 + j5)(11 - j5)} = \frac{204 + j332}{146} = 1.40 + j2.27\,\Omega$$

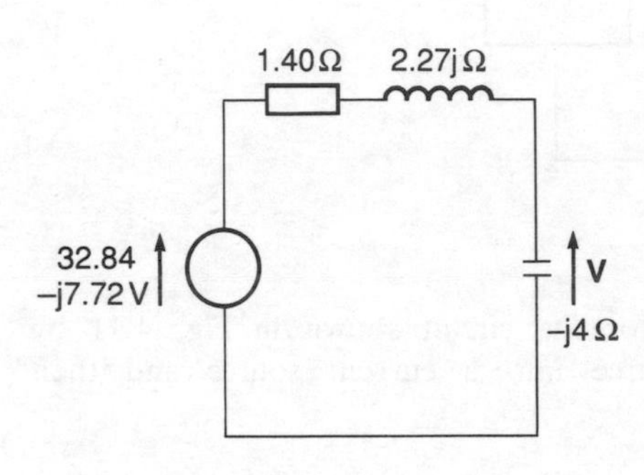

Fig. 4.13 Example 6

A source conversion to a voltage source then gives a source of voltage, given by equation [12], as

$$\mathbf{V}_s = Z_s\mathbf{I}_s = (1.40 + j2.27)(4 - j12) = (32.84 - j7.72)\,\text{V}$$

The circuit is now as shown in Fig. 4.13.

The circuit has been reduced to a simple series circuit of total impedance (1.40 − j1.73) Ω. Hence the circuit current is

$$\mathbf{I} = \frac{32.84 - j7.72}{1.40 - j1.73} = \frac{(32.84 - j7.72)(1.40 + j1.73)}{(1.40 - j1.73)(1.40 + j1.73)}$$

$$= \frac{59.3 + j46.0}{4.95} = (12.0 + j9.3)\,\text{A}$$

Thus the potential difference across the capacitor is

$$\mathbf{V} = (12.0 + j9.3)(-j4) = (37.2 - j48.0)\,\text{V}$$

In polar form this is $60.7\angle -52.2°$ V.

Superposition theorem

A circuit element is said to be *linear* if when a current i_1 gives a potential difference v_1 and a current i_2 a potential difference v_2 then a current $(i_1 + i_2)$ gives a potential difference $(v_1 + v_2)$. This principle is called the *principle of superposition*. It can also be applied to phasors and impedances. Thus a circuit element is said to be *linear* if when a current $\mathbf{I}_1$ gives a potential difference $\mathbf{V}_1$ and a current $\mathbf{I}_2$ a potential difference $\mathbf{V}_2$ then a current $(\mathbf{I}_1 + \mathbf{I}_2)$ gives a potential difference $(\mathbf{V}_1 + \mathbf{V}_2)$.

Ohm's law is really just a statement of this principle of superposition. For example, for a resistor of resistance 10 Ω a current of 1 A means a potential difference of 10 V. A current of 2 A means a potential difference of 20 V. Hence a current of (1 + 2) = 3 A means a potential difference of (10 + 20) = 30 V.

The principle of superposition can be applied to circuits provided they contain linear components and independent sources. For this situation the principle may be stated as: the potential difference across, or the current through, any element in such a circuit can be obtained by adding algebraically all the individual potential differences, or currents, for the element caused by each independent source acting alone, with all the other independent voltage sources replaced by short circuits and all other independent current sources replaced by open circuits. The procedure for using this principle is thus:

1 Select one independent source and set all other independent sources to zero, replacing voltage sources by a short circuit and current sources by an open circuit.
2 Determine the current and/or potential difference across the circuit component.
3 Now repeat the above two steps for each of the remaining independent sources.
4 Obtain the current and/or potential difference across the circuit component by algebraically summing the currents/potential differences calculated in stage 2 above due to each of the sources.

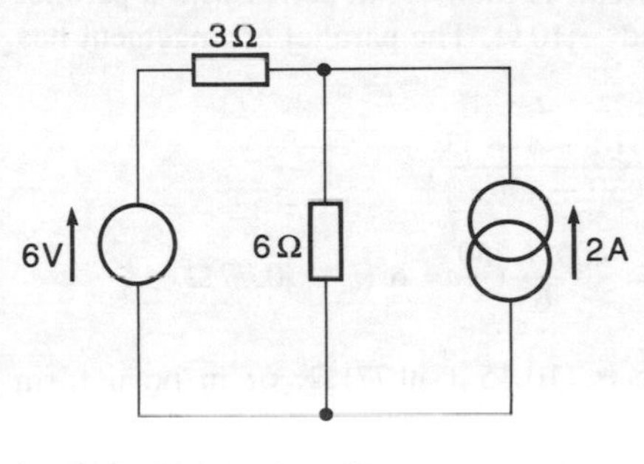

Fig. 4.14 Example 7

Example 7

Determine, using the principle of superposition, the current through the 6 Ω resistor in Fig. 4.14.

Answer

Setting the current source to zero and that branch of the circuit to open circuit results in the circuit shown in Fig. 4.15(*a*). The current through the 6 Ω resistor is thus given by

$$6 = i(3 + 6)$$

Hence $i = 0.67$ A. Setting the voltage source to zero and replacing it by a short circuit results in the circuit shown in Fig. 4.15(*b*). The 2 A current is thus divided between the 3 Ω and the 6 Ω resistor branches.

$$2 = i_1 + i_2$$

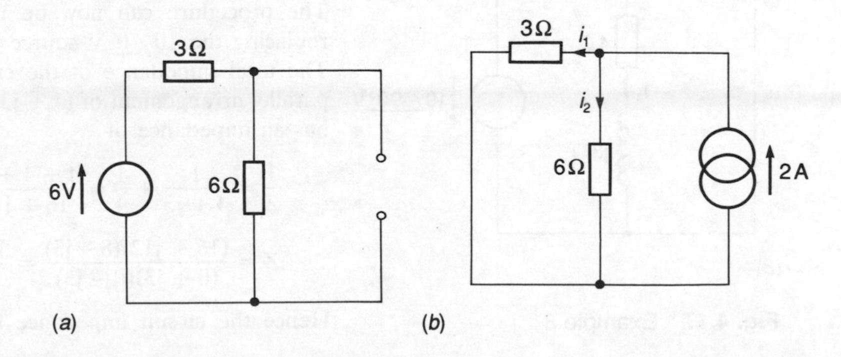

Fig. 4.15 Example 7 (*a*) (*b*)

But the potential difference v across both resistors will be the same, hence

$$v = 3i_1 = 6i_2$$

Thus

$$2 = 2i_2 + i_2$$

and so $i_2 = 0.67$ A. The current through the 6 Ω resistor will thus be 0.67 + 0.67 = 1.34 A. The currents add because the two separate currents are in the same direction through the resistor.

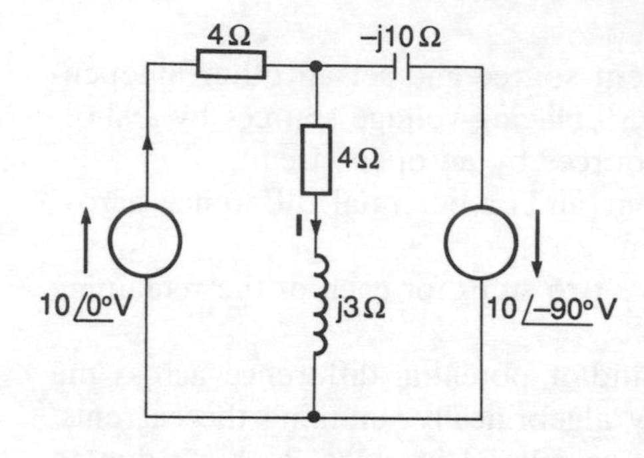

Fig. 4.16 Example 8

Example 8

Determine the current I in the circuit shown in Fig. 4.16.

Answer

Applying the superposition theorem, first the $10\angle -90°$ V source is replaced by a short circuit to give the circuit shown in Fig. 4.17(*a*). The total impedance of the circuit is then 4 Ω in series with a parallel arrangement of (4 + j3) Ω and −j10 Ω. The parallel arrangement has an impedance of

$$\frac{1}{Z} = \frac{1}{4 + j3} + \frac{1}{-j10} = \frac{-j10 + 4 + j3}{30 - j40}$$

$$Z = \frac{(30 - j40)(4 + j7)}{(4 - j7)(4 + j7)} = \frac{400 + j50}{65} = 6.15 + j0.77\,\Omega$$

Hence the circuit impedance is (10.15 + j0.77) Ω, or in polar form $10.18\angle 4.3°$. The current **I** is thus

$$\mathbf{I} = \frac{10\angle 0°}{10.18\angle 4.3°} = 0.98\angle -4.3°\,\text{A}$$

The potential difference across the parallel part of the circuit is, since (6.15 + j0.77) Ω in polar form is $6.20\angle 7.1°$ Ω,

$$\mathbf{V} = (0.98\angle -4.3°)(6.20\angle 7.1°) = 6.1\angle 2.8°\,\text{V}$$

The current through the (4 + j3) Ω or $5\angle 36.9°$ Ω is thus

$$\mathbf{I}_1 = \frac{6.1\angle 2.8°}{5\angle 36.9°} = 1.2\angle -34.1°\,\text{A}$$

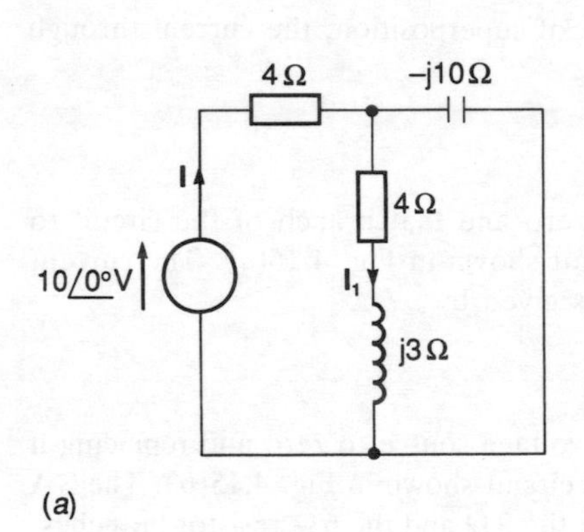

(*a*)

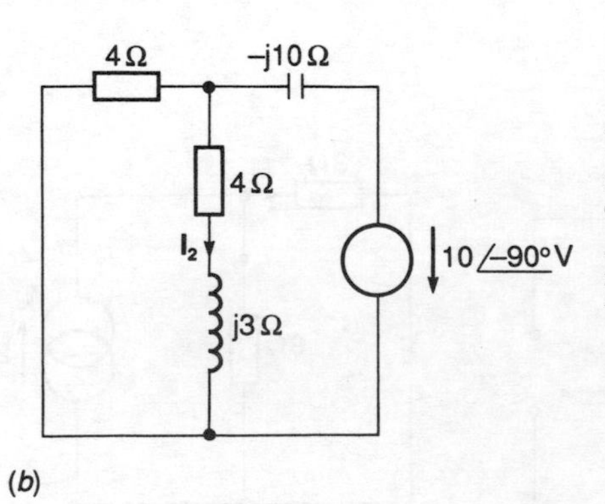

(*b*)

Fig. 4.17 Example 8

The procedure can now be repeated for the circuit produced by replacing the $10\angle 0°$ V source by a short circuit, as in Fig. 4.17(*b*). The total impedance of the circuit is then −j10 Ω in series with a parallel arrangement of (4 + j3) Ω and 4 Ω. The parallel arrangement has an impedance of

$$\frac{1}{Z} = \frac{1}{4 + j3} + \frac{1}{4} = \frac{4 + 4 + j3}{16 + j12}$$

$$Z = \frac{(16 + j12)(8 - j3)}{(8 + j3)(8 - j3)} = \frac{164 + j48}{73} = 2.25 + j0.66$$

Hence the circuit impedance is (2.25 − j9.34) Ω, or in polar form

$9.61\angle -76.5°$. The current **I** is thus

$$\mathbf{I} = \frac{10\angle -90°}{9.61\angle -76.5} = 1.04\angle -13.5° \text{ A}$$

The potential difference across the parallel part of the circuit is, since $(2.25 + j0.66)\ \Omega$ in polar form is $2.34\angle 16.3°\ \Omega$,

$$\mathbf{V} = (1.04\angle -13.5°)(2.34\angle 16.3°) = 2.43\angle -2.8° \text{ V}$$

The current through the $(4 + j3)\ \Omega$ or $5\angle 36.9°\ \Omega$ is thus

$$\mathbf{I}_2 = \frac{2.4\angle 2.8}{5\angle 36.9°} = 0.5\angle -34.1° \text{ A}$$

Thus the current through the $(4 + j3)\ \Omega$ due to both the sources is $\mathbf{I}_1 - \mathbf{I}_2$, i.e.

$$\mathbf{I} = 1.2\angle -34.1° - 0.5\angle -34.1° \text{ A}$$

Thus the current is $0.7\angle -34.1°$.

Thévenin's theorem

In many situations an electric circuit can conveniently be regarded as a source network connected to a load network. Each network can be regarded as a black box with two terminals, thc only point of concern being the voltage and current at these terminals (Fig. 4.18). Thus the source network might be a power supply with the load being some electronic circuit. *Thévenin's theorem* states that the source network may be replaced by an equivalent network consisting of an ideal independent voltage source in series with an internal impedance, the voltage of the source v_{Th} or voltage phasor of the source $\mathbf{V}_{Th}$ being the open-circuit voltage of the source network, and the internal impedance Z_{Th} being that which is measured at the terminals of the source network when all independent sources are reduced to zero, voltage sources being replaced by a short circuit and current sources by an open circuit. Figure 4.18 shows the Thévenin equivalent circuit. The Thévenin theorem is really only an application of the principle of superposition The Thévenin impedance Z_{Th} is the same as the output impedance which is frequently specified for electronic devices such as amplifiers.

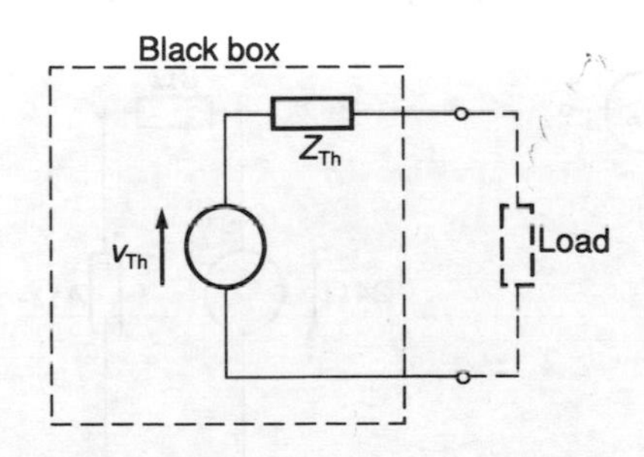

Fig. 4.18 Thévenin equivalent circuit

The procedure for finding the Thévenin equivalent circuit can be summarized as follows:

1. Divide the circuit into the source and load parts, connected at a pair of terminals.
2. Consider the source circuit in isolation from the load circuit and determine its open-circuit voltage v_{Th} or open circuit phasor voltage $\mathbf{V}_{Th}$.
3. Consider the source circuit in isolation from the load circuit and short circuit the terminals which would connect

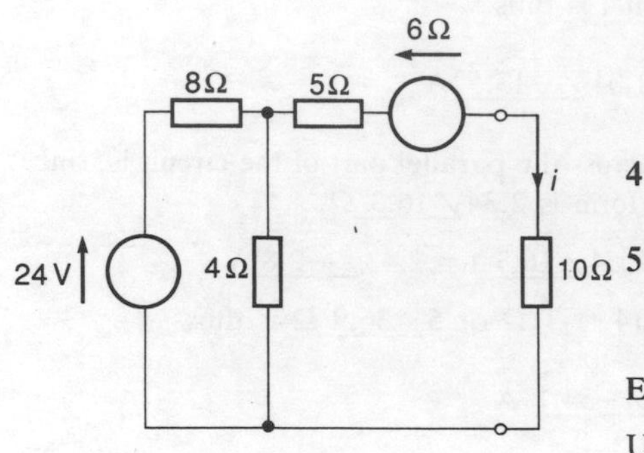

Fig. 4.19 Example 9

the load to the circuit. Determine the impedance of the circuit when all sources are made zero, voltage sources being replaced by a short circuit and current sources by an open circuit.

4 Replace the source circuit with its Thévenin equivalent circuit.

5 Now consider the Thévenin circuit connected to the load circuit and determine the variables of interest.

Example 9

Use Thévenin's theorem to determine the current through the 10 Ω resistor in Fig. 4.19.

Answer

Figure 4.20(*a*) shows the source part of the circuit. Dealing first with that part of the circuit shown in Fig. 4.20(*b*), the open-circuit voltage is that fraction of the 24 V across the 4 Ω resistor. This is the fraction 4/(8 + 4), i.e. 8 V. The resistance of this part of the circuit when the voltage source is replaced by a short circuit is $4 \times 8/(4 + 8) = 2.7\ \Omega$. Thus the circuit in Fig. 4.20(*a*) can be replaced by Fig. 4.20(*c*). This can be simplified to a circuit with an open-circuit voltage of 2 V and a resistance of 7.7 Ω. Hence the current *i* through the 10 Ω resistor is given

$$2 = i(7.7 + 10)$$

$$i = 0.11\ \text{A}$$

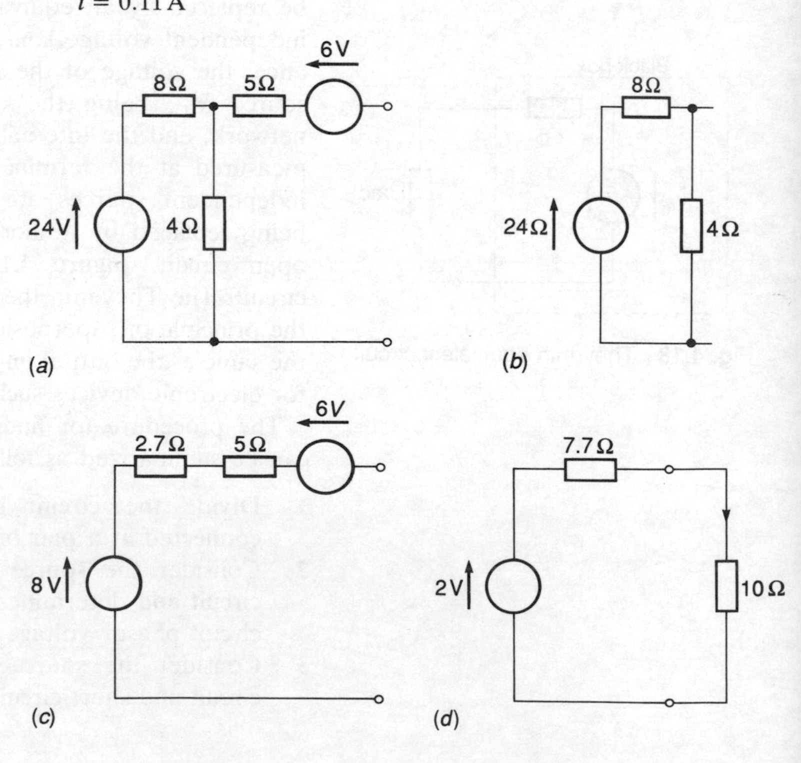

Fig. 4.20 Example 10

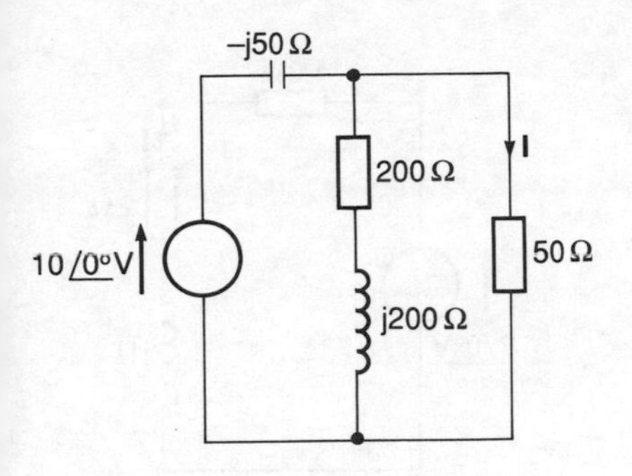

Fig. 4.21 Example 10

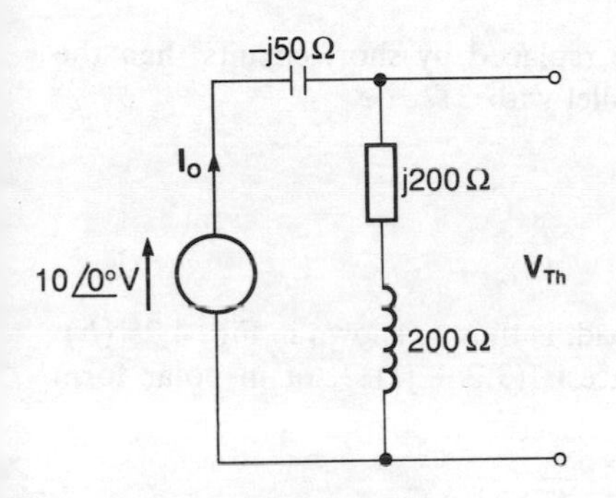

Fig. 4.22 Example 10

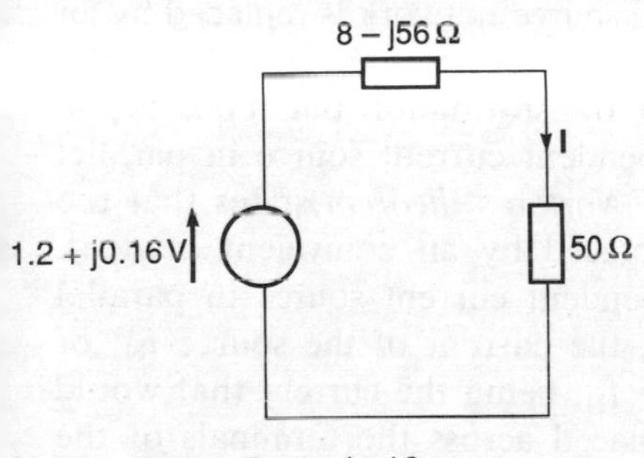

Fig. 4.23 Example 10

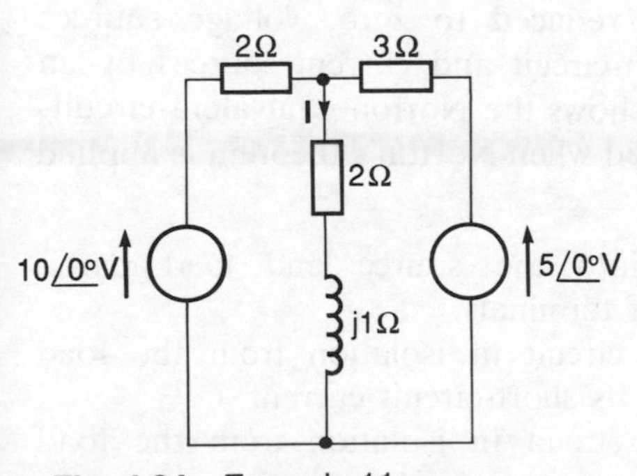

Fig. 4.24 Example 11

Example 10

Determine the current **I** through the 50 Ω resistance for the circuit shown in Fig. 4.21.

Answer

If the 50 Ω resistor is considered as the load then Thévenin's theorem can be used to simplify the rest of the circuit. The open-circuit voltage is the voltage across the impedance (200 + j200) Ω when the 50 Ω resistor is made open-circuit (Fig. 4.22). For this condition the total circuit impedance is (200 + j200 − j50) = (200 + j150) Ω. Hence the current $\mathbf{I_0}$ is

$$\mathbf{I}_o = \frac{10}{200 + j150}\,\text{A}$$

The open-circuit voltage $\mathbf{V}_{Th}$ is thus

$$\mathbf{V}_{Th} = \mathbf{I}_o(200 + j200) = \frac{10(200 + j200)}{200 + j150} = \frac{10(2 + j2)}{2 + j1.5}$$

$$= \frac{10(2 + j2)(2 - j1.5)}{(2 + j1.5)(2 - j1.5)} = \frac{10(7 + j1)}{6.25} = 11.2 + j1.6\,\text{V}$$

The Thévenin impedance is given by replacing the voltage source in Fig. 4.22 by a short circuit. The impedance is then −j50 Ω in parallel with (200 + j200) Ω. Hence

$$\frac{1}{Z_{Th}} = \frac{1}{-j50} + \frac{1}{200 + j200} = \frac{200 + j200 - j50}{-j50(200 + j200)}$$

$$Z_{Th} = \frac{50 \times 200(1 - j)}{200 + j150} = \frac{100(1 - j)(2 - j1.5)}{(2 + j1.5)(2 - j1.5)} = \frac{100(0.5 - j3.5)}{6.25}$$

$$Z_{Th} = 8 - j56\,\Omega$$

The Thévenin circuit, with the load, is thus as shown in Fig. 4.23. The current through the 50 Ω resistor is thus

$$\mathbf{I} = \frac{11.2 + j1.6}{58 - j56} = \frac{(11.2 + j1.6)(58 + j56)}{(58 - j56)(58 + j56)}$$

$$= \frac{640.64 + j636.48}{6500} = 0.099 + j0.098$$

In polar form this is $0.099\angle 45°$ A.

Example 11

Determine, using Thévenin's theorem, the current **I** in the circuit given in Fig. 4.24.

Answer

Take the (2 + j1) Ω to be the load. The circuit to which Thévenin's theorem is to be applied is thus as shown in Fig. 4.25(*a*). The open-circuit current is thus (10 − 5)/5 = 1 A. Hence the voltage $\mathbf{V}_{Th}$ is 10 V minus the potential drop across the 2 Ω resistor, i.e. (10 − 2 × 1) = 8 V.

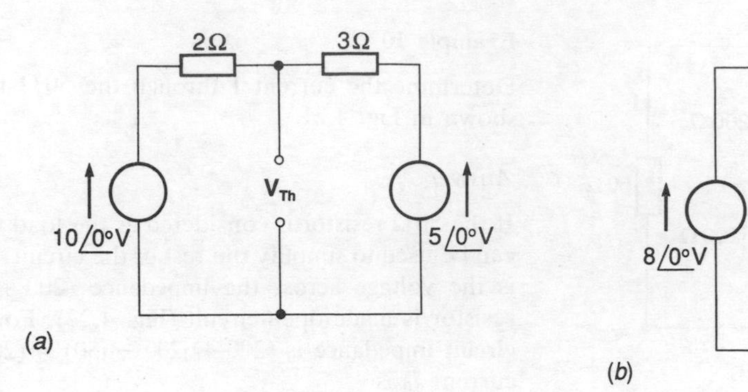

Fig. 4.25 Example 11

When the voltage sources are replaced by short circuits then the impedance is that of 2 Ω in parallel with 3 Ω, i.e.

$$\frac{1}{Z_{\text{Th}}} = \frac{1}{2} + \frac{1}{3}$$

$$Z_{\text{Th}} = 1.2\,\Omega$$

The Thévenin circuit, with the load, is thus as shown in Fig. 4.25 (*b*). Hence the total circuit impedance is $(3.2 + \text{j}1)\,\Omega$, or in polar form $3.35\angle 17.4°\,\Omega$. Thus

$$\mathbf{I} = \frac{8\angle 0°}{3.35\angle 17.4°} = 2.39\angle -17.4°\ \text{A}$$

Norton's theorem

With the Thévenin theorem a source network is replaced by an ideal voltage source in series with an internal impedance. The Norton theorem is a similar transformation but replaces the source network by an independent current source in parallel with an internal impedance. *Norton's theorem* states that the source network may be replaced by an equivalent network consisting of an ideal independent current source in parallel with an internal impedance, the current of the source i_N, or current phasor of the source $\mathbf{I}_N$, being the current that would flow if a short-circuit was placed across the terminals of the source network, and the internal impedance Z_N that which is measured at the terminals of the source network when all independent sources are reduced to zero, voltage sources being replaced by a short-circuit and current sources by an open-circuit. Figure 4.26 shows the Norton equivalent circuit.

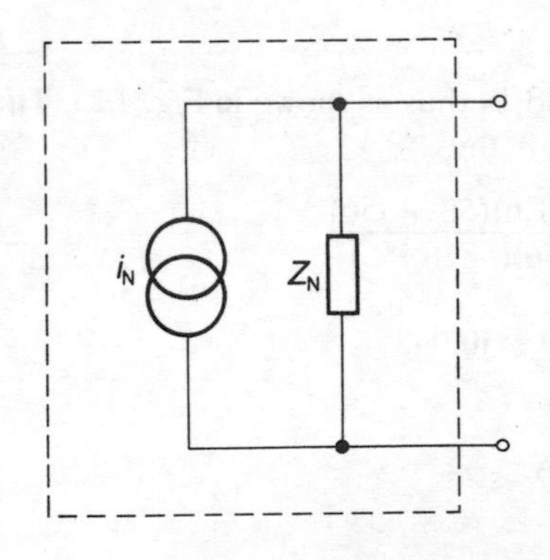

Fig. 4.26 Norton equivalent circuit

The procedure to be used when Norton's theorem is applied to a circuit is:

1 Divide the circuit into the source and load parts, connected at a pair of terminals.
2 Consider the source circuit in isolation from the load circuit and determine its short-circuit current.
3 Consider the source circuit in isolation from the load

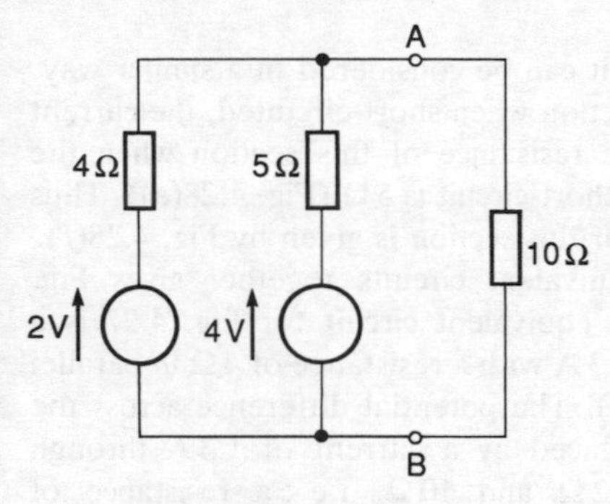

Fig. 4.27 Example 12

circuit and determine its impedance when all sources are made zero, voltage sources being replaced by a short-circuit and current sources by an open-circuit.

4 Replace the source circuit with its Norton equivalent circuit.

5 Now consider the Norton circuit connected to the load circuit and determine the variables of interest.

Example 12

Determine using Norton's theorem the potential difference across the 10 Ω resistor in Fig. 4.27.

Answer

The circuit can be broken into the source and load sections at terminals A and B. The source circuit can then be broken into two further sections. Figure 4.28(a) shows the first section when short-circuited to obtain the current i_N. This current is thus 2/4 = 0.5 A. The resistance of the section when the voltage source is replaced by a short-circuit is 4 Ω (Fig. 4.28(b)). Thus the Norton equivalent circuit for this section is given by Fig. 4.28(c).

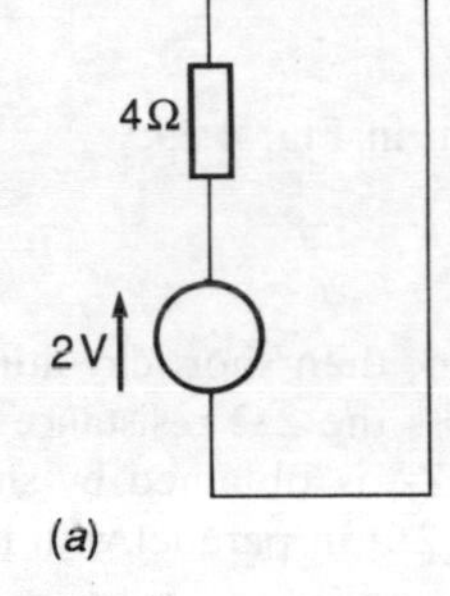

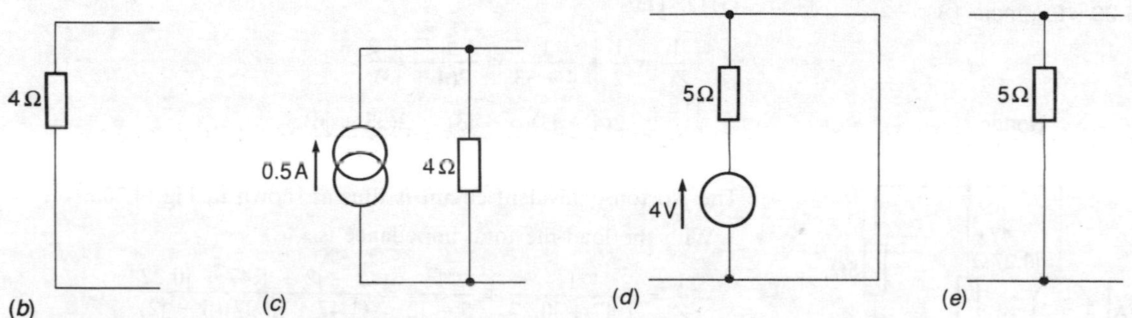

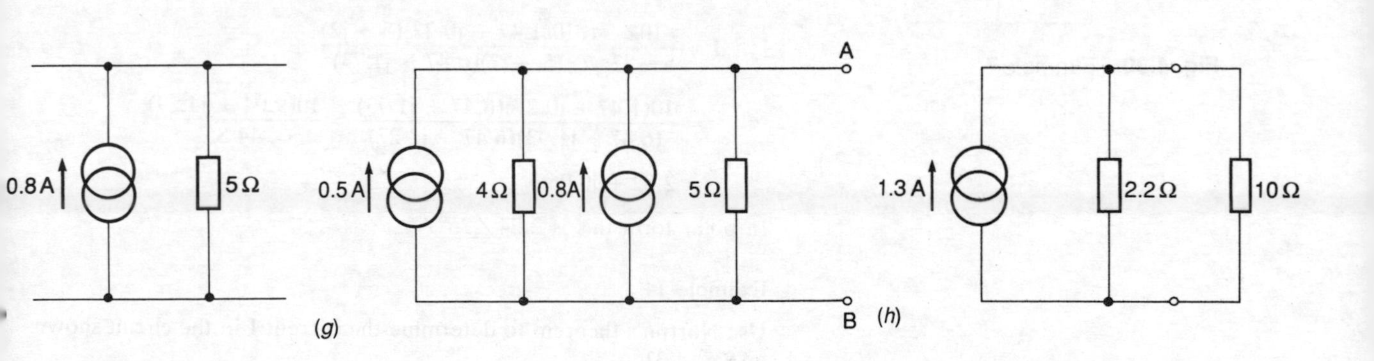

Fig. 4.28 Example 12

The next section of the circuit can be considered in a similar way. Thus Fig. 4.28(*d*) shows this section when short-circuited, the current i_N thus being $4/5 = 0.8$ A. The resistance of this section when the voltage source is replaced by a short-circuit is 5 Ω (Fig. 4.28(*e*)). Thus the Norton equivalent circuit for this section is given by Fig. 4.28(*f*).

Putting the two Norton equivalent circuits together gives Fig. 4.28(*g*) and hence the Norton equivalent circuit for Fig. 4.27 is a current source of $0.5 + 0.8 = 1.3$ A with a resistance of 4 Ω in parallel with 5 Ω, i.e. as in Fig. 4.28(*h*). The potential difference across the 10 Ω resistor is thus that produced by a current of 1.3 A through a parallel arrangement of 2.2 Ω and 10 Ω, i.e. a resistance of $[2.2 \times 10/(2.2 + 10)] = 1.8$ Ω. Thus the potential difference is $1.3 \times 1.8 = 2.3$ V.

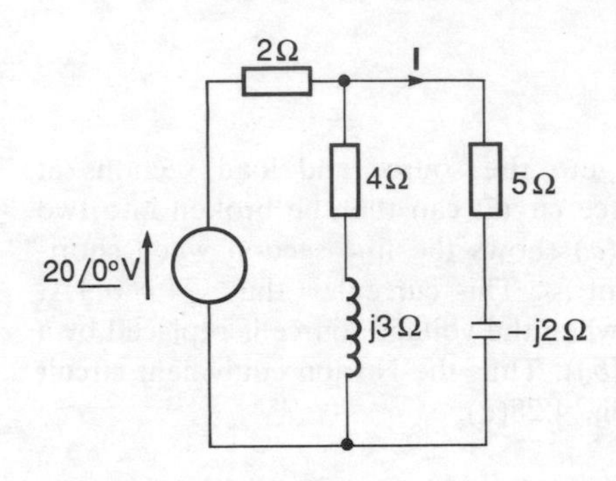

Fig. 4.29 Example 13

Example 13

Determine the current I in the circuit shown in Fig. 4.29.

Answer

If the (5 − j2) Ω is considered to be the load, then short-circuiting it means that the entire source voltage is across the 2 Ω resistance and so the current $\mathbf{I}_N$ is 10 A. The impedance Z_N is obtained by short-circuiting the voltage source. We then have 2 Ω in parallel with (4 + j3) Ω. Thus

$$\frac{1}{Z_N} = \frac{1}{2} + \frac{1}{4 + j3} = \frac{4 + j3 + 2}{2(4 + j3)}$$

$$Z_N = \frac{2(4 + j3)(6 - j3)}{(6 + j3)(6 - j3)} = \frac{2(33 + j6)}{45} = 1.47 + j0.27\,\Omega$$

The Norton equivalent circuit is thus as shown in Fig. 4.30.

With the load the total impedance is

$$\frac{1}{Z} = \frac{1}{1.47 + j0.27} + \frac{1}{5 - j2} = \frac{5 - j2 + 1.47 + j0.27}{(1.47 + j0.27)(5 - j2)}$$

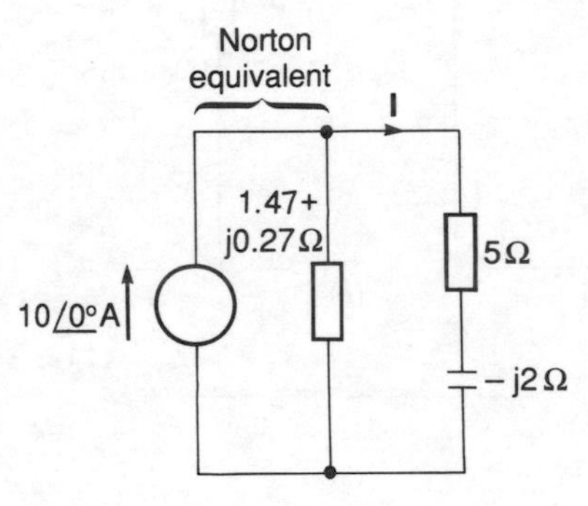

Fig. 4.30 Example 13

The potential difference across the load is $10Z$ and so the current I through it is

$$\mathbf{I} = \frac{10Z}{5 - j2} = \frac{10(1.47 + j0.27)(5 - j2)}{(5 - j2)(6.47 - j1.73)}$$

$$= \frac{10(1.47 + j0.27)(6.47 + j1.73)}{(6.47 - j1.93)(6.47 + j2.27)} = \frac{10(9.04 + j4.29)}{44.85}$$

$$= 2.02 + j0.96\,\text{A}$$

In polar form this is $2.24\angle 25°$ A.

Example 14

Use Norton's theorem to determine the current I in the circuit shown in Fig. 4.31.

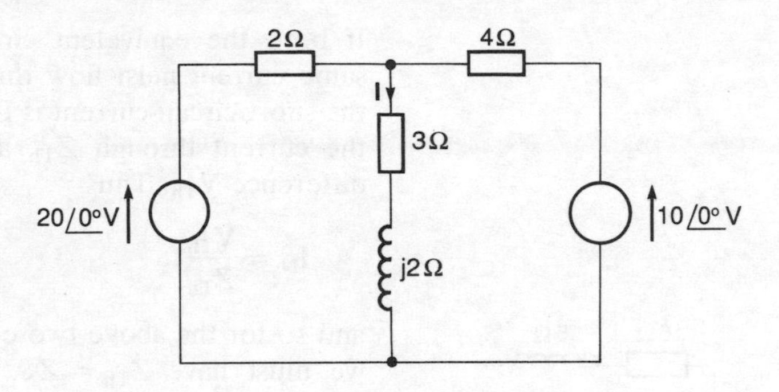

Fig. 4.31 Example 14

Answer

The load is taken to be the (3 + j2) Ω and so the circuit with the load short-circuited is as shown in Fig. 4.32(*a*). The short-circuit current $\mathbf{I_N}$ is $\mathbf{I_1} + \mathbf{I_2}$ and so is (20/2) + (10/4) = 12.5 A. If the voltages sources are replaced by short-circuits then the circuit impedance is 2 Ω in parallel with 4 Ω, i.e. 1.3 Ω. Thus the Norton equivalent circuit with the load is as shown in Fig. 4.32(*b*). The current through the 1.3 Ω is (12.5 − **I**) and so, since the potential difference across the 1.3 Ω is the same as that across the load,

$$1.3(12.5 - \mathbf{I}) = \mathbf{I}(3 + j2)$$

$$\mathbf{I} = \frac{1.3 \times 12.5}{3 + j2 + 1.3} = \frac{1.3 \times 12.5(4.3 - j2)}{(4.3 + j2)(4.3 - j2)}$$

$$= (3.1 - j1.4)\text{ A}$$

In polar form this is $3.4\angle 24°$ A.

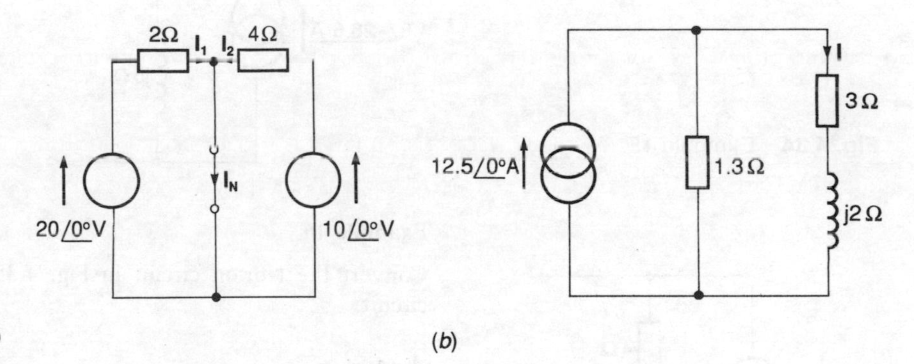

Fig. 4.32 Example 14 (*a*) (*b*)

Thévenin and Norton equivalent circuits

The Thévenin and Norton circuits are alternative equivalents to a source network. If both equivalent circuits have open-circuited terminals then both must have the same open-circuit voltage. For the Thévenin circuit the open-circuit voltage is $\mathbf{V_{Th}}$ while for the Norton equivalent circuit it is a current $\mathbf{I_N}$ through the impedance Z_N, i.e.

$$\mathbf{V_{Th}} = \mathbf{I_N} Z_N \qquad [14]$$

If both the equivalent circuits are short-circuited, then the same current must flow through each. For the Norton circuit the short-circuit current is $\mathbf{I_N}$ while for the Thévenin circuit it is the current through Z_{Th} as a result of an applied potential difference $\mathbf{V_{Th}}$. Thus

$$\mathbf{I_N} = \frac{\mathbf{V_{Th}}}{Z_{Th}} \qquad [15]$$

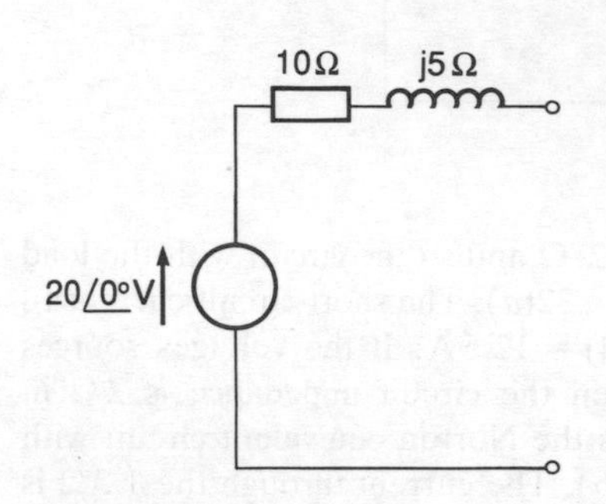

Fig. 4.33 Example 15

and so for the above two equations to be simultaneously valid we must have $Z_{Th} = Z_N$. In the solution of any particular problem, whether the Thévenin or the Norton equivalent circuit is used is purely a matter of convenience.

Example 15

Convert the Thévenin circuit in Fig. 4.33 to an equivalent Norton circuit.

Answer

Using equation [15], with (10 + j5) in polar form as $11.2\angle 26.6°\,\Omega$,

$$\mathbf{I_N} = \frac{\mathbf{V_{Th}}}{Z_{Th}} = \frac{20\angle 0°}{11.2\angle 26.6°} = 1.8\angle -26.6°\text{ A}$$

$Z_N = Z_{Th}$, hence the circuit is as shown in Fig. 4.34.

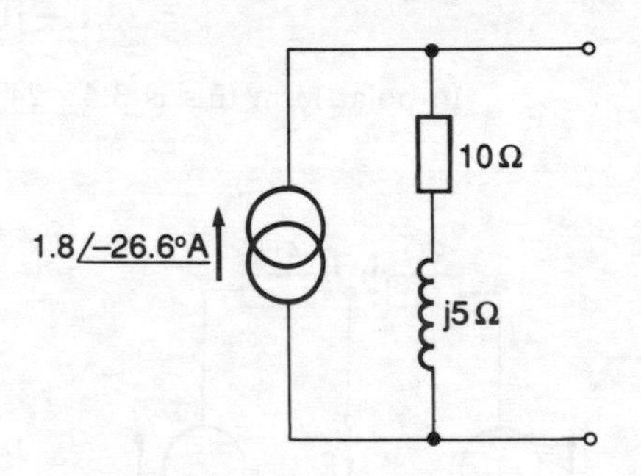

Fig. 4.34 Example 15

Example 16

Convert the Norton circuit in Fig. 4.35 to an equivalent Thévenin circuit.

Answer

Using equation [14], with Z_N in polar form as $4.5\angle 26.6°\,\Omega$, then

$$\mathbf{V_{Th}} = \mathbf{I_N}Z_N = (2\angle 30°)(4.5\angle 26.6°) = 9.0\angle 56.6°\text{ V}$$

Since $Z_{Th} = Z_N$, then the equivalent circuit is as shown in Fig. 4.36.

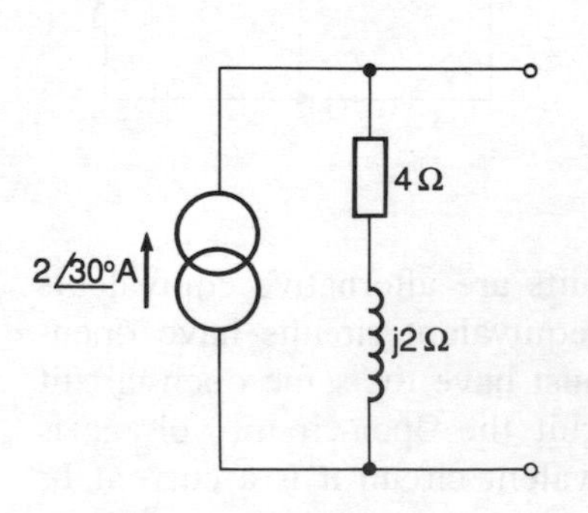

Fig. 4.35 Example 16

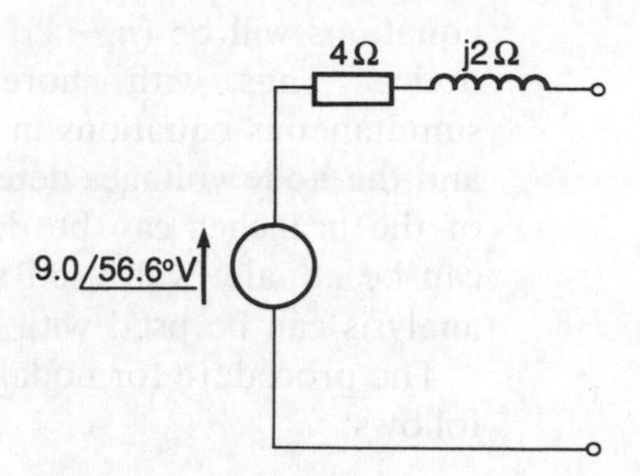

Fig. 4.36 Example 16

Node voltage analysis

A *node* is defined as being a point in an electrical circuit where two or more circuit elements are joined. If three or more circuit elements join at a node then that node is called a *principal node* or *essential node*. Thus for the circuit shown in Fig. 4.37 the nodes are points a, b, c and d, with b and d being principal nodes. The term *branch* is used for the path that connects two nodes. Thus, for example, the principal node at b in Fig. 4.37 is a junction of three branches ab, cb, and db.

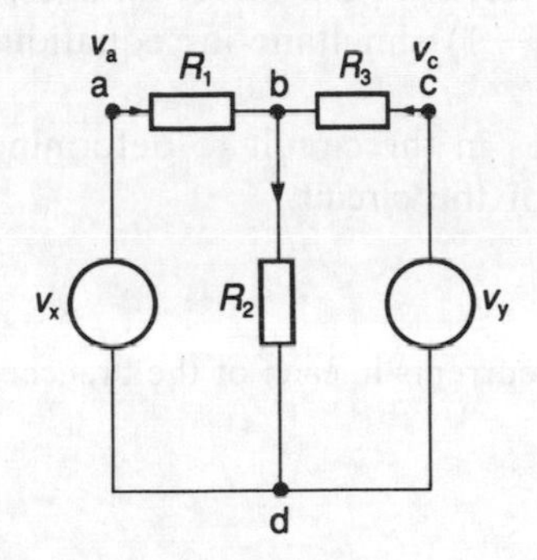

Fig. 4.37 Nodes

Node voltage analysis requires the selection of one of the principal nodes as the reference node. It does not matter which node is selected but usually it is easiest to take the node with the most branches. The voltage at any other node is then the voltage at that node relative to that at the reference node. Thus in Fig. 4.37 if node d is taken as the reference node, then the voltages v_a, v_b, and v_c are the node voltages relative to the voltage at node d. This means that the potential difference across resistor R_2 is v_b, across resistor R_1 is $(v_a - v_b)$ and resistor R_3 is $(v_c - v_b)$. Thus the current through R_2 is v_b/R_2, the current through R_1 is $(v_a - v_b)/R_1$ and through R_3 is $(v_c - v_b)$. Then Kirchoff's current law is applied to each of the principal nodes, with the exception of the reference node. Thus for node b we have

$$\frac{v_a - v_b}{R_1} + \frac{v_c - v_b}{R_3} - \frac{v_b}{R_2} = 0$$

In the example chosen the source potential difference v_x is v_a relative to the reference potential and v_c is the source potential difference v_y. Thus

$$\frac{v_x - v_b}{R_1} + \frac{v_y - v_b}{R_3} - \frac{v_b}{R_2} = 0$$

Hence the voltages at each node can be determined and so the current in each branch of the circuit. In the example chosen there are only two principal nodes and so there is only one equation. Since an equation can be produced for every principal node but the reference node, then the number of equations will be $(n - 1)$, where n is the number of principal nodes. Thus with more complex circuits a number of simultaneous equations in the node voltages can be produced and the node voltages determined. Hence the currents in each of the branches can be determined. Because Kirchoff's laws can be equally well used with phasors and impedances, node analysis can be used with phasors and impedances.

The procedure for nodal analysis can thus be summarized as follows:

1 Label the n principal nodes.
2 Select a reference node.
3 Apply Kirchoff's current law to each non-reference node, using Ohm's law or the relationship $\mathbf{V} = Z\mathbf{I}$ to express the currents through resistors in terms of the node voltages.
4 Solve the resulting set of $(n - 1)$ simultaneous equations for the node voltages.
5 Use the values of the voltages in the circuit to determine the currents in each branch of the circuit.

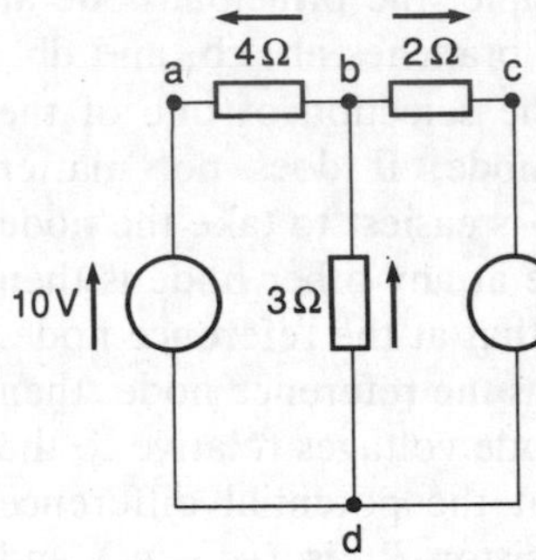

Fig. 4.38 Example 17

Example 17

Use nodal analysis to determine the currents in each of the branches of the circuit shown in Fig. 4.38.

Answer

Node d is selected as the reference node. Thus the potential difference across the 4 Ω resistor is $(v_a - v_b)$ and the current through it is thus $(v_a - v_b)/4$. The potential difference across the 3 Ω resistor is v_b and the current through it is thus $v_b/3$. The potential difference across the 2 Ω resistor is $(v_c - v_b)$ and thus the current through it is $(v_c - v_b)/2$. Thus for node b Kirchoff's current law gives

$$\frac{v_a - v_b}{4} + \frac{v_c - v_b}{2} - \frac{v_b}{3} = 0$$

But v_a is the potential difference across the voltage source and so is 10 V; likewise, $v_c = 5$ V. Thus the equation becomes

$$\frac{10 - v_b}{4} + \frac{5 - v_b}{2} - \frac{v_b}{3} = 0$$

and so

$$\frac{3(10 - v_b) + 6(5 - v_b) - 4v_b}{12} = 0$$

$$60 - 13v_b = 0$$

Hence $v_b = 4.62$ V. Since the potential difference across the 3 Ω resistor is 4.62 V then the current through that branch is 4.62/3 = 1.54 A. The potential difference across the 4 Ω resistor is (10 − 4.62) V and so the current through that branch is 5.38/4 = 1.35 A. The potential difference across the 2 Ω resistor is (5 − 4.62) V and so the current through that branch is 0.19 A.

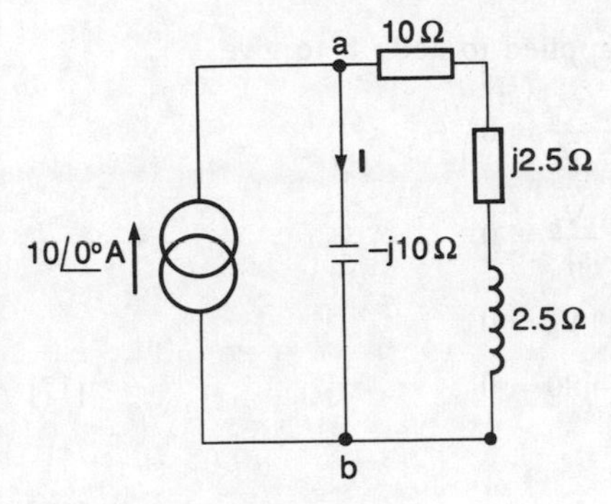

Fig. 4.39 Example 18

Example 18

Use node analysis to determine the voltages at node a and the current through the capacitor for the circuit in Fig. 4.39.

Answer

Taking b as the reference node, then applying Kirchoff's current law to node a gives

$$\frac{\mathbf{V_a}}{-j10} + \frac{\mathbf{V_a}}{10 + 2.5 + j2.5} - 10 = 0$$

$$\mathbf{V_a} = \frac{10(-j10)(12.5 + j2.5)}{12.5 + j2.5 - j10} = \frac{10(25 - j125)}{12.5 - j7.5}$$

Since $\mathbf{V_a} = (-j10)\mathbf{I}$, then

$$\mathbf{I} = \frac{10(25 - j125)}{(-j10)(12.5 - j7.5)} = \frac{25 - j125}{-7.5 - j12.5}$$

$$= \frac{(25 - j125)(-7.5 + j12.5)}{(-7.5 - j12.5)(-7.5 + j12.5)} = \frac{1375 + j1250}{212.5}$$

$$= 6.47 + j5.88\ \text{A}$$

In polar form this is $8.74\angle 42.3°$ A.

Example 19

Determine the voltages at nodes a and b in the circuit described by the circuit diagram in Fig. 4.40.

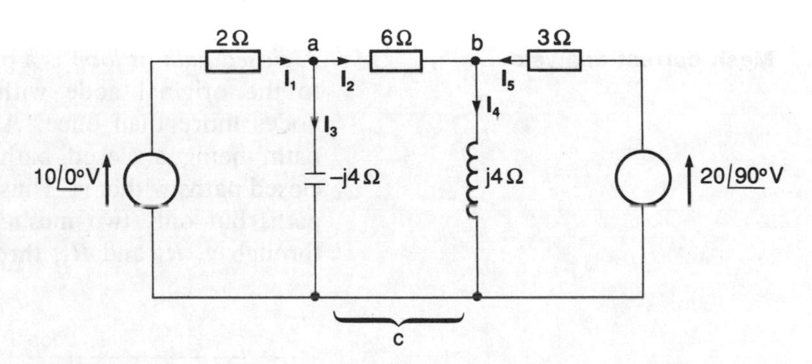

Fig. 4.40 Example 19

Answer

Taking node c as the reference node, then applying Kirchoff's current law to node a gives

$$\frac{(10 - \mathbf{V_a})}{2} - \frac{(\mathbf{V_a} - \mathbf{V_b})}{6} - \frac{\mathbf{V_a}}{-j4} = 0$$

$$\frac{(10 - \mathbf{V_a})}{2} - \frac{(\mathbf{V_a} - \mathbf{V_b})}{6} - \frac{j\mathbf{V_a}}{4} = 0$$

$$60 - 6\mathbf{V_a} - 2\mathbf{V_a} + 2\mathbf{V_b} - j3\mathbf{V_a} = 0$$

$$60 - \mathbf{V_a}(8 + j3) + 2\mathbf{V_b} = 0 \qquad [16]$$

Kirchoff's current law is then applied to node b to give

$$\frac{(\mathbf{V_a} - \mathbf{V_b})}{6} + \frac{(\mathrm{j}20 - \mathbf{V_b})}{3} - \frac{\mathbf{V_b}}{\mathrm{j}4} = 0$$

$$\frac{(\mathbf{V_a} - \mathbf{V_b})}{6} + \frac{(\mathrm{j}20 - \mathbf{V_b})}{3} + \frac{\mathrm{j}\mathbf{V_b}}{4} = 0$$

$$2\,\mathbf{V_a} - 2\,\mathbf{V_b} + \mathrm{j}80 - 4\,\mathbf{V_b} + \mathrm{j}3\,\mathbf{V_b} = 0$$

$$2\,\mathbf{V_a} - \mathbf{V_b}(6 - \mathrm{j}3) + \mathrm{j}80 = 0 \qquad [17]$$

Equation [17] gives

$$\mathbf{V_a} = 0.5\,\mathbf{V_b}(6 - \mathrm{j}3) + \mathrm{j}40 \qquad [18]$$

Hence substituting this in equation [16] gives

$$60 - [0.5\,\mathbf{V_b}(6 - \mathrm{j}3) + \mathrm{j}40](8 + \mathrm{j}3) + 2\,\mathbf{V_b} = 0$$

$$60 - \mathrm{j}40(8 + \mathrm{j}3) - \mathbf{V_b}[(3 - \mathrm{j}1.5)(8 + \mathrm{j}3) - 2] = 0$$

$$\mathbf{V_b} = \frac{180 - \mathrm{j}320}{28.5 - \mathrm{j}3} = \frac{(180 - \mathrm{j}320)(28.5 + \mathrm{j}3)}{(28.5 - \mathrm{j}3)(28.5 + \mathrm{j}3)} = \frac{6090 - \mathrm{j}8580}{821.25}$$
$$= 7.42 - \mathrm{j}10.45\,\mathrm{V}$$

In polar form this is $12.82\angle{-54.6^\circ}\,\mathrm{V}$. $\mathbf{V_a}$ can be obtained by substituting this value in equation [18]. Thus

$$\mathbf{V_a} = 0.5(7.42 - \mathrm{j}10.45)(6 - \mathrm{j}3) + \mathrm{j}40$$
$$= 6.59 - \mathrm{j}2.48\,\mathrm{V}$$

In polar form this is $7.04\angle{-20.6^\circ}\,\mathrm{V}$.

Mesh current analysis

A *closed path* or *loop* is a path that starts at a node and returns to the original node without passing through intermediate nodes more than once. A *mesh* is a special form of closed path, being a closed path that does not contain any other closed paths within it. Thus for Fig. 4.41 there are three closed paths but only two meshes. The closed paths are the paths through v, R_1 and R_3, through R_3 and R_2 and through v, R_1

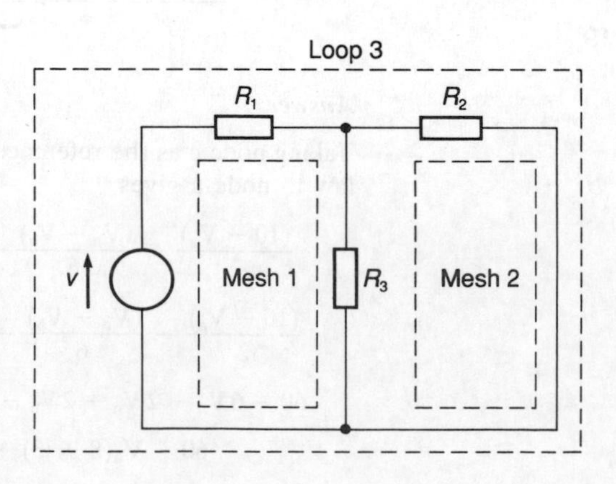

Fig. 4.41 Meshes

and R_2. Only the first two of these closed paths are meshes. The closed path through v, R_1 and R_2 is not a mesh since it contains other closed paths. Meshes can be thought of as being like the meshes of a fisherman's net.

Mesh current analysis involves defining a mesh current for each mesh. The *mesh current* is the current that is considered to circulate round the mesh path. The same direction must be chosen for all the mesh currents in a circuit. A usual convention is to make all the mesh currents clockwise currents. Figure 4.42 shows the mesh currents for a circuit with two meshes. For that circuit the current through the component R_1 is the mesh current i_1 but for the component R_3, which is common to two meshes, the current through it is the algebraic sum of the two mesh currents, i.e. $(i_1 - i_2)$ or $(i_2 - i_1)$.

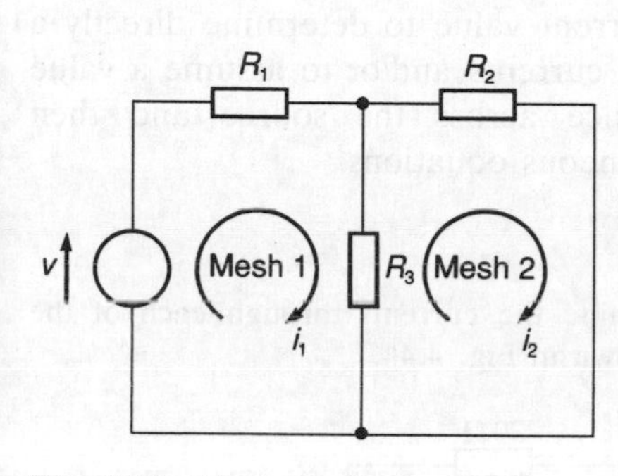

Fig. 4.42 Mesh currents

Having specified the mesh currents, Kirchoff's voltage law is applied to each mesh. Thus for Fig. 4.42 we have for mesh 1,

$$v - i_1 R_1 - (i_1 - i_2) R_3 = 0 \qquad [19]$$

and for mesh 2,

$$-R_2 i_2 - R_3(i_2 - i_1) = 0 \qquad [20]$$

The two simultaneous equations [19] and [20] can then be solved to obtain the mesh currents and hence the currents through individual components.

This technique of circuit analysis applies only to *planar circuits*. These are circuits that can be drawn on a plane so that no branches cross over each other, a branch being a path connecting two nodes (Fig. 4.43). Because Kirchoff's laws can be equally well used with phasors and impedances then mesh analysis can also be used with phasors and impedances.

The mesh analysis technique can be summarized as:

1 Label each of the m meshes with clockwise mesh currents.

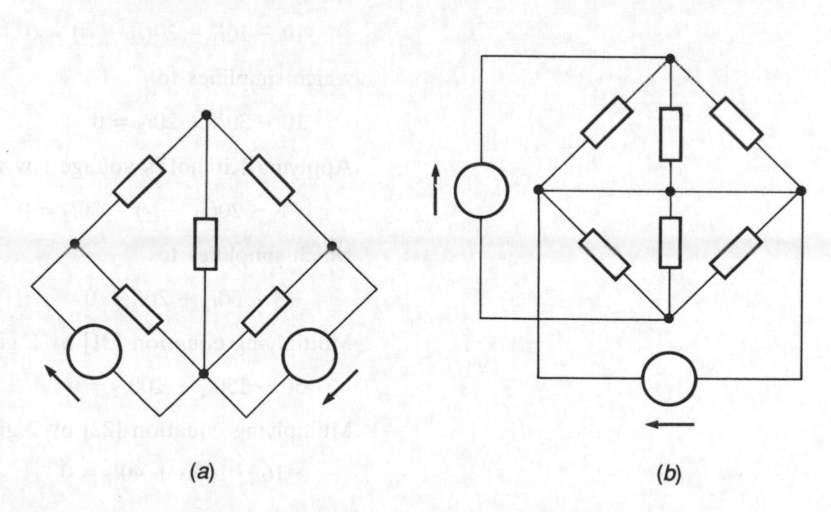

Fig. 4.43 (*a*) A planar circuit, (*b*) a non-planar circuit

2 Apply Kirchoff's voltage law to each of the m meshes by traversing the mesh in a clockwise direction, potential differences across resistor or impedance being given by Ohm's law or $V = ZI$ and in the opposite direction to the mesh current for that mesh.
3 Solve the resulting set of m simultaneous equations for the mesh currents.
4 Use the mesh currents to establish the currents through the branches of the circuit.

In considering current sources the techniques that can be adopted are to use the current value to determine directly a relationship between mesh currents and/or to assume a value for the potential difference across the source and then eliminate from the simultaneous equations.

Example 20

Use mesh analysis to determine the currents through each of the components in the circuit shown in Fig. 4.44.

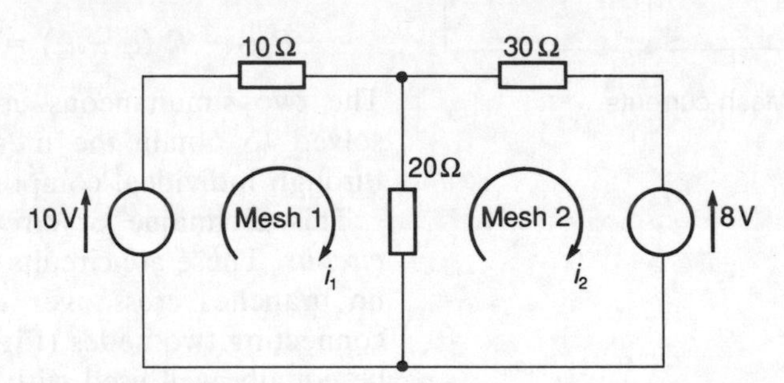

Fig. 4.44 Example 20

Answer

With the mesh currents as indicated in the circuit diagram, then applying Kirchoff's voltage law to mesh 1 gives

$$10 - 10i_1 - 20(i_1 - i_2) = 0$$

which simplifies to

$$10 - 30i_1 + 20i_2 = 0 \qquad [21]$$

Applying Kirchoff's voltage law to mesh 2 gives

$$-8 - 20(i_2 - i_1) - 30i_2 = 0$$

which simplifies to

$$-8 - 50i_2 + 20i_1 = 0 \qquad [22]$$

Multiplying equation [21] by 5 gives

$$50 - 150i_1 + 100i_2 = 0$$

Multiplying equation [22] by 2 gives

$$-16 - 100i_2 + 40i_1 = 0$$

Adding these two equations gives

$$50 - 150i_1 + 100i_2 - 16 - 100i_2 + 40i_1 = 0$$

Hence

$$110i_1 = 34$$

$$i_1 = 0.309\,\text{A}$$

Substituting this value in equation [22] gives

$$-8 - 50i_2 + 6.18 = 0$$

$$i_2 = -0.036\,\text{A}$$

The minus sign indicates that the mesh current i_2 is in the opposite direction to that drawn in Fig. 4.44. Thus the current through the 10 Ω resistor is 0.309 A, through the 30 Ω resistor 0.036 A and through the 20 Ω resistor 0.345 A.

Example 21

Use mesh analysis to determine the currents through each of the components in the circuit shown in Fig. 4.45.

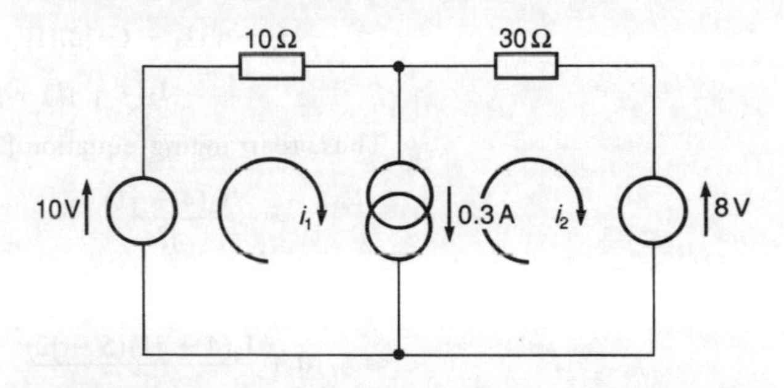

Fig. 4.45 Example 21

Answer

With the mesh currents as indicated in the circuit diagram, applying Kirchoff's voltage law to mesh 1 gives

$$10 - 10i_1 + v = 0$$

The current source is assumed to have a potential v across it. Applying Kirchoff's voltage law to mesh 2 gives

$$-8 - v + 30i_2 = 0$$

Adding these two equations gives

$$10 - 10i_1 + v - 8 - v + 30i_2 = 0$$

$$2 - 10i_1 - 30i_2 = 0 \qquad [23]$$

But the current source is 0.3 A, hence for that branch

$$i_1 - i_2 = 0.3 \qquad [24]$$

Multiplying this equation by 10 gives

$$10i_1 - 10i_2 = 3$$

Adding this to equation [23] gives

$$2 - 10i_1 - 30i_2 + 10i_1 - 10i_2 = 0 + 3$$

$$i_2 = -0.025\text{ A}$$

Hence substituting this value in equation [24] gives $i_1 = 0.275$ A. The current through the 10 Ω branch is thus 0.275 A, through the 30 Ω branch 0.025 A and through the current source 0.3 A.

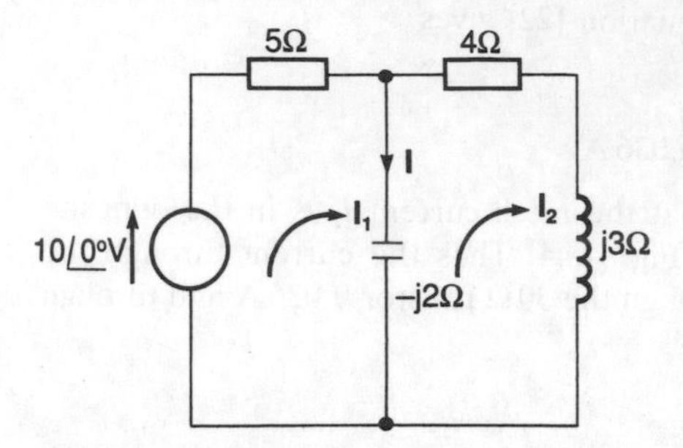

Fig. 4.46 Example 22

Example 22

Use mesh analysis to determine the current **I** in the capacitor arm of the circuit in Fig. 4.46.

Answer

There are two meshes and the mesh currents are $\mathbf{I}_1$ and $\mathbf{I}_2$. Applying Kirchoff's voltage law to the first mesh gives

$$10 - 5\mathbf{I}_1 - (-j2)(\mathbf{I}_1 - \mathbf{I}_2) = 0$$

$$10 - \mathbf{I}_1(5 - j2) - j2\mathbf{I}_2 = 0 \quad [25]$$

Applying Kirchoff's voltage law to the second mesh gives

$$-4\mathbf{I}_2 - j3\mathbf{I}_2 - (-j2)(\mathbf{I}_2 - \mathbf{I}_1) = 0$$

$$\mathbf{I}_2(4 + j1) + j2\mathbf{I}_1 = 0 \quad [26]$$

Thus, rearranging equation [26],

$$\mathbf{I}_1 = -\frac{\mathbf{I}_2(4 + j1)}{j2} \quad [27]$$

Substituting this value of $\mathbf{I}_1$ into equation [25] gives

$$10 + \frac{\mathbf{I}_2(4 + j1)(5 - j2)}{j2} - j2\mathbf{I}_2 = 0$$

$$10 - \tfrac{1}{2}j\,\mathbf{I}_2(22 - j3) - j2\mathbf{I}_2 = 0$$

$$10 - \mathbf{I}_2(1.5 + j11) - j2\mathbf{I}_2 = 0$$

Thus

$$\mathbf{I}_2 = \frac{10}{1.5 + j13} = \frac{10(1.5 - j13)}{(1.5 + j13)(1.5 - j13)} = (0.0876 - j0.759)\text{ A}$$

Substituting this value into equation [27] gives

$$\mathbf{I}_1 = \frac{-(0.0876 - j0.759)(4 + j1)}{j2}$$

$$= \tfrac{1}{2}j(0.0876 - j0.759)(4 + j1) = (1.474 - j0.555)\text{ A}$$

The current **I** is $\mathbf{I}_1 - \mathbf{I}_2$ and is thus (1.386 + j0.204) A. In polar form this is $1.40\angle 8.4°$ A.

Maximum power transfer theorem

A network can be considered as having two elements, the source and the load. The source network can then be replaced

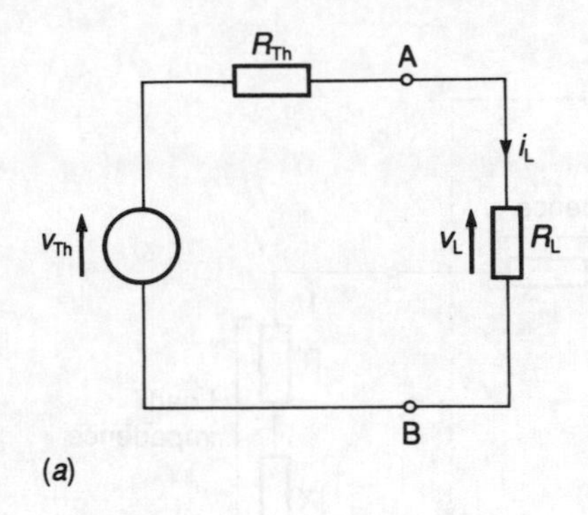

(a)

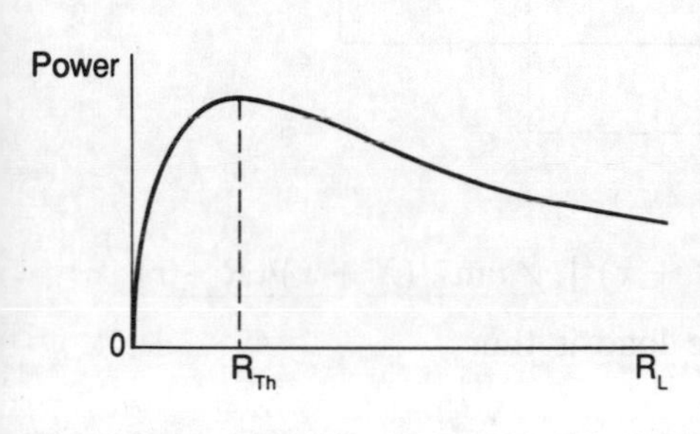

(b)

Fig. 4.47 (*a*) Thévenin circuit, (*b*) power–load resistance graph

by an equivalent Thévenin or Norton circuit. Thus, for a purely resistive network with a d.c. source, representing the source network by the Thévenin equivalent and considering the load to be represented by a resistor R_L, then the circuit is effectively as shown in Fig. 4.47(*a*). The instantaneous power p dissipated by a load network is then the product of the voltage v_L and current i_L at the load terminals A and B, i.e.

$$p = i_L v_L \qquad [28]$$

But for the circuit in Fig. 4.47(*a*),

$$v_{Th} = i_L(R_{Th} + R_L)$$

$$i_L = \frac{v_{Th}}{R_{Th} + R_L}$$

But $v_L = i_L R_L$, hence equation [28] can be written as

$$p = i_L(i_L R_L) = \frac{v_{Th}^2 R_L}{(R_{Th} + R_L)^2} \qquad [29]$$

The power p dissipated thus depends on the value of R_L, varying from zero when $R_L = 0$, i.e. a short circuit, to some maximum before falling again to zero when $R_L = \infty$, i.e. an open circuit. Figure 4.47(*b*) shows the graph of how the power depends on R_L. The maximum power occurs when $\mathrm{d}p/\mathrm{d}R_L = 0$. Differentiating equation [29] gives

$$\frac{\mathrm{d}p}{\mathrm{d}R_L} = v_{Th}^2 \left[\frac{(R_{Th} + R_L)^2 - 2R_L(R_{Th} + R_L)}{(R_{Th} + R_L)^4}\right]$$

$$= v_{Th}^2 \frac{(R_{Th} - R_L)}{(R_{Th} + R_L)^3}$$

When $\mathrm{d}p/\mathrm{d}R_L = 0$ then $R_{Th} = R_L$. Thus there is maximum power transfer from the source network to the load when the source resistance equals the load resistance. In a circuit where the source resistance equals the load resistance, the load is said to be *matched* to the source.

The maximum power p_{max} that can be dissipated in the load is thus obtained when $R_{Th} = R_L$ and thus equation [29] gives

$$p_{max} = \frac{v_{Th}^2 R_L}{(R_{Th} + R_L)^2} = \frac{v_{Th}^2 R_{Th}}{(2R_{Th})^2} = \frac{v_{Th}^2}{4R_{Th}} \qquad [30]$$

The same approach can be used with a.c. sources and loads having impedance. Figure 4.48 shows such a circuit. The source has a complex impedance of $(r + \mathrm{j}x)$ and the load a complex impedance of $(R + \mathrm{j}X)$. The total circuit impedance Z is

$$Z = (R + \mathrm{j}X) + (r + \mathrm{j}x) = (R + r) + \mathrm{j}(X + x)$$

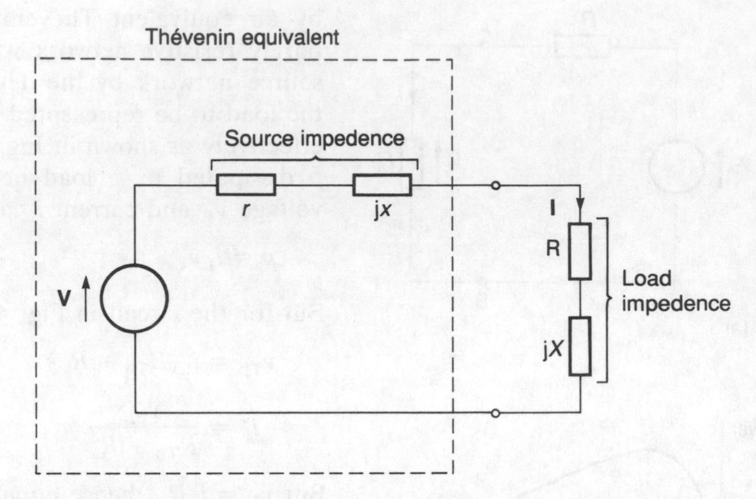

Fig. 4.48 Complex source and load

In polar form this is

$$Z = \sqrt{[(R+r)^2 + (X+x)^2]}\ \angle \tan^{-1}(X+x)/(R+r)$$

The current **I** through the load is thus

$$\mathbf{I} = \frac{\mathbf{V}}{Z} = \frac{\mathbf{V}}{\sqrt{[(R+r)^2 + (X+x)^2]}\ \angle \tan^{-1}(X+x)/(R+r)}$$

The magnitude of the current is thus

$$I = \frac{V}{\sqrt{[(R+r)^2 + (X+x)^2]}} \qquad [31]$$

The real power P dissipated in the load is I^2R, since power is only dissipated in the resistance part of the load, and is thus

$$P = \frac{V^2R}{(R+r)^2 + (X+x)^2} \qquad [32]$$

Consider the situation where both R and X can be independently varied. There are thus two variables which determine the value of the power delivered, namely X and R. Considering first the variable X, then

$$\frac{dP}{dX} = \frac{-V^2R \times 2(X+x)}{[(R+r)^2 + (X+x)^2]^2}$$

The maximum is when $dP/dX = 0$, i.e. when $(X + x) = 0$. Thus the condition for maximum power is that $X = -x$. Imposing this condition on equation [32] gives

$$P = \frac{V^2R}{(R+r)^2}$$

Since

$$\frac{dP}{dR} = \frac{(R+r)^2 \times V^2 - V^2R \times 2(R+r)}{(R+r)^4}$$

The maximum is when $\mathrm{d}P/\mathrm{d}R = 0$, i.e. when

$$(R + r)^2 \times V^2 - V^2 R \times 2(R + r) = 0$$

$$R^2 + 2Rr + r^2 - 2R^2 - 2Rr = 0$$

$$r^2 - R^2 = 0$$

The condition is thus $R = r$. Thus with a complex load and complex source, the condition for maximum power transfer is

$$R + \mathrm{j}X = r - \mathrm{j}x \qquad [33]$$

The load impedance has thus to be the conjugate of the source impedance. Since at this condition the load reactance cancels the source reactance, the voltage V and current I will be in phase. When the source and the load are both purely resistive the condition becomes $R = r$, the condition derived earlier.

The current I is, by equation [31] with $R = r$ and $X = -x$,

$$I = \frac{V}{2r} \qquad [34]$$

and thus the maximum power is

$$P_{\max} = \left(\frac{V}{2r}\right)^2 R = \frac{V^2}{4r} \qquad [35]$$

In some situations, as for when transformers are used to match a load to source, the load impedance can be adjusted through a choice of turns ratio but the power factor remains unchanged. The load impedance magnitude $|Z_L|$ is related to the load resistance R and the load reactance X by

$$\frac{R}{|Z_L|} = \cos\phi$$

$$\frac{X}{|Z_L|} = \sin\phi$$

Thus equation [32] can be written as

$$P = \frac{V^2 |Z_L| \cos\phi}{(|Z_L|\cos\phi + r)^2 + (|Z_L|\sin\phi + x)^2}$$

The maximum power is when $\mathrm{d}P/\mathrm{d}|Z_L| = 0$, i.e. when

$$0 = V^2 \cos\phi [\{(|Z_L|\cos\phi + r)^2 + (|Z_L|\sin\phi + x)^2\} \\ - |Z_L|\{2(|Z_L|\cos\phi + r)\cos\phi + 2(|Z_L|\sin\phi + x)\sin\phi\}]$$

Thus the condition is

$$|Z_L|^2\cos^2\phi + 2|Z_L|r\cos\phi + r^2 + |Z_L|^2\sin^2\phi \\ + 2|Z_L|x\sin\phi + x^2 - 2|Z_L|^2\cos^2\phi - 2|Z_L|r\cos\phi \\ - 2|Z_L|^2\sin^2\phi - 2|Z_L|x\sin\phi = 0$$

$$r^2 + x^2 - |Z_L|^2\cos^2\phi - |Z_L|^2\sin^2\phi = 0$$

But $\sin^2\phi + \cos^2\phi = 1$, hence

$$r^2 + x^2 = |Z_L|^2$$

But the impedance magnitude of the source $|Z_s| = \sqrt{(r^2 + x^2)}$. Thus the condition for maximum power is that

$$|Z_s| = |Z_L| \qquad [36]$$

Now consider the condition for maximum power when there is a complex source delivering power to a load for which only the resistance can be varied. As before we have for the power equation [32]

$$P = \frac{V^2R}{(R + r)^2 + (X + x)^2}$$

and the condition for maximum power is $dP/dR = 0$. Thus

$$0 = V^2[\{(R + r)^2 + (X + x)^2\} - R\{2(R + r)\}]$$

$$0 = R^2 + 2Rr + r^2 + (X + x)^2 - 2R^2 - 2Rr$$

$$R^2 = r^2 + (X + x)^2$$

$$R = \sqrt{[r^2 + (X + x)^2]} \qquad [37]$$

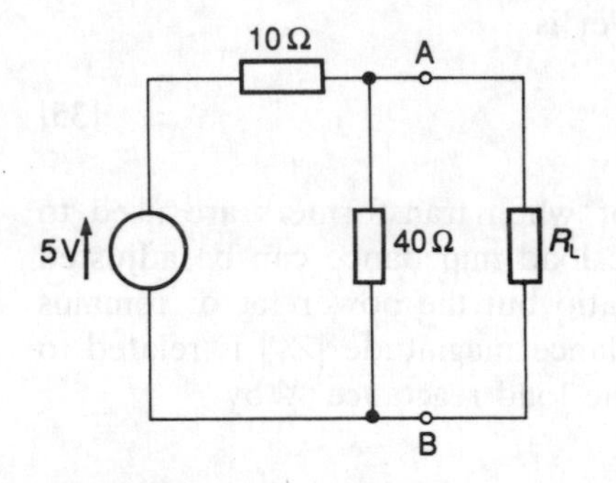

Fig. 4.49 Example 22

Example 22

Determine the value of the load resistor R_L in Fig. 4.49 which will give maximum power transfer from source to load.

Answer

The first step is to find the Thévenin equivalent circuit for the source part of Fig. 4.49. Thus the open-circuit voltage for Fig. 4.50(*a*) is the fraction of the 5 V that is across the 40 Ω resistor. This fraction is 40/(10 + 40) and thus the open circuit voltage v_{Th} is 4 V. The resistance R_{Th} is given by replacing the independent voltage source by a short circuit and so by circuit (Fig. 4.50(*b*)). The resistance is thus 10 Ω in parallel with 40 Ω, i.e. 10 × 40/(10 + 40) = 8 Ω. Hence the circuit is as in Fig. 2.50(*c*). Maximum power transfer thus occurs when $R_L = R_{Th} = 8\ \Omega$.

Fig. 4.50 Example 22

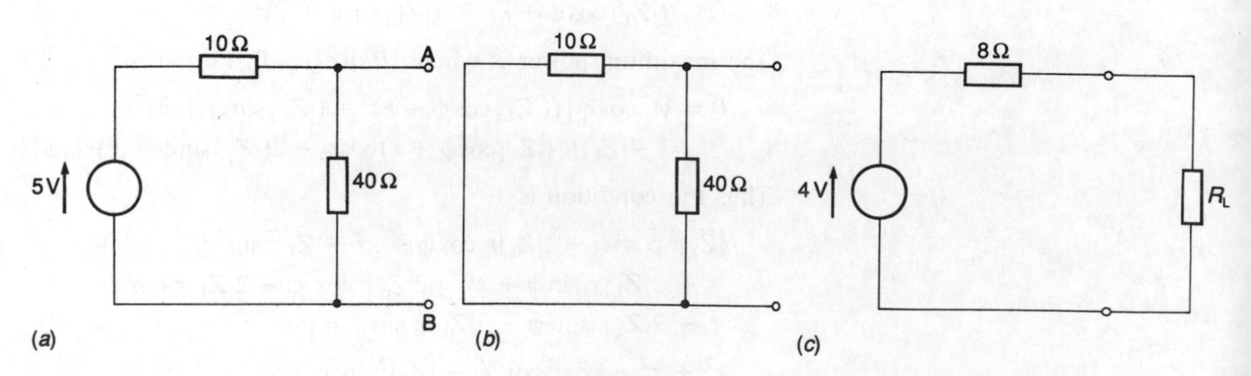

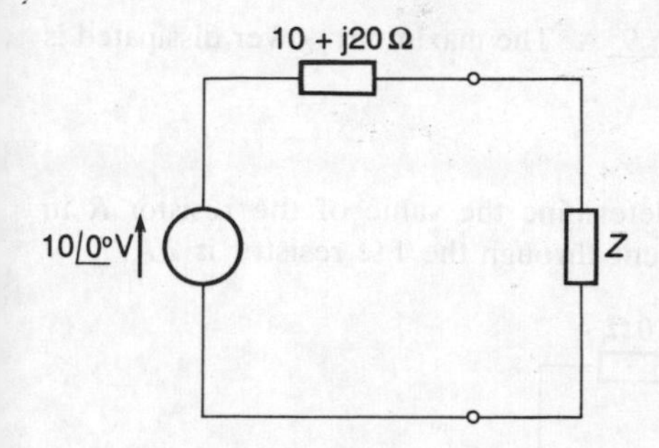

Fig. 4.51 Example 23

Example 23

Determine, for the circuit in Fig. 4.51, the impedance of the load that is necessary for maximum power transfer, and the value of the maximum power.

Answer

The maximum power is delivered when the load has an impedance which is the complex conjugate of the source, i.e. when

$$Z = 10 - j20\,\Omega$$

The maximum power is (equation [35])

$$P_{max} = \frac{V^2}{4r} = \frac{100}{40} = 2.5\,W$$

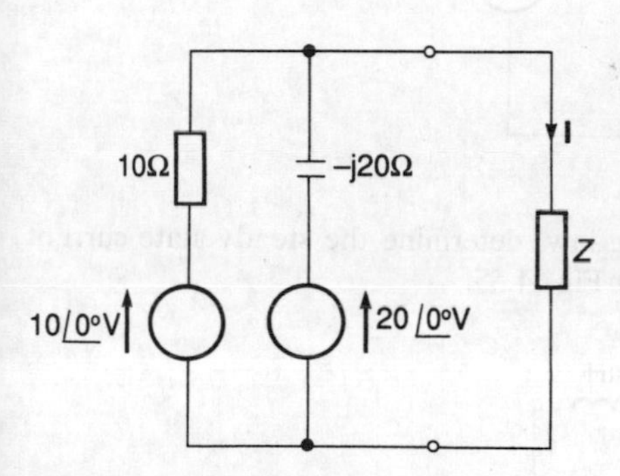

Fig. 4.52 Example 24

Example 24

Determine for the circuit in Fig. 4.52 the load impedance for maximum power, the current through the load **I** and the power at maximum power.

Answer

The source part of the circuit needs to be put into the Thévenin circuit form. The open-circuit voltage can be determined by using the superposition theorem. With the $20\angle 0°$ V source replaced by a short circuit then the circuit current is 10/(10 − j20) A and so the output potential difference is

$$\mathbf{V}_1 = \frac{10(-j20)}{10 - j20} = \frac{-j20}{1 - j2}$$

With the $10\angle 0°$ source replaced by a short circuit then the circuit current is 20/(10 − j20) A and so the output potential difference is

$$\mathbf{V}_2 = \frac{20 \times 20}{10 - j20} = \frac{40}{1 - j2}$$

Hence the open circuit voltage is

$$\frac{40 - j20}{1 - j2} = \frac{(40 - j20)(1 + j2)}{(1 - j2)(1 + j2)} = 16 + j12\,V$$

The Thévenin impedance Z is

$$\frac{1}{Z} = \frac{1}{10} + \frac{1}{-j20} = \frac{1 - j2}{-j20}$$

$$Z = \frac{-j20}{1 - j2} = \frac{-j20(1 + j2)}{(1 - j2)(1 + j2)} = 8 - j4\,\Omega$$

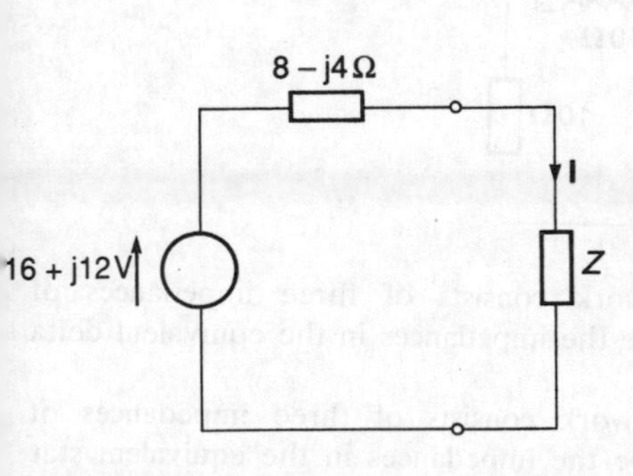

Fig. 4.53 Example 24

Figure 4.53 shows the Thévenin equivalent circuit.

Thus for maximum power the load must have an impedance of (8 + j4) Ω. The current **I** at maximum power is given by

$$\mathbf{I} = \frac{16 + j12}{(8 - j4) + (8 + j4)} = 1.0 + j0.75\,A$$

In polar form this is $1.25\angle 36.9°$ A. The maximum power dissipated is $1.25^2 \times 8 = 12.5$ W.

Problems

1 Using Kirchoff's laws determine the value of the resistor R in Fig. 4.54 when the current through the 1 Ω resistor is 2 A.

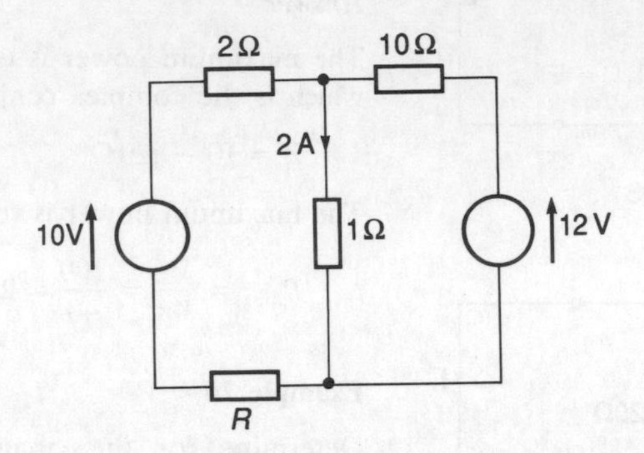

Fig. 4.54 Problem 1

2 Using Kirchoff's voltage law, determine the steady state current for the circuit shown in Fig. 4.55.

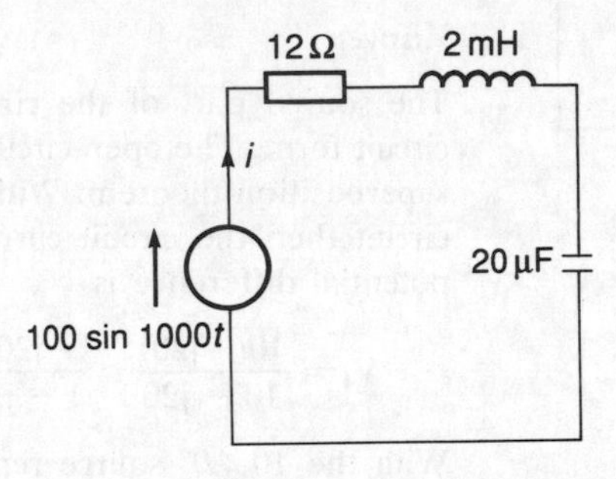

Fig. 4.55 Problem 2

3 Use the delta–star transformation to simplify the circuit in Fig. 4.56 and so determine the current taken from the supply.

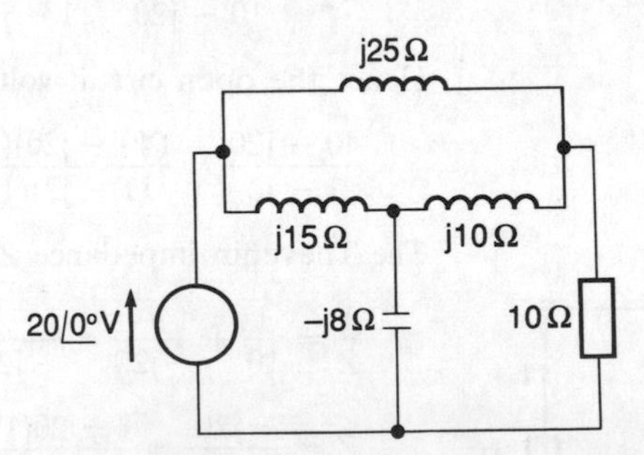

Fig. 4.56 Problem 3

4 A star-connected network consists of three impedances of (2 + j4) Ω. What will be the impedances in the equivalent delta network?

5 A delta-connected network consists of three impedances of (3 + j6) Ω. What will be the impedances in the equivalent star network?

6 Determine, by means of a star–delta transformation, the current I in the circuit shown in Fig. 4.57.

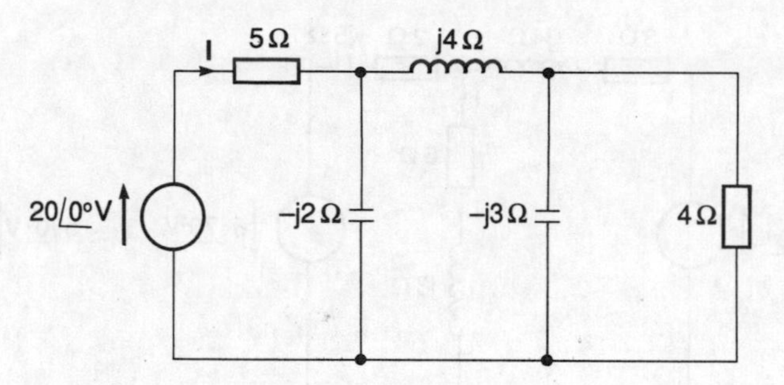

Fig. 4.57 Problem 6

7 Convert the circuit shown in Fig. 4.58 into a current source circuit.

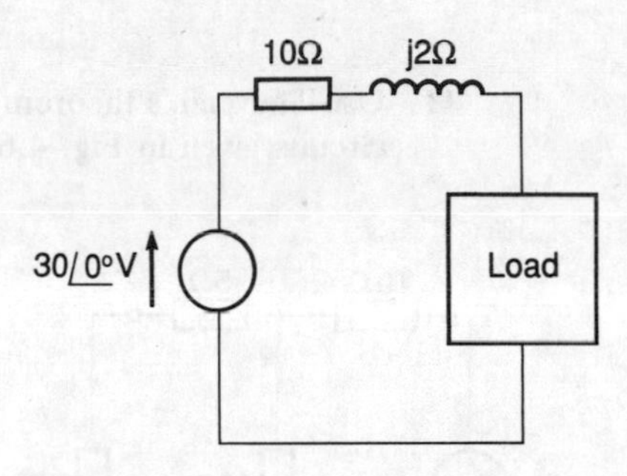

Fig. 4.58 Problem 7

8 Convert the circuit shown in Fig. 4.59 into a voltage source circuit.

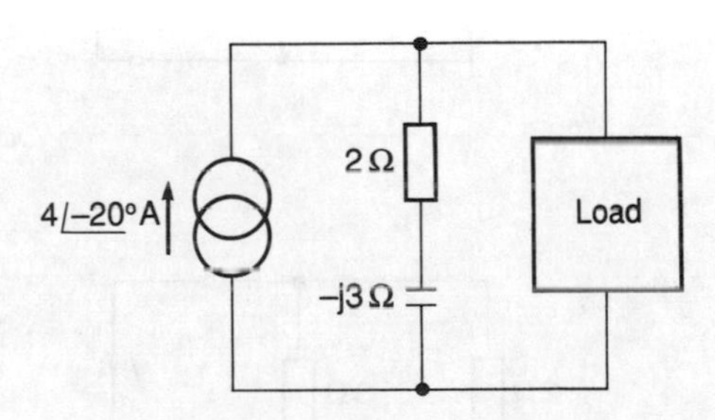

Fig. 4.59 Problem 8

9 Use the superposition principle to obtain for the circuits shown in Fig. 4.60 the current through the 10 Ω resistor in (*a*), through the 20 Ω resistor in (*b*) and the 40 Ω resistor in (*c*).

Fig. 4.60 Problem 9

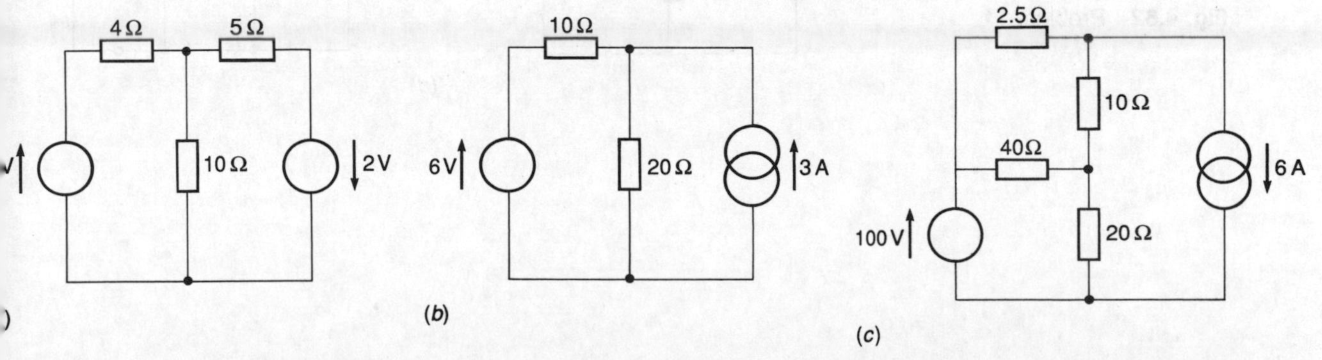

10 Use the superposition theorem to determine the current **I** for the circuits shown in Fig. 4.61.

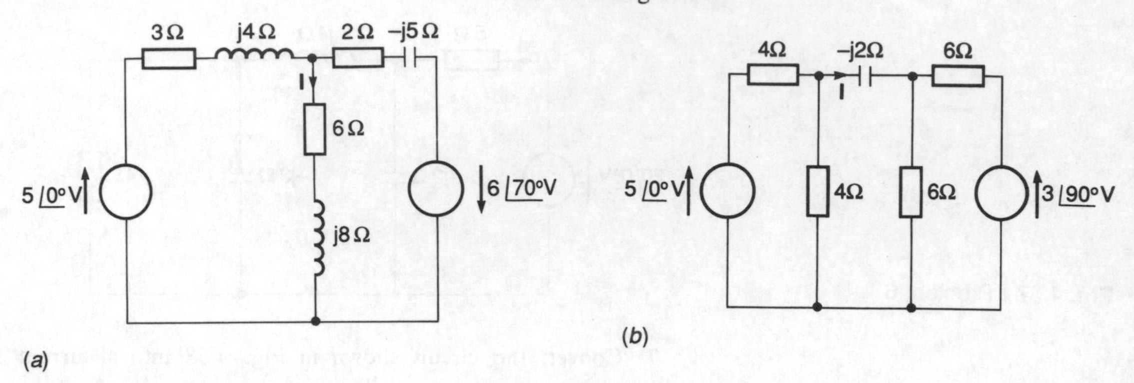

Fig. 4.61 Problem 10

11 Use Thévenin's theorem to determine the current i in each of the circuits given in Fig. 4.62.

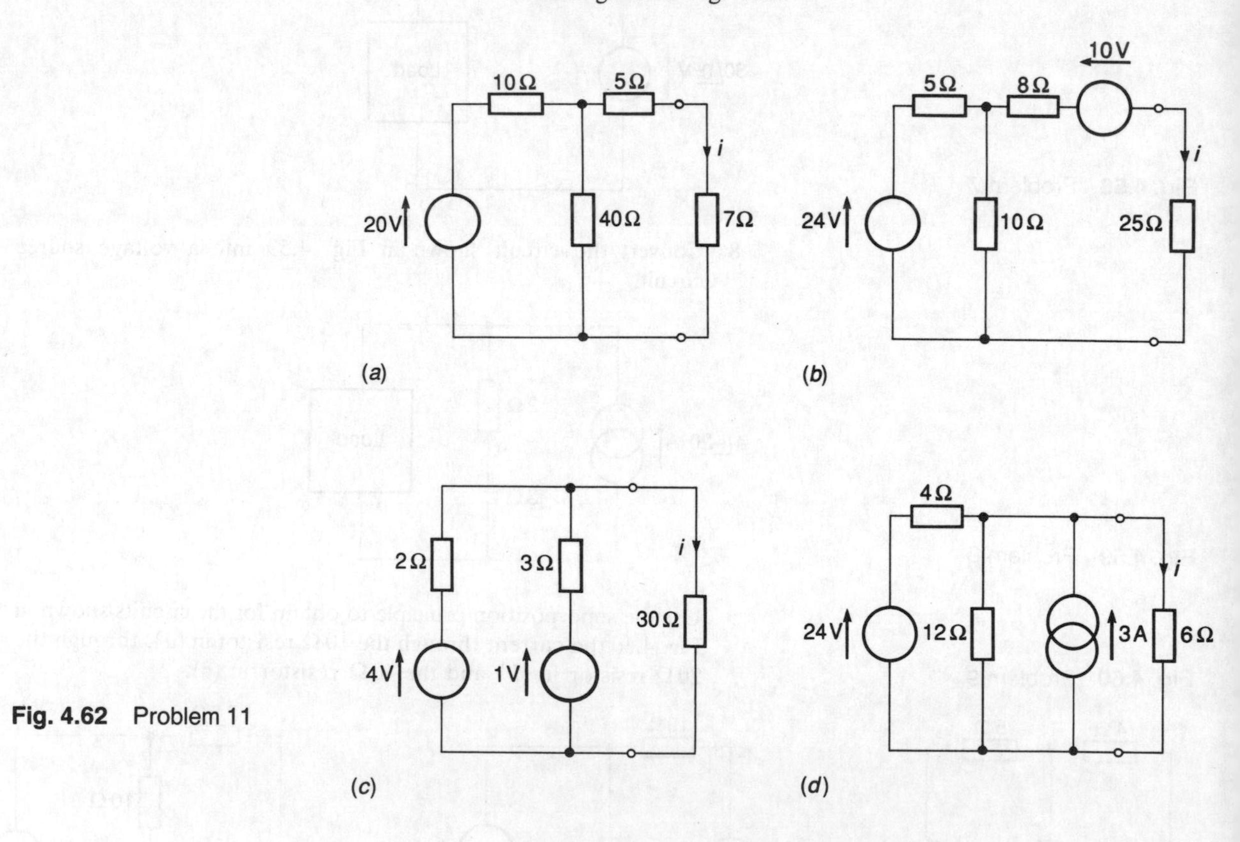

Fig. 4.62 Problem 11

12 Determine the Thévenin equivalent circuits for the circuits shown in Fig. 4.63.

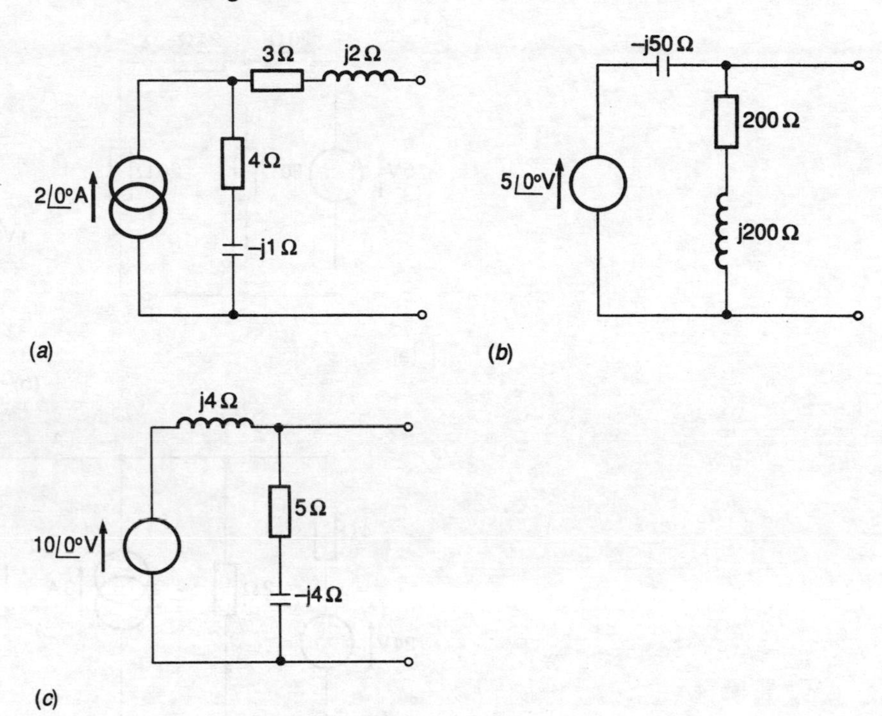

Fig. 4.63 Problem 12

13 Use Thévenin's theorem to determine the current **I** through the capacitor in the circuit given in Fig. 4.64.

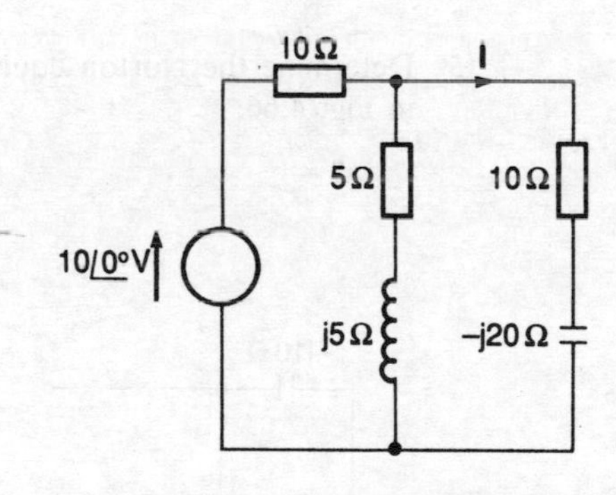

Fig. 4.64 Problem 13

14 Use Norton's theorem to obtain the potential difference v between A and B in each of the circuits given in Fig. 4.65.

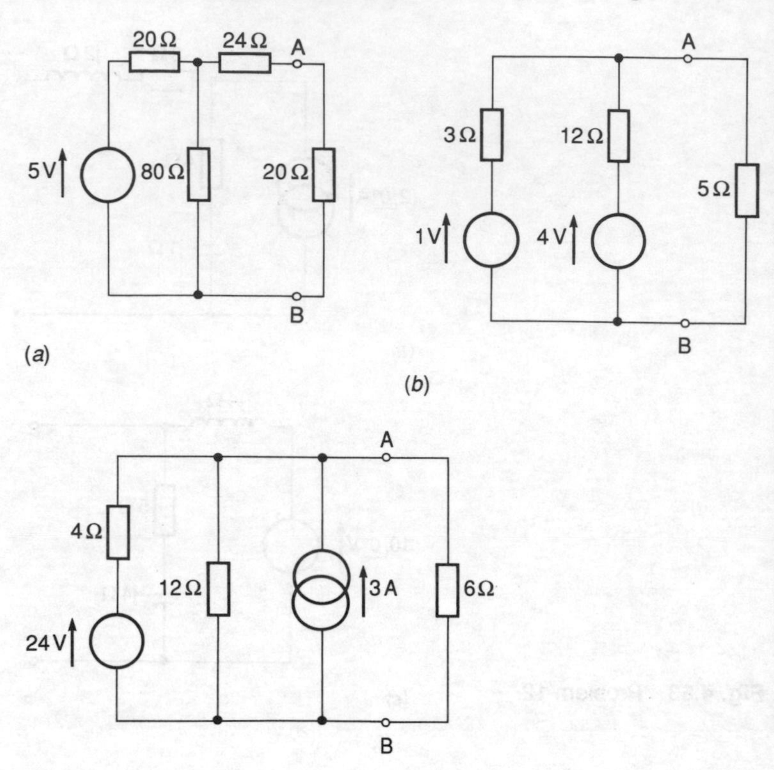

Fig. 4.65 Problem 14

15 Determine the Norton equivalent circuits for the circuits shown in Fig. 4.66.

Fig. 4.66 Problem 15

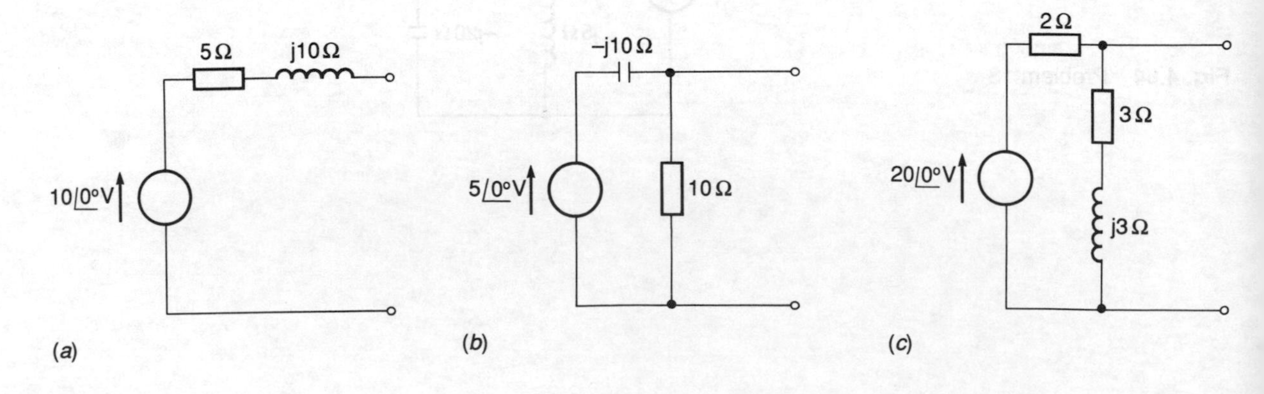

16 Determine the equivalent Norton circuit for the Thévenin circuit shown in Fig. 4.67.

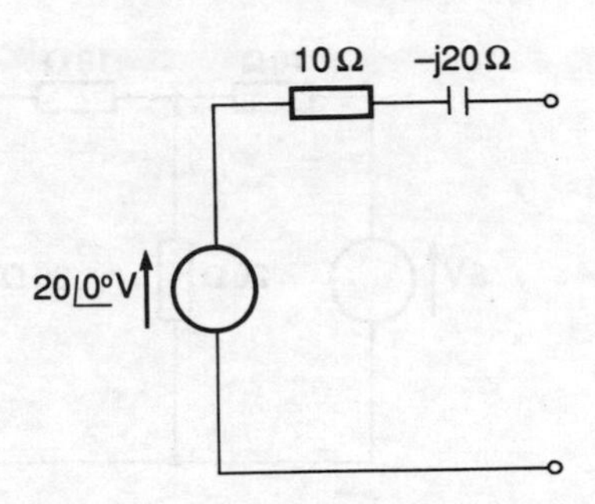

Fig. 4.67 Problem 16

17 Determine the equivalent Thévenin circuit for the Norton circuit shown in Fig. 4.68.

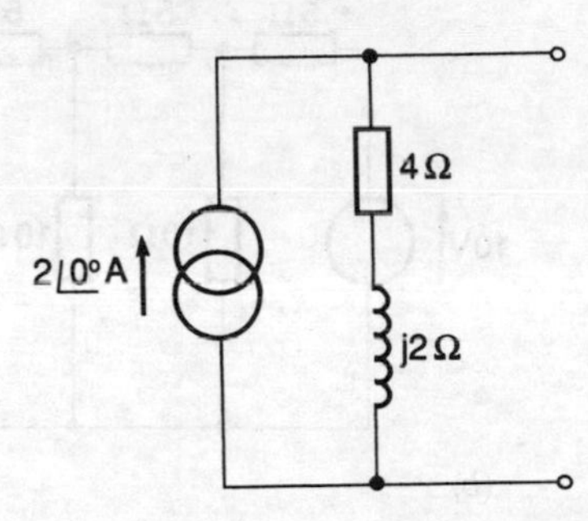

Fig. 4.68 Problem 17

18 Use either Thévenin's theorem or Norton's theorem to determine the current **I** in Fig. 4.69.

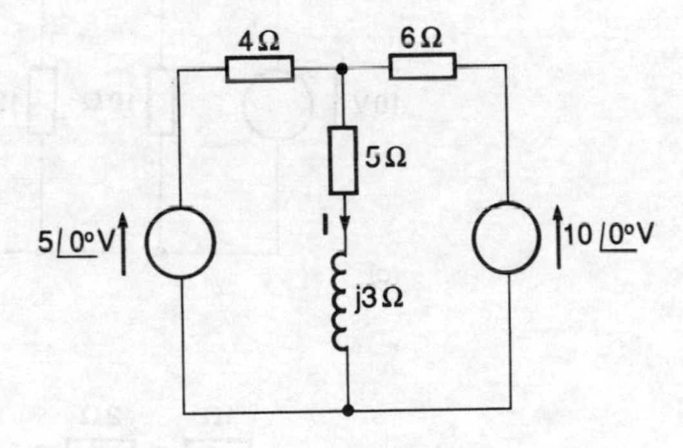

Fig. 4.69 Problem 18

19 Use node analysis to determine the currents in each of the branches of the circuits shown in Fig. 4.70.

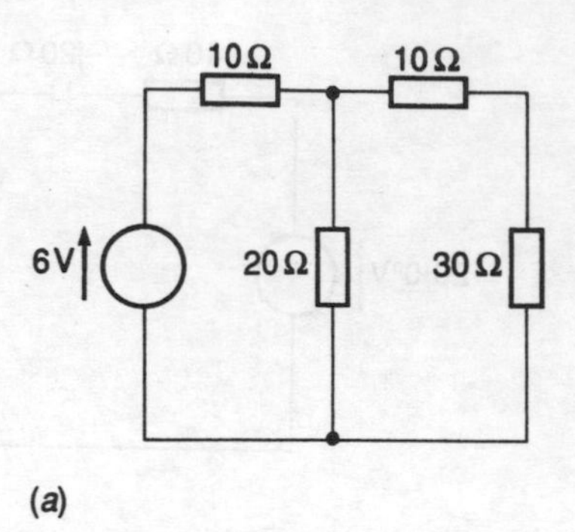

(*a*)

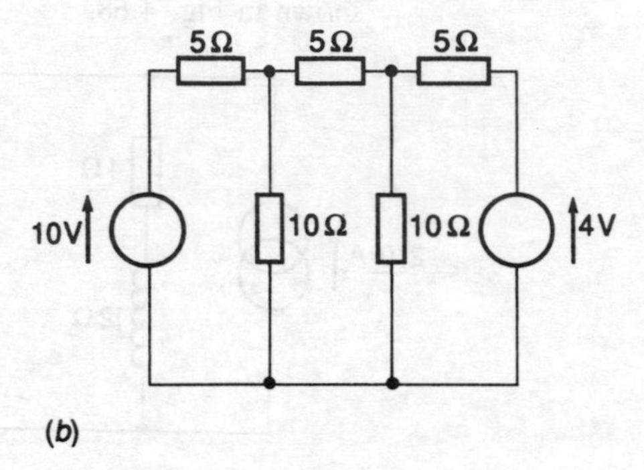

(*b*)

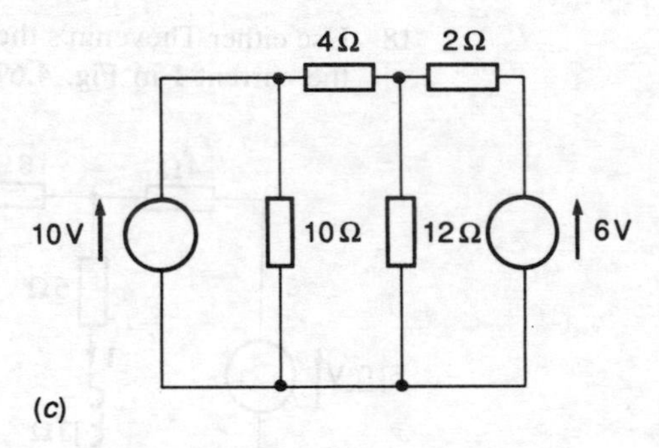

(*c*)

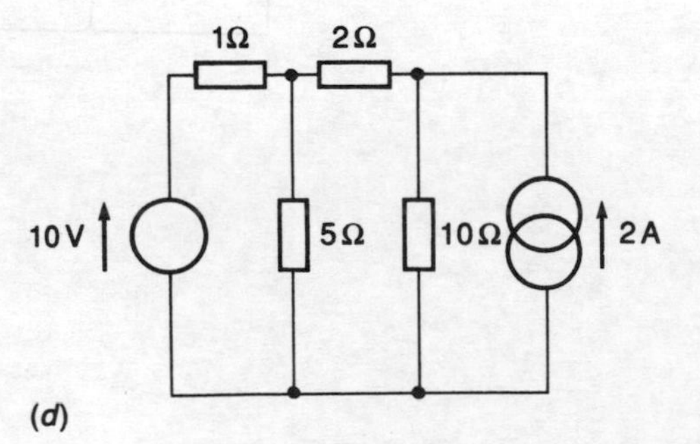

Fig. 4.70 Problem 19 (*d*)

20 Use node analysis to determine the current **I** in each of the circuits in Fig. 4.71.

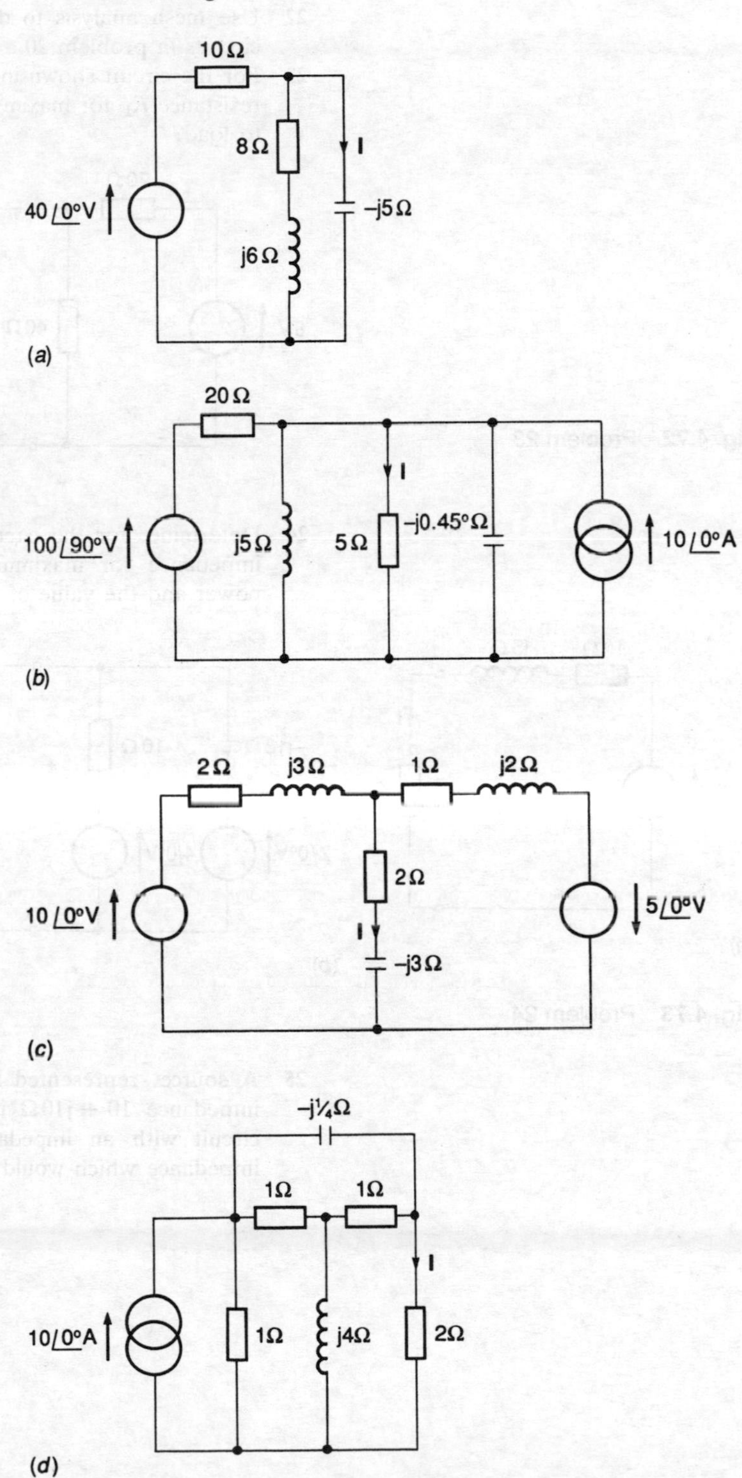

Fig. 4.71 Problem 20

21 Use mesh analysis to determine the currents through each of the components in the circuits in problem 19.

22 Use mesh analysis to determine the current **I** in each of the circuits in problem 20.

23 For the circuit shown in Fig. 4.72 what is the value of the load resistance R_L for maximum power transfer to occur from source to load?

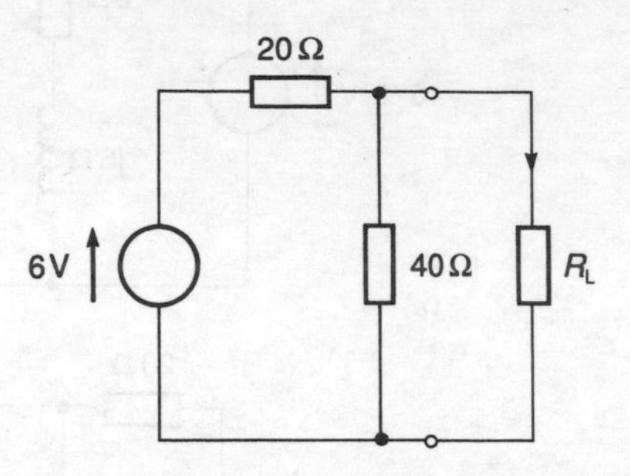

Fig. 4.72 Problem 23

24 Determine, for the circuits in Fig. 4.73, the value of the load impedance for maximum power, the current **I** at maximum power and the value of the maximum power.

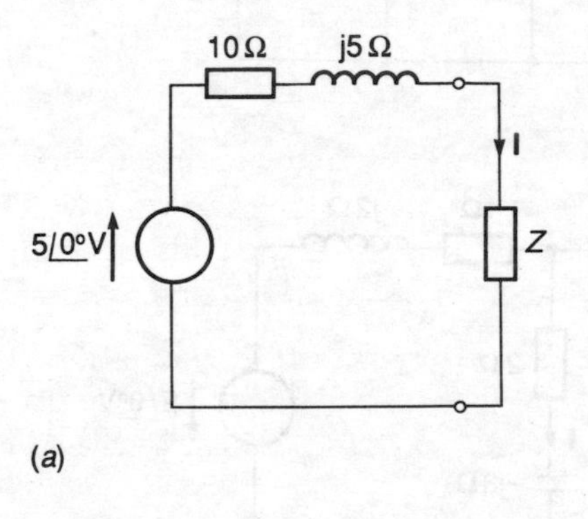

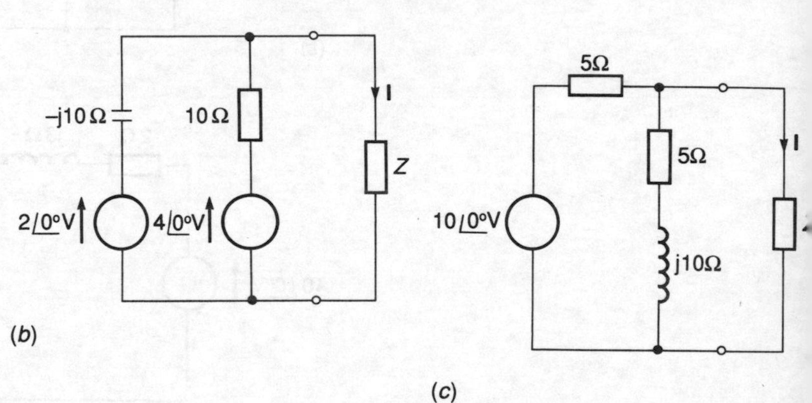

Fig. 4.73 Problem 24

25 A source, represented by a voltage of $100\angle 0°$ V and internal impedance $10 + j10\,\Omega$ is connected to a load by means of a circuit with an impedance of $2 + j0.5\,\Omega$. What is the load impedance which would give maximum power transfer?

5 Magnetically coupled circuits

Introduction

Coupled circuits are circuits which are linked in some way such that changes in one circuit result in changes in the other circuit. A common way of obtaining such coupling is by mutual inductance between coils in the circuits. Such coupling occurs as a result of the changing magnetic field of one coil inducing an e.m.f. in the other coil. This chapter is concerned with coupling as a result of mutual inductance and the transformer as a circuit element, the transformer being an example of mutual inductance.

Mutual inductance

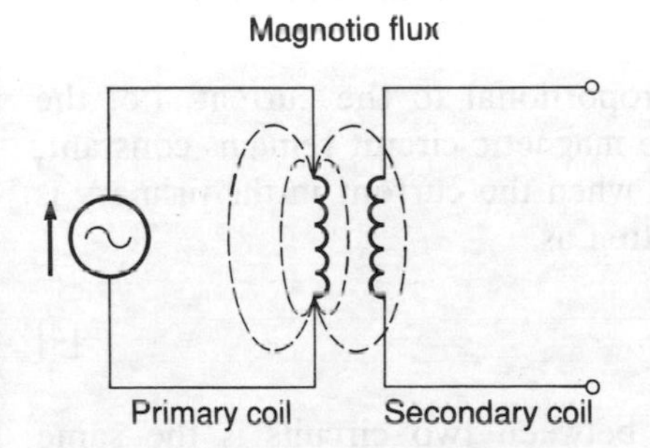

Fig. 5.1 Magnetic flux linking two coils

When the current in a coil changes, the magnetic flux of that coil changes. If this flux links another coil then an e.m.f. will be induced in it (Fig. 5.1). In such a situation the coils are said to be *magnetically coupled* and the effect is called *mutual inductance*.

The flux linked by the secondary coil depends on the flux produced by the primary coil and this in turn depends on the current I_p in the primary coil. The e.m.f. induced in the secondary, e_s, is proportional to the rate of change of flux linked by it. Hence the e.m.f. induced in the secondary coil must be proportional to the rate of change of current in the primary coil.

$$\text{Induced e.m.f. } e_s \text{ is proportional to } \frac{-dI_p}{dt}$$

$$e_s = -M\frac{dI_p}{dt} \qquad [1]$$

M is a constant for the arrangement of coils concerned and is called the *mutual inductance*. The unit of mutual inductance is the henry (H). The minus sign is because the induced e.m.f. tends to circulate a current in the secondary coil in such a direction as to itself produce magnetic flux which opposes the change in flux produced by the current change in the primary coil.

A change in magnetic flux of $\delta\phi$ in a time δt in a turn of wire produces an induced e.m.f. of $-\delta\phi/\delta t$ (the laws of electromagnetic induction). If the secondary coil has N_s turns then the e.m.f. e_s induced as a result of such a flux change in each turn is

$$e_s = -N_s \frac{\delta\phi}{\delta t}$$

If this flux change is produced by the current in the primary coil changing by δI_p in a time δt, then

$$e_s = -M \frac{\delta I_p}{\delta t}$$

Hence

$$M \frac{\delta I_p}{\delta t} = N_s \frac{\delta\phi}{\delta t}$$

$$M = N_s \frac{\delta\phi}{\delta I_p} \qquad [2]$$

$N_s\delta\phi$ is the change of flux linkages with the secondary coil. Hence

$$M = \frac{\text{change of flux linkages with the secondary coil}}{\text{change in current in the primary}} \qquad [3]$$

If the flux is directly proportional to the current, i.e. the relative permeability of the magnetic circuit remains constant, then if ϕ is the linked flux when the current in the primary is I_p equation [2] can be written as

$$M = N_2 \frac{\phi}{I_p} \qquad [4]$$

The mutual inductance between two circuits is the same whichever one produces the flux change. Thus we could have derived the above relationships for the e.m.f. induced in the primary as a result of a current change in the secondary. Thus

$$M = N_p \frac{\delta\phi}{\delta I_s}$$

Example 1

If the mutual inductance of a pair of coils is 200 μH, what will be the e.m.f. induced in the secondary coil when the current in the primary coil changes at the rate of 2000 A/s?

Answer

Using equation [1]

$$\text{induced e.m.f.} = M\frac{dI_p}{dt} = 200 \times 10^{-6} \times 2000 = 0.4\,\text{V}$$

Coefficient of magnetic coupling

The *magnetic coupling coefficient* k is used to describe the degree of magnetic coupling that occurs between circuits.

$$k = \frac{\text{flux linking the two circuits}}{\text{total flux produced}} \qquad [5]$$

If there is no magnetic coupling then k is zero. If the magnetic coupling is perfect and all the flux produced in the primary links with the secondary then k is 1. Where k is low circuits are said to be loosely coupled, where k is high tightly coupled.

The e.m.f. induced in a coil of N turns when the flux changes at the rate $d\phi/dt$ is

$$\text{e.m.f.} = -N\frac{d\phi}{dt}$$

This e.m.f. is related to the inductance L of the coil and the rate of change of current by

$$\text{e.m.f.} = -L\frac{dI}{dt}$$

Hence

$$N\frac{d\phi}{dt} = L\frac{dI}{dt}$$

$$L = N\frac{d\phi}{dI} = \frac{\text{change in flux linkages}}{\text{change in current}}$$

For a coil where the flux is directly proportional to the current, then

$$L = \frac{N\phi}{I} \qquad [6]$$

We can consider the inductance to be the flux linkages produced per ampere.

Consider a primary coil of inductance L_p which is coupled with a secondary coil of inductance L_s. If each has a core of constant relative permeability, then a current of I_p in the primary coil will produce a flux of ϕ_p through its turns.

$$L_p = \frac{N_p\phi_p}{I_p}$$

and thus

$$\phi_p = \frac{L_p I_p}{N_p}$$

The flux that links the secondary coil will be $k\phi_p$. Thus if the coils have a mutual inductance M, equation [4] gives

$$M = N_s \frac{k\phi_p}{I_p}$$

$$= kN_s \frac{L_p}{N_p}$$

Hence, with rearrangement we have

$$\frac{N_s}{N_p} = \frac{M}{kL_p}$$

Similarly, since we could reverse the two coils, we can also write

$$M = kN_p \frac{L_s}{N_s}$$

and so

$$\frac{N_s}{N_p} = \frac{kL_s}{M}$$

Thus

$$\frac{N_s}{N_p} = \frac{M}{kL_p} = \frac{kL_s}{M}$$

and so

$$k = \frac{M}{\sqrt{(L_1 L_2)}} \qquad [7]$$

Example 2

What is the magnetic coupling coefficient of a pair of coils if they have inductances of 200 mH and 400 mH and the mutual inductance is 50 mH?

Answer

Using equation [7]

$$k = \frac{M}{\sqrt{(L_1 L_2)}} = \frac{0.050}{\sqrt{(0.200 \times 0.400)}} = 0.18$$

Coupled circuits

Consider a coupled pair of coils with a mutual inductance M, as in Fig. 5.2. If the primary coil is supplied with a sinusoidal

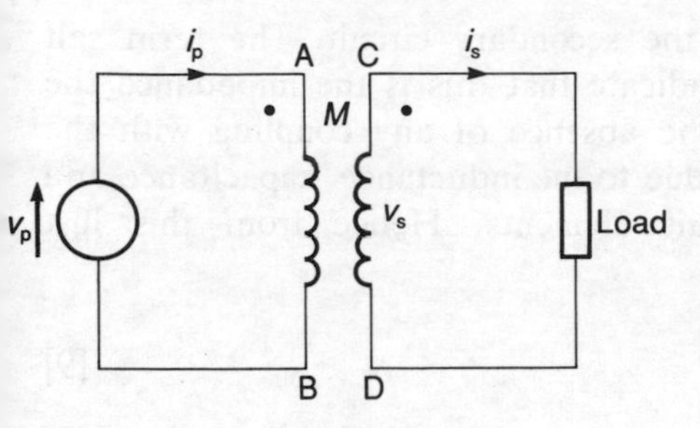

Fig. 5.2 Coupled circuits

alternating current then the current i_p at any instant can be represented by

$$i_p = I_{pm} \sin \omega t$$

where I_{pm} is the maximum value of the current and ω is the angular frequency. The e.m.f. v_s induced in the secondary coil is

$$v_s = -M\frac{di_p}{dt} = -M\frac{d(I_{pm}\sin\omega t)}{dt}$$

$$v_s = -MI_{pm}\omega\cos\omega t = -MI_{pm}\omega\sin(\omega t - 90°)$$

This e.m.f. is 90° out of phase with the current in the primary circuit. We can thus represent this e.m.f. as

$$\mathbf{V_s} = -j\omega M\mathbf{I_p} \qquad [8]$$

and so produce an equivalent secondary which consists of a voltage generator of size $j\omega M\mathbf{I_p}$ in series with the coil. Similarly, if the secondary is not open circuit and there is a secondary current, then for a changing current in the secondary coil an e.m.f. will be induced in the primary coil, and so we can produce an equivalent circuit which consists of a voltage generator of size $j\omega M\mathbf{I_s}$ in series with the coil.

The direction of these e.m.f.s is indicated by *dot notation*. A dot is put near one end of the primary coil and then another dot is placed at the end of the magnetically coupled secondary coil which has the same instantaneous polarity as the dotted end of the primary coil. With this notation, if a current enters the end of the coil marked with a dot then the induced e.m.f. in the coupled coil will make that end of the coupled coil marked with a dot positive. If a current leaves the end of a coil marked with a dot then the induced e.m.f. in the end of the coupled coil marked with a dot will be negative. Thus for the dot notation used in Fig. 5.2, current $\mathbf{I_p}$ enters point A, the dot end of the primary coil, and so the induced e.m.f. $j\omega M\mathbf{I_p}$ in the dot end of the secondary coil, point C, will be positive. Current $\mathbf{I_s}$ leaves the dot end of the secondary coil, C, and so the dot end of the primary coil, A, will be negative. Hence the equivalent circuits for the two coils are as shown in Fig. 5.3.

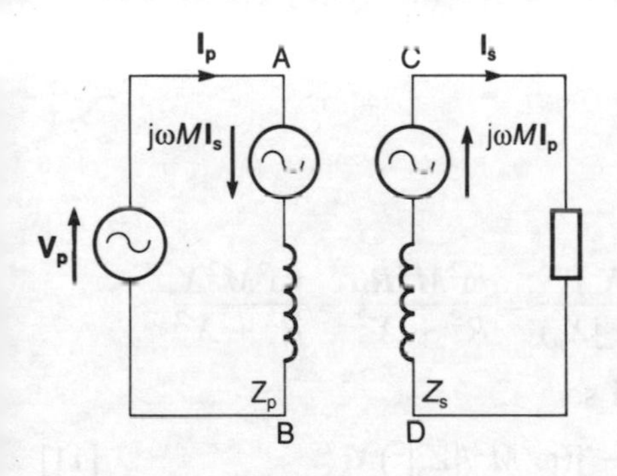

Fig. 5.3 Equivalent circuits for coupled coils

Reflected impedance

For the primary circuit in Fig. 5.3 Kirchoff's voltage law gives

$$\mathbf{V_p} - \mathbf{I_p}Z_p + j\omega M\mathbf{I_s} = 0$$

and for the secondary circuit

$$-j\omega M\mathbf{I_p} + \mathbf{I_s}Z_s = 0$$

where Z_p is the self impedance of the primary circuit and Z_s

the self impedance of the secondary circuit. The term self impedance is used to indicate that this is the impedance the circuit would have in the absence of any coupling with the other circuit and is just due to the inductance, capacitance and resistance of the circuit elements. Hence from this last equation

$$\mathbf{I}_s = \frac{j\omega M \mathbf{I}_p}{Z_s} \qquad [9]$$

and substituting this value in the previous equation gives

$$\mathbf{V}_p = \mathbf{I}_p(Z_p + \omega^2 M^2/Z_s) \qquad [10]$$

If there had been no secondary coil then the primary circuit would have just had

$$\mathbf{V}_p = \mathbf{I}_p Z_p$$

Thus the presence of a current through the secondary coil has resulted in an extra impedance of $\omega^2 M^2/Z_s$ appearing in the primary circuit. This extra impedance is called the *reflected impedance*, the term reflected being used because it represents the impedance reflected back into the primary side by the presence of the secondary circuit.

Since, in general, the self secondary impedance Z_s is complex and can be considered to be

$$Z_s = R_s + jX_s$$

then the reflected impedance Z_r is

$$Z_r = \frac{\omega^2 M^2}{R_s + jX_s}$$

and so

$$Z_r = \frac{\omega^2 M^2 (R_s - jX_s)}{(R_s + jX_s)(R_s - jX_s)} = \frac{\omega^2 M^2 R_s}{R_s^2 + X_s^2} - \frac{j\omega^2 M^2 X_s}{R_s^2 + X_s^2}$$

But $R_s^2 + X_s^2 = |Z_s|^2$, and so

$$Z_r = (\omega^2 M^2/|Z_s|^2)R_s - j(\omega^2 M^2/|Z_s|^2)X_s \qquad [11]$$

The reflected impedance has thus a resistive component of $(\omega^2 M^2/|Z_s|^2)R_s$ and a reactive component of $-j(\omega^2 M^2/|Z_s|^2)X_s$. This reactive component has the opposite sign to the reactance in the secondary circuit. Thus if X_s is inductive then the reflected reactance will be capacitive, and vice versa.

Example 3

For the coupled circuits shown in Fig. 5.4 determine (*a*) the self impedance of the primary circuit, (*b*) the self impedance of the secondary circuit, (*c*) the impedance reflected into the primary circuit, (*d*) the effective primary impedance, (*e*) the primary current, (*f*) the secondary current. The input has an angular frequency of 1000 rad/s.

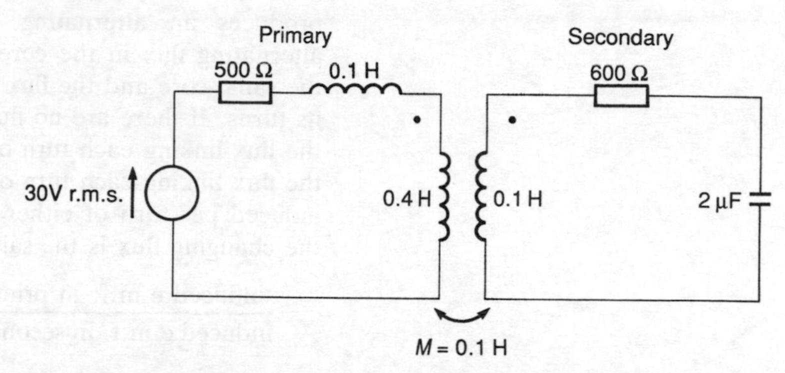

Fig. 5.4 Example 3

Answer

(*a*) The primary circuit has a resistance of 500 Ω in series with a total inductance of 0.5 H, i.e. an impedance of $j\omega L = j500\,\Omega$. Hence the self impedance Z_p is

$$Z_p = 500 + j500\,\Omega$$

(*b*) The secondary circuit has a resistance of 600 Ω in series with a capacitor of 2 μF, i.e. an impedance of $1/j\omega C = -j500\,\Omega$, and an inductor of 0.1 H, i.e. an impedance of $j\omega L = j100\,\Omega$. Hence the self impedance Z_s is

$$Z_s = 600 - j500 + j100 = 600 - j400\,\Omega$$

(*c*) The impedance reflected into the primary circuit is

$$Z_r = \frac{\omega^2 M^2}{Z_s} = \frac{1000^2 \times 0.1^2}{600 - j400}$$

$$= \frac{10\,000(600 + j400)}{(600 - j400)(600 + j400)} = 11.5 + j7.7\,\Omega$$

(*d*) The effective impedance of the primary circuit is thus

$$500 + j500 + 11.5 + j7.7 = 511.5 + j507.7\,\Omega$$

(*e*) The primary current $\mathbf{I}_p$ is

$$\mathbf{I}_p = \frac{\mathbf{V}_p}{511.5 + j507.7} = \frac{30\angle 0^\circ}{721\angle 44.8^\circ} = 0.042\angle -44.8^\circ\,\text{A r.m.s.}$$

(*f*) The secondary circuit can be considered to be a source of $j\omega M\mathbf{I}_p$ directing a current of $\mathbf{I}_s$ through an impedance of Z_s (as in Fig. 5.3). Thus

$$\mathbf{I}_s = \frac{j\omega M\mathbf{I}_p}{Z_s} = j\left(\frac{1000 \times 0.1}{600 - j400}\right) \times 0.0142\angle -44.8^\circ$$

$$= \frac{j(100 \times 0.042\angle -44.8^\circ)}{721\angle -33.7^\circ} = j0.058\angle -11.1^\circ$$

$$= 0.0058\angle 78.9^\circ\,\text{A}$$

Transformers

The principle behind the operation of a transformer is mutual inductance between two coils wound on the same core (Fig. 5.5). An alternating voltage is applied to the primary coil. This produces an alternating current in the coil and so an alternating flux in the core. The secondary coil is wound on the same core and the flux produced by the primary coil links its turns. If there are no flux losses from the magnetic circuit, the flux linking each turn of the secondary coil is the same as the flux linking each turn of the primary coil. Thus the e.m.f. induced per turn of either the secondary or primary coils by the changing flux is the same. Hence

$$\frac{\text{induced e.m.f. in primary}}{\text{induced e.m.f. in secondary}} = \frac{N_p}{N_s}$$

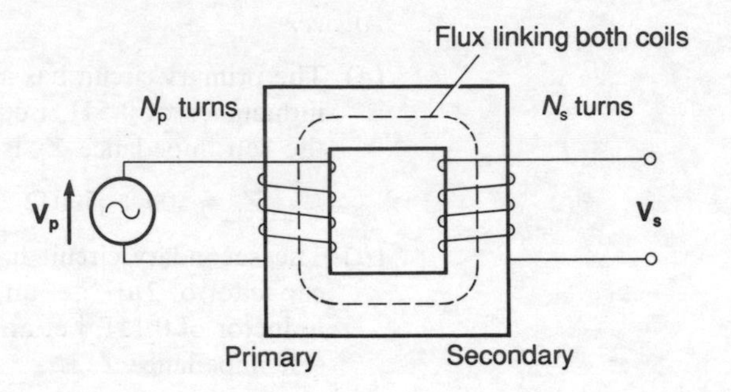

Fig. 5.5 Transformer

This condition that all the flux produced by the primary coil links the secondary coils means that the coefficient of coupling k is 1. A transformer based on this assumption is said to be *ideal*.

If there is no significant secondary current, i.e. the secondary is effectively open circuit, then the voltage between the terminals of the coil is the same as the induced e.m.f. If there is no current then no energy is taken from the secondary coil. This means therefore that no energy is taken from the primary coil and so there will be no current in the primary circuit (the primary circuit has to have also an infinite inductance so that negligible current is needed to produce the magnetic flux, since $L_p = N_p\phi/I_p$). This can only be the case if the induced e.m.f. is equal to and opposing the input voltage V_p. Thus

$$\frac{V_p}{V_s} = \frac{N_p}{N_s} \qquad [12]$$

If V_s is less than V_p the transformer is said to have a step-down voltage ratio, if V_s is more than V_p a step-up voltage ratio.

For an ideal transformer we can also postulate that there

will be no power loss and thus

$$I_pV_p = I_sV_s$$

$$\text{apparent power supplied to the primary} = \text{apparent power supplied to the load}$$

Thus

$$\frac{V_p}{V_s} = \frac{I_s}{I_p} = \frac{N_p}{N_s}$$

and so

$$I_pN_p = I_sN_s \qquad [13]$$

Thus, the number of ampere-turns on the secondary winding equals the number of ampere-turns on the primary winding. In terms of the current phasors it is usual to write the equation with a minus sign (see later this chapter) to indicate that the primary and secondary phasors are in opposite directions.

$$\frac{\mathbf{I_p}}{\mathbf{I_s}} = -\frac{N_s}{N_p} \qquad [14]$$

The *ideal transformer* discussed above can be summarized as having:

1. a coupling coefficient of 1
2. infinite primary inductance
3. coil power losses are zero.

In specifying a transformer the term *rating* is used. The rating is quoted in volt-amperes (V A) at the full rated voltage. For example a transformer might be rated as 2 kV A, 100/10 V. This means that a primary voltage of 100 V will produce a secondary voltage of 10 V and that at these values the apparent power is 2 kV A and so the primary current is 2000/100 = 20 A and the secondary current 2000/10 = 200 A. It in effect thus specifies how much current the transformer windings can take without overheating.

Example 4

What will be the secondary voltage produced if a single phase transformer, with 200 primary turns and 50 secondary turns, has an a.c. input of 240 V (r.m.s.)?

Answer

Assuming the transformer to be ideal

$$\frac{V_p}{V_s} = \frac{N_p}{N_s}$$

$$V_s = \frac{240 \times 50}{200} = 60\,\text{V}$$

Example 5

What will be the rated current in the primary and secondary windings of a 20 kV A, 1100 : 240 V single phase transformer?

Answer

For the primary windings

$$I = \frac{20\,000}{1100} = 18.2\,\text{A}$$

For the secondary windings

$$I = \frac{20\,000}{240} = 83.3\,\text{A}$$

The lossless transformer

Consider a lossless transformer as being represented by a primary circuit with a sinusoidal source and inductance L_p and a secondary circuit containing an inductance L_s, coupled to the first circuit by mutual inductance M, and a complex load Z_L (Fig. 5.6). Applying Kirchoff's voltage law to each circuit gives

$$\mathbf{V_p} - j\omega L_p\mathbf{I_p} - j\omega M\,\mathbf{I_s} = 0 \qquad [15]$$

$$j\omega M\,\mathbf{I_p} + j\omega L_s\mathbf{I_s} + Z_L\mathbf{I_s} = 0 \qquad [16]$$

Also

$$\mathbf{V_s} = -Z_L\mathbf{I_s} \qquad [17]$$

Equation [15] gives

$$\mathbf{V_p} = j\omega L_p\mathbf{I_p} + j\omega M\,\mathbf{I_s}$$

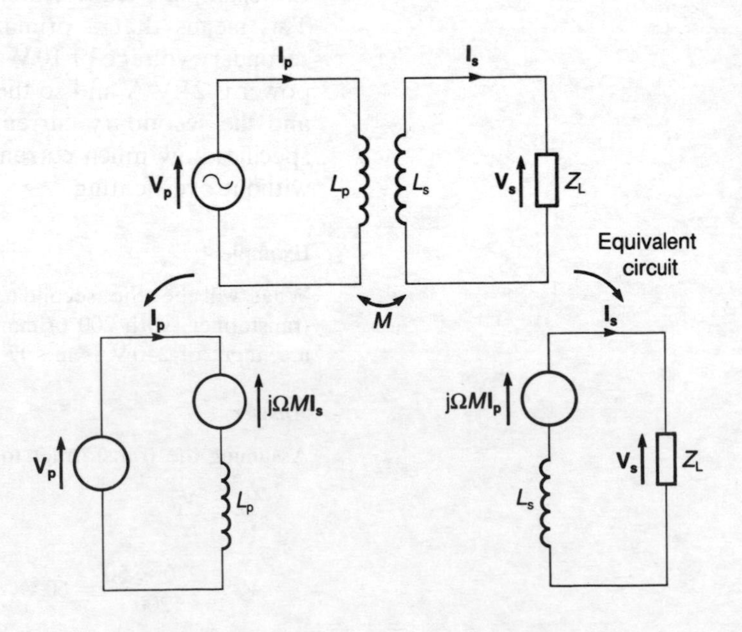

Fig. 5.6 Lossless transformer

and dividing equation [17] by this gives

$$\frac{\mathbf{V_s}}{\mathbf{V_p}} = -\frac{Z_L\mathbf{I_s}}{j\omega L_p\mathbf{I_p} + j\omega M\mathbf{I_s}}$$

Equation [16], with rearrangement, gives

$$Z_L\mathbf{I_s} = -j\omega M\mathbf{I_p} - j\omega L_s\mathbf{I_s}$$

and so

$$\frac{\mathbf{V_s}}{\mathbf{V_p}} = \frac{j\omega M\mathbf{I_p} + j\omega L_s\mathbf{I_s}}{j\omega L_p\mathbf{I_p} + j\omega M\mathbf{I_s}} = \frac{M\mathbf{I_p} + L_s\mathbf{I_s}}{L_p\mathbf{I_p} + M\mathbf{I_s}}$$

But $M = k\sqrt{(L_pL_s)}$, where k is the coupling coefficient. Thus

$$\frac{\mathbf{V_s}}{\mathbf{V_p}} = \frac{k\sqrt{L_p}\sqrt{L_s}\mathbf{I_p} + L_s\mathbf{I_s}}{L_p\mathbf{I_p} + k\sqrt{L_p}\sqrt{L_s}\mathbf{I_s}}$$

$$= \sqrt{\left(\frac{L_s}{L_p}\right)}\left(\frac{k\sqrt{L_p}\mathbf{I_p} + \sqrt{L_s}\mathbf{I_s}}{\sqrt{L_p}\mathbf{I_p} + k\sqrt{L_s}\mathbf{I_s}}\right) \qquad [18]$$

Only if $k = 1$ does

$$\frac{\mathbf{V_s}}{\mathbf{V_p}} = \sqrt{\left(\frac{L_s}{L_p}\right)}$$

The inductance of a coil is N^2/S, where S is the reluctance. Since both coils are wound on the same core then S is the same for both the primary and secondary coils. Hence $L_s/L_p = N_s^2/N_p^2$. Thus if $k = 1$ then

$$\frac{\mathbf{V_s}}{\mathbf{V_p}} = \frac{N_s}{N_p}$$

and so equation [12], derived earlier, is obtained. For all other values of k the above relationship is not valid. For transformers k is usually very close to 1.

Rearranging equation [16] gives

$$j\omega L_s\mathbf{I_s} + Z_L\mathbf{I_s} = -j\omega M\mathbf{I_p}$$

$$\frac{\mathbf{I_s}}{\mathbf{I_p}} = -\frac{j\omega M}{j\omega L_s + Z_L} \qquad [19]$$

Since $M = k\sqrt{(L_pL_s)}$ then equation [19] becomes

$$\frac{\mathbf{I_s}}{\mathbf{I_p}} = -\frac{j\omega k\sqrt{(L_pL_s)}}{j\omega L_s + Z_L} = -\frac{j\omega kL_s}{j\omega L_s + Z_L}\sqrt{\frac{L_p}{L_s}}$$

Since $L_p/L_s = N_p^2/N_s^2$ then

$$\frac{\mathbf{I_s}}{\mathbf{I_p}} = -\frac{j\omega kL_s}{j\omega L_s + Z_L}\left(\frac{N_p}{N_s}\right) \qquad [20]$$

To obtain the relationship $\mathbf{I_s}/\mathbf{I_p} = -N_p/N_s$ (equation [14]) and so have an ideal transformer, then we must have $k = 1$ and

$|Z_L|$ must be insignificant in comparison with ωL_s, i.e. an infinitely large secondary inductance is required.

Equation [15] gives

$$\mathbf{V_p} = j\omega L_p \mathbf{I_p} + j\omega M \mathbf{I_s}$$

and thus substituting for $\mathbf{I_s}$ using equation [19] gives

$$\mathbf{V_p} = j\omega L_p \mathbf{I_p} + \frac{\omega^2 M^2 \mathbf{I_p}}{j\omega L_s + Z_L}$$

$$\frac{\mathbf{V_p}}{\mathbf{I_p}} = j\omega L_p + \frac{\omega^2 M^2}{j\omega L_s + Z_L} \qquad [21]$$

But $\mathbf{V_p}/\mathbf{I_p} = Z_p$, the effective impedance of the primary circuit. The effective impedance thus comprises the inductive reactance of the primary coil plus an impedance due to the presence of the secondary coil, this being called the reflected impedance (see earlier this chapter). But $M = k\sqrt{(L_p L_s)}$ and so equation [21] can be written as

$$Z_p = j\omega L_p + \frac{\omega^2 k^2 L_p L_s}{j\omega L_s + Z_L}$$

$$Z_p = \frac{j\omega L_p(j\omega L_s + Z_L) + \omega^2 k^2 L_p L_s}{j\omega L_s + Z_L}$$

When $k = 1$ then

$$Z_p = \frac{j\omega L_p Z_L}{j\omega L_s + Z_L}$$

When we have $Z_L \ll j\omega L_s$ then the equation simplifies to

$$Z_p \approx \frac{L_p}{L_s} Z_L$$

and since $L_p/L_s = N_p^2/N_s^2$ then

$$Z_p \approx \frac{N_p^2}{N_s^2} Z_L \qquad [22]$$

Thus for an ideal transformer the primary has an effective impedance of (N_p^2/N_s^2) times that of the load. Thus positioning a transformer between the source and the load alters the apparent impedance of the load. The impedance seen at the primary terminals is increased or decreased by a factor of $(N_p/N_s)^2$.

According to the *maximum power transfer theorem* (see Chapter 4, equation [36]), when maximum power is to be transferred from one circuit to another, without any change in power factor, then the impedance of the load circuit must equal the impedance of the supply circuit. A transformer can be used to achieve this by adjustment of the turns ratio and is then said to be *impedance matching*.

Example 6

What is the input impedance of an ideal transformer with a turns ratio 10:1 if the load connected to the secondary terminals is (*a*) a resistance of 2 kΩ and (*b*) a 2 kΩ resistance in series with a 1 μF capacitor? The angular frequency ω is 1000 rad/s.

Answer

(*a*) Using equation [22]

$$Z_p = \frac{N_p^2}{N_s^2} Z_L = 100 \times 2000 = 200\,\text{k}\Omega$$

(*b*) Using equation [22]

$$Z_p = \frac{N_p^2}{N_s^2} Z_L = 100\left(2000 + \frac{1}{\text{j}1000 \times 10^{-6}}\right)$$

$$= 100(2000 - \text{j}1000)$$

$$= 200 - \text{j}100\,\text{k}\Omega.$$

Example 7

What should be the turns ratio of a transformer which is to be used to match an amplifier with an output impedance of 40 Ω with a circuit of input impedance 600 Ω and so give maximum power transfer?

Answer

Using equation [22]

$$Z_r = \left(\frac{N_p}{N_s}\right)^2 Z_l$$

$$40 = \left(\frac{N_p}{N_s}\right)^2 \times 600$$

$$\frac{N_p}{N_s} = 0.26$$

The e.m.f. equation for a transformer

The alternating current in the primary coil of a transformer gives rise to alternating flux. If the current is sinusoidal of frequency f then since the flux φ will be proportional to the current the flux varies with time according to an equation of the form

$$\phi = \phi_m \sin 2\pi f t$$

where ϕ_m is the maximum value of the flux, φ is the flux at time t. The flux is in phase with the primary current. The induced e.m.f. per turn of both primary and secondary coils is the rate of change of flux and is thus

$$\text{e.m.f. per turn} = -\frac{d\phi}{dt} = -\frac{d(\phi_m \sin 2\pi ft)}{dt}$$

$$\text{e.m.f. per turn} = -2\pi f\phi_m \cos 2\pi ft$$

The e.m.f. is thus 90° out of phase with the flux, and hence with the primary current. The maximum value of the e.m.f. will be when the cosine term has its maximum value of 1. Hence

$$\text{max. e.m.f. per turn} = 2\pi f\phi_m$$

The r.m.s. value of the e.m.f. per turn is (max. e.m.f.)/√2, hence

$$\text{r.m.s. e.m.f. per turn} = \frac{2\pi f\phi_m}{\sqrt{2}} = 4.44 f\phi_m \qquad [23]$$

Example 8

What is the maximum value of the flux in the core of an ideal transformer if the primary has 500 turns and the input is 240 V at 50 Hz?

Answer

The e.m.f. per turn is 240/500 = 0.48 V. Hence using the equation [23] developed above

$$\text{r.m.s. e.m.f. per turn} = 4.44 f\phi_m$$

$$\phi_m = \frac{0.48}{4.44 \times 50} = 2.2 \times 10^{-3}\,\text{Wb}$$

Real transformers

Real transformers, unlike ideal transformers, do have power losses and do have coupling coefficients which are less than 1. The power losses with a transformer can be considered to be classified as *iron losses* (or *core losses*) and *copper losses*.

Iron losses are the power dissipated in the magnetic material used for the transformer core. The iron loss is essentially constant and independent of the currents in the coils. The iron loss consists of two forms of losses, eddy current losses and hysteresis losses. The term *eddy currents* is given to the currents induced in a conductor as a result of flux changes within that conductor. Thus for a magnetic circuit, as for example a transformer, when the flux in the core changes then an e.m.f. is induced which gives rise to eddy currents in the core material. These induced currents circulate in the core in such a direction as to produce flux which opposes the change responsible for their production. Since the core material is generally a good conductor of electricity the size of the induced currents can be large. The power loss, the I^2R loss, associated with these currents can thus be quite large. The size of the eddy currents can be reduced by using a laminated core instead of a solid one. *Hysteresis losses* result from the magnetization of the core. When the current through a coil is reversed the state of magnetization of the core is reversed. Each time this occurs energy is expended, this being known as the *hysteresis loss*. The hysteresis power loss is proportional to the volume of material for which the magnetization is being continually reversed.

Copper losses are the I^2R losses in the copper conductors of the primary and secondary windings which result from currents flowing through them.

The term *leakage flux* is used for the flux produced by the primary coil which does not link with turns of the secondary coil. The greater the leakage flux the greater the departure of the coupling coefficient from the ideal value of 1.

Equivalent circuits for transformers

An ideal transformer is one that has no losses and no flux leakage. Real transformers have both. It is however useful to represent a real transformer by an ideal one in a circuit such that the ideal transformer plus the circuit gives the same results as the real transformer.

Copper losses can be accounted for by considering there to be a resistance in series with the primary coil R_p and one in series with the secondary coil R_s of an ideal transformer. The copper loss in the primary is thus $I_p^2R_p$ and in the secondary $I_s^2R_s$. These resistances R_p and R_s are really the resistances of the primary and secondary coils.

Iron losses are due to losses associated with the production of flux in the transformer core. When there is no load in the secondary of a transformer the current in the primary $\mathbf{I_p}$, which we will denote by $\mathbf{I_o}$ to indicate that it is the open-circuit current, can be considered to be composed of two parts. One part, called the magnetizing current $\mathbf{I_{mag}}$, is responsible for producing the flux. This component is thus in phase with the flux. The other part of the primary current is responsible for supplying the hysteresis and eddy current losses in the core. This component $\mathbf{I_c}$ is in phase with the primary voltage $\mathbf{V_p}$ and so 90° out of phase with the magnetizing component. Figure 5.7 shows the phasor diagram for this no-load condition. Thus

$$I_0 = \surd(I_c^2 + I_{mag}^2) \qquad [24]$$

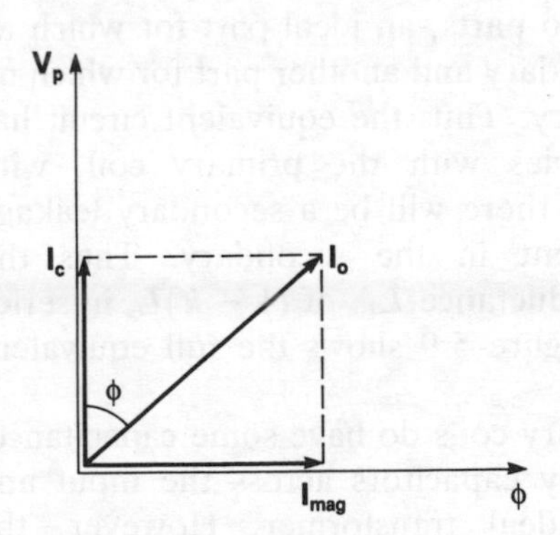

Fig. 5.7 Phasor diagram for a transformer on no-load

With the ideal transformer there is no net current through the primary coil (see earlier this chapter) and thus in an equivalent circuit the current $\mathbf{I_0}$ is considered to by-pass the primary coil. The phasor relationship can be represented by the magnetizing current $\mathbf{I_{mag}}$ being considered to pass through an inductor L_{mag}

and the current $\mathbf{I}_c$ through a resistor R_c, as in Fig. 5.8(*a*). When there is a load in the secondary then there is a current in the secondary $\mathbf{I}_s$ and as a consequence a current of $(N_s/N_p)\mathbf{I}_s$ in the primary. Figure 5.8(*b*) thus shows the equivalent circuit with such a current.

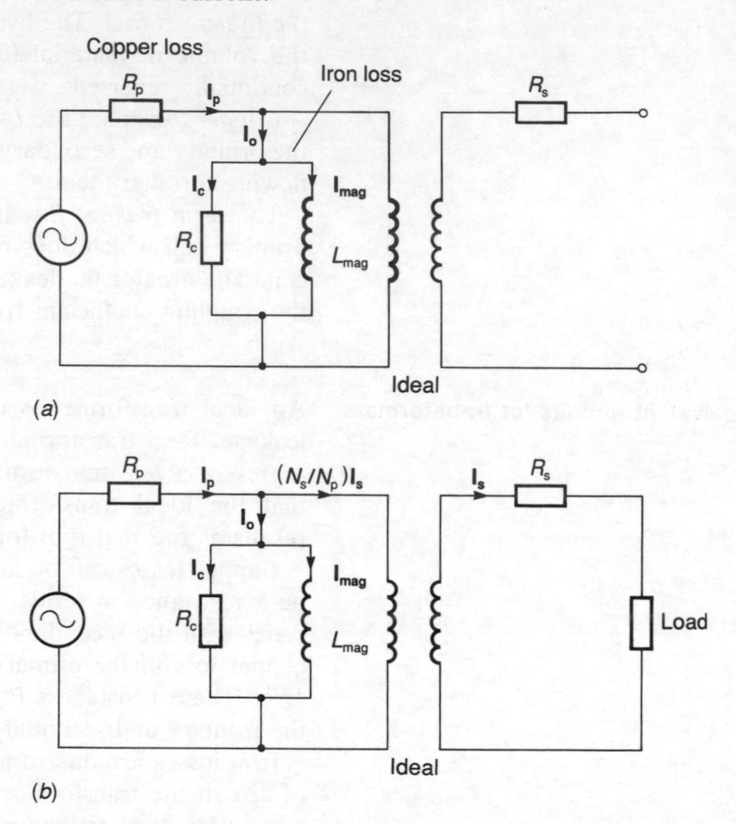

Fig. 5.8 Equivalent circuits with copper and iron losses

Leakage flux means that of the total flux produced only a fraction k links with the secondary coil. This is equivalent to having a primary coil in two parts, an ideal part for which all the flux links with the secondary and another part for which no flux links with the secondary. Thus the equivalent circuit has an inductance L_{Lp} in series with the primary coil, with $L_{Lp} = (1 - k)L_p$. Similarly there will be a secondary leakage flux as a result of current in the secondary. Thus the equivalent circuit has an inductance L_{Ls} of $(1 - k)L_s$ in series with the secondary coil. Figure 5.9 shows the full equivalent circuit.

The primary and secondary coils do have some capacitance. This can be represented by capacitors across the input and output terminals of an ideal transformer. However, the capacitance is only of significance at high audio frequencies and is generally neglected at all lower frequencies.

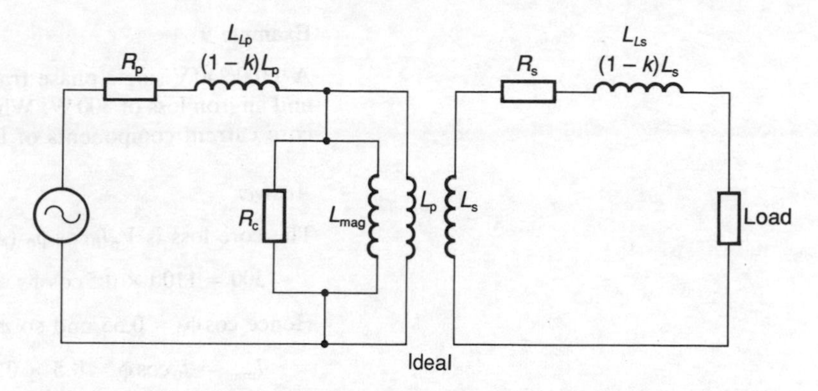

Fig. 5.9 Equivalent circuit for a real transformer

Since the value of the no-load current is generally less than about 5 per cent of the full-load primary current the circuit branch for R_c and L_{mag} is often neglected and the equivalent circuit has just the resistances representing the copper losses and the inductances representing the leakage flux. This equivalent circuit can be further simplified by replacing R_p and R_s by just a single resistance in the primary circuit. The resistance element needed in the primary circuit to give the same effect as R_s in the secondary circuit is $(N_p/N_s)^2R_s$. Thus the total resistance in the primary circuit is $R_p + (N_p/N_s)^2R_s$. The inductance element needed in the primary circuit to give the same effect as the L_{Ls} reactance in the secondary is $(N_p/N_s)^2L_{L}s$. Thus the total inductance in the primary circuit is $L_{Lp} + (N_p/N_s)^2L_{Ls}$. An alternative to this simplification is to replace R_p and L_{Lp} by equivalent resistance and inductance in the secondary circuit. Figure 5.10 shows both these forms of simplified circuits.

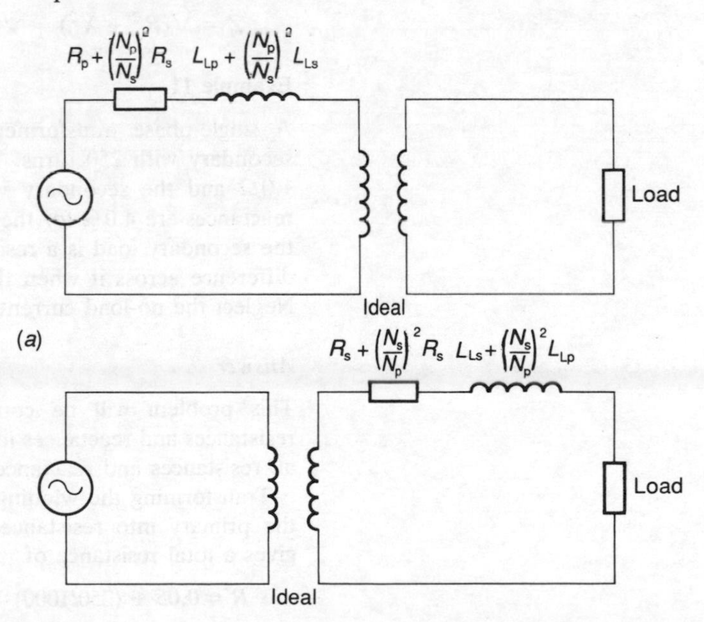

Fig. 5.10 Simplified equivalent circuits with all components referred to (a) the primary circuit, (b) the secondary circuit

Example 9

A 1100/240 V single phase transformer has a no-load current of 0.5 A and an iron loss of 300 W. What are the values of the magnetizing and core current components of the no-load current?

Answer

The core loss is $V_p I_0 \cos \phi_0$ (see Fig. 5.7). Thus

$$300 = 1100 \times 0.5 \cos \phi_0$$

Hence $\cos \phi_0 = 0.55$ and so $\phi_0 = 56.9°$.

$$I_{mag} = I_0 \cos \phi = 0.5 \times 0.55 = 0.28 \text{ A}$$

$$I_c = I_0 \sin \phi = 0.5 \times 0.84 = 0.42 \text{ A}$$

Example 10

A transformer has 500 primary turns and 100 secondary turns. The primary resistance is 0.40 Ω and the secondary 0.02 Ω. The primary leakage reactance is 1.20 Ω and the secondary 0.03 Ω. What is the equivalent impedance which could be placed in the primary circuit with an ideal transformer to represent all these factors?

Answer

The resistance R required in the primary circuit is

$$R = R_p + R_s(N_p/N_s)^2 = 0.40 + 0.02(500/100)^2 = 0.90 \,\Omega$$

The reactance X needed in the primary circuit is

$$X = X_{Lp} + X_{Ls}(N_p/N_s)^2 = 1.20 + 0.03(500/100)^2 = 1.95 \,\Omega$$

The impedance Z that has thus to be put in the primary circuit is

$$Z = \sqrt{(R^2 + X^2)} = \sqrt{(0.90^2 + 1.9^5)} = 2.15 \,\Omega$$

Example 11

A single-phase transformer has a primary with 1000 turns and a secondary with 250 turns. The primary winding has a resistance of 1.0 Ω and the secondary winding a resistance of 0.05 Ω. Leakage reactances are 4.0 Ω for the primary and 0.25 Ω for the secondary. If the secondary load is a resistance of 10 Ω what will be the potential difference across it when the input to the primary circuit is 200 V? Neglect the no-load current.

Answer

This problem will be considered in two ways: transforming all resistances and reactances into the secondary circuit and transforming all resistances and reactances into the primary circuit.

Transforming the winding resistance and the leakage reactance in the primary into resistance and reactance in the secondary circuit gives a total resistance of

$$R = 0.05 + (250/1000)^2 \times 1.0 = 0.1125 \,\Omega$$

and a reactance X of

$$X = 0.25 + (250/1000)^2 \times 4.0 = 0.50\,\Omega$$

Thus the total impedance in the secondary circuit is

$$10 + 0.1125 + j0.50\,\Omega$$

The induced voltage in the secondary is

$$(N_p/N_s)V_p = (250/1000) \times 200 = 50\,\text{V}.$$

Hence

$$\mathbf{I_s} = \frac{50\angle 0°}{10.1125 + j0.50} = \frac{50\angle 0°}{10.1\angle 2.8°} = 4.9\angle -2.8°\,\text{A}$$

Transforming the winding resistance and the leakage reactance in the secondary into resistance and reactance in the primary circuit gives a total resistance R of

$$R = 1.0 + (1000/250)^2 \times 0.05 = 1.8\,\Omega$$

and a reactance X of

$$X = 4.0 + (1000/250)^2 \times 0.25 = 8.0\,\Omega$$

The load is transformed into a resistance R_L in the primary circuit of

$$R_L = (1000/250)^2 \times 10 = 160\,\Omega$$

Thus the total impedance Z in the equivalent primary circuit is

$$Z = 160 + 1.8 + j8.0\,\Omega$$

Hence the current $\mathbf{I_p}$ is

$$\mathbf{I_p} = \frac{200\angle 0°}{161.8 + j8.0} = \frac{200\angle 0°}{162\angle 2.8°} = 1.23\angle -2.8°\,\text{A}$$

Since this is the effective current with an ideal transformer, then $\mathbf{I_s}/\mathbf{I_p} = N_p/N_s = 1000/250$ and so

$$\mathbf{I_s} = (1000/250) \times 1.23\angle -2.8° = 4.9\angle -2.8°\,\text{A}$$

and so the potential difference across the load $\mathbf{V_L}$ is

$$\mathbf{V_L} = 10 \times 4.92\angle -2.8° = 49.2\angle -2.8°\,\text{V}$$

Problems

1 If two coils have a mutual inductance of 200 μH, what will be the e.m.f. induced in one coil when the current in the other coil changes at the rate of 20 000 A/s?

2 What is the coupling coefficient for a pair of coils if they have inductances of 12 mH and 24 mH and a mutual inductance of 6 mH?

3 For the coupled circuits shown in Fig. 5.11 determine (*a*) the self impedance of the primary circuit, (*b*) the self impedance of the secondary circuit, (*c*) the impedance reflected into the primary circuit, (*d*) the effective primary impedance, (*e*) the primary current, (*f*) the secondary current.

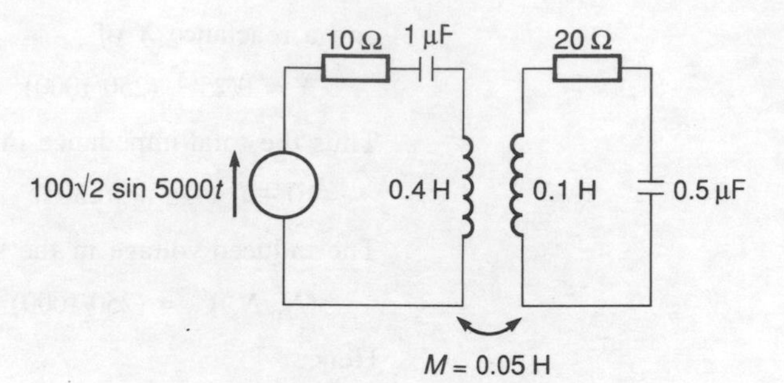

Fig. 5.11 Problem 3

4 For the coupled circuits shown in Fig. 5.12 determine (*a*) the current in the primary circuit and (*b*) the current in the secondary circuit.

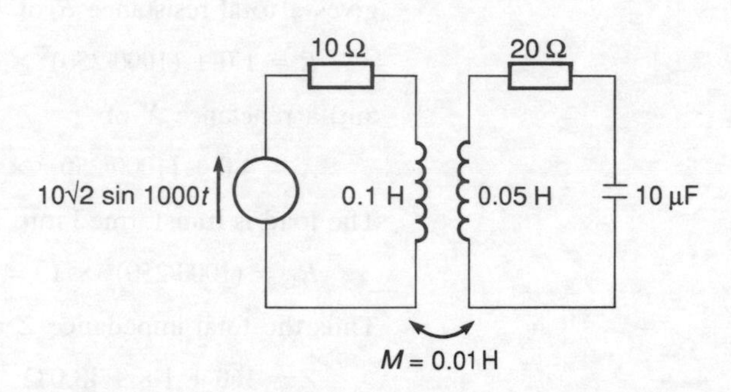

Fig. 5.12 Problem 4

5 A single-phase transformer has a step-down ratio of 10 : 1. (*a*) How many turns will there be on the secondary coil if the primary has 200 turns? (*b*) What will be the primary current if the secondary current is 30 A?

6 What will be the rated current in the primary and secondary windings of a 10 kVA, 1100 : 240 V single phase transformer?

7 What is the input impedance of an ideal transformer with a turns ratio 5 : 1 if the load connected to the secondary terminals is (*a*) a resistance of 500 Ω and (*b*) a 1 kΩ resistance in series with a 1 μF capacitor? The angular frequency ω is 1000 rad/s.

8 What should be the turns ratio of a transformer which is to be used to match an amplifier with an output impedance of 100 Ω with a circuit of input impedance 50 Ω and so give maximum power transfer?

9 A single-phase transformer has 600 primary turns and 1000 secondary turns. What will be (*a*) the maximum value of the flux produced in the transformer core, (*b*) the secondary induced e.m.f., if the input to the primary turns is 240 V, 50 Hz?

10 A 240/12 V single phase transformer has a no-load current of 0.5 A and a core loss of 50 W. What are the values of the magnetizing and core current components of the no-load current?

11 Sketch the equivalent circuit of a single-phase transformer that includes the effects of copper loss and leakage reactance.

12 A transformer has 600 primary turns and 120 secondary turns. The primary resistance is 0.42 Ω and the secondary 0.02 Ω. The primary leakage reactance is 1.30 Ω and the secondary 0.04 Ω. What is the equivalent impedance which could be placed in the primary circuit with an ideal transformer?

6 Tuned circuits

Introduction

This chapter is about the frequency response characteristics of circuits. Circuits containing resistance, inductance and capacitance have one or more resonant frequencies at which the impedance becomes purely resistive and the phase angle between the current and the applied voltage zero. Such circuits are said to be *tuned circuits* because they exhibit discriminatory characteristics as far as frequency is concerned. They will transmit signals at some frequencies better than at other frequencies and so can be used to filter out, or eliminate, signals in an unwanted frequency range. They can thus be used to produce frequency-selective circuits, a vital ingredient in radio and television communications.

Series *RLC* circuits

For a *series circuit* of a resistor, inductor and capacitor (Fig. 6.1) the total impedance Z is

$$Z = R + j\omega L - \frac{j}{\omega C} = R + j\left(\omega L - \frac{1}{\omega C}\right)$$

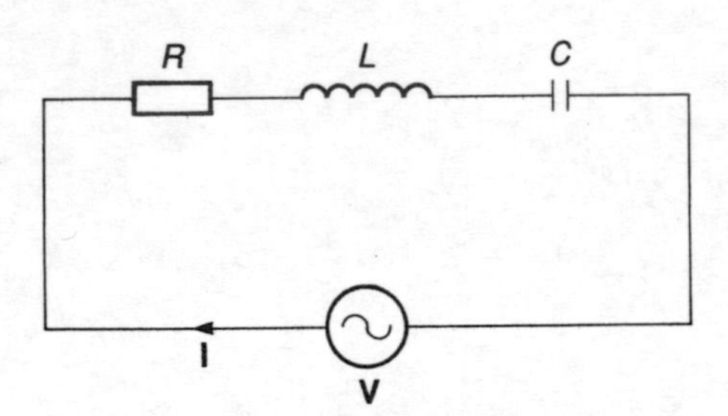

Fig. 6.1 Series *RLC* circuit

Thus, in polar notation,

$$Z = \sqrt{\left[R^2 + \left(\omega L - \frac{1}{\omega C}\right)^2\right]} \angle\phi \qquad [1]$$

with

$$\phi = \tan^{-1}\left[\frac{\omega L - (1/\omega C)}{R}\right] \qquad [2]$$

The phase angle ϕ is zero and the impedance Z a minimum of just R when $\omega L = 1/\omega C$. This condition is called *resonance* and so the resonant angular frequency ω_0 is

$$\omega_0 = \frac{1}{\sqrt{LC}} \qquad [3]$$

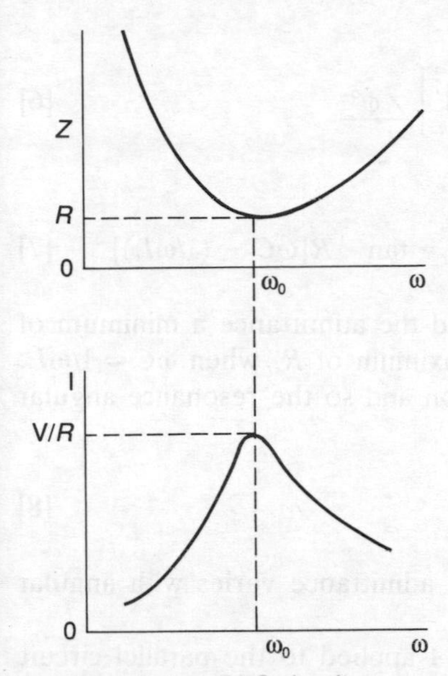

Fig. 6.2 Series *RLC* circuit, (*a*) impedance and (*b*) current as a function of frequency

Figure 6.2(*a*) shows how the impedance varies with angular frequency.

For a sinusoidal voltage of V applied to the series arrangement, the current I is given by

$$\mathbf{I} = \frac{\mathbf{V}}{Z} = \frac{V\angle 0°}{\sqrt{[R^2 + (\omega L - 1/\omega C)^2]}\angle\phi} \qquad [4]$$

At resonance this becomes

$$\mathbf{I} = \frac{V\angle 0°}{R\angle 0°} = \frac{V}{R}\angle 0° \qquad [5]$$

Figure 6.2(*b*) shows how the current varies with angular frequency.

Example 1

An inductor having an inductance of 50 mH and resistance 10 Ω is connected in series with a 12 μF capacitor and a 20 V r.m.s. sinusoidal supply. What is (*a*) the resonant frequency, (*b*) the current at resonance and (*c*) the potential difference across the capacitor at resonance?

Answer

(*a*) The resonant angular frequency is given by equation [3] as

$$\omega_0 = \frac{1}{\sqrt{LC}}$$

and so, since $\omega_0 = 2\pi f_0$, then

$$f_0 = \frac{1}{2\pi\sqrt{(0.050 \times 12 \times 10^{-6})}} = 205\text{ Hz}$$

(*b*) At resonance the impedance of the circuit will be just the resistance of 10 Ω. Thus

$$\mathbf{I} - \frac{20\angle 0°}{10} = 2.0\angle 0°\text{ A}$$

The current is in phase with the supplied voltage.

(*c*) The potential difference across the capacitor at resonance is

$$\mathbf{V} = \mathbf{I}\mathbf{Z} = (2.0\angle 0°)\left(\frac{1}{\mathrm{j}\omega_0 C}\right) = (2.0\angle 0°)\left(-\frac{\mathrm{j}}{\omega_0 C}\right)$$

$$= (2.0\angle 0°)\left(\frac{1}{2\pi \times 205 \times 12 \times 10^{-6}}\angle -90°\right)$$

$$= 129\angle -90°\text{ V r.m.s.}$$

Parallel *RLC* circuits

For a *parallel circuit* of resistor, inductor and capacitor (Fig. 6.3) the total admittance *Y* of the circuit is

$$Y = \frac{1}{R} - \frac{\mathrm{j}}{\omega L} + \mathrm{j}\omega C = \frac{1}{R} + \mathrm{j}\left(\omega C - \frac{1}{\omega L}\right)$$

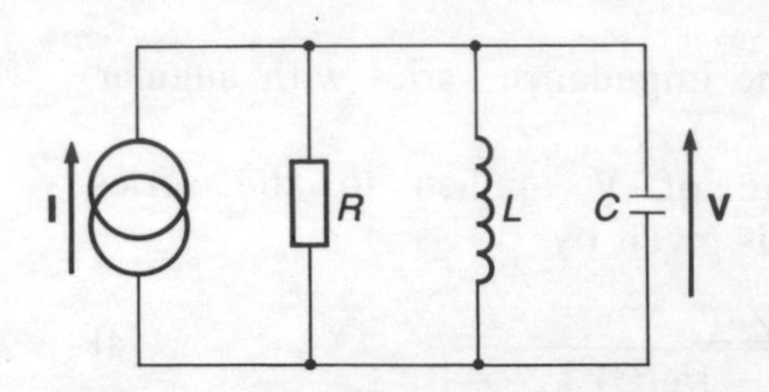

Fig. 6.3 Parallel *RLC* circuit

Thus in polar notation

$$Y = \sqrt{\left[\frac{1}{R^2} + \left(\omega C - \frac{1}{\omega L}\right)^2\right]} \angle \phi^\circ \qquad [6]$$

with

$$\phi = \tan^{-1}\left[\frac{\omega C - (1/\omega L)}{(1/R)}\right] = \tan^{-1} R[\omega C - (1/\omega L)] \qquad [7]$$

The phase angle ϕ is zero and the admittance a minimum of just $1/R$, i.e. impedance a maximum of R, when $\omega C = 1/\omega L$. This is the resonance condition and so the resonance angular frequency ω_0 is

$$\omega_0 = \frac{1}{\sqrt{LC}} \qquad [8]$$

Figure 6.4(*a*) shows how the admittance varies with angular frequency.

For a sinusoidal current of **I** applied to the parallel circuit the potential difference **V** across the arrangement is

$$\mathbf{V} = \frac{\mathbf{I}}{Y} = \frac{I \angle 0^\circ}{\sqrt{\left[\frac{1}{R^2} + \left(\omega C - \frac{1}{\omega L}\right)^2\right]} \angle \phi} \qquad [9]$$

At resonance this becomes

$$\mathbf{V} = \frac{I \angle 0^\circ}{(1/R) \angle 0^\circ} = RI \qquad [10]$$

Figure 6.4(*b*) shows how the voltage varies with angular frequency.

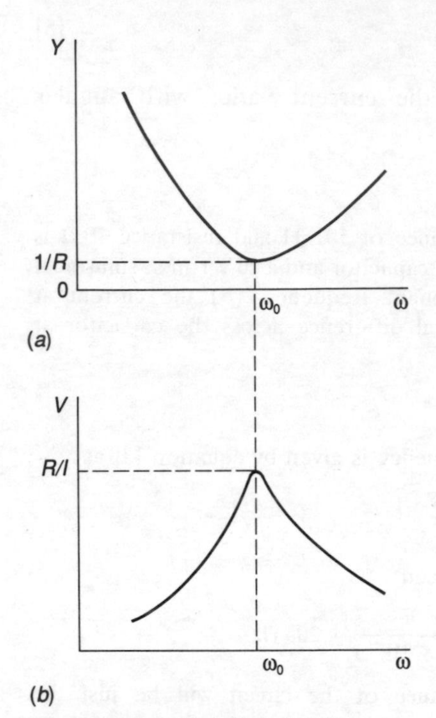

Fig. 6.4 Parallel *RLC* circuit, (*a*) admittance and (*b*) voltage as a function of frequency

Parallel *RL* and *C* circuits

For a circuit of a resistance plus inductance, e.g. an inductor having both resistance and inductance, in parallel with a capacitor (Fig. 6.5), the impedance of the resistor plus inductor arm is $(R + j\omega L)$. Thus the admittance Y of the circuit is

$$Y = \frac{1}{R + j\omega L} + j\omega C$$

$$= \frac{R - j\omega L}{R^2 + \omega^2 L^2} + j\omega C$$

$$= \frac{R}{R^2 + \omega^2 L^2} + j\left(\omega C - \frac{\omega L}{R^2 + \omega^2 L^2}\right)$$

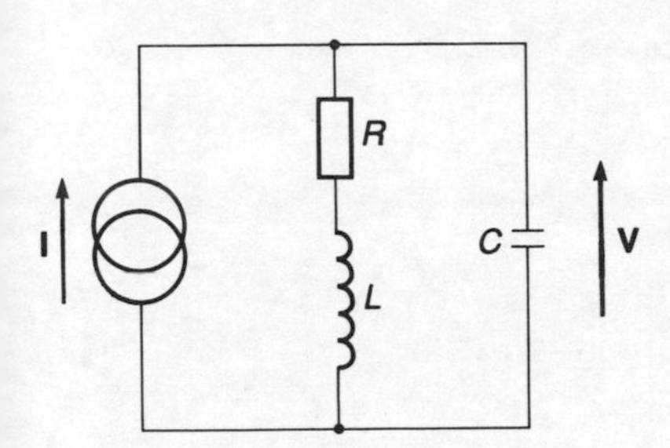

Fig. 6.5 Parallel RL and C circuit

In polar form this is

$$Y = \sqrt{\left[\frac{R^2}{(R^2+\omega^2L^2)^2} + \left(\omega C - \frac{\omega L}{R^2+\omega^2L^2}\right)^2\right]} \angle\phi \qquad [11]$$

with

$$\phi = \tan^{-1}\left[\left(\omega C - \frac{\omega L}{R^2+\omega^2L^2}\right)\left(\frac{R^2+\omega^2L^2}{R^2}\right)\right] \qquad [12]$$

The phase angle is zero when

$$\omega C = \frac{\omega L}{R^2+\omega^2L^2}$$

$$R^2+\omega^2L^2 = \frac{L}{C}$$

Thus the resonant angular frequency ω_0 is

$$\omega_0 = \sqrt{\left(\frac{1}{LC} - \frac{R^2}{L^2}\right)} \qquad [13]$$

At this condition the admittance, given by equation [11] becomes

$$Y = \frac{R}{R^2+\omega_0^2L^2} = \frac{R}{L/C} \qquad [14]$$

The impedance is thus L/RC and is purely resistive. It is usually referred to as the *dynamic resistance*.

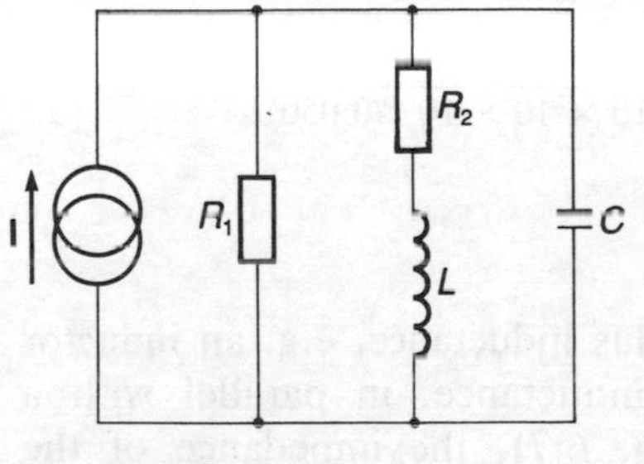

Fig. 6.6 Example 1

Example 2

Determine the resonant frequency for the circuit in Fig. 6.6.

Answer

The admittance Y of the circuit is

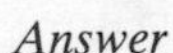

$$Y = \frac{1}{R_1} + \frac{1}{R_2 + j\omega L} + j\omega C$$
$$= \frac{1}{R_1} + \frac{R_2 - j\omega L}{R_2^2+\omega^2L^2} + j\omega C$$
$$= \left[\frac{1}{R_1} + \frac{R_2}{R_2^2+\omega^2L^2}\right] + j\left[\omega C - \frac{\omega L}{R_2^2+\omega^2L^2}\right]$$

In polar form this is

$$Y = \sqrt{\left[\left(\frac{1}{R_1} + \frac{R_2}{R_2^2+\omega^2L^2}\right)^2 + \left(\omega C - \frac{\omega L}{R^2+\omega^2L^2}\right)^2\right]} \angle\phi$$

with

$$\phi = \tan^{-1}\frac{\left[\omega C - \dfrac{\omega L}{R_2^2+\omega^2L^2}\right]}{\left[\dfrac{1}{R_1} + \dfrac{R_2}{R_2^2+\omega^2L^2}\right]}$$

Resonance is when $\phi = 0°$, hence

$$\omega C = \frac{\omega L}{R_2^2 + \omega^2 L^2}$$

$$R_2^2 + \omega^2 L^2 = (L/C)$$

$$\omega_0 = \sqrt{\left(\frac{1}{LC} - \frac{R_2^2}{L^2}\right)}$$

Example 3

An inductor of resistance 10 Ω and inductance 50 mH is in parallel with a 10 μF capacitor. What will be (*a*) the resonant frequency of the circuit and (*b*) the potential difference across the capacitor at resonance if the circuit is supplied by a sinusoidal current of 10 mA r.m.s.?

Answer

(*a*) Using equation [13], and since $\omega_0 = 2\pi f_0$,

$$f_0 = \frac{1}{2\pi}\sqrt{\left(\frac{1}{LC} - \frac{R^2}{L^2}\right)}$$

$$= \frac{1}{2\pi}\sqrt{\left(\frac{1}{0.050 \times 10 \times 10^{-6}} - \frac{10^2}{0.050^2}\right)}$$

$$= 223\,\text{Hz}$$

(*b*) At resonance the admittance Y will be, by equation [14], RC/L. Thus the potential difference **V** is

$$\mathbf{V} = \frac{\mathbf{I}}{Y} = \frac{0.010\angle 0°}{(RC/L)} = \frac{0.010\angle 0°}{(10 \times 10 \times 10^{-6}/0.050)}$$

$$= 5.0\angle 0°\,\text{V r.m.s.}$$

Parallel *RL* and *RC* circuits

For a circuit of a resistance plus inductance, e.g. an inductor having both resistance and inductance, in parallel with a capacitor plus resistance (Fig. 6.7), the impedance of the resistor plus inductor arm is $(R_1 + j\omega L)$ and that of the capacitor arm $(R_2 - j/\omega C)$. Thus the admittance Y of the circuit is

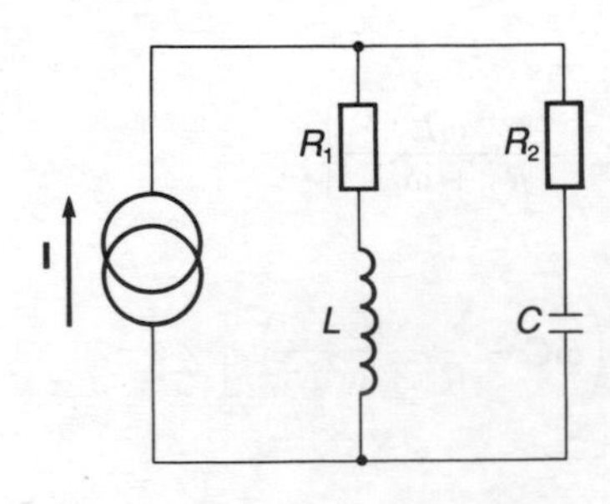

Fig. 6.7 Parallel *RL* and *RC* circuit

$$Y = \frac{1}{R_1 + j\omega L} + \frac{1}{R_2 - j/\omega C}$$

$$= \frac{R_1 - j\omega L}{R_1^2 + \omega^2 L^2} + \frac{R_2 + j/\omega C}{R_2^2 + (1/\omega C)^2}$$

$$= \frac{R_1}{R_1^2 + \omega^2 L^2} + \frac{R_2}{R_2^2 + (1/\omega C)^2} + j\left(\frac{1/\omega C}{R_2^2 + (1/\omega C)^2} - \frac{\omega L}{R_1^2 + \omega^2 L^2}\right)$$

In polar form this is

$$Y = \sqrt{\left[\left(\frac{R_1}{R_1^2 + \omega^2 L^2} + \frac{R_2}{R_2^2 + (1/\omega C)^2}\right)^2 + \left(\frac{1/\omega C}{R_2^2 + (1/\omega C)^2} - \frac{\omega L}{R_1^2 + \omega^2 L^2}\right)\right]} \angle 0° \quad [15]$$

with

$$\phi = \tan^{-1}\left[\frac{\dfrac{1/\omega C}{R_2^2 + (1/\omega C)^2} - \dfrac{\omega L}{R_1^2 + \omega^2 L^2}}{\dfrac{R1}{R_2^2 + \omega^2 L^2} + \dfrac{R_2}{R_2^2 + (1/\omega C)^2}}\right] \quad [16]$$

The phase angle is zero when

$$\frac{1/\omega C}{R_2^2 + (1/\omega C)^2} = \frac{\omega L}{R_1^2 + \omega^2 L^2}$$

$$\omega^2 LCR_2^2 + (L^2/C) = R_1^2 + \omega^2 L^2$$

$$\omega^2 = \frac{R_1^2 - (L^2/C)}{LCR_2^2 - L^2}$$

$$= \frac{CR_1^2 - L^2}{LC(CR_2^2 - L)}$$

and hence the resonance angular frequency ω_0 is

$$\omega_0 = \frac{1}{\sqrt{LC}}\sqrt{\left[\frac{R_1^2 - (L/C)}{R_2^2 - (L/C)}\right]} \quad [17]$$

At this condition the admittance, given by equation [15] becomes

$$Y = \frac{R_1}{R_1^2 + \omega_0^2 L^2} + \frac{R_2}{R_2^2 + (1/\omega_0 C)^2} \quad [18]$$

The impedance is purely resistive. It is usually referred to as the *dynamic resistance*.

Example 4

What is the resonant frequency for a network consisting of an inductor of resistance 10 Ω and inductance 50 mH in parallel with a 10 μF capacitor and a 2 Ω resistor?

Answer

The resonant angular frequency ω_0 is given by equation [17] as

$$\omega_0 = \frac{1}{\sqrt{LC}}\sqrt{\left[\frac{R_1^2 - (L/C)}{R_2^2 - (L/C)}\right]} \quad [17]$$

$$= \frac{1}{\sqrt{(0.050 \times 10^{-5})}}\sqrt{\left[\frac{10^2 - (0.050/10^{-5})}{2^2 - (0.050/10^{-5})}\right]}$$

$= 1401$ rad/s

Hence $f_0 = \omega_0/2\pi = 223$ Hz.

Q-factor with series *RLC* circuit

The *Q-factor*, or quality factor, is defined as being

$$Q = 2\pi \times \frac{\text{maximum energy stored in a component}}{\text{energy dissipated per cycle}} \qquad [19]$$

at resonance.

For a *series RLC circuit* the average power dissipated at resonance is $I_{rms}^2 R$ or $(I_m/\sqrt{2})^2 R$. Thus since the duration of 1 cycle is the periodic time T and $T = 1/f_0$, then the power dissipated in one cycle is $(1/f_0)(I_m/\sqrt{2})^2 R$. The maximum energy stored in the capacitor is $\frac{1}{2}CV_m^2$. But $V_m = I_m X_C$, where X_C is the reactance of the capacitor at the resonant frequency f_0. Hence the Q-factor for the capacitor is

$$Q = 2\pi \times \frac{\frac{1}{2}C(I_m X_C)^2}{(1/f_0)(I_m/\sqrt{2})^2 R} = \frac{2\pi f_0 C X_C^2}{R}$$

Since $X_C = 1/2\pi f_0 C$, then

$$Q = \frac{X_C}{R} \qquad [20]$$

But since the capacitor is in series in the circuit the current through it is the same as the circuit current. Thus, at resonance,

$$\frac{V_C}{V} = \frac{IX_C}{IR}$$

Hence equation [20] can be written as

$$Q = \frac{V_C}{V} \qquad [21]$$

and hence is the *voltage magnification* produced by the component.

Since $X_C = 1/\omega_0 C$, equation [20] can also be expressed as

$$Q = \frac{1}{\omega_0 CR} \qquad [22]$$

Likewise for the inductor in the series *RLC* circuit similar equations can be developed.

$$Q = \frac{X_L}{R} = \frac{V_L}{V} = \frac{\omega_0 L}{R} \qquad [23]$$

At resonance, since $\omega_0 L = 1/\omega_0 C$, then we must have the Q-factor for the capacitor the same as the Q-factor for the inductor.

Bandwidth with series *RLC* circuit

At resonance the power dissipated in the series RCL circuit is I_0^2R, where I_0 is the r.m.s. current at resonance. In this discussion the subscript r.m.s. has been omitted for simplification; it must however be realised that all the phasors have magnitudes which are r.m.s. values. I_0 is given by (equation [5])

$$I_0 = \frac{V}{R} \qquad [24]$$

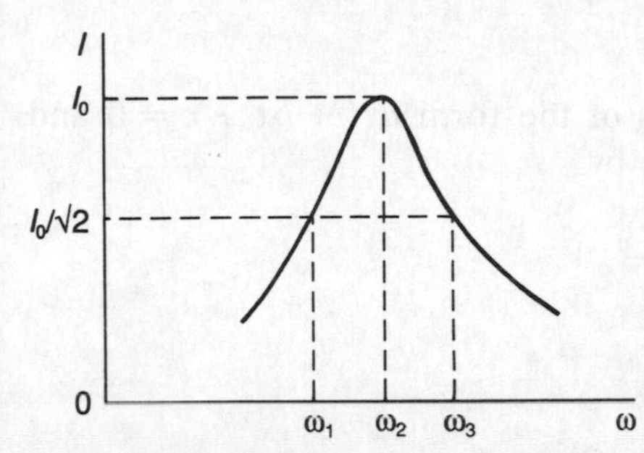

Fig. 6.8 Response of *RLC* circuit

At frequencies other than the resonant frequency the current will be smaller. Consider the frequencies on either side of the resonant frequency ω_0 at which the power has dropped by a half (Fig. 6.8). Since power is only dissipated in the resistance, then for half the power we must have, for a current I,

$$I^2R = \tfrac{1}{2}I_0^2R$$

Hence substituting for I_0 using equation [24],

$$I = \frac{I_0}{\sqrt{2}} = \frac{V}{R\sqrt{2}} \qquad [25]$$

But the current at an angular frequency ω is given by equation [4] as

$$I = \frac{V}{Z} = \frac{V}{\sqrt{[R^2 + (\omega L - 1/\omega C)^2]}} \qquad [26]$$

for frequencies above ω_0 when the inductive reactance is greater than the capacitive reactance. For frequencies below ω_0 when the inductive reactance is less than the capacitive reactance,

$$I = \frac{V}{Z} = \frac{V}{\sqrt{[R^2 + (1/\omega C - \omega L)^2]}} \qquad [27]$$

Thus we must have, because of equation [25],

$$R\sqrt{2} = \sqrt{[R^2 + (\omega L - 1/\omega C)^2]}$$

or

$$R\sqrt{2} = \sqrt{[R^2 + (1/\omega C - \omega L)^2]}$$

Hence

$$2R^2 = R^2 \pm (\omega L - 1/\omega C)^2$$

$$R = \pm\left(\omega L - \frac{1}{\omega C}\right) \qquad [28]$$

$$\pm 1 = \frac{\omega L}{R} - \frac{1}{\omega CR}$$

But for a capacitor $Q = 1/\omega_0 CR$ (equation [22]) and for an inductor $Q = \omega_0 L/R$ (equation [23]), hence

$$\pm 1 = \frac{\omega Q}{\omega_0} - \frac{\omega_0 Q}{\omega}$$

$$\pm \omega_0 \omega = \omega^2 Q - \omega_0^2 Q$$

$$\omega^2 \pm \frac{\omega_0 \omega}{Q} - \omega_0^2 = 0$$

This quadratic equation is of the form $ax^2 + bx + c = 0$ and the roots of this are given by

$$x = \frac{-b \pm \sqrt{(b^2 - 4ac)}}{2a}$$

Hence

$$\omega = \frac{\pm(\omega_0/Q) \pm \sqrt{[(\omega_0/Q)^2 + 4\omega_0^2]}}{2}$$

$$\omega = \pm \frac{\omega_0}{2Q} \pm \frac{\omega_0}{2Q} \sqrt{(1 + 4Q^2)} \quad [29]$$

There are four solutions to this equation; however, only two give positive values for ω. These are

$$\omega_1 = -\frac{\omega_0}{2Q} + \frac{\omega_0}{2Q} \sqrt{(1 + 4Q^2)} \quad [30]$$

$$\omega_2 = +\frac{\omega_0}{2Q} + \frac{\omega_0}{2Q} \sqrt{(1 + 4Q^2)} \quad [31]$$

Equations [30] and [31] can often be approximated to

$$\omega_1 \approx -\frac{\omega_0}{2Q} + \omega_0$$

$$\omega_1 \approx -\frac{\omega_0}{2Q} + \omega_0$$

The frequencies ω_1 and ω_2 are the frequencies at which the power is half that at the resonant frequency and are known as the *half-power*, or the *band edge* or *the −3 dB frequencies*. This latter term is because a power decrease by $\frac{1}{2}$ is in decibels

$$10 \log \tfrac{1}{2} = -3\,\text{dB}$$

At these frequencies the circuit impedance Z, given by equation [25] is

$$Z = R\sqrt{2} \quad [32]$$

The *3 dB bandwidth* is the difference between these two frequencies, i.e.

$$\text{bandwidth} = \omega_2 - \omega_1$$

Using equations [27] and [28], this becomes

$$\text{bandwidth} = \frac{\omega_0}{2Q} + \frac{\omega_0}{2Q} = \frac{\omega_0}{Q} \qquad [33]$$

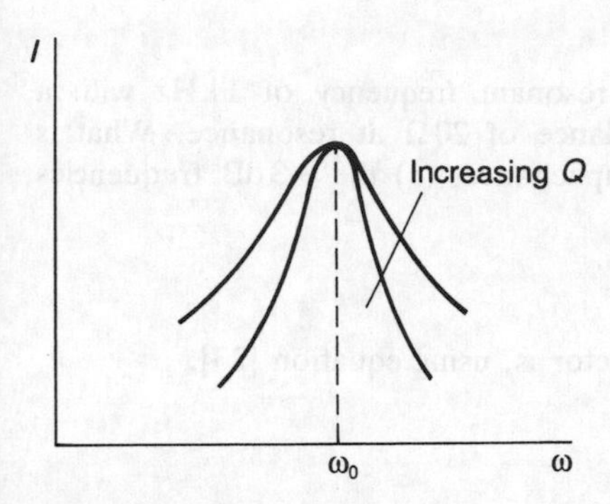

Fig. 6.9 Effect of Q-factor on sharpness of response

Thus the Q-factor is a measure of the sharpness or *selectivity* of the response curve, the larger the Q-factor the sharper the response and the more selective the circuit (Fig. 6.9). In a series circuit a high Q-factor, though giving the circuit high selectivity, can however result in high voltages across components (Q is the voltage magnification) and hence the possibility of electrical breakdown. Thus if a capacitor with a working voltage of 1 kV is used in a series RLC circuit with a Q-factor of 10, then the maximum value of the voltage supply in that circuit must not exceed 100 V if the working voltage of the capacitor is not to be exceeded.

Example 5

An inductor of resistance 10 Ω and inductance 0.10 H is connected in series with a 0.5 μF capacitor. What is (*a*) the resonant frequency, (*b*) the Q-factor, (*c*) the −3 dB frequencies, (*d*) the bandwidth, (*e*) the impedance at the resonance frequency, (*f*) the impedance at the −3 dB frequencies?

Answer

(*a*) The resonant frequency is given by equation [3] as

$$f_0 = \frac{1}{2\pi\sqrt{LC}} = \frac{1}{2\pi\sqrt{(0.10 \times 0.5 \times 10^{-6})}} = 712\,\text{Hz}$$

(*b*) Using equation [23],

$$Q = \frac{\omega_0 L}{R} = \frac{2\pi \times 712 \times 0.10}{10} = 44.7$$

(*c*) Using equations [30] and [31]

$$\omega_1 = -\frac{\omega_0}{2Q} + \frac{\omega_0}{2Q}\sqrt{(1 + 4Q^2)}$$

$$= -\frac{2\pi \times 712}{2 \times 44.7} + \frac{2\pi \times 712}{2 \times 44.7}\sqrt{(1 + 4 \times 44.7^2)}$$

Hence $f_1 = 704\,\text{Hz}$

$$\omega_2 = +\frac{\omega_0}{2Q} + \frac{\omega_0}{2Q}\sqrt{(1 + 4Q^2)}$$

and so $f_2 = 720\,\text{Hz}$

(*d*) The bandwidth is $f_2 - f_1 = 720 - 704 = 16\,\text{Hz}$.

(*e*) The impedance at the resonant frequency is just the resistance 10 Ω.

(*f*) The impedance at the −3 dB frequencies is $R\sqrt{2}$ (equation [32]) and so is 14.1 Ω.

Example 6

A series *RLC* circuit has a resonant frequency of 1 kHz with a *Q*-factor of 50 and an impedance of 20 Ω at resonance. What is (*a*) the inductance, (*b*) the capacitance, (*c*) the −3 dB frequencies and (*d*) the bandwidth?

Answer

(*a*) The *Q*-factor for the inductor is, using equation [23],

$$Q = \frac{\omega_0 L}{R}$$

The resistance in the circuit is the value of the impedance at resonance, i.e. 20 Ω. Hence

$$L = \frac{QR}{\omega_0} = \frac{50 \times 20}{2\pi \times 1000} = 0.159\,\text{H}$$

(*b*) Using equation [22]

$$Q = \frac{1}{\omega_0 CR}$$

$$C = \frac{1}{\omega_0 RQ} = \frac{1}{2\pi \times 1000 \times 20 \times 50} = 1.59 \times 10^{-7}\,\text{F}$$

(*c*) Using equations [30] and [31]

$$\omega_1 = -\frac{\omega_0}{2Q} + \frac{\omega_0}{2Q}\sqrt{(1 + 4Q^2)}$$

$$= -\frac{2\pi \times 1000}{2 \times 50} + \frac{2\pi \times 1000}{2 \times 50}\sqrt{(1 + 4 \times 50^2)}$$

Hence $f_1 = 990\,\text{Hz}$

$$\omega_2 = +\frac{\omega_0}{2Q} + \frac{\omega_0}{2Q}\sqrt{(1 + 4Q^2)}$$

and so $f_2 = 1010\,\text{Hz}$

(*d*) The bandwidth is $f_2 - f_1 = 1010 - 990 = 20\,\text{Hz}$.

Example 7

A series *RLC* circuit when connected to a sinusoidal 4 V supply gives a maximum current of 0.1 A at a frequency of 2 kHz with a bandwidth of 200 Hz. What is (*a*) the *Q*-factor, (*b*) the capacitance and (*c*) the potential difference across the capacitor?

Answer

(*a*) The bandwidth in rad/s is given by equation [33] as ω_0/Q. Thus

$$Q = \frac{2000}{200} = 10$$

(*b*) The impedance of the circuit at resonance is just the circuit resistance. This is thus 4/0.1 = 40 Ω. Hence, using equation [22],

$$Q = \frac{1}{\omega_0 CR}$$

$$C = \frac{1}{\omega_0 RQ} = \frac{1}{2\pi \times 2000 \times 40 \times 10} = 1.99 \times 10^{-7}\,\text{F}$$

(*c*) The Q-factor can be expressed as the voltage magnification (equation [21]), hence

$$Q = \frac{V_C}{V}$$

$$V_C = QV = 10 \times 4 = 40\,\text{V}.$$

Voltage magnification with the series *RLC* circuit

With the series resonant circuit the potential difference across the capacitor $\mathbf{V_C}$ has an amplitude which is Q times the amplitude of the input source voltage $\mathbf{V}$ for the circuit. But $\mathbf{V_C} = \mathbf{I}(1/\text{j}\omega C)$ and $\mathbf{V} = \mathbf{I}Z$ where Z is the total circuit impedance. Thus

$$\mathbf{V_C} = \frac{\mathbf{V}(1/\text{j}\omega C)}{Z}$$

Since

$$Z = R + \text{j}[\omega L - (1/\omega C)]$$

then

$$\mathbf{V_C} = \frac{\mathbf{V}}{\text{j}\omega C\{R + \text{j}[\omega L - (1/\omega C)]\}}$$

$$= \frac{\mathbf{V}}{1 - \omega^2 LC + \text{j}\omega RC}$$

$$= \frac{\mathbf{V}[(1 - \omega^2 LC) - \text{j}\omega RC]}{[(1 - \omega^2 LC) + \text{j}\omega RC][(1 - \omega^2 LC) - \text{j}\omega RC]}$$

The magnitude of $\mathbf{V_C}$ is thus

$$V_C = \frac{V\sqrt{[(1 - \omega^2 LC)^2 + \omega^2 R^2 C^2]}}{[(1 - \omega^2 LC)^2 + \omega^2 R^2 C^2]}$$

$$V_C = \frac{V}{\sqrt{[(1 - \omega^2 LC)^2 + \omega^2 R^2 C^2]}} \qquad [34]$$

The maximum magnitude for $\mathbf{V_C}$ is when $dV_C/d\omega = 0$.

$$\frac{dV_C}{d\omega} = \frac{-\frac{1}{2}V[2(1 - \omega^2 LC)(-2\omega LC) + 2\omega R^2 C^2]}{[(1 - \omega^2 LC)^2 + \omega^2 R^2 C^2]^{3/2}}$$

Hence the maximum value occurs when

$$0 = -\tfrac{1}{2}V[2(1 - \omega^2 LC)(-2\omega LC) + 2\omega R^2 C^2]$$

$$(1 - \omega^2 LC)\omega LC = \omega R^2 C^2$$

$$\omega^2 = \frac{1}{LC} - \frac{1}{2}\left(\frac{R}{L}\right)^2 \quad [35]$$

The resonant frequency ω_0 is $1/\sqrt{LC}$, thus

$$\omega^2 = \omega_0^2 - \frac{1}{2}\left(\frac{R}{L}\right)^2$$

But $Q = \omega_0 L/R$, and so

$$\omega^2 = \omega_0^2 - \frac{\omega_0^2}{2Q^2}$$

$$\omega = \omega_0\sqrt{\left[1 - \frac{1}{2Q^2}\right]} \quad [36]$$

The maximum value of the potential difference across the capacitor does not therefore occur at the resonant frequency. The frequency is always less than the resonant frequency. The maximum value of the potential difference V_{Cm} can be obtained by substituting for ω, using equation [36] in equation [34].

$$V_{Cm} = \frac{V}{\sqrt{[(1 - \omega^2 LC)^2 + \omega^2 R^2 C^2]}}$$

$$= \frac{V}{\sqrt{\left[\left(1 - \omega_0^2 LC + \frac{\omega_0^2 LC}{2Q^2}\right)^2 + \omega_0^2 R^2 C^2 - \frac{\omega_0^2 R^2 C_2}{2Q^2}\right]}}$$

Since $\omega_0 = 1/\sqrt{LC}$, then

$$V_{Cm} = \frac{V}{\sqrt{\left[\frac{1}{4Q^4} + \frac{R^2 C}{L} - \frac{R^2 C}{2Q^2 L}\right]}}$$

But $Q = \omega_0 L/R = 1/\omega_0 CR$, hence $Q^2 = L/R^2 C$. Thus

$$V_{Cm} = \frac{V}{\sqrt{\left[\frac{1}{4Q^4} + \frac{1}{Q^2} - \frac{1}{2Q^4}\right]}}$$

$$V_{Cm} = \frac{QV}{\sqrt{[1 - (1/2Q^2)]}} \quad [37]$$

Similarly the condition for the maximum value of the potential difference across the inductor can be derived. It is

$$\omega = \frac{\omega_0}{\sqrt{\left[1 - \frac{1}{2Q^2}\right]}} \quad [38]$$

This gives a frequency which is always greater than the resonant frequency.

Example 8

A series RLC circuit has a sinusoidal voltage input of maximum value 10 V. The resistance 60 Ω, the inductance 10 mH and the capacitance 0.5 μF. What is (*a*) the resonance angular frequency, (*b*) the value of the voltage across the capacitor at the resonance frequency, (*c*) the angular frequency at which the voltage across the capacitor is a maximum, and (*d*) the value of this maximum voltage?

Answer

(*a*) The resonance frequency is

$$\omega_0 = \frac{1}{\sqrt{LC}} = \frac{1}{\sqrt{(10 \times 10^{-3} \times 0.5 \times 10^{-6})}}$$
$$= 1.41 \times 10^4 \text{ rad/s}$$

(*b*) The voltage V_C across the capacitor at the resonance frequency is QV. Since

$$Q = \frac{1}{\omega_0 CR} = \frac{1}{1.41 \times 10^4 \times 0.5 \times 10^{-6} \times 60} = 2.36$$

then

$$V_C = QV = 2.36 \times 10 = 23.6 \text{ V}$$

(*c*) The angular frequency at which V_C is a maximum is given by equation [36] as

$$\omega = \omega_0 \sqrt{\left[1 - \frac{1}{2Q^2}\right]} = 1.41 \times 10^4 \sqrt{\left[1 - \frac{1}{2 \times 2.36^2}\right]}$$
$$= 1.35 \times 10^4 \text{ rad/s}$$

(*d*) The maximum value of the voltage across the capacitor V_{Cm} is given by equation [37] as

$$V_{Cm} = \frac{QV}{\sqrt{\left[1 - \left(\frac{1}{2Q^2}\right)\right]}} = \frac{2.36 \times 10}{\sqrt{\left[1 - \left(\frac{1}{2 \times 2.36^2}\right)\right]}} = 24.7 \text{ V}$$

The universal series resonance equation

Consider a series RLC circuit and a frequency ω_1 which is below the resonant frequency ω_0 and another frequency ω_2 which is the same amount above the resonant frequency as ω_1 is below (Fig. 6.10), i.e.

$$\omega_0 - \omega_1 = \omega_2 - \omega_0$$

The fractional deviation from the resonant frequency δ is

$$\delta = \frac{\omega_0 - \omega_1}{\omega_0} = \frac{\omega_2 - \omega_0}{\omega_0}$$

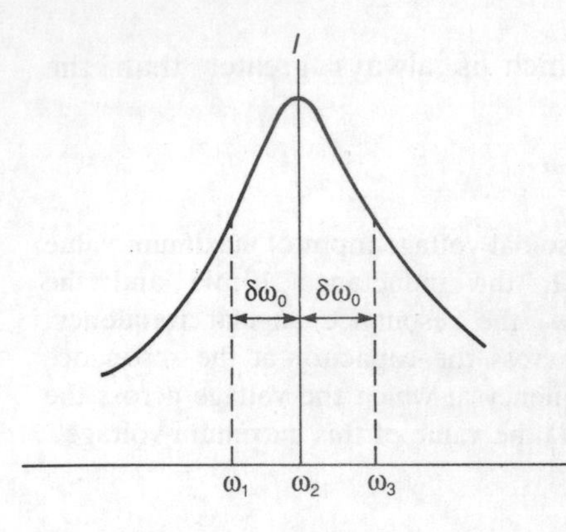

Fig. 6.10 Deviations from resonance

and thus

$$\omega_1 = \omega_0(1 - \delta) \qquad [39]$$

$$\omega_2 = \omega_0(1 + \delta) \qquad [40]$$

At resonance $\mathbf{I_0} = \mathbf{V}/R$ and at other frequencies $\mathbf{I} = \mathbf{V}/Z$, where Z is the circuit impedance. Thus

$$\frac{\mathbf{I}}{\mathbf{I_0}} = \frac{R}{Z} = \frac{R}{R + \mathrm{j}(\omega L - 1/\omega C)}$$

Hence at frequency ω_1 we can write, using equation [39],

$$\frac{\mathbf{I}}{\mathbf{I_0}} = \frac{R}{R + \mathrm{j}\left[\omega_0 L(1 - \delta) - \dfrac{1}{\omega_0 C(1 - \delta)}\right]}$$

At resonance $\omega_0 L = 1/\omega_0 C$, hence

$$\frac{\mathbf{I}}{\mathbf{I_0}} = \frac{1}{1 + \mathrm{j}\dfrac{\omega_0 L}{R}\left[(1 - \delta) - \dfrac{1}{(1 - \delta)}\right]}$$

Since $Q = \omega_0 L/R$ then

$$\frac{\mathbf{I}}{\mathbf{I_0}} = \frac{1}{1 + \mathrm{j}Q\left[\dfrac{(1 - \delta)^2 - 1}{1 - \delta}\right]}$$

$$\frac{\mathbf{I}}{\mathbf{I_0}} = \frac{1}{1 + \mathrm{j}Q\delta\left[\dfrac{\delta - 2}{1 - \delta}\right]}$$

For small deviations from the resonant frequency such that $\delta \ll 1$, then

$$\frac{\mathbf{I}}{\mathbf{I_0}} = \frac{1}{1 - \mathrm{j}2Q\delta}$$

$$= \frac{1 + \mathrm{j}2Q\delta}{(1 - \mathrm{j}2Q\delta)(1 + \mathrm{j}2Q\delta)}$$

Thus, for the magnitudes of the currents,

$$\frac{I}{I_0} = \sqrt{\left[\frac{1 + (2Q\delta)^2}{(1 + 4Q^2\delta^2)^2}\right]}$$

$$\frac{I}{I_0} = \frac{1}{\sqrt{(1 + 4Q^2\delta^2)}} \qquad [41]$$

The phase difference between $\mathbf{I}$ and $\mathbf{I_0}$ is

$$\phi = \tan^{-1} 2Q\delta \qquad [42]$$

Figure 6.11 shows a graph of how the ratio I/I_0 varies with the

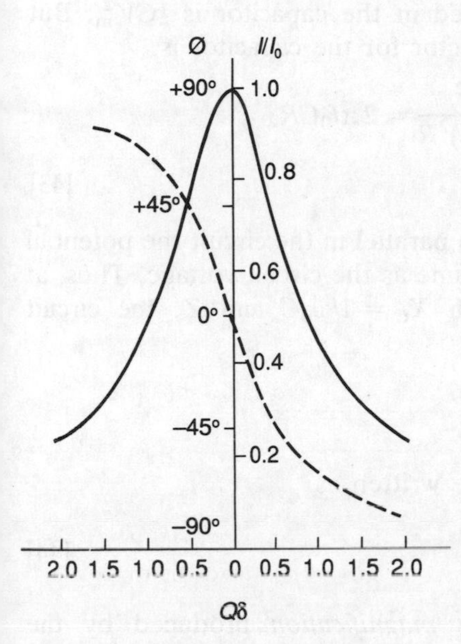

Fig. 6.11 Universal series *RLC* resonance graph

value of $Q\delta$. This graph is the same for all series resonant circuits. Also superimposed on Fig. 6.11 is the graph showing how the phase angle ϕ varies with $Q\delta$. This graph is also the same for all series resonant circuits.

Example 9

A series *RLC* circuit has $R = 10\,\Omega$, $L = 5\,\text{mH}$ and $C = 0.20\,\mu\text{F}$. Calculate the current through the circuit when the input voltage is 2.0 V and the frequency is (*a*) the resonant frequency and (*b*) a frequency 2% below the resonant frequency.

Answer

(*a*) At the resonant frequency the circuit impedance is just the resistance of $10\,\Omega$. Hence the current is $2.0/10 = 0.20$ A. The current is in phase with the voltage. Thus if the voltage is taken as being $10\angle 0°$ V then the current is $0.20\angle 0°$ A.

(*b*) For a frequency 2% below the resonant frequency $\delta = 0.02$. Since the *Q*-factor is $\omega_0 L/R$ and $\omega_0 = 1/\sqrt{LC}$,

$$Q = \frac{L}{R\sqrt{LC}} = \frac{\sqrt{L}}{R\sqrt{C}} = \frac{\sqrt{0.005}}{10 \times \sqrt{(0.20 \times 10^{-6})}} = 15.8$$

Thus, using equation [41]

$$\frac{I}{I_0} = \frac{1}{\sqrt{(1 + 4Q^2\delta^2)}}$$

$$= \frac{1}{\sqrt{(1 + 4 \times 15.8^2 \times 0.02^2)}}$$

$$= 0.845$$

Hence $I = 0.169$ A. The phase difference between $\mathbf{I}$ and $\mathbf{I_0}$ is given by equation [42] as

$$\phi = \tan^{-1} 2Q\delta = \tan^{-1}(2 \times 15.8 \times 0.02) = 32.3°$$

If the frequency is less than the resonance frequency then ϕ leads $\mathbf{I_0}$, if greater then it lags. Thus for the lesser frequency the current is $0.169\angle 32.3°$ A.

Q-factor with parallel *RLC* circuits

The *Q-factor*, or quality factor, is defined (equation [19]) as being

$$Q = 2\pi \times \frac{\text{maximum energy stored in a component}}{\text{energy dissipated per cycle}}$$

at resonance.

For a *parallel RLC circuit*, with R in parallel with L which in turn is in parallel with C (as in Fig. 6.3), the average power dissipated at resonance is $I_{\text{rms}}^2 R$ or $(I_m/\sqrt{2})^2 R$. Thus since the duration of 1 cycle is the periodic time T and $T = 1/f_0$, then the power dissipated in one cycle is $(1/f_0)(I_m/\sqrt{2})^2 R$.

The maximum energy stored in the capacitor is $\frac{1}{2}CV_m^2$. But $V_m = I_m R$. Hence the Q-factor for the capacitor is

$$Q = 2\pi \times \frac{\frac{1}{2}C(I_m R)^2}{(1/f_0)(I_m/\sqrt{2})^2 R} = 2\pi f_0 CR$$

$$Q = \omega_0 CR \qquad [43]$$

But since the capacitor is in parallel in the circuit the potential difference across it is the same as the circuit voltage. Thus, at resonance, $I_C X_C = IZ$ with $X_C = 1/\omega_0 C$ and Z, the circuit impedance, is R. Thus

$$\frac{I_C}{I} = \frac{R}{1/\omega_0 C} = \omega_0 CR$$

Hence equation [43] can be written as

$$Q = \frac{I_C}{I} \qquad [44]$$

and hence is the *current magnification* produced by the component.

Likewise for the inductor in the parallel RLC circuit similar equations can be developed.

$$Q = \frac{R}{\omega_0 L} \qquad [45]$$

At resonance, since $\omega_0 L = 1/\omega_0 C$, then we must have the Q-factor for the capacitor the same as the Q-factor for the inductor. Note that the Q-factors in the parallel RLC circuit are the inverse of those in the series RLC circuit.

For a *parallel RL and C circuit*, with R and L in series and the arrangement in parallel with C (as in Fig. 6.5) at resonance the dynamic impedance is L/RC (equation [14]). Thus the average power dissipated at resonance is $I_{rms}^2(L/RC)$ or $(I_m/\sqrt{2})^2(L/RC)$. Thus since the duration of 1 cycle is the periodic time T and $T = 1/f_0$, then the power dissipated in one cycle is $(1/f_0)(I_m/\sqrt{2})^2(L/RC)$. The maximum energy stored in the capacitor is $\frac{1}{2}CV_m^2$. But $V_m = I_m(L/RC)$. Hence the Q-factor for the capacitor is

$$Q = 2\pi \times \frac{\frac{1}{2}C(I_m L/RC)^2}{(1/f_0)(I_m/\sqrt{2})^2(L/RC)} = \frac{2\pi f_0 L}{R}$$

$$Q = \frac{\omega_0 L}{R} \qquad [46]$$

But since the capacitor is in parallel in the circuit the potential difference across it is the same as the circuit voltage. Thus, at resonance, $I_C X_C = IZ$ with $X_C = 1/\omega_0 C$ and Z, the circuit impedance, is the dynamic impedance L/RC. Thus

$$\frac{I_C}{I} = \frac{L/RC}{1/\omega_0 C} = \frac{\omega_0 L}{R}$$

Hence equation [46] can be written as

$$Q = \frac{I_C}{I} \qquad [47]$$

and hence is the *current magnification* produced by the component.

Likewise, by considering the energy stored in the inductor in the parallel RL and C circuit we can derive

$$Q = \frac{1}{\omega_0 CR} \qquad [48]$$

Note that the Q-factors in the parallel RL and C circuit are the same as those in the series RLC circuit. For the parallel RL and C circuit the resonant angular frequency ω_0 is given by equation [13] as

$$\omega_0 = \sqrt{\left(\frac{1}{LC} - \frac{R^2}{L^2}\right)}$$

This can be rearranged to give

$$\omega_0 = \frac{1}{\sqrt{LC}}\sqrt{\left(1 - \frac{R^2 C}{L}\right)}$$

But $Q = \omega_0 L/R = 1/\omega_0 CR$ and so $\omega_0^2 = L/C^2$. Hence

$$\omega_0 = \frac{1}{\sqrt{LC}}\sqrt{\left(1 - \frac{1}{Q^2}\right)} \qquad [49]$$

When Q is large then this equation approximates to

$$\omega_0 \approx \frac{1}{\sqrt{LC}}$$

and is the same as the equation for the series RLC or the parallel RLC circuit.

Example 10

A circuit consists of an inductor of resistance 10 Ω and inductance 1 mH in parallel with a 1 nF capacitor. What is (*a*) the resonant frequency and (*b*) the Q-factor of the circuit?

Answer

(*a*) Using equation [13]

$$\omega_0 = \sqrt{\left(\frac{1}{LC} - \frac{R^2}{L^2}\right)} = \sqrt{\left(\frac{1}{10^{-3} \times 10^{-9}} - \frac{10^2}{10^{-6}}\right)}$$

$$= 1.0 \times 10^6 \text{ rad/s}$$

Note that the R^2/L^2 term is insignificant compared with the $1/LC$ term and so that effectively the resonant frequency is $1/\sqrt{LC}$, the same as the parallel RLC or series RLC circuit. This approximation can be made whenever there is a large Q-factor.

(*b*) Using equation [46]

$$Q = \frac{\omega_0 L}{R} = \frac{1.0 \times 10^6 \times 1 \times 10^{-3}}{10} = 100$$

Bandwidth with parallel *RLC* circuits

At resonance the power dissipated in the parallel RLC circuit is V_0^2/R_d, where V_0 is the r.m.s. voltage at resonance and R_d the dynamic resistance. In this discussion the subscript r.m.s. has been omitted for simplification, it must however be realised that all the phasors have magnitudes which are r.m.s. values. V_0 is given by

$$V_0 = IR_d \qquad [50]$$

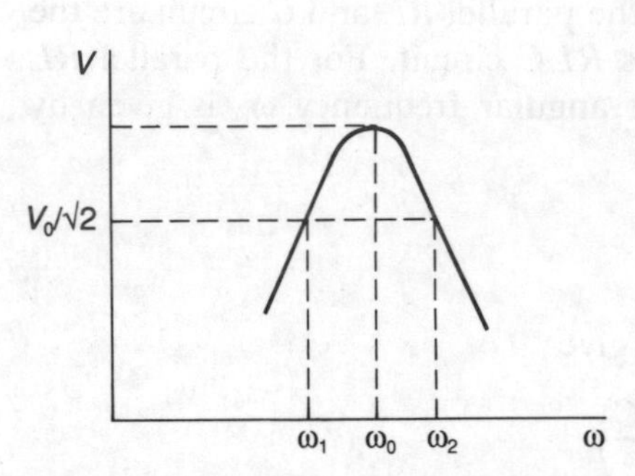

Fig. 6.12 Response of parallel *RLC* circuit

At frequencies other than the resonant frequency the voltage will be smaller. Consider the frequencies on either side of the resonant frequency ω_0 at which the power has dropped by a half (Fig. 6.12). Since power is only dissipated in the dynamic resistance, then for half the power we must have, for a voltage V,

$$V^2/R_d = \tfrac{1}{2}V_0^2/R_d$$

Hence substituting for V_0 using equation [50],

$$V = \frac{V_0}{\sqrt{2}} = \frac{IR_d}{\sqrt{2}} \qquad [51]$$

For a parallel circuit consisting of a resistance R in parallel with inductance L and this then in parallel with capacitance C (as in Fig. 6.3), equation [6] gives for the admittance magnitude Y at angular frequency ω

$$Y = \sqrt{\left[\frac{1}{R^2} + \left(\omega C - \frac{1}{\omega L}\right)^2\right]}$$

Since $V = I/Y$ then

$$V = \frac{I}{\sqrt{\left[\frac{1}{R^2} + \left(\omega C - \frac{1}{\omega L}\right)^2\right]}}$$

Thus we must have, because of equation [51] and since $R_d = R$,

$$\frac{\sqrt{2}}{R} = \sqrt{\left[\frac{1}{R^2} + \left(\omega C - \frac{1}{\omega L}\right)^2\right]}$$

$$\frac{2}{R^2} = \frac{1}{R^2} + \left(\omega C - \frac{1}{\omega L}\right)^2$$

Hence

$$\frac{1}{R^2} = \left(\omega C - \frac{1}{\omega L}\right)^2$$

But $Q = \omega_0 CR = R/\omega_0 L$ for such a circuit (equations [43] and [45]), thus

$$\frac{1}{R^2} = \left(\frac{\omega Q}{\omega_0 R} - \frac{Q\omega_0}{\omega R}\right)^2$$

$$\pm 1 = \frac{\omega Q}{\omega_0} - \frac{\omega_0 Q}{\omega}$$

$$\pm \omega_0 \omega = \omega^2 Q - \omega_0^2 Q$$

$$\omega^2 \pm \frac{\omega_0 \omega}{Q} - \omega_0^2 = 0$$

This quadratic equation is of the form $ax^2 + bx + c = 0$ and the roots of this are given by

$$x = \frac{-b \pm \sqrt{(b^2 - 4ac)}}{2a}$$

Hence

$$\omega = \frac{\pm(\omega_0/Q) \pm \sqrt{[(\omega_0/Q)^2 + 4\omega_0^2]}}{2}$$

$$\omega = \pm\frac{\omega_0}{2Q} \pm \frac{\omega_0}{2Q}\sqrt{(1 + 4Q^2)} \qquad [52]$$

There are four solutions to this equation, however only two give positive values for ω, these being

$$\omega_1 = -\frac{\omega_0}{2Q} + \frac{\omega_0}{2Q}\sqrt{(1 + 4Q)} \qquad [53]$$

$$\omega_2 = +\frac{\omega_0}{2Q} + \frac{\omega_0}{2Q}\sqrt{(1 + 4Q)} \qquad [54]$$

Equations [53] and [54] can often be approximated to

$$\omega_1 \approx -\frac{\omega_0}{2Q} + \omega_0$$

$$\omega_2 \approx +\frac{\omega_0}{2Q} + \omega_0$$

The frequencies ω_1 and ω_2 are the same as those obtained for series resonance and are the frequencies at which the power is half that at the resonant frequency and are known as the *half-power*, or the *band edge* or *the −3 dB frequencies*. This latter term is because a power decrease by ½ is in decibels

$$10 \log \tfrac{1}{2} = -3\,\text{dB}$$

At these frequencies the circuit admittance Y is, by equation [51],

$$Y = \frac{\sqrt{2}}{R} \qquad [55]$$

The *3 dB bandwidth* is the difference between these two frequencies, i.e. bandwidth = $\omega_2 - \omega_1$. Using equations [53] and [54], this becomes

$$\text{bandwidth} = \frac{\omega_0}{2Q} + \frac{\omega_0}{2Q} = \frac{\omega_0}{2Q} \qquad [56]$$

The bandwidth equation is thus the same as that of the series RLC circuit.

A parallel circuit consisting of an inductor, having resistance and inductance, in parallel with a capacitor gives the same equations as [53], [54] and [56] provided the Q-factor is taken as $\omega_0 L/R$ or $1/\omega_0 CR$. The circuit admittance at the 3 dB frequencies is $\sqrt{2}/R_d$, where R_d is the dynamic resistance L/CR.

Example 11

A parallel RLC circuit has $R = 2\,\text{k}\Omega$, $L = 40\,\text{mH}$ and $C = 0.25\,\mu\text{F}$. What is (*a*) the resonance angular frequency, (*b*) the Q-factor and (*c*) the bandwidth of the circuit?

Answer

(*a*) The resonance frequency is given by

$$\omega_0 = \frac{1}{\sqrt{LC}} = \frac{1}{\sqrt{(40 \times 10^{-3} \times 0.25 \times 10^{-6})}} = 10\,000\,\text{rad/s}$$

(*b*) Using equation [43]

$$Q = \omega_0 CR = 10\,000 \times 0.25 \times 10^{-6} \times 2000 = 5$$

(*c*) Using equation [55]

$$\text{bandwidth} = \frac{\omega_0}{Q} = \frac{10\,000}{5} = 2000\,\text{rad/s}$$

Example 12

An inductor of resistance $10\,\Omega$ and inductance 100 mH is in parallel with a 10 nF capacitor. What is (*a*) the resonance angular frequency, (*b*) the Q-factor and (*c*) the bandwidth?

Answer

(*a*) The resonant frequency is given by equation [13] as

$$\omega_0 = \sqrt{\left(\frac{1}{LC} - \frac{R^2}{L^2}\right)}$$

$$= \sqrt{\left(\frac{1}{100 \times 10^{-3} \times 10 \times 10^{-9}} - \frac{10^2}{0.100^2}\right)}$$

$$= 3.16 \times 10^4 \text{ rad/s}$$

(*b*) The Q-factor is given by equation [46] as

$$Q = \frac{\omega_0 L}{R} = \frac{3.16 \times 10^4 \times 100 \times 10^{-3}}{10} = 316$$

(*c*) The bandwidth is given by equation [56] as

$$\text{bandwidth} = \frac{\omega_0}{Q} = \frac{3.16 \times 10^4}{316} = 100 \text{ rad/s}$$

The universal parallel resonance equation

Consider a parallel RLC circuit and a frequency ω_1 which is below the resonant frequency ω_0 and another frequency ω_2 which is the same amount above the resonant frequency as ω_1 is below, i.e.

$$\omega_0 - \omega_1 = \omega_2 - \omega_0$$

The fractional deviation from the resonant frequency, δ, is

$$\delta = \frac{\omega_0 - \omega_1}{\omega_0} = \frac{\omega_2 - \omega_0}{\omega_0}$$

and thus

$$\omega_1 = \omega_0(1 - \delta) \qquad [57]$$

$$\omega_2 = \omega_0(1 + \delta) \qquad [58]$$

At resonance $\mathbf{I_0} = \mathbf{V}/R_D$ where R_D is the dynamic resistance, and at other frequencies $\mathbf{I} = \mathbf{V}Y$, where Y is the circuit admittance. The circuit admittance is

$$Y = \frac{1}{R} + \mathrm{j}\left(\omega C - \frac{1}{\omega L}\right)$$

Thus

$$\frac{\mathbf{I}}{\mathbf{I_0}} = R_D Y = R_D\left[\frac{1}{R} + \mathrm{j}\left(\omega C - \frac{1}{\omega L}\right)\right]$$

Hence at frequency ω_1 we can write, using equation [57],

$$\frac{\mathbf{I}}{\mathbf{I_0}} = R_D\left[\frac{1}{R} + \mathrm{j}\left(\omega_0\{1 - \delta\}C - \frac{1}{\omega_0\{1 - \delta\}L}\right)\right]$$

At resonance $\omega_0 L = 1/\omega_0 C$, hence

$$\frac{\mathbf{I}}{\mathbf{I_0}} = R_D\left[\frac{1}{R} + \mathrm{j}\omega_0 C\left(\{1 - \delta\} - \frac{1}{\{1 - \delta\}}\right)\right]$$

Since $Q = \omega_0 CR$ (equation [43]) and $R_D = R$, then

$$\frac{\mathbf{I}}{\mathbf{I_0}} = R\left[\frac{1}{R} + \mathrm{j}\frac{Q}{R}\left(\{1-\delta\} - \frac{1}{\{1-\delta\}}\right)\right]$$

$$= R\left[\frac{1}{R} + \mathrm{j}\frac{Q}{R}\left(\frac{1-2\delta+\delta^2-1}{1-\delta}\right)\right]$$

$$= R\left[\frac{1}{R} + \mathrm{j}\frac{Q\delta}{R}\left(\frac{\delta-2}{1-\delta}\right)\right]$$

For small deviations from the resonant frequency such that $\delta \ll 1$, then

$$\frac{\mathbf{I}}{\mathbf{I_0}} = \frac{R}{R}[1 - \mathrm{j}2Q\delta]$$

Thus, for the magnitudes of the currents,

$$\frac{I}{I_0} = \sqrt{[1 + (2Q\delta)^2]} \qquad [59]$$

The phase difference between $\mathbf{I}$ and $\mathbf{I_0}$ is

$$\phi = \tan^{-1} 2Q\delta \qquad [60]$$

The graph of how the ratio I/I_0 varies with the value of $Q\delta$ is of the same form as that for the series circuit in Fig. 6.11.

Since, in general, $V = IZ$ and $V = I_0R$ then equation [59] can be expressed as the ratio of the impedance Z at frequency ω to the impedance R_D at resonance, i.e.

$$\frac{R_D}{Z} = \sqrt{[1 + (2Q\delta)^2]} \qquad [61]$$

Equations [59], [60] and [61] apply in general to parallel circuits, though it must be remembered that the dynamic resistance R_D and the Q-factor will depend on the type of parallel circuit involved.

Example 13

A circuit consists of an inductor, having both resistance and inductance, in parallel with a 5 nF capacitor. The circuit has a bandwidth of 4 krad/s and a resonant angular frequency of 1 Mrad/s. What is (*a*) the Q-factor, and (*b*) the impedance when the applied frequency is 0.5% off the tuned value?

Answer

(*a*) Equation [56] gives bandwidth = ω_0/Q. Hence

$$Q = \frac{1 \times 10^6}{4 \times 10^3} = 250$$

(*b*) For the circuit

$$R_D = \frac{L}{CR} = \frac{\omega_0 L}{\omega_0 CR}$$

But $Q = \omega_0 L/R$, hence

$$R_D = \frac{Q}{\omega_0 C} = \frac{250}{10^6 \times 5 \times 10^{-9}} = 50\,\text{k}\Omega$$

Thus equation [61] gives, with $\delta = 0.005$,

$$\frac{R_D}{Z} = \sqrt{[1 + (2Q\delta)^2]}$$

$$Z = \frac{50 \times 10^3}{\sqrt{[1 + (2 \times 250 \times 0.005)^2]}} = 18.6\,\text{k}\Omega$$

The effect of loading

There are situations where the potential difference across the capacitor in a series *RLC* circuit is used to 'drive' some other circuit or device, i.e. there is a load across the capacitor (Fig. 6.13). For a resistive load R_L in parallel with capacitance C the impedance Z is

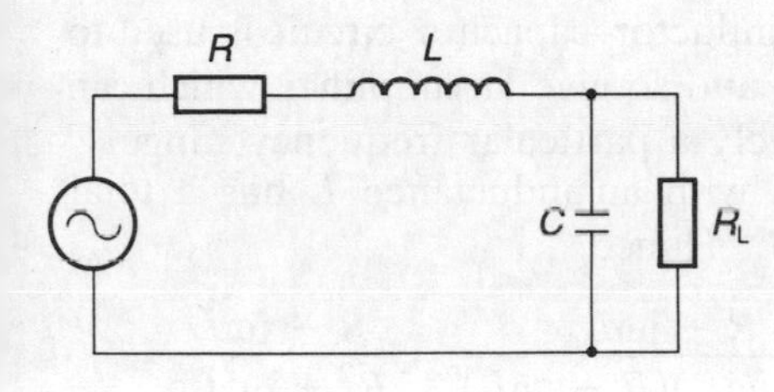

Fig. 6.13 Loading a series *RLC* circuit

$$\frac{1}{Z} = \frac{1}{R_L} + \frac{1}{1/j\omega C} = \frac{1 + j\omega C R_L}{R_L}$$

$$Z = \frac{R_L}{1 + j\omega C R_L} = \frac{R_L(1 - j\omega C R_L)}{(1 + j\omega C R_L)(1 - j\omega C R_L)}$$

$$= \frac{R_L - j\omega C R_L^2}{1 + \omega^2 C^2 R_L^2}$$

$$= \frac{R_L}{1 + \omega^2 C^2 R_L^2} + \frac{(-j\omega)(+j\omega) C R_L^2}{(+j\omega)(1 + \omega^2 C^2 R_L^2)}$$

$$= \frac{R_L}{1 + \omega^2 C^2 R_L^2} + \frac{\omega^2 C R_L^2}{j\omega(1 + \omega^2 C^2 R_L^2)}$$

We can consider this impedance to be equivalent to a resistance R_e in series with a capacitor C_e (Fig. 6.14), where

$$R_e = \frac{R_L}{1 + \omega^2 C^2 R_L^2} \qquad [62]$$

$$C_e = \frac{1 + \omega^2 C^2 R_L^2}{\omega^2 C R_L^2} \qquad [63]$$

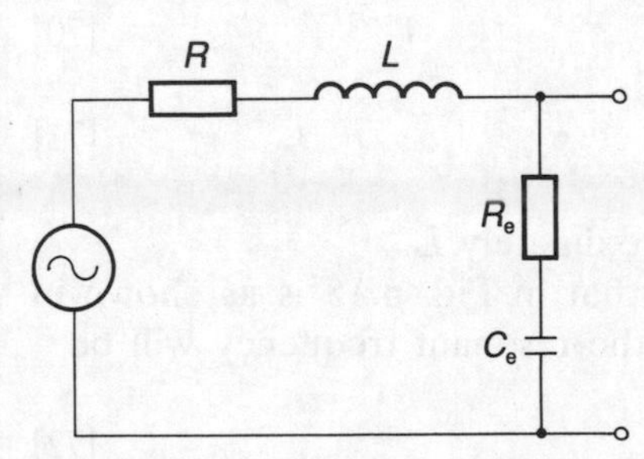

Fig. 6.14 Equivalent circuit of Fig. 6.13

The Q-factor of the original parallel circuit was $Q = \omega_0 R_L C$ and if we restrict ω to values close to ω_0 then equation [62] can be approximated to

$$R_e \approx \frac{R_L}{1 + Q^2} \qquad [64]$$

and, for equation [63],

$$C_e \approx C\left(1 + \frac{1}{Q^2}\right) \qquad [65]$$

For high Q values C_e is approximately equal to C.

The resonant frequency of the equivalent circuit is thus

$$\omega_{0L} = \frac{1}{\sqrt{LC_e}} \qquad [66]$$

and the Q-factor.

$$Q_L = \frac{\omega_{0L}L}{R + R_e} = \frac{1}{\omega_{0L}C_e(R + R_e)} \qquad [67]$$

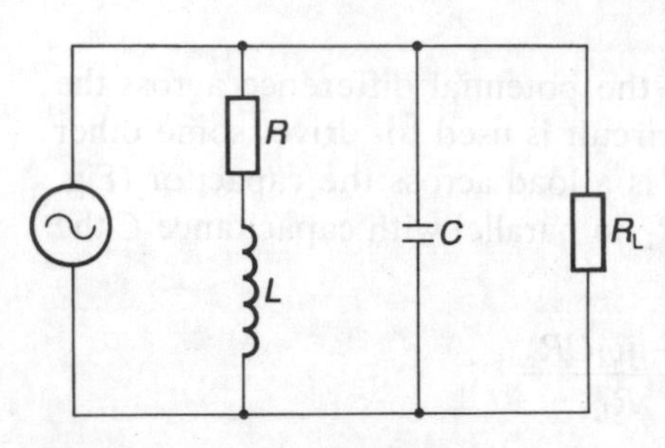

Fig. 6.15 Loading a parallel inductor–capacitor circuit

Another example of loading is that occurring when a resistive load is placed across the capacitor in a parallel inductor–capacitor circuit, the inductor having both inductance and resistance (Fig. 6.15). Such a situation can occur with a tuned amplifier where a parallel inductor–capacitor circuit is used to link stages of the amplifier and so give an amplifier which can be used to amplify selectively a particular frequency range.

A resistance R in series with an inductance L has a total impedance of $Z = R + j\omega L$. Hence

$$\frac{1}{Z} = \frac{1}{R + j\omega L} = \frac{R - j\omega L}{(R + j\omega L)(R - j\omega L)} = \frac{R - j\omega L}{R^2 + \omega^2 L^2}$$

We can consider this to be equivalent to an inductance L_e in parallel with a resistance R_e, where

$$R_e = \frac{R^2 + \omega^2 L^2}{R} \qquad [68]$$

and

$$j\omega L_e = -\left(\frac{R^2 + \omega^2 L^2}{j\omega L}\right)$$

$$L_e = \frac{R^2 + \omega^2 L^2}{\omega^2 L} \qquad [69]$$

Equations [68] and [69] can be expressed in terms of the Q-factor that existed in the series arrangement of L and R, i.e. $Q = \omega_0 L/R$. Then if ω is close to ω_0

$$R_e \approx R + Q^2 R \qquad [70]$$

$$L_e \approx L\left(1 + \frac{1}{Q^2}\right) \qquad [71]$$

For high Q values L_e is approximately L.

The equivalent circuit for that in Fig. 6.15 is as shown in Fig. 6.16. For such a circuit the resonant frequency will be

$$\omega_{0L} = \frac{1}{\sqrt{L_e C}} \qquad [72]$$

and the Q-factor

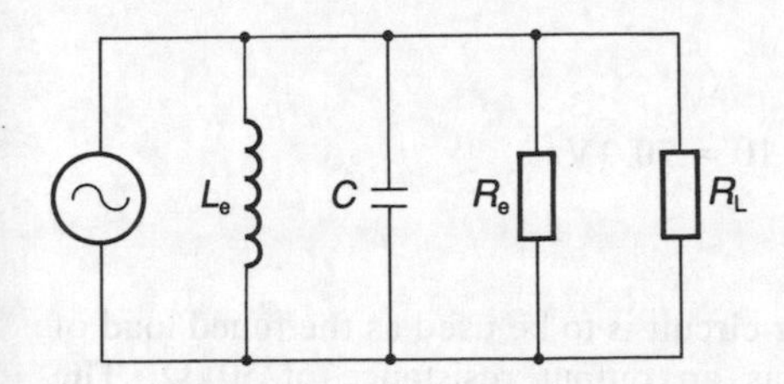

Fig. 6.16 Equivalent circuit for Fig. 6.15

$$Q_L = \omega_0 LCR_T = \frac{R_T}{\omega_{0L} L} \qquad [73]$$

where R_T is the total resistance and is given by

$$\frac{1}{R_T} = \frac{1}{R_L} + \frac{1}{R_e}$$

In tackling problems involving loaded tuned circuits the conversions of segments of such circuits from series to parallel or parallel to series is a useful simplifying routine. The equations derived earlier in this section and others are summarized in Fig. 6.17.

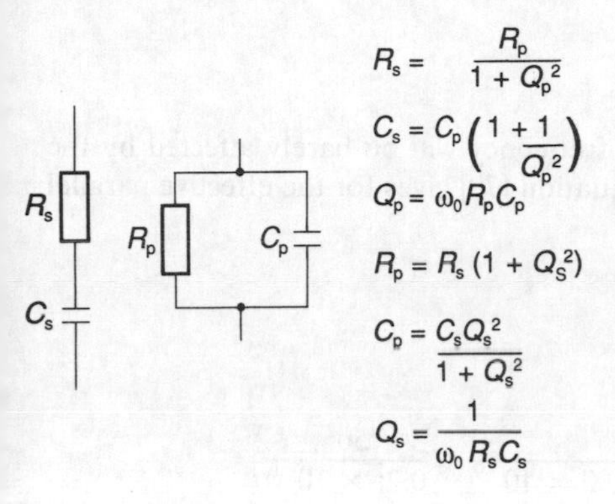

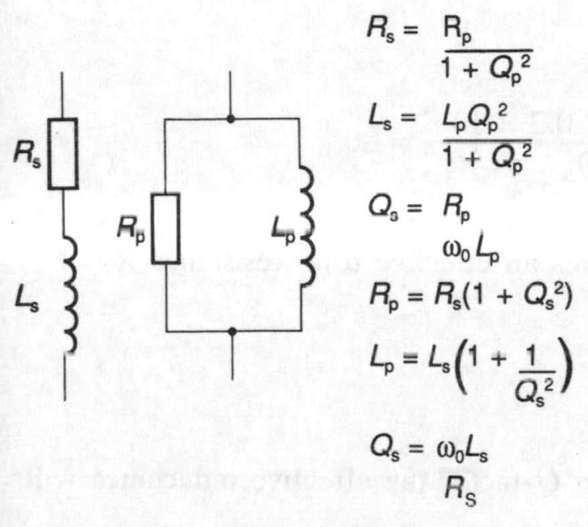

Fig. 6.17 Series and parallel equivalents

Example 14

A series RLC circuit has $R = 10\,\Omega$, $L = 5\,\text{mH}$ and $C = 0.5\,\mu\text{F}$ and a sinusoidal voltage input of 10 V. What is the potential difference across the capacitor when the circuit is (*a*) unloaded and (*b*) with a load of $10\,\text{k}\Omega$ across the capacitor?

Answer

(*a*) The resonant frequency ω_0 is

$$\omega_0 = \frac{1}{\sqrt{LC}} = \frac{1}{\sqrt{(5 \times 10^{-3} \times 0.5 \times 10^{-6})}} = 20\,\text{krad/s}$$

The Q-factor is

$$Q = \frac{\omega_{0L}}{R} = \frac{20 \times 10^3 \times 5 \times 10^{-3}}{10} = 10$$

Hence the potential difference across the capacitor V_C has the magnitude

$$V_C = QV = 10 \times 10 = 100\,\text{V}$$

(*b*) Making the assumption that the new resonant frequency is little changed from the one without the loading then, since

$$Q = \omega_0 R_L C = 20 \times 10^3 \times 10 \times 10^3 \times 0.5 \times 10^{-6} = 100$$

equation [64] gives

$$R_e \approx \frac{R_L}{1 + Q^2} \approx \frac{10 \times 10^3}{1 + 100^2} \approx 1\,\Omega$$

and equation [65] gives

$$C_e \approx C\left(1 + \frac{1}{Q^2}\right) \approx 0.5 \times 10^{-6}\left(1 + \frac{1}{100^2}\right) \approx 0.5 \times 10^{-6}\,\text{F}$$

Thus, since C_e is almost equal to C, the resonant frequency is virtually unchanged. The change in the total resistance in the circuit to $(10 + 1)\,\Omega$ will, however, result in the equivalent circuit having a Q-factor of

$$Q = \frac{\omega_0 L}{R} = \frac{20 \times 10^3 \times 5 \times 10^{-3}}{11} = 9.09$$

Hence

$$V_C = QV = 9.09 \times 10 = 90.9\,\text{V}.$$

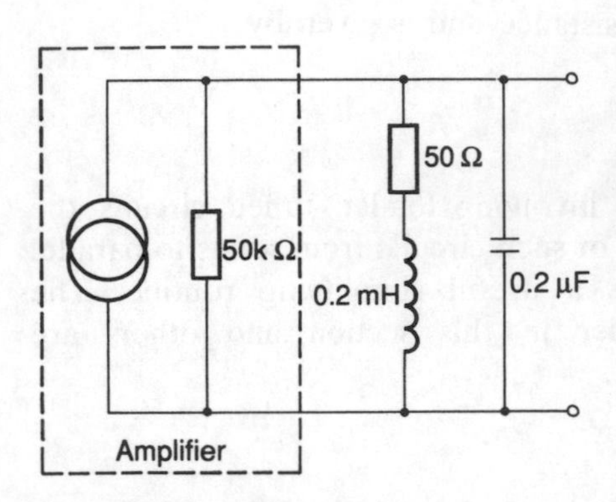

Fig. 6.18 Example 15

Example 15

A parallel inductor–capacitor circuit is to be used as the tuned load of an amplifier stage which has an output resistance of 50 kΩ. The inductor has an inductance of 0.2 mH and a resistance of 50 Ω and the capacitance is 0.2 nF. Figure 6.18 shows the circuit. What is the amplifier bandwidth?

Answer

Assuming that the resonant frequency will be barely affected by the presence of the load, then equation [70] gives for the effective parallel resistance of the inductor

$$R_e \approx R + Q^2R$$

$$\omega_0 = \sqrt{\left(\frac{1}{LC} - \frac{R^2}{L^2}\right)}$$

$$= \sqrt{\left(\frac{1}{0.2 \times 10^{-3} \times 0.2 \times 10^{-9}} - \frac{50^2}{0.2^2 \times 10^{-6}}\right)}$$

$$= 4.99 \times 10^6\,\text{rad/s}$$

Thus

$$Q = \frac{\omega_0 L}{R} = \frac{4.99 \times 10^6 \times 0.2 \times 10^{-3}}{50} = 20$$

Hence $R_e = 50 + 20^2 \times 50 = 20\,\text{k}\Omega$. This resistance is in parallel with the load resistance and so gives an effective total resistance R_T of

$$\frac{1}{R_T} = \frac{1}{20} + \frac{1}{50}$$

$$R_T = 14.3\,\text{k}\Omega$$

Because of the relatively high Q-factor the effective inductance will be unchanged as 0.2 mH.

The effective Q-factor of the circuit with load is thus

$$Q_e = \frac{R_T}{\omega_0 L} = \frac{14.3 \times 10^3}{4.99 \times 10^6 \times 0.2 \times 10^{-3}} = 14.3$$

Hence the bandwidth is

$$\text{bandwidth} = \frac{\omega_0}{Q} = \frac{4.99 \times 10^6}{14.3} = 3.49 \times 10^5\,\text{rad/s}$$

Example 16

A parallel inductor–capacitor circuit is to be used to link two stages of an amplifier. The input stage has a capacitance of 500 pF and a resistance of 1 kΩ and the output stage 200 pF and 2 kΩ. The resulting tuned amplifier is to have a resonance frequency of 500 krad/s. If the inductor has an inductance of 1 mH and a resistance of 5 Ω what should be the value of the capacitance used and the resulting bandwidth?

Answer

The circuit can be considered to be as in Fig. 6.19. The problem can be tackled by replacing the inductor with its associated resistance by an equivalent inductance in parallel with resistance and by also replacing the series capacitor and resistor of each amplifier stage by equivalent parallel capacitance and resistance.

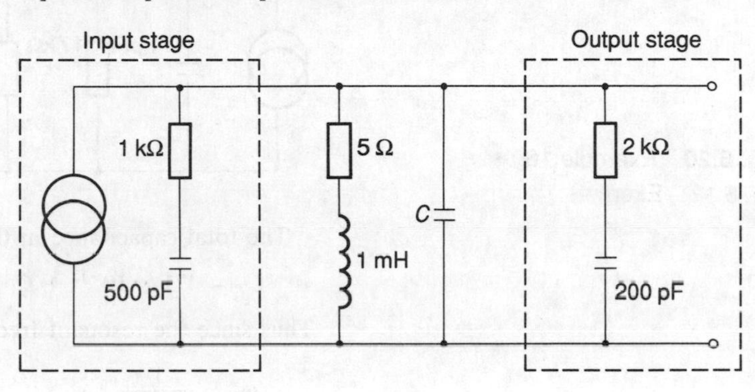

Fig. 6.19 Example 16

Assuming that the resonant frequency will be barely affected by the presence of the load, then equation [70] gives for the effective parallel resistance of the inductor

$$R_e \approx R + Q^2 R$$

For the inductor

$$Q = \frac{\omega_0 L}{R} = \frac{500 \times 10^3 \times 1 \times 10^{-3}}{5} = 100$$

Thus $R_e = 5 + 100^2 \times 5 = 50\,\text{k}\Omega$. Because of the large Q-factor the inductance is essentially unchanged and so is 1 mH.

For the input stage, the effective parallel resistance is

$$R_e \approx R + Q^2 R$$

For the capacitor

$$Q = \frac{1}{\omega_0 CR} = \frac{1}{500 \times 10^3 \times 500 \times 10^{-12} \times 1000} = 4$$

Thus $R_e = 1000 + 4^2 \times 1000 = 17\,\text{k}\Omega$. The effective parallel capacitance is

$$C_e \approx \frac{Q^2 C}{1 + Q^2} = \frac{4^2 \times 500 \times 10^{-12}}{1 + 4^2} = 471\,\text{pF}$$

For the output stage, the effective parallel resistance is

$$R_e \approx R + Q^2 R$$

For the capacitor

$$Q = \frac{1}{\omega_0 CR} = \frac{1}{500 \times 10^3 \times 200 \times 10^{-12} \times 2000} = 5$$

Thus $R_e = 2000 + 5^2 \times 2000 = 52\,\text{k}\Omega$. The effective parallel capacitance is

$$C_e \approx \frac{Q^2C}{1+Q^2} = \frac{5^2 \times 200 \times 10^{-12}}{1+5^2} = 192\,\text{pF}$$

The result of the above calculations is that the circuit is effectively that shown in Fig. 6.20.

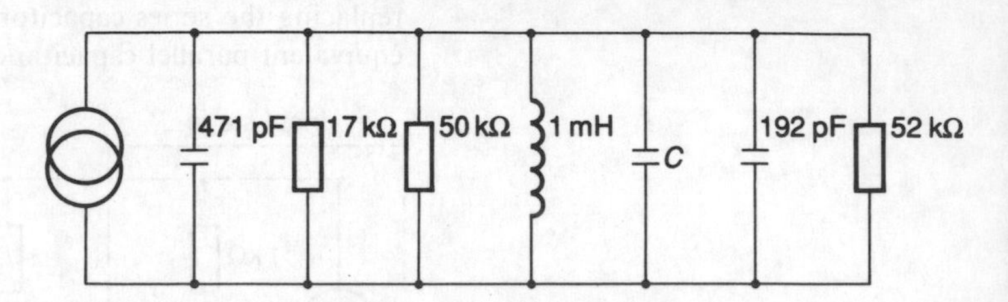

Fig. 6.20 Example 16

The total capacitance in the circuit is thus

$$C_T = 471 \times 10^{-12} + C + 192 \times 10^{-12} = 663 \times 10^{-12} + C$$

Thus since the resonant frequency ω_0 is given by

$$\omega_0 = \frac{1}{\sqrt{LC_T}}$$

$$C_T = \frac{1}{\omega_0^2 L} = \frac{1}{(500 \times 10^3)^2 \times 1 \times 10^{-3}} = 4000\,\text{pF}$$

Hence C must be $(4000 - 663) = 3337\,\text{pF}$

The total resistance R_T of the circuit is given by

$$\frac{1}{R_T} = \frac{1}{17} + \frac{1}{50} + \frac{1}{52}$$

$$R_T = 10.2\,\text{k}\Omega$$

Thus the effective Q-factor of the circuit is

$$Q = \frac{R}{\omega_0 L} = \frac{10.2 \times 10^3}{500 \times 10^3 \times 1 \times 10^{-3}} = 20.4$$

Hence the bandwidth is

$$\text{bandwidth} = \frac{\omega_0}{Q} = \frac{500 \times 10^3}{20.4} = 24.5\,\text{krad/s}$$

Mutually coupled tuned circuits

Consider two mutually coupled, identical circuits (Fig. 6.21) (see Chapter 5 for a discussion of coupled circuits). Each circuit is separately a series RLC circuit and is tuned to the same frequency. For the primary circuit, Kirchoff's voltage law gives

$$\mathbf{V} = Z\mathbf{I}_p + j\omega M\mathbf{I}_s \qquad [74]$$

and for the secondary circuit

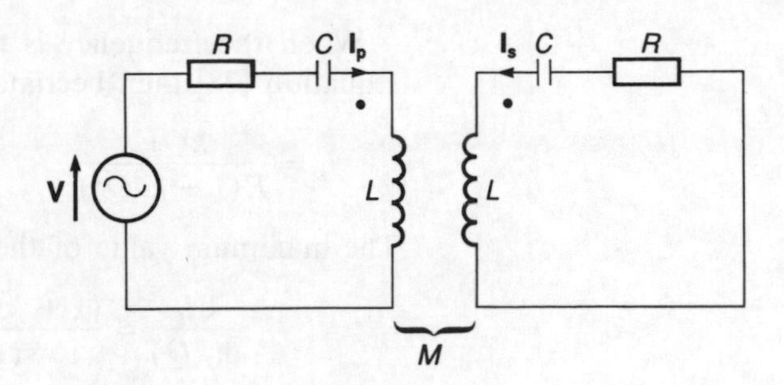

Fig. 6.21 Coupled, identical, resonant circuits

$$0 = Z\mathbf{I}_s + j\omega M\mathbf{I}_p \qquad [75]$$

with

$$Z = R + j\omega L + \frac{1}{j\omega C} = R + j\left(\omega L - \frac{1}{\omega C}\right) \qquad [76]$$

Equation [75] gives $\mathbf{I}_p = -Z\mathbf{I}_s/j\omega M$, hence equation [74] gives

$$\mathbf{V} = -\frac{Z^2\mathbf{I}_s}{j\omega M} + j\omega M\mathbf{I}_s = \frac{-Z^2\mathbf{I}_s - \omega^2 M^2\mathbf{I}_s}{j\omega M}$$

$$\mathbf{I}_s = \frac{-j\omega M\mathbf{V}}{Z^2 + \omega^2 M^2}$$

But $M = k\sqrt{L_1 L_2}$ and since $L_1 = L_2 = L$ then $M = kL$, with k being the coefficient of coupling. Thus

$$\mathbf{I}_s = \frac{-j\omega kL\mathbf{V}}{Z^2 + \omega^2 k^2 L^2} \qquad [77]$$

The resonant frequency ω_0 is $1/\sqrt{LC}$ and since the Q-factor is $\omega_0 L/R$ then $Q = 1/(\omega_0 CR)$. Hence the impedance Z, equation [76] can be written as

$$Z = R + j\left(\frac{RQ\omega}{\omega_0} - \frac{Q\omega_0 R}{\omega}\right)$$

Writing x for ω/ω_0 and substituting for Z in equation [77],

$$\mathbf{Is} = \frac{-j\omega kL\mathbf{V}}{[R + jRQ(x - 1/x)]^2 + \omega^2 k^2 L^2}$$

$$= -\frac{j}{R}\left[\frac{kQx}{\{1 + jQ(x - 1/x)\}^2 + k^2Q^2x^2}\right]\mathbf{V}$$

$$= -\frac{j}{R}\left[\frac{kQx}{1 - Q^2(x - 1/x)^2 + k^2Q^2x^2 + j2Q(x - 1/x)}\right]\mathbf{V}$$

$$= -\frac{j}{R}\left[\frac{kQx\{1 - Q^2(x - 1/x)^2 + kQ^2x^2 - j2Q(x - 1/x)\}}{\{1 - Q^2(x - 1/x)^2 + k^2Q^2x^2\}^2 + \{2Q(x - 1/x)\}^2}\right]\mathbf{V}$$

The magnitude of the secondary current is thus

$$I_s = \frac{kQxV}{R\sqrt{[\{1 - Q^2(x - 1/x)^2 + k^2Q^2x^2\}^2 + \{2Q(x - 1/x)\}^2}} \qquad [78]$$

When the frequency is the resonant frequency then $x = 1$. Equation [78] then becomes

$$I_s = \frac{kQV}{R(1 + k^2Q^2)}$$

The maximum value of the secondary current is given by

$$\frac{dI_s}{d(kQ)} = \frac{(1 + k^2Q^2)R - kQ\,2kQR}{(1 + k^2Q^2)^2R^2} = 0$$

$$(1 + k^2Q^2)R = kQ\,2kQR$$

$$k^2Q^2 = 1 \qquad [79]$$

This condition is known as *critical coupling*. Thus, for example, if $Q = 10$ then the maximum secondary current is produced with a coupling coefficient of 0.1. For $kQ < 1$ the circuits are said to be *undercoupled* and for $kQ > 1$ *overcoupled*.

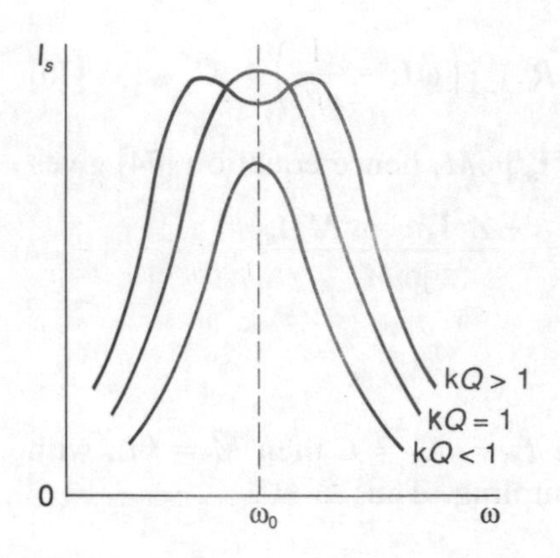

Fig. 6.22 Frequency response curves for coupled circuits

Figure 6.22 shows how the secondary current varies with frequency for different values of kQ. At the critical coupling the secondary current shows a single maximum at the resonant frequency. With undercoupling there is also only one maximum at the resonance frequency. This is a reversion towards the response of a single series resonant circuit as the coupling between the circuits becomes looser. The looser coupling results in smaller currents in the secondary but a circuit which is more highly selective. With overcoupling there are two resonant current peaks.

For the overcoupled situation when we have $kQ > 1$, equation [78],

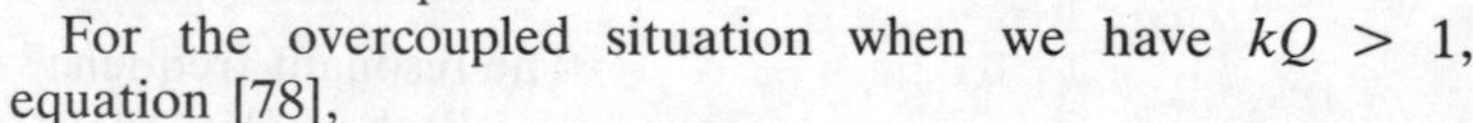

$$I_s = \frac{kQxV}{R\surd]\{1 - Q^2(x - 1/x)^2 + k^2Q^2x^2\}^2 + \{2Q(x - 1/x)\}^2]}$$

gives two maxima, these being when the $\{1 - Q^2(x - 1/x)^2 + k^2Q^2x^2\}^2$ term is zero. Since this is a quadratic in x there are two values of x which will make it zero and hence two values of x which will give maximum values for I_s.

$$1 - Q^2(x - 1/x)^2 + k^2Q^2x^2 = 0$$

$$\frac{1}{Q^2} - (x - 1/x)^2 + k^2x^2 = 0$$

For Q greater than about 10, the term $1/Q^2$ is insignificant and so

$$(x - 1/x)^2 = k^2x^2$$

$$x(1 \pm k) = 1/x$$

$$x^2 = \frac{1}{1 \pm k}$$

Thus

$$\omega = \frac{\omega_0}{\sqrt{(1 \pm k)}} \qquad [80]$$

Thus the frequency separation between the two current peaks is

$$\omega_2 - \omega_1 = \frac{\omega_0}{\sqrt{(1-k)}} - \frac{\omega_0}{\sqrt{(1+k)}} \qquad [81]$$

An approximate value for this frequency separation can be obtained as follows.

$$\omega_2^2 - \omega_1^2 = \frac{\omega_0^2}{1-k} - \frac{\omega_0^2}{1+k}$$

$$= \frac{\omega_0^2[(1+k)-(1-k)]}{(1-k)(1+k)}$$

$$= \frac{2\omega_0^2 k}{1-k^2}$$

Thus since k is generally much less than 1 we can neglect k^2 and so

$$\omega_2^2 - \omega_1^2 \approx 2\omega_0^2 k$$

$$(\omega_2 - \omega_1)(\omega_2 + \omega_1) \approx 2\omega_0^2 k$$

Since the peaks are symmetrical about ω_0 then $\omega_0 = \frac{1}{2}(\omega_1 + \omega_2)$, thus

$$\omega_2 - \omega_1 \approx \omega_0 k \qquad [82]$$

For coupling with kQ having values between 1.0 and about 1.5 the twin peaks give only a shallow minima between them and can be considered to give almost a flat-topped response (Fig. 6.23). This enables an appreciable bandwidth to be obtained with reasonable selectivity, i.e. a bandpass characteristic. The 3 dB bandwidth is

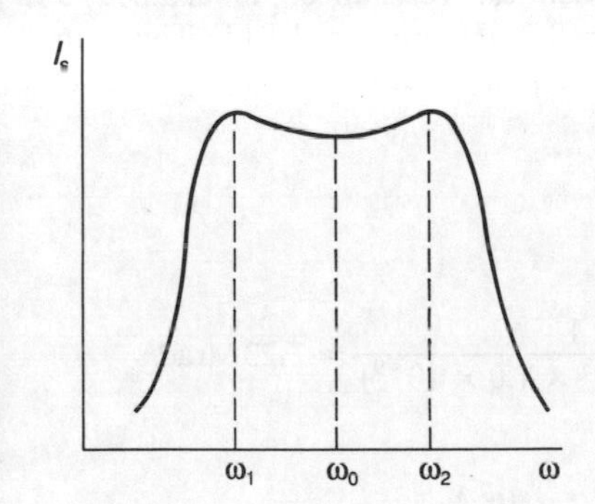

Fig. 6.23 A bandpass characteristic

$$3\text{ dB bandwidth} = (\omega_2 - \omega_1)\sqrt{2} \qquad [83]$$

For kQ values close to 1 then $k \approx 1/Q$ and so equation [82] gives

$$\omega_2 - \omega_1 \approx \frac{\omega_0}{Q}$$

and so equation [83] becomes

$$3\text{ dB bandwidth} \approx \frac{\omega_0\sqrt{2}}{Q} \qquad [84]$$

This is a greater bandwidth than a single resonant frequency having the same Q-factor and is a more flat-topped response.

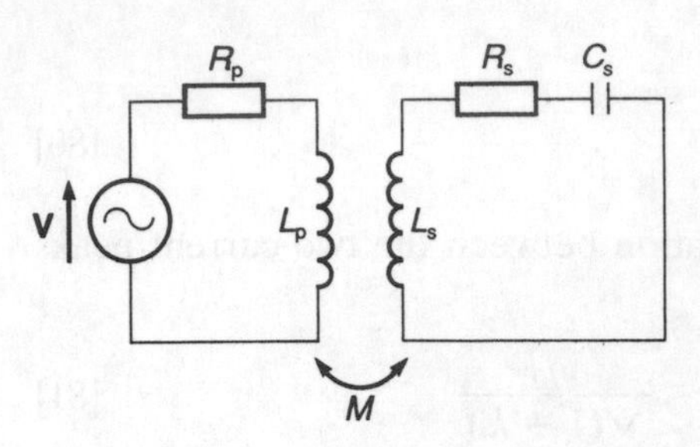

Fig. 6.24 Example 17

Example 17

Derive an expression for the secondary current $\mathbf{I}_s$ in terms of the voltage input to the primary $\mathbf{V}$ and the mutual inductance for the circuit shown in Fig. 6.24, where the primary is an untuned circuit and the secondary a tuned circuit.

Answer

For the primary circuit, Kirchoff's voltage law gives

$$\mathbf{V} = R_p\mathbf{I}_p + j\omega M\mathbf{I}_s \qquad [85]$$

and for the secondary circuit

$$0 = Z_s\mathbf{I}_s + j\omega M\mathbf{I}_p \qquad [86]$$

with

$$Z_s = R + j\omega L + \frac{1}{j\omega C} = R + j\left(\omega L - \frac{1}{\omega C}\right) \qquad [87]$$

Equation [86] gives $\mathbf{I}_p = -Z_s\mathbf{I}_s/j\omega M$, hence equation [85] gives

$$\mathbf{V} = -\frac{R_pZ_s\mathbf{I}_s}{j\omega M} + j\omega M\,\mathbf{Is} = \frac{-R_pZ_s\mathbf{I}_s - \omega^2M^2\mathbf{I}_s}{j\omega M}$$

$$\mathbf{Is} = \frac{-j\omega M\mathbf{V}}{R_pZ_s + \omega^2M^2}$$

Example 18

Two coupled circuits have identical resistance, inductance and capacitance, with $R = 20\,\Omega$, $L = 0.2\,\text{mH}$ and $C = 1.0\,\text{nF}$. The mutual inductance is $25\,\mu\text{H}$. Will the response curve be single or double humped?

Answer

The resonant frequency ω_0 is

$$\omega_0 = \frac{1}{\sqrt{LC}} = \frac{1}{\sqrt{(0.2 \times 10^{-3} \times 1.0 \times 10^{-9})}} = 2.23\,\text{Mrad/s}$$

The circuits have Q-factors of

$$Q = \frac{\omega_0 L}{R} = \frac{2.23 \times 10^6 \times 0.2 \times 10^{-3}}{20} = 22.3$$

Since $M = k\sqrt{L_pL_s} = kL$, then

$$k = \frac{25 \times 10^{-6}}{0.2 \times 10^{-3}} = 0.125$$

Thus kQ has the value $0.125 \times 22.3 = 2.8$. Since this is greater than one, the circuit is overcoupled and so the response will be double-humped.

Example 19

Two identical coupled circuits have inductances of 0.1 mH and are each tuned to give a resonant frequency of 1 Mrad/s by series connected capacitors in each circuit. The 3 dB bandwidth of the secondary current is 20 krad/s. What is (*a*) the *Q*-factor, (*b*) the coefficient of coupling for critical coupling, (*c*) the separation between the secondary current peaks when the coupling coefficient is 0.02?

Answer

(*a*) Equation [84] gives

$$3\,\text{dB bandwidth} \approx \frac{\omega_0\sqrt{2}}{Q}$$

Hence

$$Q = \frac{1 \times 10^6 \times \sqrt{2}}{20 \times 10^3} = 70.7$$

(*b*) For critical coupling $kQ = 1$, hence

$$k = \frac{1}{Q} = \frac{1}{70.7} = 0.014$$

(*c*) The separation is given by equation [82] as

$$\omega_2 - \omega_1 \approx \omega_0 k \approx 1 \times 10^6 \times 0.02 \approx 20\,\text{krad/s}$$

Problems

1 A circuit consists of a variable capacitor in series with an inductor of resistance 10 Ω and inductance 40 mH. What will be the value of the capacitance required if the circuit is to have a resonance frequency of 1 kHz?

2 A circuit consists of a 20 μF capacitor in series with an inductor of resistance 8 Ω and inductance 100 mH and a sinusoidal voltage supply of 10 V r.m.s.. What is (*a*) the resonant frequency, (*b*) the current at resonance and (*c*) the potential difference across the capacitor at resonance?

3 An inductor of resistance 10 Ω and inductance 100 mH is in series with a 20 Ω resistance and this arrangement is in parallel with a 40 μF capacitor. What will be (*a*) the resonant frequency of the circuit and (*b*) the potential difference across the capacitor at resonance if the circuit is supplied by a sinusoidal current of 100 mA r.m.s.?

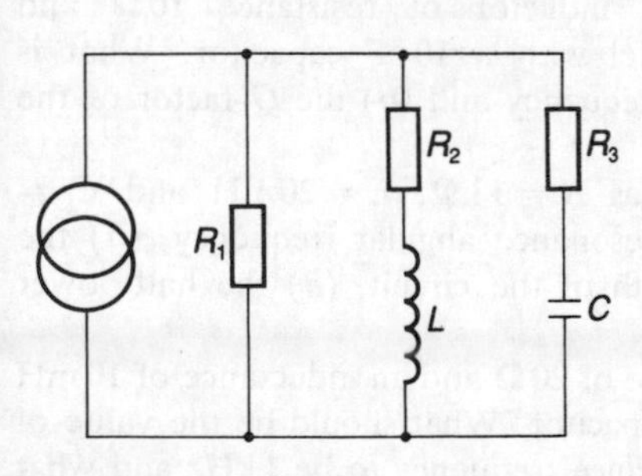

Fig. 6.25 Problem 4

4 Determine the resonant frequency for the circuit in Fig. 6.25.

5 What is the resonant frequency for a network consisting of an inductor of resistance 4 Ω and inductance 10 mH in parallel with a 10 μF capacitor and a 5 Ω resistor?

6 An inductor of resistance 8 Ω and inductance 0.10 H is connected in series with a 0.05 μF capacitor. What is (*a*) the resonant frequency, (*b*) the *Q*-factor, (*c*) the −3 dB frequencies, (*d*) the bandwidth, (*e*) the impedance at the resonant frequency, (*f*) the impedance at the −3 dB frequencies?

7 A series *RLC* circuit is to be designed to have a resonant frequency of 2 kHz, with a *Q*-factor of 40 and an impedance of 20 Ω at the resonant frequency. What are the values of the inductance and capacitance and the resulting bandwidth of the circuit?

8 A series *RLC* circuit when connected to a 2 V sinusoidal supply has a maximum current of 40 mA at a frequency of 2 kHz and a bandwidth of 200 Hz. (*a*) What are the values of the capacitance and inductance? (*b*) What is the potential difference across the capacitor?

9 A series *RLC* circuit is designed to have a resonant angular frequency of 5 Mrad/s and a bandwidth of 1 Mrad/s. What will be the resistance and the inductance of the circuit if the capacitance is 0.01 μF?

10 A series *RLC* circuit has a resistance of 10 Ω, an inductance of 200 μH and a capacitance of 5 nF. What is (*a*) the resonant frequency, (*b*) the *Q*-factor, (*c*) the bandwidth?

11 A variable capacitor is in series with an inductor having an inductance of 10 mH and a resistance of 5 Ω. The circuit is designed to operate as a filter with a resonant frequency of 10 kHz. What is (*a*) the capacitance, (*b*) the bandwidth of the filter?

12 A series *RLC* circuit has a sinusoidal voltage input of maximum value 10 V. The resistance is 50 Ω, the inductance 5 mH and the capacitance 0.5 μF. What is (*a*) the resonance angular frequency, (*b*) the value of the voltage across the capacitor at the resonance frequency, (*c*) the angular frequency at which the voltage across the capacitor is a maximum, and (*d*) the value of this maximum voltage?

13 A series *RLC* circuit has $R = 8\,\Omega$, $L = 0.5\,\text{mH}$ and $C = 500\,\text{pF}$. Calculate the current through the circuit when the input voltage is 0.5 V and the frequency is (*a*) the resonant frequency and (*b*) a frequency 2% below the resonant frequency?

14 A circuit consists of a 10 Ω resistance in parallel with an inductor of 1 mH in parallel with a 1 nF capacitor. What is (*a*) the resonant angular frequency and (*b*) the *Q*-factor of the circuit?

15 A circuit consists of an inductor of resistance 10 Ω and inductance 5 mH in parallel with a 10 nF capacitor. What is (*a*) the resonant angular frequency and (*b*) the *Q*-factor of the circuit?

16 A parallel *RLC* circuit has $R = 1\,\text{k}\Omega$, $L = 20\,\text{mH}$ and $C = 0.50\,\mu\text{F}$. What is (*a*) the resonance angular frequency, (*b*) the *Q*-factor, (*c*) the bandwidth of the circuit, (*d*) the half-power frequencies?

17 An inductor has a resistance of 20 Ω and an inductance of 10 mH and is in parallel with a capacitor. What should be the value of the capacitor for the resonance frequency to be 1 kHz and what are the *Q*-factor and bandwidth that then result?

18 A circuit consists of an inductor, having both resistance and inductance, in parallel with a 0.2 nF capacitor. The circuit has a bandwidth of 30 kHz and a resonant frequency of 5 MHz. What is (*a*) the *Q*-factor, (*b*) the dynamic resistance, and (*c*) the

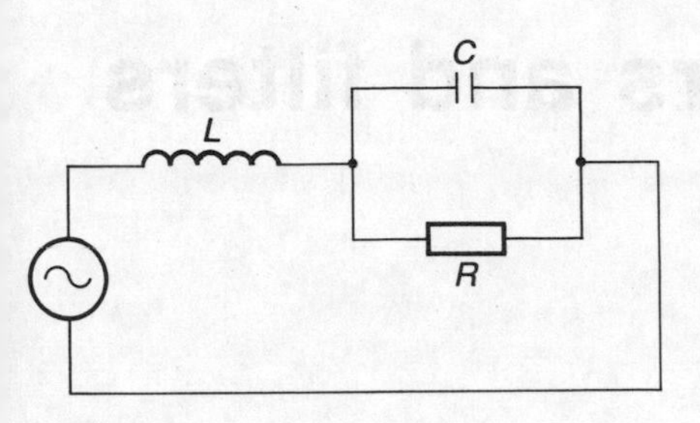

Fig. 6.26 Problem 19

impedance when the applied frequency is 0.1% off the tuned value?

19 Derive equations for the resonance frequency and the Q-factor for the circuit in Fig. 6.26.

20 A series RLC circuit has $R = 50\,\Omega$, $L = 50\,\text{mH}$ and $C = 0.2\,\mu\text{F}$ and a sinusoidal voltage input of 10 V. What is the potential difference across the capacitor when the circuit is (*a*) unloaded and (*b*) measured by a voltmeter of resistance 200 kΩ placed across it?

21 A parallel inductor–capacitor circuit is to be used as the tuned load of an amplifier stage which has an output resistance of 20 kΩ. The inductor has an inductance of 0.1 mH and a resistance of 10 Ω and the capacitance is 0.5 nF. What is the amplifier bandwidth?

22 A parallel inductor–capacitor circuit has an inductor with an inductance of 25 mH and resistance 100 Ω and a 10 nF capacitor. If the circuit is supplied by a sinusoidal current source of 5 mA and internal resistance 10 kΩ, what is (*a*) the resonance frequency of the circuit, and (*b*) the bandwidth?

23 A parallel inductor–capacitor circuit is to be used as the load of an amplifier. The amplifier has an output impedance equivalent to a capacitance of 500 pF and a resistance of 1 kΩ. The resulting tuned amplifier is to have a resonance frequency of 500 krad/s. If the inductor has an inductance of 2 mH and a resistance of 10 Ω what should be the value of the capacitance used and the resulting bandwidth?

24 Two coupled circuits have identical resistance, inductance and capacitance, with $R = 30\,\Omega$, $L = 0.1\,\text{mH}$ and $C = 2.0\,\text{nF}$. The mutual inductance is 10 μH. Will the response curve be single or double humped?

25 Two identical coupled circuits have $L = 0.3\,\text{mH}$ and are each tuned to give a resonance frequency of 2 Mrad/s by series connected capacitors in each circuit. The 3 dB bandwidth of the secondary current is 50 krad/s. What is (*a*) coefficient of coupling for critical coupling, (*b*) the separation between the secondary current peaks when the coupling coefficient is 0.04?

26 Two identical coupled circuits have each a coil of inductance 1 mH and resistance 6 Ω connected in series with a capacitor and are each tuned to give a resonance frequency of 100 kHz. What is (*a*) the coefficient of coupling needed to give a secondary current with a bandwidth of 5 kHz, (*b*) the coefficient of coupling to give critical coupling?

7 Attenuators and filters

Introduction

Attenuators are networks which reduce the magnitude of an electrical signal to a fraction which ideally is constant irrespective of frequency. A fixed attenuation network is called a *pad*. *Filters* are networks which attenuate certain ranges of frequency and pass others without loss. They can be considered to be attenuators which have an attenuation which depends on the frequency of the input signal. If the network contains only components such as resistors, inductors and capacitors then it is said to be *passive*, if it contains a source of e.m.f. then it is referred to as *active*. All such networks are examples of what are termed *two-port networks*, though this term does include more than attenuators and filters. Chapter 8 is a more general consideration of two-port networks, this chapter being only a consideration of them as passive attenuators and filters.

Attenuators

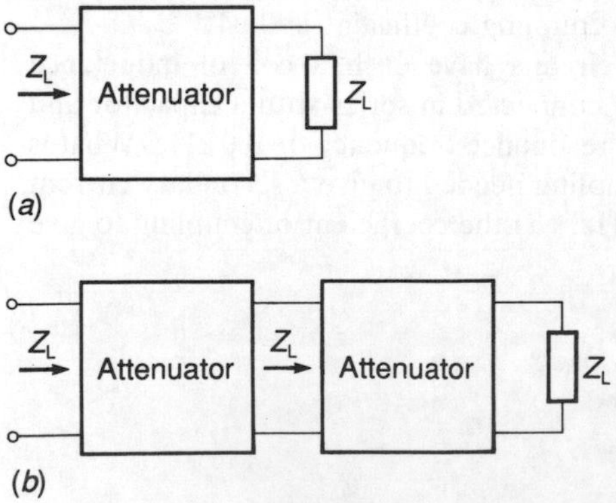

An attenuator has generally to provide not only the required amount of attenuation but also, when inserted between a source and a load, it must not change the loading of the source (Fig. 7.1(*a*)). This has the consequence that, whatever the impedance of the source, the input to the attenuator is the same as the input that would have occurred to the load prior to the insertion of the attenuator. This characteristic means that the input impedance of the attenuator, when connected across the load, is the same as that of the load and thus enables a sequence of identical attenuator networks to be connected together to give a compound attenuation and still present the same input impedance (Fig. 7.1(*b*)).

Fig. 7.1 Input impedance equals load impedance, (*a*) single attenuator network, (*b*) sequence of identical attenuator networks

Most attenuators are made up of either repeated T or repeated π configurations of impedances (Fig. 7.2). Such configurations may be *symmetrical* or *asymmetrical*. Thus the T-network in Fig. 7.2(*a*) is symmetrical because the two series impedances are the same; with the asymmetrical form this would not be the case. The π-network in Fig. 7.2(*b*) is symmetrical because both the shunt impedances are the same. Combinations of T-networks or π-networks lead to what is termed a *ladder network* (Fig. 7.2(*c*)). The T and the π sections can be considered to be just different ways of subdividing a ladder network. It is to enable this identity between T and π networks to be maintained that the impedances indicated in Fig. 7.2(*a*) and (*b*) are in the form they are.

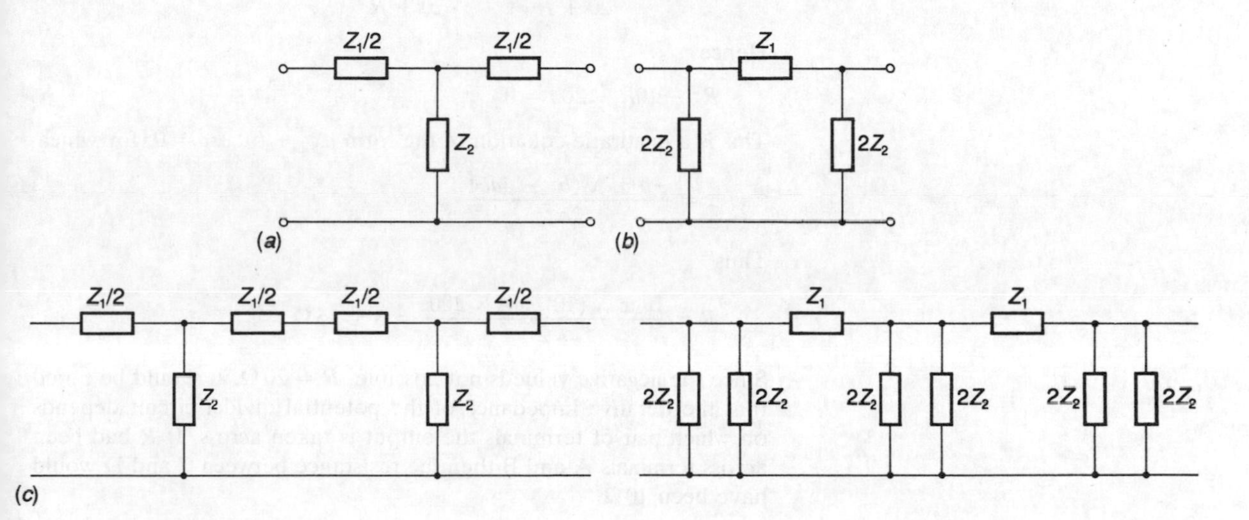

Fig. 7.2 Networks, (*a*) T, (*b*) π, (*c*) ladder

Characteristic impedance

The *input impedance* of a two-port network is the ratio of the voltage to current at the input terminals. The term *iterative impedance* is used for the impedance which if connected across one pair of the two-port network terminals leads to the same value of impedance across the other. Thus a network has an iterative impedance of, say, 100 Ω if when an impedance of 100 Ω is connected across a pair of its terminals the impedance across the other pair becomes 100 Ω.

For a symmetrical network it does not matter across which pair of terminals the impedance is connected, the same impedance appears across the other. For an asymmetrical network this is not the case; the value of the impedance which is an iterative impedance depends on which pair of terminals is taken as the input and which the output. For a symmetrical two-port network there is thus just one value of iterative impedance, this value being called the *characteristic impedance*. For an asymmetrical network there are two values of iterative impedance (see later this chapter).

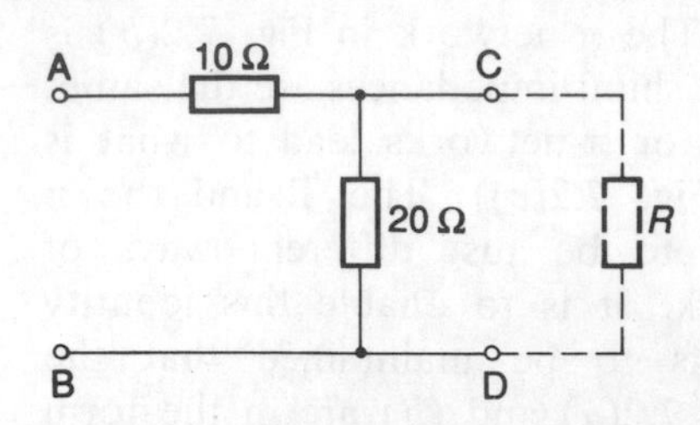

Fig. 7.3 Example 1

Example 1

For the simple potential divider circuit shown in Fig. 7.3, what is the iterative impedance R?

Answer

The resistance between the terminals A and B must be R. But this is the sum of the 10 Ω resistance and the parallel arrangement of R and the 20 Ω. Thus

$$R = 10 + \frac{20R}{20 + R} = \frac{200 + 10R + 20R}{20 + R}$$

Hence

$$R^2 - 10R - 200 = 0$$

This is a quadratic equation of the form $ax^2 + bx + c = 0$, for which

$$x = \frac{-b \pm \sqrt{(b^2 - 4ac)}}{2a}$$

Thus

$$R = \frac{10 \pm \sqrt{(10^2 + 4 \times 200)}}{2} = 5 \pm 15\,\Omega$$

Since the negative value is not possible, $R = 20\,\Omega$. It should be noted that the iterative impedance of the potential divider circuit depends on which pair of terminals the output is taken across. If R had been across terminals A and B then the resistance between C and D would have been 10 Ω.

Attenuation

The attenuation produced by a network can be described in a number of ways. One way is in terms of the ratio of the input and output powers (Fig. 7.4),

$$\text{attenuation} = \frac{P_1}{P_2}$$

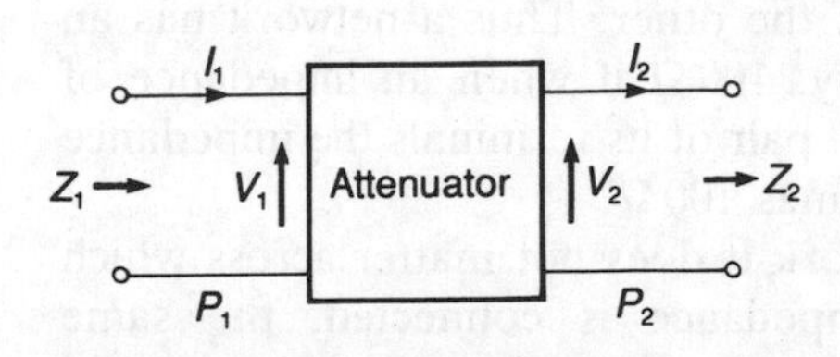

Fig. 7.4 Input and output to an attenuator

This ratio is usually expressed in logarithmic form. Thus, with logarithms to base 10, written usually as lg,

$$\text{attenuation in bels} = \lg\left(\frac{P_1}{P_2}\right)$$

The units for this power ratio are bels. Since the bel is a large unit, the decibel (dB) is generally used,

$$\text{attenuation in dB} = 10\lg\left(\frac{P_1}{P_2}\right) \qquad [1]$$

Since the input power $P_1 = V_1^2/Z_1$ and the output power $P_2 = V_2^2/Z_2$, where Z_1 is the input impedance and Z_2 the output impedance, then equation [1] can be written as

$$\text{attenuation in dB} = 10\lg\left(\frac{V_1^2/Z_1}{V_2^2/Z_2}\right)$$

For a symmetrical attenuator $Z_1 = Z_2$, and so

$$\text{attenuation in dB} = 20\lg\left(\frac{V_1}{V_2}\right) \quad [2]$$

Since the input power $P_1 = I_1^2 Z_1$ and the output power $P_2 = I_2^2 Z_0$, then the equation can also be written, for a symmetrical attenuator, as

$$\text{attenuation in dB} = 20\lg\left(\frac{I_1}{I_2}\right) \quad [3]$$

An alternative way of specifying the attenuation is in terms of logarithms to base e, indicated by ln. When these are used then the attenuation is in units of nepers (Np). This form of definition is discussed in more detail later in this chapter. The definition used is

$$\text{attenuation in Np } \alpha = \ln\left(\frac{I_1}{I_2}\right) \quad [4]$$

where I_1 and I_2 are the magnitudes of the currents. This equation can also be written in the form

$$e^{\alpha} = \frac{I_1}{I_2} \quad [5]$$

Since $P_1 = I_1^2/Z_1$ and $P_2 = I_2^2/Z_2$ then for a symmetrical attenuator when $Z_1 = Z_2$ the attenuation in nepers can be written as

$$\text{attenuation in Np } \alpha = \ln\left(\frac{P_1 Z_1}{P_2 Z_2}\right)^{\frac{1}{2}}$$

$$\text{attenuation in Np } \alpha = \tfrac{1}{2}\ln\left(\frac{P_1}{P_2}\right) \quad [6]$$

Alternatively, since $I_1 = V_1/Z_1$ and $I_2 = V_2/Z_2$ then with a symmetrical attenuator when $Z_1 = Z_2$ equation [4] can be written as

$$\text{attenuation in Np } \alpha = \ln\left(\frac{V_1}{V_2}\right) \quad [7]$$

The attenuation in dB, as specified by equation [1], can be related to the attenuation in Np by using equation [5].

$$\text{attenuation in dB} = 20\lg\left(\frac{I_1}{I_2}\right) = 20\lg e^{\alpha}$$

$$= 20\alpha \lg e = 20\alpha \times 0.4343$$

Thus

$$\text{attenuation in dB} = 8.686 \times \text{attenuation in Np} \qquad [8]$$

In all the above definitions the attenuation has been described in terms of the ratio of input to output powers, voltages or currents. When the output is smaller than the input this leads to positive values for the attenuation when expressed in dB or Np. Sometimes, however, attenuation is described in terms of the ratio of the output to input. This leads to the same numerical value in dB or Np, but the values are negative.

Example 2

If the ratio of input power to output power for an attenuator is 5, what is the attenuation in (*a*) dB, (*b*) Np?

Answer

(*a*) Using equation [1]

$$\text{attenuation} = 10 \lg \left(\frac{P_1}{P_2}\right) = 10 \lg 5 = 7.0\,\text{dB}$$

Using equation [6]

$$\text{attenuation in } \tfrac{1}{2} \ln \left(\frac{P_1}{P_2}\right) = \tfrac{1}{2} \ln 5 = 0.80\,\text{Np}$$

Example 3

A symmetrical attenuator has a current input of 100 mA. What will be the current output if the attenuation is 10 dB?

Answer

Using equation [3],

$$\text{attenuation in dB} = 20 \lg \left(\frac{I_1}{I_2}\right)$$

$$10 = 20 \lg (100/I_2)$$

$$10^{0.5} = 100/I_2$$

$$I_2 = 31.6\,\text{mA}$$

The propagation coefficient

The propagation coefficient γ is defined as

$$\gamma = \ln \left(\frac{\mathbf{I}_1}{\mathbf{I}_2}\right) \qquad [9]$$

where the currents are both phasors having both a magnitude and phase. This equation can also be written in the form

$$e^{\gamma} = \frac{\mathbf{I}_1}{\mathbf{I}_2} \qquad [10]$$

In general, since both the currents can be complex quantities then γ can be a complex quantity. It can thus be represented by

$$\gamma = \alpha + j\beta \qquad [11]$$

where α is called the *attenuation coefficient* and is

$$\alpha = \ln \left| \frac{I_1}{I_2} \right|$$

i.e. the definition introduced earlier as equation [4]. β is the *phase change coefficient* and gives the phase difference between $\mathbf{I}_1$ and $\mathbf{I}_2$.

In polar form equation [9] thus becomes

$$\gamma = \ln \left(\left| \frac{I_1}{I_2} \right| \angle \beta \right) \qquad [12]$$

Attenuators in cascade

Consider a number of attenuators connected in cascade, i.e. the output from the first attenuator becomes the input to the second and its output becomes the input to the third attenuator, and so on (Fig. 7.5). Assuming the attenuators are symmetrical and that each has the same characteristic impedance Z_0 and the final attenuator has a load Z_0, then the attenuation of the first attenuator is, by equation [2],

$$\text{attenuation in dB} = 20 \lg \left(\frac{V_1}{V_2} \right)$$

and the attenuation of the second attenuator is

$$\text{attenuation in dB} = 20 \lg \left(\frac{V_2}{V_3} \right)$$

and the third, assuming there to be just three, has an attenuation

$$\text{attenuation in dB} = 20 \lg \left(\frac{V_3}{V_4} \right)$$

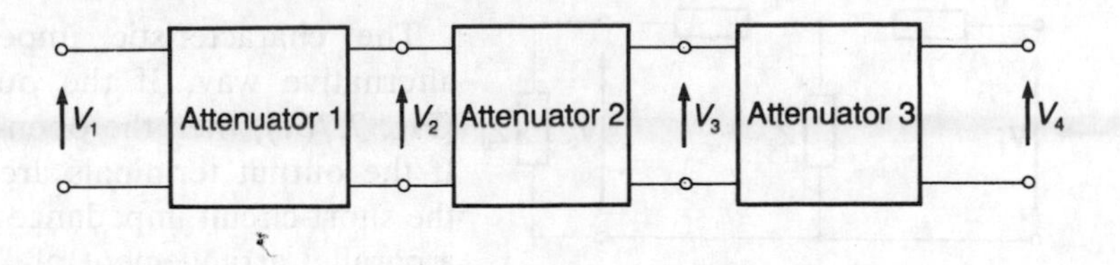

Fig. 7.5 Attenuators in cascade

The overall attenuation is

$$\text{attenuation in dB} = 20 \lg \left(\frac{V_1}{V_4} \right)$$

and this can be written as

$$\text{attenuation in dB} = 20\lg\left(\frac{V_1}{V_2} \times \frac{V_2}{V_3} \times \frac{V_3}{V_4}\right)$$

$$= 20\lg\left(\frac{V_1}{V_2}\right) + 20\lg\left(\frac{V_2}{V_3}\right)$$

$$+ 20\lg\left(\frac{V_3}{V_4}\right) \qquad [13]$$

Thus the total attenuation is the sum of the attenuations, in dB, of the separate attenuators.

Example 4

Three identical attenuator sections are connected in cascade. If each section has an attenuation of 10 dB, what is the overall attenuation?

Answer

The total attenuation is the sum of the attenuations of the separate attenuator sections and is thus 3 × 10 dB = 30 dB.

Symmetrical T-section attenuator

Consider a symmetrical T-section attenuator with a characteristic impedance of Z_0, as in Fig. 7.6. The impedance looking into the input port is thus Z_0. Then Z_0 must be equal to the impedance due to $\frac{1}{2}Z_1$ in series with a parallel arrangement of Z_2 with $(\frac{1}{2}Z_1 + Z_0)$, i.e.

$$Z_0 = \tfrac{1}{2}Z_1 + \frac{Z_2(\frac{1}{2}Z_1 + Z_0)}{Z_0 + \frac{1}{2}Z_1 + Z_2}$$

$$= \frac{\frac{1}{2}Z_1(Z_0 + \frac{1}{2}Z_1 + Z_2) + Z_2(\frac{1}{2}Z_1 + Z_0)}{Z_0 + \frac{1}{2}Z_1 + Z_2}$$

$$Z_0(Z_0 + \tfrac{1}{2}Z_1 + Z_2) = \tfrac{1}{2}Z_1(Z_0 + \tfrac{1}{2}Z_1 + Z_2) + Z_2(\tfrac{1}{2}Z_1 + Z_0)$$

$$Z_0^2 = (\tfrac{1}{2}Z_1)^2 + Z_1Z_2$$

$$Z_0 = \sqrt{(\tfrac{1}{4}Z_1^2 + Z_1Z_2)} \qquad [14]$$

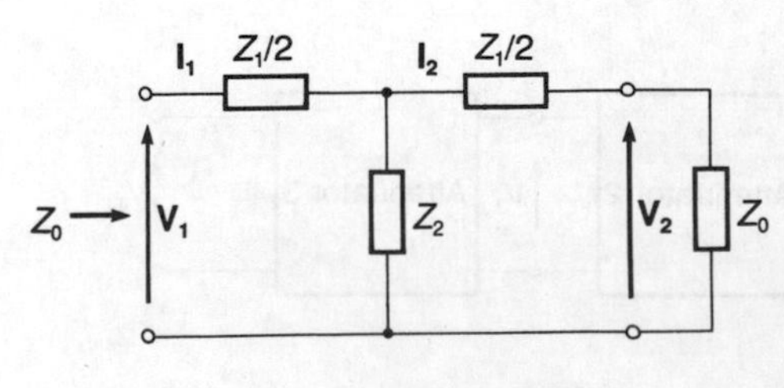

Fig. 7.6 Symmetrical T-section

The characteristic impedance can be expressed in an alternative way. If the output terminals are open-circuited (Fig. 7.7(*a*)) then the open-circuit impedance Z_{oc} is $\frac{1}{2}Z_1 + Z_2$. If the output terminals are short-circuited (Fig. 7.7(*b*)) then the short-circuit impedance Z_{sc} is that due to $\frac{1}{2}Z_1$ in series with a parallel arrangement of $\frac{1}{2}Z_1$ and Z_2, i.e.

$$Z_{sc} = \tfrac{1}{2}Z_1 + \frac{\frac{1}{2}Z_1Z_2}{\frac{1}{2}Z_1 + Z_2} = \frac{\frac{1}{4}Z_1^2 + \frac{1}{2}Z_1Z_2 + \frac{1}{2}Z_1Z_2}{\frac{1}{2}Z_1 + Z_2}$$

$$Z_{sc}(\tfrac{1}{2}Z_1 + Z_2) = \tfrac{1}{4}Z_1^2 + Z_1Z_2$$

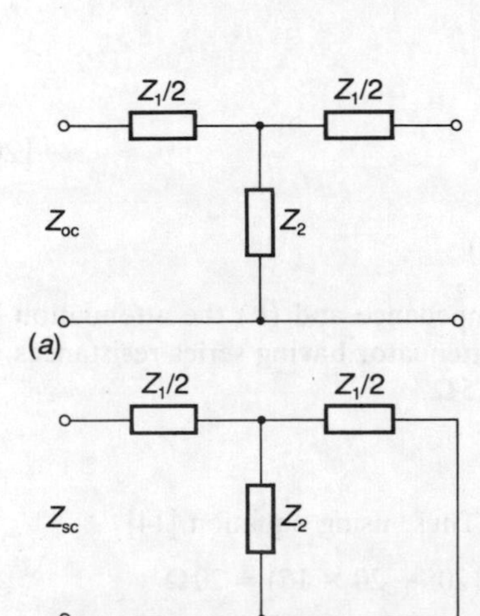

Fig. 7.7 Symmetrical T-section, (*a*) Z_{oc}, (*b*) Z_{sc}

But $Z_{oc} = \frac{1}{2}Z_1 + Z_2$, hence

$$Z_{sc}Z_{oc} = \tfrac{1}{4}Z_1^2 + Z_1Z_2$$

Hence, using equation [14],

$$Z_0 = \sqrt{(Z_{sc}Z_{oc})} \qquad [15]$$

A problem that is often encountered is as follows: given the characteristic impedance and attenuation required, what are the values of Z_1 and Z_2? Consider the T-section circuit. For the source side of the T-section Kirchoff's voltage law gives

$$\mathbf{V}_1 - \mathbf{I}_1\tfrac{1}{2}Z_1 - (\mathbf{I}_1 - \mathbf{I}_2)Z_2 = 0$$

and for the load side of the T-section

$$\mathbf{I}_2Z_0 + \mathbf{I}_2\tfrac{1}{2}Z_1 + (\mathbf{I}_2 - \mathbf{I}_1)Z_2 = 0$$

Subtracting the two equations eliminates Z_2 and gives

$$\mathbf{V}_1 - \mathbf{I}_1\tfrac{1}{2}Z_1 - \mathbf{I}_2Z_0 - \mathbf{I}_2\tfrac{1}{2}Z_1 - 0$$

But $\mathbf{V}_1 = \mathbf{I}_1Z_0$ and so

$$\mathbf{I}_1Z_0 - \mathbf{I}_1\tfrac{1}{2}Z_1 = \mathbf{I}_2Z_0 + \mathbf{I}_2\tfrac{1}{2}Z_1$$

$$\frac{\mathbf{I}_1}{\mathbf{I}_2} = \frac{Z_0 + \frac{1}{2}Z_1}{Z_0 - \frac{1}{2}Z_1} \qquad [16]$$

Alternatively, since $\mathbf{V}_1 = \mathbf{I}_1Z_0$ and $\mathbf{V}_2 = \mathbf{I}_2Z_0$ we could write

$$\frac{\mathbf{V}_1}{\mathbf{V}_2} = \frac{Z_0 + \frac{1}{2}Z_1}{Z_0 - \frac{1}{2}Z_1} \qquad [17]$$

If we denote this ratio $\mathbf{I}_1/\mathbf{I}_2$ or $\mathbf{V}_1/\mathbf{V}_2$ by N then

$$N = \frac{Z_0 + \frac{1}{2}Z_1}{Z_0 - \frac{1}{2}Z_1} \qquad [18]$$

and so

$$NZ_0 - N\tfrac{1}{2}Z_1 = Z_0 + \tfrac{1}{2}Z_1$$

$$Z_1 = 2Z_0\left(\frac{N-1}{N+1}\right) \qquad [19]$$

We can obtain an equation giving Z_2 by using equation [14].

$$Z_0 = \sqrt{(\tfrac{1}{4}Z_1^2 + Z_1Z_2)}$$

$$Z_0^2 = \tfrac{1}{4}Z_1^2 + Z_1Z_2$$

Substituting for Z_1 using equation [19]

$$Z_0^2 = Z_0^2\left(\frac{N-1}{N+1}\right)^2 + 2Z_0\left(\frac{N-1}{N+1}\right)Z_2$$

$$Z_0(N+1)^2 = Z_0(N-1)^2 + 2(N-1)(N+1)Z_2$$

$$Z_0(N^2 + 2N + 1) = Z_0(N^2 - 2N + 1) + 2(N^2 - 1)Z_2$$

Hence

$$Z_2 = Z_0\left(\frac{2N}{N^2 - 1}\right) \qquad [20]$$

Example 5

What is (*a*) the characteristic impedance and (*b*) the attenuation in dB of a symmetrical T-section attenuator having series resistances of 10 Ω and a shunt resistance of 15 Ω?

Answer

(*a*) $\frac{1}{2}Z_1 = 10\,\Omega$ and $Z_2 = 15\,\Omega$. Thus, using equation [14]

$$Z_0 = \sqrt{(\tfrac{1}{4}Z_1^2 + Z_1Z_2)} = \sqrt{(\tfrac{1}{4}20^2 + 20 \times 15)} = 20\,\Omega$$

(*b*) Using equation [18]

$$N = \frac{Z_0 + \frac{1}{2}Z_1}{Z_0 - \frac{1}{2}Z_1} = \frac{20 + 10}{20 - 10} = 3$$

Thus

$$\text{attenuation} = 20\lg N = 20\lg 3 = 9.5\,\text{dB}$$

Example 6

Design a symmetrical T-section attenuator to give an attenuation of 20 dB and have a characteristic impedance of 500 Ω.

Answer

The attenuation in dB is given by

$$\text{attenuation} = 20\lg N = 20\,\text{dB}$$

Hence $N = 10$. Thus equation [19] gives

$$Z_1 = 2Z_0\left(\frac{N-1}{N+1}\right) = 2 \times 500\left(\frac{10-1}{10+1}\right) = 2 \times 409\,\Omega$$

and equation [20] gives

$$Z_2 = Z_0\left(\frac{2N}{N^2-1}\right) = 500\left(\frac{2 \times 10}{10^2 - 1}\right) = 101\,\Omega$$

Thus the resistors are $\frac{1}{2}Z_1 = 409\,\Omega$ and $Z_2 = 101\,\Omega$.

Symmetrical π-section attenuator

Consider a symmetrical π-section attenuator with a characteristic impedance Z_0 (Fig. 7.8). The impedance looking into the input port is thus Z_0. The load Z_0 is in parallel with $2Z_2$, i.e. an impedance of $Z_0 2Z_2/(Z_0 + 2Z_2)$. This is in series with Z_1, to give an impedance of $Z_1 + Z_0 2Z_2/(Z_0 + 2Z_2)$. This is in parallel with $2Z_2$. Thus, since the total impedance is Z_0,

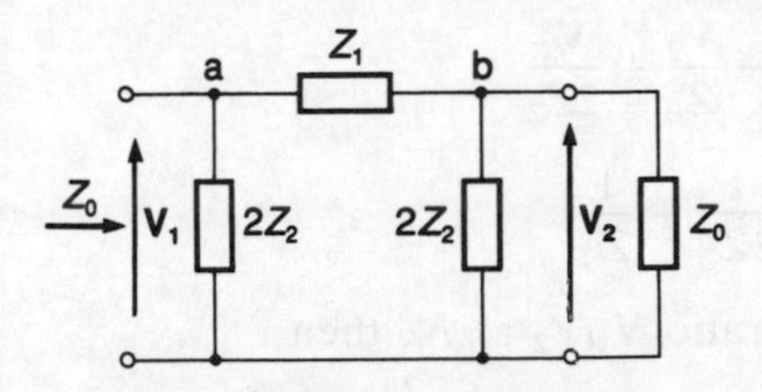

Fig. 7.8 Symmetrical π-section

$$Z_0 = \frac{2Z_2\left(Z_1 + \dfrac{Z_0 2Z_2}{Z_0 + 2Z_2}\right)}{2Z_2 + \left(Z_1 + \dfrac{Z_0 2Z_2}{Z_0 + 2Z_2}\right)}$$

$$= \frac{2Z_2(Z_1Z_0 + Z_1 2Z_2 + Z_0 2Z_2)}{2Z_2Z_0 + 4Z_2^2 + Z_1Z_0 + 2Z_1Z_2 + 2Z_0Z_2}$$

Hence

$$Z_0(4Z_2Z_0 + 4Z_2^2 + Z_1Z_0 + 2Z_1Z_2) = 2Z_0Z_1Z_2 + 4Z_1Z_2^2 + 4Z_0Z_2^2$$

$$Z_0 = \sqrt{\left(\frac{4Z_1Z_2^2}{Z_1 + 4Z_2}\right)} \qquad [21]$$

The characteristic impedance can be expressed in an alternative way. If the output terminals are open-circuited then the open-circuit impedance Z_{oc} is that of $2Z_2$ in parallel with $Z_1 + 2Z_2$. Thus

$$Z_{oc} = \frac{2Z_2(Z_1 + 2Z_2)}{2Z_2 + Z_1 + 2Z_2} = \frac{2Z_2(Z_1 + 2Z_2)}{Z_1 + 4Z_2}$$

If the output terminals are short-circuited then the short-circuit impedance Z_{sc} is that of $2Z_2$ in parallel with Z_1.

$$Z_{sc} = \frac{2Z_2Z_1}{Z_1 + 2Z_2}$$

Thus

$$Z_{oc}Z_{sc} = \frac{4Z_1Z_2^2(Z_1 + 2Z_2)}{(Z_1 + 4Z_2)(Z_1 + 4Z_2)} = \frac{4Z_1Z_2^2}{Z_1 + 4Z_2}$$

Thus the characteristic impedance Z_0, given by equation [21], can be expressed as

$$Z_0 = \sqrt{(Z_{oc}Z_{sc})} \qquad [22]$$

This is the same equation as for the symmetrical T-section (equation [15]).

A problem that is often encountered is that, given the characteristic impedance and attenuation required, what are the values of Z_1 and Z_2? Consider the π-section circuit. Kirchoff's current law gives for node a

$$\mathbf{V}_1/Z_0 = (\mathbf{V}_{ab}/Z_1) + (\mathbf{V}_1/2Z_2)$$

and for node b

$$\mathbf{V}_{ab}/Z_1 = (\mathbf{V}_2/Z_0) + (\mathbf{V}_2/2Z_2)$$

Hence

$$\frac{\mathbf{V}_1}{Z_0} = \frac{\mathbf{V}_2}{2Z_2} + \frac{\mathbf{V}_2}{Z_0} + \frac{\mathbf{V}_1}{2Z_2}$$

$$\mathbf{V}_1\left(\frac{1}{Z_0} - \frac{1}{2Z_2}\right) = \mathbf{V}_2\left(\frac{1}{2Z_2} + \frac{1}{Z_0}\right)$$

If we write the voltage ratio $\mathbf{V}_1/\mathbf{V}_2$ as N, then

$$N\left(\frac{1}{Z_0} - \frac{1}{2Z_2}\right) = \left(\frac{1}{2Z_2} + \frac{1}{Z_0}\right)$$

$$N(2Z_2 - Z_0) = Z_0 + 2Z_2$$

Thus

$$N = \frac{Z_0 + 2Z_2}{2Z_2 - Z_0} \qquad [23]$$

or

$$Z_2 = \frac{Z_0}{2}\left(\frac{N+1}{N-1}\right) \qquad [24]$$

By substituting this value of Z_2 into equation [21] then a value for Z_1 can be obtained. Equation [21] gives

$$Z_0 = \sqrt{\left(\frac{4Z_1Z_2^2}{Z_1 + 4Z_2}\right)}$$

and so

$$Z_0^2 = \frac{Z_1Z_0^2\left(\frac{N+1}{N-1}\right)^2}{Z_1 + 2Z_0\left(\frac{N+1}{N-1}\right)}$$

$$= \frac{Z_1Z_0^2(N+1)^2}{Z_1(N-1)^2 + 2Z_0(N+1)(N-1)}$$

Thus

$$Z_1(N-1)^2 + 2Z_0(N+1)(N-1) = Z_1(N+1)^2$$

$$Z_1(N^2 + 2N + 1 - N^2 + 2N - 1) = 2Z_0(N^2 - 1)$$

$$Z_1 = Z_0\left(\frac{N^2 - 1}{2N}\right) \qquad [25]$$

Example 7

A symmetrical π-attenuator has a series resistance of 100 Ω and shunt resistances of 1000 Ω. What is (*a*) the characteristic impedance and (*b*) the attenuation in dB?

Answer

(*a*) $Z_1 = 100\,\Omega$ and $2Z_2 = 1000\,\Omega$. Thus using equation [21]

$$Z_0 = \sqrt{\left(\frac{4Z_1Z_2^2}{Z_1 + 4Z_2}\right)} = \sqrt{\left(\frac{4 \times 100 \times 500^2}{100 + 4 \times 500}\right)} = 218\,\Omega$$

(*b*) Using equation [23]

$$N = \frac{Z_0 + 2Z_2}{2Z_2 - Z_0} = \frac{218 + 1000}{1000 - 218} = 1.56$$

Thus the attenuation in dB is

$$\text{attenuation} = 20\lg N = 20\lg 1.56 = 3.86\,\text{dB}$$

Example 8

Design a symmetrical π-section attenuator to have a characteristic resistance of $400\,\Omega$ and an attenuation of 20 dB.

Answer

The attenuation in dB is given by

$$\text{attenuation} = 20\lg N = 20\,\text{dB}$$

Hence $N = 10$. Thus equation [24] gives

$$Z_2 = \frac{Z_0}{2}\left(\frac{N+1}{N-1}\right) = \frac{400}{2}\left(\frac{10+1}{10-1}\right) = \frac{489}{2}\,\Omega$$

Thus $2Z_2$, the resistor to be included, is $489\,\Omega$. Equation [25] gives

$$Z_1 = Z_0\left(\frac{N^2 - 1}{2N}\right) = 400\left(\frac{10^2 - 1}{2 \times 100}\right) = 198\,\Omega$$

Insertion loss

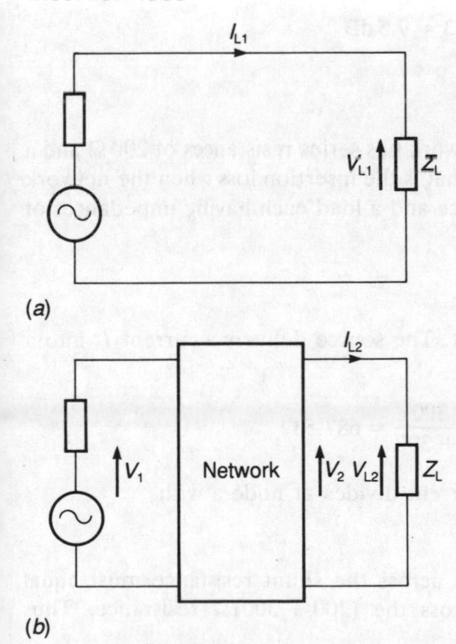

Fig. 7.9 Insertion loss

The *insertion loss* A_L of a network is defined as being the ratio of the powers, expressed in dB, dissipated in a load before and after the insertion of the network between the source and load. Thus if the current through a load Z_L is initially I_{L1} and after the introduction of the network becomes I_{L2} (Fig. 7.9) then

$$A_L = \frac{10\lg I_{L1}^2 Z_L}{10\lg I_{L2}^2 Z_L}$$

$$A_L = 20\lg\left(\frac{I_{L1}}{I_{L2}}\right) \qquad [26]$$

Similarly we can derive the relationship

$$A_L = 20\lg\left(\frac{V_{L1}}{V_{L2}}\right) \qquad [27]$$

When the load of the inserted two-port network has the characteristic impedance Z_0 then the input impedance of the network is also Z_0. In such a situation V_{L1} is the same as the voltage V_1 across the input of the network and thus, since V_{L2} is the voltage V_2 across the output,

$$A_L = 20\lg\left(\frac{V_1}{V_2}\right) \quad [28]$$

or alternatively

$$A_L = 20\lg\left(\frac{I_1}{I_2}\right) \quad [29]$$

Example 9

What is the insertion loss of a symmetrical T-section network, with series resistances of 300 Ω and a shunt resistance of 450 Ω, when it is connected to a matched load?

Answer

The network has $\frac{1}{2}Z_1 = 300\,\Omega$ and $Z_2 = 450\,\Omega$. The characteristic impedance is thus, by equation [14],

$$Z_0 = \surd(\tfrac{1}{4}Z_1^2 + Z_1Z_2) = \surd(\tfrac{1}{4} \times 600^2 + 600 \times 450) = 600\,\Omega$$

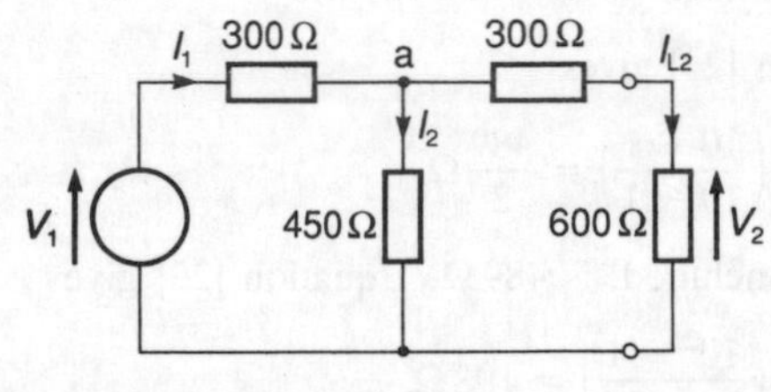

Fig. 7.10 Example 9

Figure 7.10 shows the circuit. For node a we have

$$I_1 = I_2 + I_{L2}$$

But the potential difference across the 450 Ω must be equal to that across the (300 + 600) Ω. Thus $450I_2 = 900I_{L2}$. Hence

$$I_1 = (900/450)I_{L2} + I_{L2}$$

$$\frac{I_1}{I_{L2}} = 3$$

Thus, using equation [29],

$$A_L = 20\lg\left(\frac{I_1}{I_{L2}}\right) = 20\lg 3 = 9.5\,\text{dB}$$

Example 10

A symmetrical T-section network has series resistances of 200 Ω and a shunt resistance of 300 Ω. What is the insertion loss when the network is connected between a source and a load each having impedances of 300 Ω?

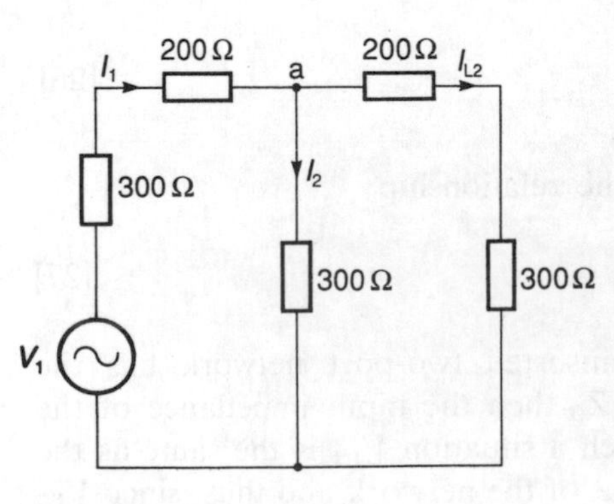

Fig. 7.11 Example 10

Answer

Figure 7.11 shows the circuit. The source delivers a current I_1 into a total impedance of

$$300 + 200 + \frac{300(200 + 300)}{300 + 200 + 300} = 687.5\,\Omega$$

Thus I_1 is $V_1/687.5$. This current divides at node a with

$$I_1 = I_2 + I_{L2}$$

But the potential difference across the shunt resistance must equal the potential difference across the (200 + 300) Ω resistance. Thus $300I_2 = 500I_{L2}$. Hence

$$I_1 = (5/3)I_{L2} + I_{L2} = (8/3)I_{L2}$$

Hence

$$I_{L2} = \frac{3V_1}{8 \times 687.5}$$

In the absence of the T-section the source would have been delivering a current I_{L1} through $(300 + 300)\,\Omega$. Thus the current would have been $I_{L1} = V_1/600$. The insertion loss is thus, by equation [26],

$$A_L = 20\lg\left(\frac{I_{L1}}{I_{L2}}\right) = 20\lg\left(\frac{8 \times 687.5}{3 \times 600}\right) = 9.7\,\text{dB}$$

Filters

A filter is a network designed to pass certain bands of frequencies and introduce considerable attenuation for other frequencies. The band, or bands, of frequencies for which the attenuation is effectively zero is called the *pass band*. The frequencies at which the attenuation changes from finite to zero are called the *cut-off frequencies*. There are four basic types of filter, the low pass, the high pass, the band pass and the band stop. The *low-pass* filter passes signals up to some limiting frequency but not above it, the *high-pass* filter passes signals down to some limiting frequency but not below it, the *band-pass* filter passes signals over a range of frequencies but not outside it, and the *band-stop* filter only passes signals outside a range of frequencies. Fig. 7.12 shows the ideal characteristics of these filters and the symbols used for them. In reality the transition in attenuation at the cut-off frequencies is not the abrupt change shown in the figures but a gradual change.

L-section filters

Figure 7.13 shows two basic forms of an L-section filter section. The circuits are potential divider circuits with one of the components, the capacitor, having an impedance which varies with frequency. Thus for Fig. 7.13(*a*), an increase in frequency produces a decrease in the reactance of the capacitor ($X_C = 1/\omega C$) and hence a smaller potential drop across the output. The attenuation of the circuit has thus increased. The arrangement is a low-pass filter. With Fig. 7.13(*b*) the converse occurs and as the frequency increases the potential drop across the output increases. It is a high-pass filter. This simple type of filter finds widespread use in electronics, e.g. blocking a.c. interference or blocking a d.c. signal.

With a load R_L the total circuit impedance for the circuit in Fig. 7.13(*a*) is

$$Z = R + \frac{(1/j\omega C)R_L}{R_L + (1/j\omega C)}$$

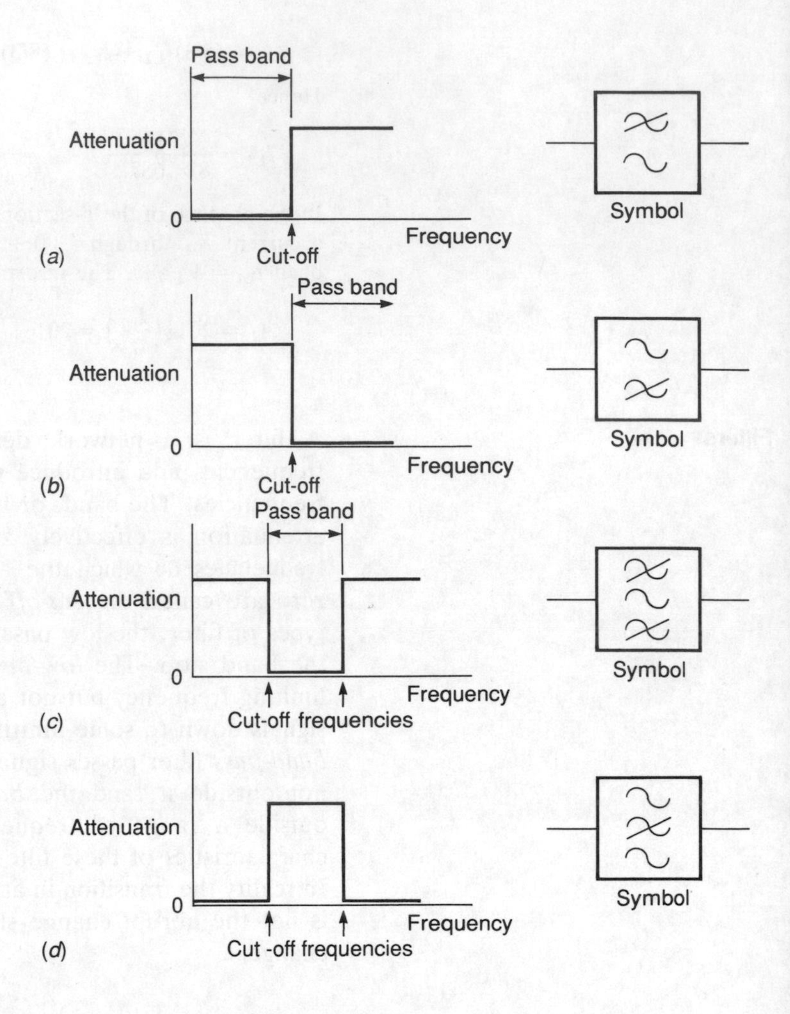

Fig. 7.12 (*a*) Low-pass filter, (*b*) high-pass filter, (*c*) band-pass filter, (*d*) band-stop filter

Fig. 7.13 L-section filters, (*a*) low-pass, (*b*) high-pass

The current $\mathbf{I_1}$ is thus given by

$$\mathbf{V_1} = Z\mathbf{I_1} = \mathbf{I_1}\left[R + \frac{(1/\mathrm{j}\omega C)R_\mathrm{L}}{R_\mathrm{L} + (1/\mathrm{j}\omega C)}\right]$$

$$= \mathbf{I_1}\left[R + \frac{R_\mathrm{L}}{1 + \mathrm{j}\omega CR_\mathrm{L}}\right]$$

The output voltage $\mathbf{V_2}$ is across the parallel arrangement of C and R_L and is thus

$$\mathbf{V_2} = \mathbf{I_1}\left[\frac{(1/\mathrm{j}\omega C)R_\mathrm{L}}{R_\mathrm{L} + (1/\mathrm{j}\omega C)}\right] = \mathbf{I_1}\left[\frac{R_\mathrm{L}}{1 + \mathrm{j}\omega CR_\mathrm{L}}\right]$$

Thus

$$\frac{\mathbf{V_2}}{\mathbf{V_1}} = \frac{\left[\dfrac{R_\mathrm{L}}{1 + \mathrm{j}\omega CR_\mathrm{L}}\right]}{\left[R + \dfrac{R_\mathrm{L}}{1 + \mathrm{j}\omega CR_\mathrm{L}}\right]} = \frac{1}{R + \mathrm{j}\omega CR_\mathrm{L}R + R_\mathrm{L}}$$

With $R_L \gg R$, then

$$\frac{\mathbf{V}_2}{\mathbf{V}_1} \approx \frac{1}{1 + j\omega CR} \approx \frac{1 - j\omega CR}{1 + \omega^2 C^2 R^2} \qquad [30]$$

Thus the magnitudes of the voltages have the ratio

$$\left|\frac{V_2}{V_1}\right| = \sqrt{\left[\frac{1 + \omega^2 C^2 R^2}{(1 + \omega^2 C^2 R^2)^2}\right]} = \frac{1}{\sqrt{(1 + \omega^2 C^2 R^2)}} \qquad [31]$$

and the phase difference between $\mathbf{V}_2$ and $\mathbf{V}_1$ is

$$\phi = \tan^{-1}(-\omega CR)$$

With $\omega CR \ll 1$ then $\mathbf{V}_2/\mathbf{V}_1 \approx 1$, and thus there is effectively no attenuation. With $\omega CR \gg 1$ then $\mathbf{V}_2/\mathbf{V}_1 \approx -\mathrm{j}/\omega CR$ and the circuit behaves as a purely reactive circuit. As a purely reactive circuit no power is delivered to the load and thus the attenuation is effectively infinite. The circuit is thus a low-pass filter. The transition between the resistive or reactive nature dominating is when $\omega CR = 1$. Thus the cut-off frequency is defined by

$$\omega_c = \frac{1}{CR} \qquad [32]$$

For the circuit in Fig. 7.13(*b*) a similar analysis leads to

$$\frac{\mathbf{V}_2}{\mathbf{V}_1} \approx \frac{\omega^2 C^2 R^2 + j\omega CR}{1 + \omega^2 C^2 R^2} \qquad [33]$$

and so

$$\left|\frac{V_2}{V_1}\right| = \sqrt{\left[\frac{(\omega^2 C^2 R^2)^2 + (\omega CR)^2}{(1 + \omega^2 C^2 R^2)^2}\right]}$$

$$= \sqrt{\left[\frac{(\omega^2 C^2 R^2)^2\ (1 + 1/\omega^2 C^2 R^2)}{(\omega^2 C^2 R^2)^2\ (1 + 1/\omega^2 C^2 R^2)^2}\right]}$$

$$\left|\frac{V_2}{V_1}\right| = \frac{1}{\sqrt{(1 + 1/\omega^2 C^2 R^2)}} \qquad [34]$$

and

$$\phi = \tan^{-1}\left(\frac{\omega CR}{\omega^2 C^2 R^2}\right) = \tan^{-1}\left(\frac{1}{\omega CR}\right) \qquad [35]$$

With $\omega CR \ll 1$ then $\mathbf{V}_2/\mathbf{V}_1 \approx \mathrm{j}\omega CR$ and the circuit behaves as a purely reactive circuit. With $\omega CR \gg 1$ then $\mathbf{V}_2/\mathbf{V}_1 \approx 1$, and thus there is effectively no attenuation. The circuit is thus a high-pass filter. The transition between the reactive and resistive nature dominating is when $\omega CR = 1$. Thus the cut-off frequency is defined by

$$\omega_c = \frac{1}{CR} \qquad [36]$$

Example 11

A simple RC filter is to be used to give a low-pass filter with a cut-off frequency of 1 kHz. What capacitance will be required if the resistance R is 1 kΩ and what will be the attenuation in dB at the cut-off frequency?

Answer

Using equation [32] $\omega_c = 1/CR$, hence

$$C = \frac{1}{2\pi \times 1000 \times 1000} = 0.16\,\mu\text{F}$$

Equation [31] gives

$$\left|\frac{V_2}{V_1}\right| = \frac{1}{\sqrt{(1 + \omega^2 C^2 R2)}}$$

When $\omega = 1/CR$ then

$$\left|\frac{V_2}{V_1}\right| = \frac{1}{\sqrt{[1 + 1]}}$$

Hence, equation [2],

$$\text{attenuation} = 20\lg\left(\frac{V_1}{V_2}\right) = 20\lg(\sqrt{2}) = 3.01\,\text{dB}$$

Low-pass T-section filter

Figure 7.14 shows a *T-section low-pass filter*. The filter is known as a *constant k filter* since the product of total series and shunt impedances is a constant which is independent of frequency.

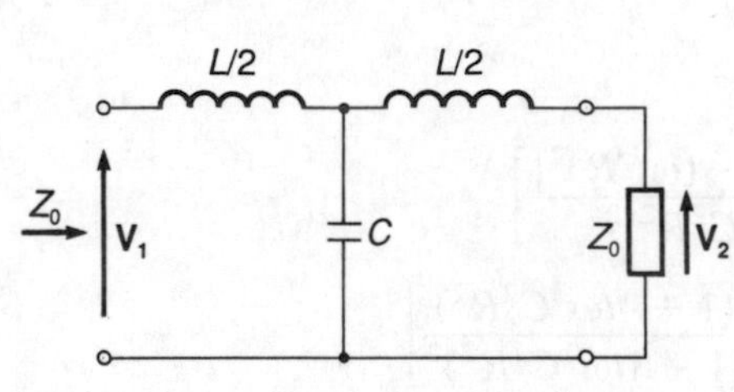

Fig. 7.14 Low-pass T-section filter

$$\text{j}2\,\omega(L/2) \times \frac{1}{\text{j}\omega C} = \frac{L}{C}$$

The constant k T-section filter is often referred to as a *prototype filter*.

The characteristic impedance Z_0 of the low-pass filter is given by equation [14], with $\frac{1}{2}Z_1 = \text{j}\omega\frac{1}{2}L$ and $Z_2 = 1/\text{j}\omega C$, as

$$Z_0 = \sqrt{(\tfrac{1}{4}Z_1^2 + Z_1Z_2)} = \sqrt{[\tfrac{1}{4}(\text{j}\omega L)^2 + (\text{j}\omega L)(1/\text{j}\omega C)]}$$

$$= \sqrt{\left[\frac{L}{C} - \frac{\omega^2 L^2}{4}\right]} \qquad [37]$$

Z_0 is a real quantity if $(L/C) > (\omega^2 L^2/4)$ and thus effectively purely resistive. Because all the elements in the T-section are reactive no power will be dissipated in the filter and thus all the input power will be transmitted through to the load. Thus the attenuation of the filter operating under this condition is zero. Z_0 is an imaginary quantity if $(L/C) < (\omega^2 L^2/4)$ and thus purely reactive. When this occurs there is no power dissipation in the load. Thus the attenuation is infinite. The T-section is thus operating as a low-pass filter. The cut-off frequency is when

$$\frac{L}{C} = \frac{\omega^2 L^2}{4}$$

and thus

$$f_c = \frac{1}{\pi\sqrt{(LC)}} \qquad [38]$$

When the frequency ω is zero then equation [37] gives for the characteristic impedance

$$Z_0 = R_0 = \sqrt{(L/C)} \qquad [39]$$

This impedance is called the *design impedance* R_0.

Example 12

What is the cut-off frequency and the design impedance for a constant k T-section filter which has series inductances of 20 mH and a shunt capacitor of 0.1 μF?

Answer

Since $L/2$ = 20 mH, then using equation [38]

$$f_c = \frac{1}{\pi\sqrt{(LC)}} = \frac{1}{\pi\sqrt{(40 \times 10^{-3} \times 0.1 \times 10^{-6})}} = 5.03\,\text{kHz}$$

Using equation [39]

$$R_0 = \sqrt{\left(\frac{L}{C}\right)} = \sqrt{\left(\frac{40 \times 10^{-3}}{0.1 \times 10^{-6}}\right)} = 632\,\Omega$$

Low-pass π-section filter

Figure 7.15 shows a *low-pass* π-section filter. The filter is a *constant k filter* since the product of the series and total shunt impedances is a constant which is independent of frequency. The impedance of the two $C/2$ capacitors in parallel is half the impedance of one and so $1/j\omega C$. Thus the product is

$$j\omega L \times \frac{1}{j\omega C} = \frac{L}{C}$$

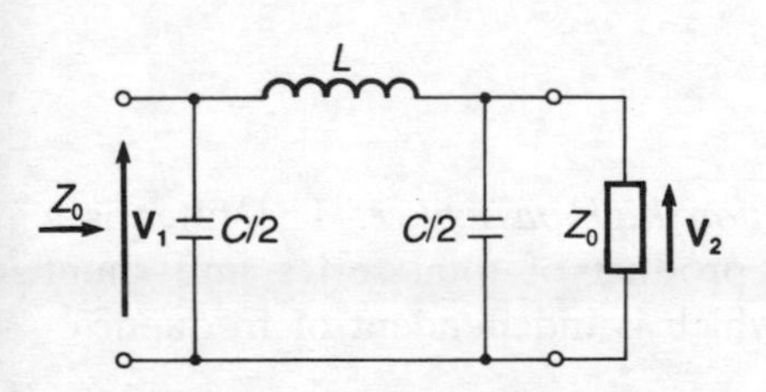

Fig. 7.15 Low-pass π-section filter

The constant k π-section is often referred to as a *prototype filter*.

The characteristic impedance Z_0 of the π-section filter is given by equation [21], with $Z_1 = j\omega L$ and $2Z_2 = 1/j\omega\frac{1}{2}C$, as

$$Z_0 = \sqrt{\left(\frac{4Z_1Z_2^2}{Z_1 + 4Z_2}\right)} = \sqrt{\left[\frac{4j\omega L/(j\omega C)^2}{j\omega L + 4/(j\omega C)}\right]}$$

$$= \sqrt{\left[\frac{4j\omega L}{j\omega C(4 - \omega^2 LC)}\right]} = \sqrt{\left[\frac{(L/C)}{1 - \omega^2 LC/4}\right]} \qquad [40]$$

Z_0 is a real quantity if $1 > (\omega^2 LC/4)$ and thus effectively purely resistive. Because all the elements in the T-section are reactive no power will be dissipated in the filter and thus all the input power will be transmitted through to the load. Thus the attenuation of the filter operating under this condition is zero. Z_0 is an imaginary quantity if $1 < (\omega^2 LC/4)$ and thus purely reactive. When this occurs there is no power dissipation in the load. Thus the attenuation is infinite. The π-section is thus operating as a low-pass filter. The cut-off frequency is when

$$1 = \frac{\omega^2 LC}{4}$$

and thus

$$f_c = \frac{1}{\pi\sqrt{(LC)}} \qquad [41]$$

When the frequency ω is zero then equation [40] gives for the characteristic impedance

$$Z_0 = R_0 = \sqrt{(L/C)} \qquad [42]$$

Example 13

Determine the component values for a constant k low-pass π filter if the design impedance is to be 600 Ω and the cut-off frequency 2 kHz.

Answer

Using equation [42] $R_0 = 600 = \sqrt{(L/C)}$ and so $\sqrt{L} = 600\sqrt{C}$. Using equation [41]

$$f_c = 2000 = \frac{1}{\pi\sqrt{(LC)}} = \frac{1}{\pi \times 600C}$$

$$C = 0.27\,\mu\text{F}$$

Hence the required shunt capacitance $C/2$ is 0.135 μF.

$$\sqrt{L} = 600\sqrt{(0.27 \times 10^{-6})}$$

and so

$$L = 0.097\,\text{H}.$$

High-pass T-section filter

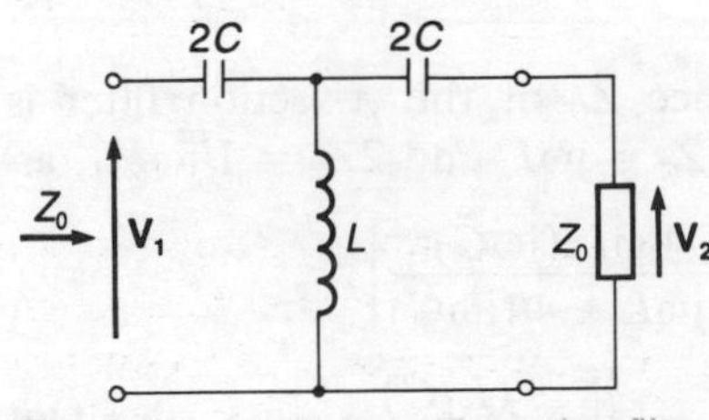

Fig. 7.16 High-pass T-section filter

Figure 7.16 shows a *T-section high-pass filter*. The filter is a *constant k filter* since the product of total series and shunt impedances is a constant which is independent of frequency.

$$\frac{2}{j\omega 2C} \times j\omega L = \frac{L}{C}$$

The constant k T-section filter is also referred to as a *prototype filter*.

The characteristic impedance Z_0 of the high-pass filter is given by equation [14], with $\frac{1}{2}Z_1 = 1/j\omega 2C$ and $Z_2 = j\omega L$, as

$$Z_0 = \sqrt{(\tfrac{1}{4}Z_1^2 + Z_1Z_2)} = \sqrt{[\tfrac{1}{4}(1/j\omega C)^2 + (1/j\omega C)j\omega L]}$$

$$= \sqrt{\left[\frac{L}{C} - \frac{1}{4\omega^2C^2}\right]} \qquad [43]$$

Z_0 is a real quantity if $(L/C) > (1/4\omega^2C^2)$, i.e. $\omega^2 > 1/4LC$, and thus effectively purely resistive. Because all the elements in the T-section are reactive no power will be dissipated in the filter and thus all the input power will be transmitted through to the load. Thus the attenuation of the filter operating with $\omega^2 > 1/4LC$ is zero. Z_0 is an imaginary quantity if $(L/C) < (4\omega^2C^2)$, i.e. $\omega^2 < 1/4LC$, and thus purely reactive. When this occurs there is no power dissipation in the load. Thus the attenuation is infinite. The T-section is thus operating as a high-pass filter with a cut-off frequency given by

$$\omega^2 = \frac{1}{4LC}$$

and thus

$$f_c = \frac{1}{4\pi\sqrt{(LC)}} \qquad [44]$$

When the frequency ω is infinite then equation [43] gives for the characteristic impedance

$$Z_0 = R_0 = \sqrt{(L/C)} \qquad [45]$$

This impedance is called the *design impedance* R_0.

Example 14

Design a high-pass T-section filter to have a design impedance of 600 Ω and a cut-off frequency of 2 kHz.

Answer

The design impedance is given by equation [45] as $R_0 = 600 = \sqrt{(L/C)}$. Thus $\sqrt{L} = 600\sqrt{C}$. The cut-off frequency is given by equation [44] as

$$f_c = 2000 = \frac{1}{4\pi\sqrt{(LC)}} = \frac{1}{4\pi \times 600C}$$

Thus $C = 66.3$ nF and so a capacitor of $2C = 132.6$ nF. This leads to $L = 60^2C = 23.9$ mH.

High-pass π-section filter

Figure 7.17 shows a π-section high-pass filter. The filter is a *constant k filter* since the product of series and total shunt impedances is a constant which is independent of frequency. The impedance of the two $2L$ inductances in parallel is half the

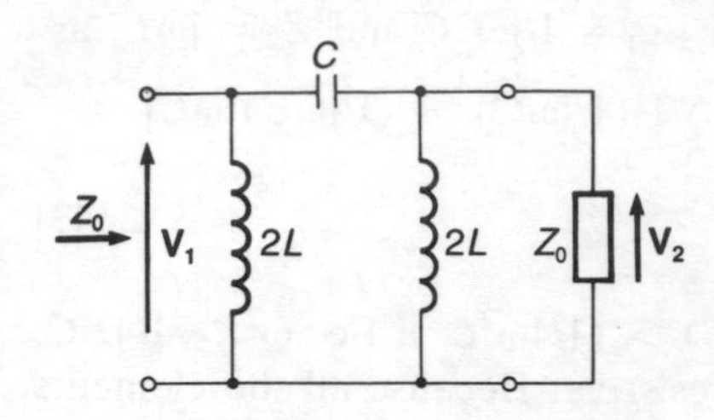

Fig. 7.17 High-pass π-section filter

impedance of one and so $j\omega L$. Thus the product is

$$\frac{1}{j\omega C} \times j\omega L = \frac{L}{C}$$

The constant k T-section filter is also referred to as a *prototype filter*.

The characteristic impedance Z_0 of the high-pass filter is given by equation [21], with $Z_1 = 1/j\omega C$ and $2Z_2 = j\omega 2L$, as

$$Z_0 = \sqrt{\left(\frac{4Z_1Z_2^2}{Z_1 + 4Z_2}\right)} = \sqrt{\left[\frac{4(1/j\omega C)(j\omega L)^2}{(1/j\omega C) + 4(j\omega L)}\right]}$$

$$= \frac{\sqrt{[L/C]}}{\sqrt{\left[1 - \dfrac{1}{4\omega^2 LC}\right]}} \qquad [46]$$

Z_0 is a real quantity if $1 > (1/4\omega^2 LC)$, i.e. $\omega^2 > 1/4LC$, and thus effectively purely resistive. Because all the elements in the π-section are reactive no power will be dissipated in the filter and thus all the input power will be transmitted through to the load. Thus the attenuation of the filter operating with $\omega^2 > 1/4LC$ is zero. Z_0 is an imaginary quantity if $1 < (4\omega^2 LC)$, i.e. $\omega^2 < 1/4LC$, and thus purely reactive. When this occurs there is no power dissipation in the load. Thus the attenuation is infinite. The π-section is thus operating as a high-pass filter with a cut-off frequency given by

$$\omega^2 = \frac{1}{4LC}$$

and thus

$$f_c = \frac{1}{4\pi\sqrt{(LC)}} \qquad [47]$$

When the frequency ω is infinite then equation [46] gives for the characteristic impedance

$$Z_0 = R_0 = \sqrt{(L/C)} \qquad [48]$$

This impedance is called the *design impedance* R_0.

Example 15

A high-pass π-section filter has a series capacitance of 0.2 μF and shunt inductances of 20 mH. What is the design impedance and the cut-off frequency?

Answer

Using equation [48], since $2L = 20$ mH and $C = 0.2$ μF, the design impedance is

$$R_0 = \sqrt{\left(\frac{L}{C}\right)} = \sqrt{\left(\frac{10 \times 10^{-3}}{0.2 \times 10^{-6}}\right)} = 223.6\,\Omega$$

The cut-off frequency is given by equation [47] as

$$f_c = \frac{1}{4\pi\sqrt{(LC)}} = \frac{1}{4\pi\sqrt{(10 \times 10^{-3} \times 0.2 \times 10^{-6})}} = 1.8\,\text{kHz}$$

Attenuation of T and π-section filters

The attenuation for a T-section filter will be considered here, though the argument applies equally well to a π-section filter and yields the same result. If we apply Kirchoff's voltage law to the source end of the T-section in Fig. 7.18, then

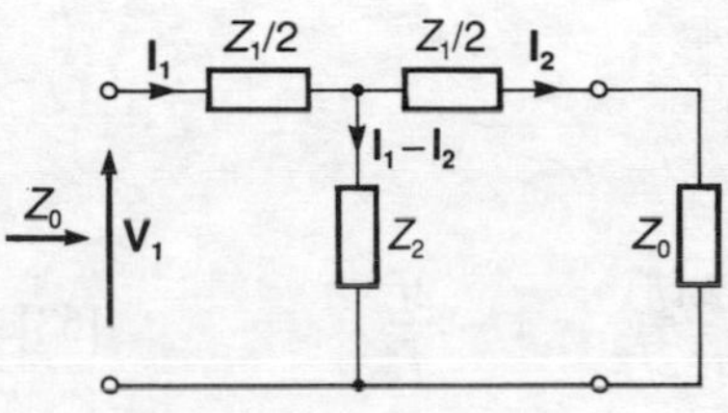

$$\mathbf{V}_1 = \mathbf{I}_1\tfrac{1}{2}Z_1 + (\mathbf{I}_1 - \mathbf{I}_2)Z_2$$

Fig. 7.18 T-section filter

But $\mathbf{V}_1 = Z_0\mathbf{I}_1$, thus

$$\mathbf{I}_2 Z_2 = \mathbf{I}_1[\tfrac{1}{2}Z_1 + Z_2 - Z_0]$$

$$\frac{\mathbf{I}_2}{\mathbf{I}_1} = \frac{Z_1}{2Z_2} + 1 - \frac{Z_0}{Z_2} \qquad [49]$$

The propagation coefficient γ is, however, given by equation [10] as

$$\frac{\mathbf{I}_1}{\mathbf{I}_2} = \mathrm{e}^{\gamma}$$

Thus

$$\mathrm{e}^{-\gamma} = \frac{Z_1}{2Z_2} + 1 - \frac{Z_0}{Z_2} \qquad [50]$$

and

$$\mathrm{e}^{\gamma} = \frac{1}{\dfrac{Z_1}{2Z_2} + 1 - \dfrac{Z_0}{Z_2}} = \frac{\dfrac{Z_1}{2Z_2} + 1 + \dfrac{Z_0}{Z_2}}{\left[\dfrac{Z_1}{2Z_2} + 1 - \dfrac{Z_0}{Z_2}\right]\left[\dfrac{Z_1}{2Z_2} + 1 + \dfrac{Z_0}{Z_2}\right]}$$

Substituting for Z_0 in the denominator results in

$$\mathrm{e}^{\gamma} = \frac{Z_1}{2Z_2} + 1 + \frac{Z_0}{Z_2} \qquad [51]$$

The expression $\tfrac{1}{2}(\mathrm{e}^{x} + \mathrm{e}^{-x})$ is called $\cosh x$. Thus, adding equations [50] and [51] gives

$$\cosh\gamma = \frac{1}{2}\left[\frac{Z_1}{2Z_2} + 1 - \frac{Z_0}{Z_2} + \frac{Z_1}{2Z_2} + 1 + \frac{Z_0}{Z_2}\right]$$

$$\cosh\gamma = 1 + \frac{Z_1}{2Z_2} \qquad [52]$$

Consider this equation applied to a *low-pass filter*, either a T-section or π-section. For a T-section when used as a low-pass filter $\frac{1}{2}Z_1 = j\omega\frac{1}{2}L$ and $Z_2 = 1/j\omega C$ (see Fig. 7.14). Thus equation [52] gives

$$\cosh\gamma = 1 + \frac{Z_1}{2Z_2} = 1 + \frac{j\omega L}{2/j\omega C} = 1 - \frac{\omega^2 LC}{2}$$

But the cut-off frequency f_c is given by equation [38] as

$$f_c = \frac{1}{\pi\sqrt{(LC)}}$$

Thus

$$LC = \frac{1}{\pi^2 f_c^2}$$

and

$$\cosh\gamma = 1 - \frac{\omega^2}{\pi^2 2f_c^2} = 1 - \frac{(2\pi f)^2}{\pi^2 2f_c^2} = 1 - \frac{2f^2}{f_c^2} \qquad [53]$$

But $\gamma = \alpha + j\beta$, where α is the attenuation coefficient and β the phase change coefficient.

$$\begin{aligned}\cosh\gamma &= \cosh(\alpha + j\beta)\\ &= \tfrac{1}{2}[e^{(\alpha+j\beta)} + e^{-(\alpha+j\beta)}]\\ &= \tfrac{1}{2}[e^{\alpha}e^{j\beta} + e^{-\alpha}e^{-j\beta}]\\ &= \tfrac{1}{2}(e^{\alpha} + e^{-\alpha})\,\tfrac{1}{2}(e^{j\beta} + {}^{e-j\beta})\\ &\quad - \tfrac{1}{2}(e^{\alpha} - e^{-\alpha})\,\tfrac{1}{2}j(e^{j\beta} - e^{-j\beta})\\ &= \cosh\alpha\cos\beta + j\sinh\alpha\sin\beta \qquad [54]\end{aligned}$$

with $\sinh x$ being given by

$$\sinh x = \tfrac{1}{2}(e^{x} - e^{-x})$$

and

$$\cos x = \tfrac{1}{2}(e^{jx} + e^{-jx})$$

$$\sin x = -\tfrac{1}{2}j(e^{jx} - e^{-jx})$$

For the low-pass filter when f is less than the critical frequency the attenuation coefficient is zero. This means that

$$\cosh\alpha = \tfrac{1}{2}(e^{0} + e^{-0}) = 1$$

$$\sinh\alpha = \tfrac{1}{2}(e^{0} - e^{-0}) = 0$$

Thus equation [54] becomes

$$\cosh\gamma = \cos\beta$$

and so equation [53] gives, for the pass band,

$$\cos\beta = 1 - \frac{2f^2}{f_c^2} \qquad [55]$$

For the low-pass filter in the attenuation band when $f > f_c$ there is no power dissipation in the load. This can be explained by the phase difference being 180° or π. With this value for β then $\sin\beta = 0$ and $\cos\beta = -1$. Thus equation [54] becomes

$$\cosh\gamma = -\cosh\alpha$$

and so equation [53] gives

$$-\cosh\alpha = 1 - \frac{2f^2}{f_c^2}$$

$$\cosh\alpha = \frac{2f^2}{f_c^2} - 1 \qquad [56]$$

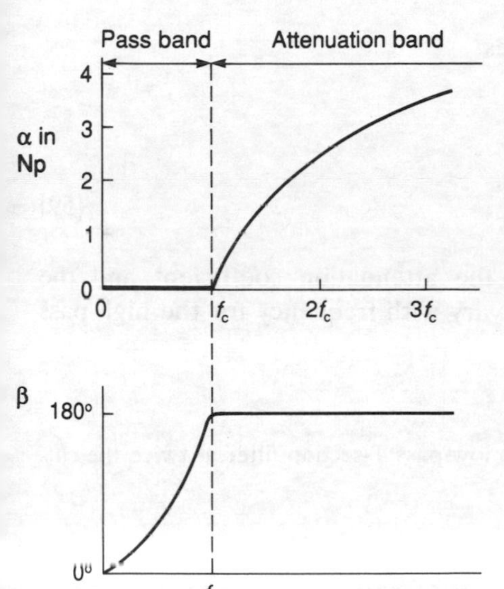

Fig. 7.19 Low-pass constant *k* filter

Figure 7.19 shows how the attenuation coefficient and the phase change coefficient vary with frequency for the low-pass filter.

Now consider equation [52] applied to a *high-pass filter*, either a T-section or π-section. For a T-section when used as a low-pass filter $\frac{1}{2}Z_1 = 1/j\omega 2C$ and $Z_2 = j\omega L$ (see Fig. 7.16). Thus equation [52] gives

$$\cosh\gamma = 1 + \frac{Z_1}{2Z_2} = 1 + \frac{1/j\omega C}{2j\omega L} = 1 - \frac{1}{2\omega^2 LC}$$

But the cut-off frequency f_c is given by equation [44] as

$$f_c = \frac{1}{4\pi\sqrt{(LC)}}$$

Thus

$$LC = \frac{1}{16\pi^2 f_c^2}$$

Thus

$$\cosh\gamma = 1 - \frac{16\pi^2 f_c^2}{2(2\pi f)^2} = 1 - \frac{2f_c^2}{f^2} \qquad [57]$$

For the high-pass filter when f is greater than the critical frequency then the attenuation coefficient is zero. This means that

$$\cosh\alpha = \tfrac{1}{2}(e^0 + e^{-0}) = 1$$

$$\sinh\alpha = \tfrac{1}{2}(e^0 - e^{-0}) = 0$$

Thus equation [54] becomes

$$\cosh\gamma = \cos\beta$$

and so equation [57] gives, in the pass band,

$$\cos\beta = 1 - \frac{2f_c^{\,2}}{f^2} \qquad [58]$$

For the high-pass filter in the attenuation band when $f < f_c$ there is no power dissipation in the load. This can be explained by the phase difference being 180° or π. With this value for β then $\sin\beta = 0$ and $\cos\beta = -1$. Thus equation [54] becomes

$$\cosh\gamma = -\cosh\alpha$$

and so equation [57] gives

$$-\cosh\alpha = 1 - \frac{2f_c^2}{f^2}$$

$$\cosh\alpha = \frac{2f_c^2}{f^2} - 1 \qquad [59]$$

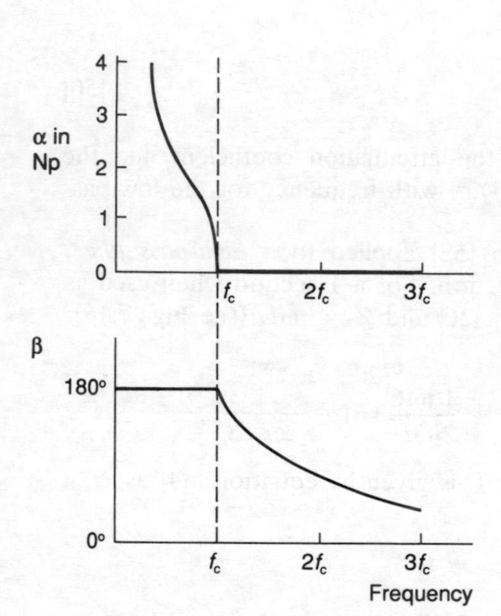

Fig. 7.20 High-pass constant k filter

Figure 7.20 shows how the attenuation coefficient and the phase change coefficient vary with frequency for the high-pass filter.

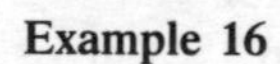

Example 16

What is the attenuation of a low-pass T-section filter at twice the cut-off frequency?

Answer

Using equation [56]

$$\cosh\alpha = \frac{2f^2}{f_c^2} - 1 = 2 \times 2^2 - 1 = 7$$

Hence

$$\tfrac{1}{2}(e^{\alpha} + e^{-\alpha}) = 7$$

$$e^{\alpha} + \frac{1}{e^{\alpha}} = 14$$

$$e^{2\alpha} - 14\,e^{\alpha} + 1 = 0$$

This quadratic equation can be solved for e^{α}.

$$e^{\alpha} = \frac{14 \pm \sqrt{(14^2 - 4)}}{2}$$

$$e^{\alpha} = +13.9 \text{ or } +0.0718$$

Hence $\alpha = +2.63$ or -2.63 Np. This can be converted to dB using equation [8]

$$\text{attenuation in dB} = 8.686 \times \text{attenuation in Np}$$

$$= 8.686 \times 2.63 = 22.8\,\text{dB}$$

Band-pass and band-stop filters

A *band-pass filter* can be produced by the cascade connection of a low-pass and a high-pass filters. However, the more usual method is to use a T-section or π-section with the series and shunt branches being series or parallel resonant circuits. For such a T-section filter (Fig. 7.21) we have, since $2C_1$ in series with $2C_1$ is equivalent to a capacitance of C_1,

$$\text{total series impedance} = j\omega L_1 + \frac{1}{j\omega C_1} = \left(\frac{1 - \omega^2 L_1 C_1}{j\omega C_1}\right)$$

$$\text{total shunt impedance} = \frac{j\omega L_2(1/j\omega C_2)}{j\omega L_2 + (1/j\omega C_2)} = \frac{j\omega L_2}{1 - \omega^2 L_2 C_2}$$

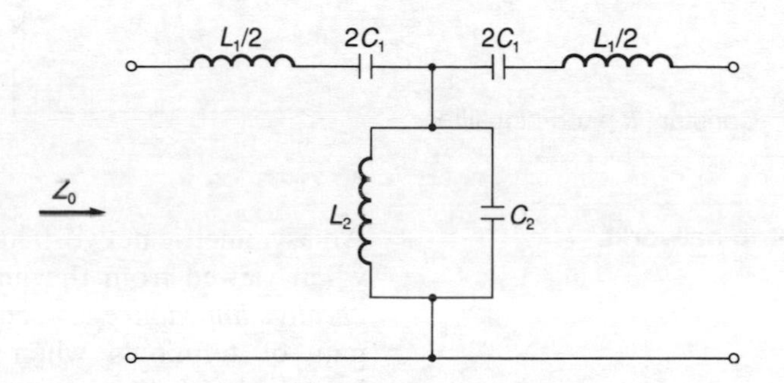

Fig. 7.21 Constant *k* band-pass filter

Thus the product of the series and shunt impedances is

$$\left(\frac{1 - \omega^2 L_1 C_1}{j\omega C_1}\right)\left(\frac{j\omega L_2}{1 - \omega^2 L_2 C_2}\right)$$

For the filter to be constant *k* we must have the product of the series and shunt impedances to be a constant which does not change with frequency. With

$$L_1 C_1 = L_2 C_2 \qquad [60]$$

we have the product becoming L_2/C_1 and so a constant. When this occurs both the resonant circuits resonate at the same frequency.

At frequencies below the resonant frequency the series arms of the T-section, which are series resonant circuits, have high impedances while the impedance of the shunt arm, a parallel resonant circuit, is low. Thus the output is low, i.e. the attenuation is high. At, and close to, the resonant frequency the series arms have zero, or very low, impedance and the shunt arms have high impedance. Thus the output is high, i.e. the attenuation is low. At frequencies above the resonant frequency the series arms have high impedances while the shunt arm has a low impedance. The output is thus low, i.e. the attenuation is again high. Figure 7.22 shows the general characteristics of such a constant *k* band-pass filter.

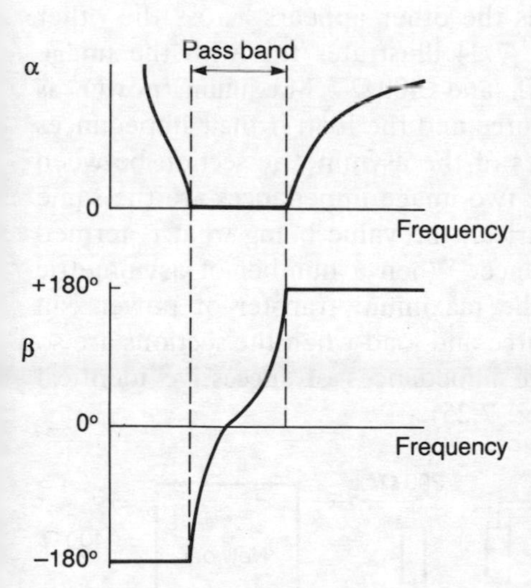

Fig. 7.22 Constant *k* band-pass filter

A *band-stop filter* is produced by interchanging the series and shunt arms of the band-pass filter. Figure 7.23 shows the circuit.

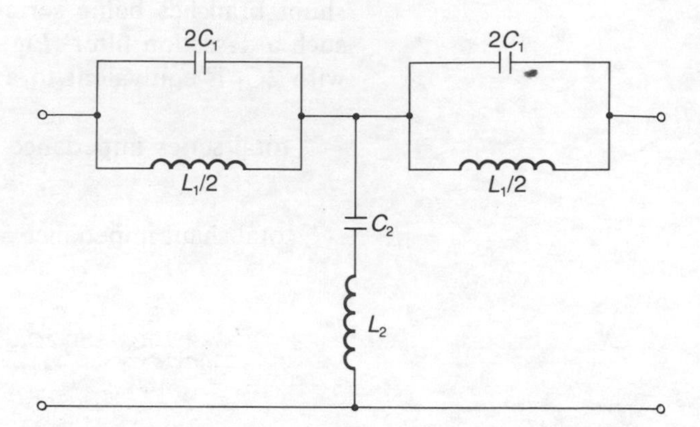

Fig. 7.23 Constant *k* band-stop filter

Asymmetric networks

An asymmetric network has different characteristic impedances when viewed from the input and output terminals. The term *iterative impedance* is used for the impedance measured at one pair of terminals when the other is terminated with an impedance of the same value. With symmetrical networks the iterative impedance viewed from the input terminals is the same as the one viewed from the output terminals, i.e. there is thus just one value of characteristic impedance. With asymmetric networks this is not the case and there are two values of the characteristic impedance.

The term *image impedance* is used for those impedances which have values such that when one of them is connected across a pair of terminals the other appears across the other pair of terminals. Figure 7.24 illustrates this with the image impedances being 200 Ω and 400 Ω. Maximum power is transferred between a source and the load if their impedances are the image impedances of the asymmetric section between the source and load. The two image impedances are the same if the network is symmetrical, the value being what is termed the characteristic impedance. When a number of asymmetric sections are cascaded, the maximum transfer of power can only occur between a source and load when the sections are so connected that the image impedances of successive identical sections are reversed (Fig. 7.25).

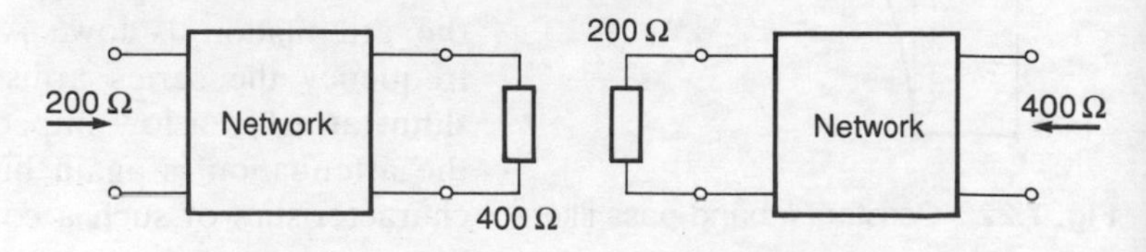

Fig. 7.24 Image impedances

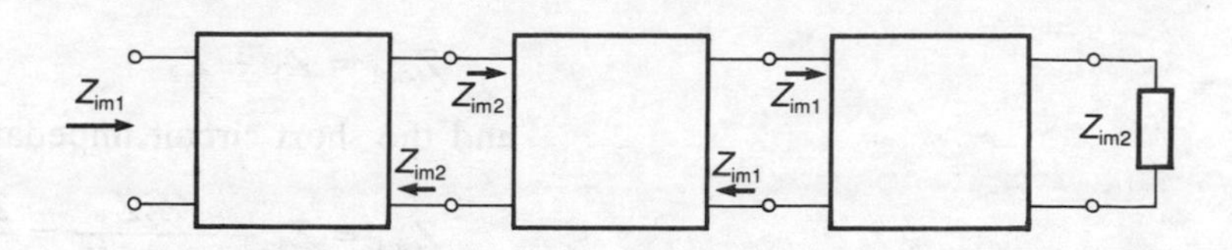

Fig. 7.25 Cascaded asymmetric sections

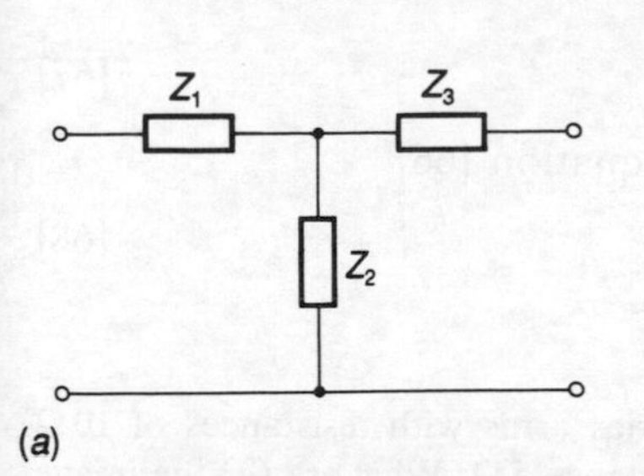

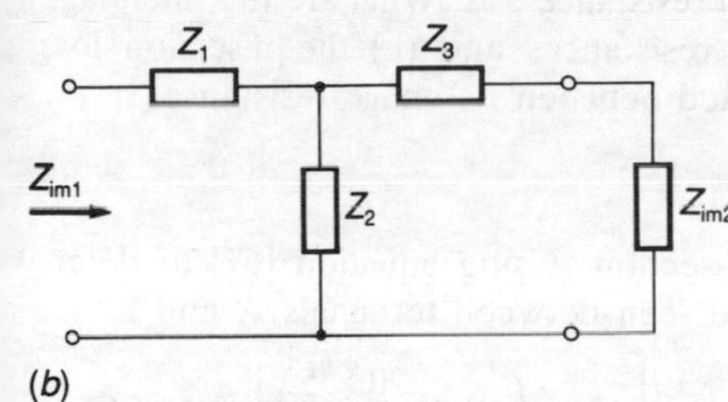

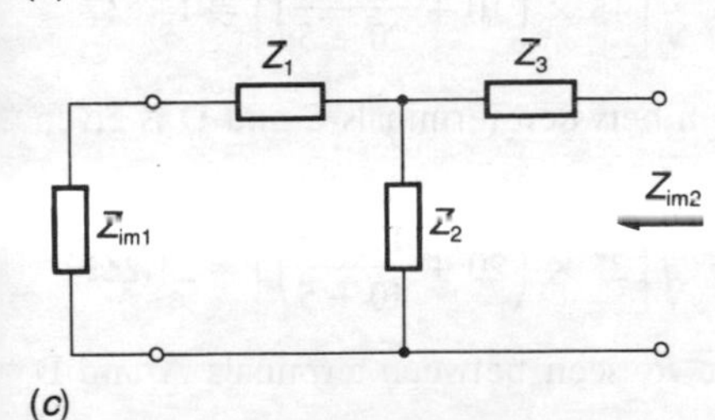

Fig. 7.26 (*a*) An asymmetric T-section, (*b*) terminated by Z_{im1}, (*c*) terminated by Z_{im2}

Figure 7.26(*a*) shows an asymmetric T-section. One of the image impedances Z_{im1} is the impedance of the T-section with the load Z_{im2}, as in Fig. 7.26(*b*). For this circuit

$$Z_{im1} = Z_1 + \frac{Z_2(Z_3 + Z_{im2})}{Z_2 + Z_3 + Z_{im2}} \qquad [61]$$

The other image impedance Z_{im2} is the impedance of the T-section with the load Z_{im1}, as in Fig. 7.26(*c*). For this circuit

$$Z_{im2} = Z_3 + \frac{Z_2(Z_1 + Z_{im1})}{Z_2 + Z_1 + Z_{im1}} \qquad [62]$$

Rearrangement of equations [61] and [62] gives

$$Z_{im1}(Z_2 + Z_3) + Z_{im1}Z_{im2} = Z_1Z_2 + Z_1Z_3 + Z_1Z_{im2} + Z_2Z_3 + Z_2Z_{im2}$$

$$Z_{im2}(Z_1 + Z_2) + Z_{im1}Z_{im2} = Z_2Z_3 + Z_1Z_3 + Z_3Z_{im1} + Z_1Z_2 + Z_2Z_{im1}$$

Subtracting these two equations gives

$$Z_{im1}(Z_2 + Z_3) - Z_{im2}(Z_1 + Z_2) = Z_{im2}(Z_1 + Z_2) - Z_{im1}(Z_2 + Z_3)$$

$$\frac{Z_{im1}}{Z_{im2}} = \frac{Z_1 + Z_2}{Z_2 + Z_3} \qquad [63]$$

Adding the two equations gives

$$Z_{im1}Z_{im2} = Z_1Z_2 + Z_2Z_3 + Z_1Z_3 \qquad [64]$$

Multiplying [63] by [64] eliminates Z_{im2} and gives

$$Z^2_{im1} = \frac{(Z_1 + Z_2)(Z_1Z_2 + Z_2Z_3 + Z_1Z_3)}{Z_2 + Z_3} \qquad [65]$$

Dividing [64] by [63] eliminates Z_{im1} and gives

$$Z^2_{im2} = \frac{(Z_2 + Z_3)(Z_1Z_2 + Z_2Z_3 + Z_1Z_3)}{Z_1 + Z_2} \qquad [66]$$

Equations [65] and [66] can be used to determine the image impedances.

A simpler way of expressing the image impedances is in terms of the open-circuit and short-circuit impedances. For Fig. 7.26(*b*) the open-circuit impedance Z_{oc1} is

$$Z_{oc1} = Z_1 + Z_2$$

and the short-circuit impedance Z_{sc1} is

$$Z_{sc1} = Z_1 + \frac{Z_2 Z_3}{Z_2 + Z_3} = \frac{Z_1 Z_2 + Z_2 Z_3 + Z_1 Z_3}{Z_2 + Z_3}$$

Thus equation [65] gives

$$Z_{im1} = \sqrt{(Z_{oc1} Z_{sc1})} \qquad [67]$$

Likewise for Fig. 7.26(*c*) and equation [66]

$$Z_{im2} = \sqrt{(Z_{oc2} Z_{sc2})} \qquad [68]$$

Example 17

An asymmetrical T-section has series arms with resistances of 10 Ω and 20 Ω and a shunt arm with resistance 5 Ω. What are (*a*) the image resistances, (*b*) the iterative resistances and (*c*) the insertion loss when the T-section is connected between its image resistances?

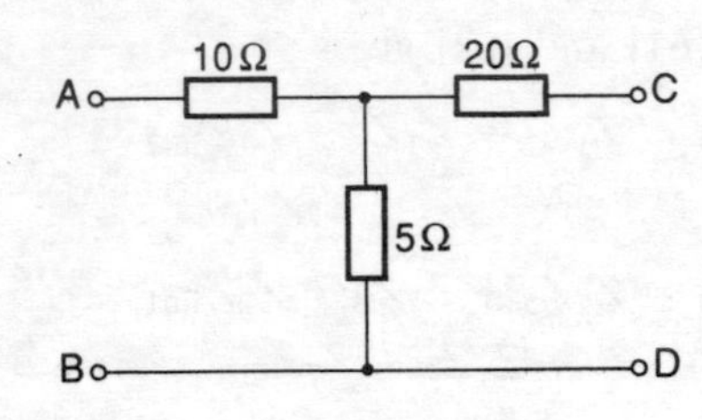

Fig. 7.27 Example 18

Answer

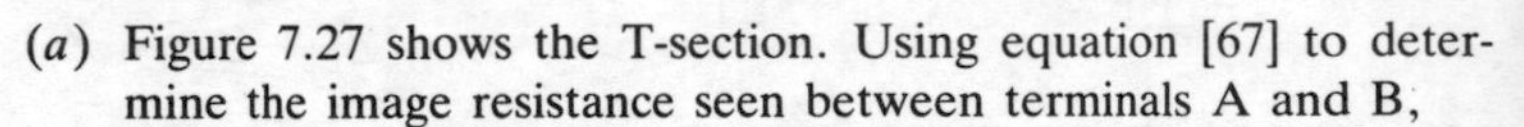

(*a*) Figure 7.27 shows the T-section. Using equation [67] to determine the image resistance seen between terminals A and B,

$$Z_{im1} = \sqrt{(Z_{oc1} Z_{sc1})} = \sqrt{\left[15 \times \left(10 + \frac{20 \times 5}{20 + 5}\right)\right]} = 14.5\,\Omega$$

The image resistance seen between terminals C and D is given by equation [68] as

$$Z_{im2} = \sqrt{(Z_{oc2} Z_{sc2})} = \sqrt{\left[25 \times \left(20 + \frac{10 \times 5}{10 + 5}\right)\right]} = 24.2\,\Omega$$

(*b*) For the iterative resistance R_1 seen between terminals A and B

$$R_1 = 10 + \frac{5(20 + R_1)}{5 + 20 + R_1}$$

Thus

$$R_1^2 + 25R_1 = 250 + 10R_1 + 100 + 5R_1$$
$$R_1^2 + 10R_1 - 350 = 0$$

This is a quadratic equation and thus

$$R_1 = \frac{-10 \pm \sqrt{(100 + 4 \times 350)}}{2}$$

Hence $R_1 = -5 \pm 19.4 = 14.4\,\Omega$ or $-24.4\,\Omega$. Thus the positive solution gives 14.4 Ω.

For the iterative resistance R_2 seen between terminals C and D

$$R_2 = 20 + \frac{5(10 + R_2)}{5 + 10 + R_2}$$

Thus

$$R_2^2 + 15R_2 = 300 + 20R_2 + 50 + 5R_2$$

$$R_2^2 - 10R_2 - 350 = 0$$

Thus

$$R_2 = \frac{10 \pm \sqrt{(100 + 4 \times 350)}}{2}$$

Hence $R_2 = +5 \pm 19.4 = 24.4$ or $-14.4\,\Omega$. Thus the positive solution gives $24.4\,\Omega$. Note that this is the negative solution obtained when R_1 was being determined.

(*c*) The insertion loss is given by equation [26] as

$$A_L = 20\lg\left(\frac{I_{L1}}{I_{L2}}\right)$$

where I_{L1} is the current that flows through the load when the source is connected directly to the load and I_{L2} is the current when the network is between the source and the load. With the image resistance of $24.2\,\Omega$ between C and D, then in the absence of the network the source voltage V will give the current

$$I_{L1} = \frac{V}{24.2}$$

With the network present the total resistance of the network plus load is $14.4\,\Omega$. The current taken from the source is thus $V/14.4$ and so the current through the load I_{L2} is given by

$$5(I - I_{L2}) = (20 + 24.4)I_{L2}$$

$$I_{L2} = \frac{5I}{49.2} = \frac{5V}{14.4 \times 49.2}$$

Thus the insertion loss is

$$A_L = 20\lg\left[\frac{(V/24.4)}{5V/(14.4 \times 49.2)}\right] = 15.3\,\text{dB}$$

Problems

1 For the simple potential divider circuit shown in Fig. 7.28, what are the iterative impedances?

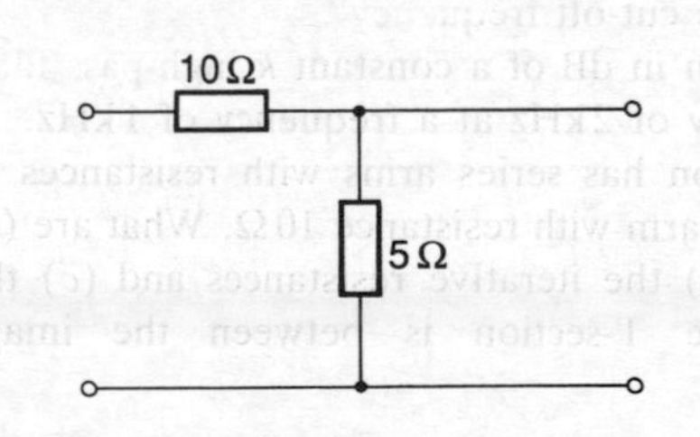

Fig. 7.28 Problem 1

2 If the ratio of input power to output power for an attenuator is 2, what is the attenuation in (*a*) dB, (*b*) Np?

3 What is the attenuation in dB produced by a cable if 10% of the input power appears at the output?

4 A symmetrical attenuator has a current input of 100 mA. What will be the current output if the attenuation is 4 dB?

5 Four identical attenuator sections are connected in cascade. If each has an attenuation of 5 dB, what will be the overall attenuation?

6 Four identical attenuator sections are connected in cascade to give an overall attenuation of 40 dB. What is (*a*) the attenuation of each section and (*b*) the output of the final stage if the input to the first is 20 mV?

7 What is (*a*) the characteristic impedance and (*b*) the attenuation

in dB of a symmetrical T-section attenuator having series resistances of 100 Ω and a shunt resistance of 80 Ω?

8 Design a symmetrical T-section attenuator to give an attenuation of 10 dB and have a characteristic impedance of 200 Ω.

9 A symmetrical π-attenuator has a series resistance of 400 Ω and shunt resistances of 1 kΩ. What is (*a*) the characteristic impedance and (*b*) the attenuation in dB?

10 Design a symmetrical π-section attenuator to have a characteristic resistance of 500 Ω and an attenuation of 15 dB.

11 Design a symmetrical π-section attenuator to have a characteristic resistance of 600 Ω and an attenuation of 20 dB.

12 What is the insertion loss of a symmetrical T-section network, with series resistances of 100 Ω and a shunt resistance of 200 Ω, when it is connected to a matched load?

13 What is the insertion loss of a symmetrical π-section network, with series resistance of 100 Ω and shunt resistances of 500 Ω, when it is connected to a matched load?

14 A symmetrical T-section network has series resistances of 400 Ω and a shunt resistance of 600 Ω. What is the insertion loss when the network is connected between a source and a load each having impedances of 600 Ω?

15 Explain how an L-section *RC* circuit can be used in some circumstances to give a low-pass filter and others a high-pass filter.

16 An L-section *RC* filter has R = 100 Ω and C = 10 μF and is used as a low pass filter. What is the cut off frequency and the attenuation in dB at that frequency?

17 What is the cut-off frequency and the design impedance for a constant k T-section filter having series inductances of 50 mH and a shunt capacitor of 0.1 μF?

18 Determine the component values for a constant k low-pass π filter if the design impedance is to be 600 Ω and the cut-off frequency 3 kHz.

19 Design a high-pass T-section filter to have a design impedance of 600 Ω and a cut-off frequency of 3 kHz.

20 Design a high-pass π-section filter to have a design impedance of 600 Ω and a cut-off frequency of 1.5 kHz.

21 What is the attenuation in dB of a low-pass π-section filter at (*a*) twice, (*b*) three times the cut-off frequency?

22 Determine the attenuation in dB of a constant k high-pass filter having a cut-off frequency of 2 kHz at a frequency of 1 kHz.

23 An asymmetrical T-section has series arms with resistances of 20 Ω and 5 Ω and a shunt arm with resistance 10 Ω. What are (*a*) the image resistances, (*b*) the iterative resistances and (*c*) the insertion loss when the T-section is between the image resistances?.

8 Two-port networks

Introduction

The term *two-port network* is used to describe a network that has two input and two output terminals, though very often one input and one output terminal are common. It is sometimes also called a *four-terminal network*. Such networks are of importance as transmission elements. In Chapter 7 the use of such devices as attenuators and filters was discussed. Other examples of two-port networks are transformers (see Chapter 5), transmission lines (see Chapter 9), and amplifiers. For all such networks there is a specific relationship between the input and output currents and voltages. This chapter considers briefly the range of such parameters but then concentrates on the application of the *ABCD*, or transmission, parameters.

Network parameters

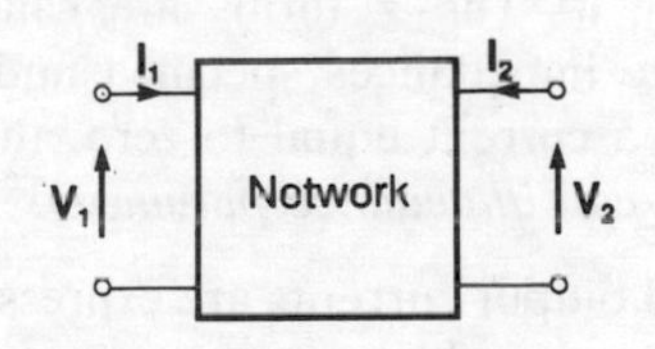

Fig. 8.1 Two-port network

A two-port network has a pair of input terminals and a pair of output terminals. When a voltage V_1 is applied to the input terminals then input and output currents, I_1 and I_2, flow and an output voltage V_2 is produced (Fig. 8.1). Note the directions of the voltages and currents specified in Fig. 8.1. Often I_2 is specified in the opposite direction; when this occurs it just assumes a negative value in equations. The reason for specifying both I_1 and I_2 as currents entering the network is that it really does not matter which of the ports is taken as the input and which as the output and so it is generally more convenient to indicate both ports as potential input ports and let the signs that develop for the currents indicate which one is.

There are thus four variable quantities V_1, I_1, V_2 and I_2 for a two-port network. Any two of these variables may be taken as independent variables; the other two then depend on them and can be expressed in terms of them. There are six different ways we can select two of the four variables and thus six different ways of describing a two-port network by means of writing equations for two of the terminal variables in terms of the other two. If the elements in the networks are assumed to be linear then we can write these six sets of relationships as:

1 *ABCD parameters* The input voltage and current are specified in terms of the output voltage and current. Since the relationships will be linear we can write them as

$$\mathbf{V}_1 = A\mathbf{V}_2 - B\mathbf{I}_2 \quad [1]$$

$$\mathbf{I}_1 = C\mathbf{V}_2 - D\mathbf{I}_2 \quad [2]$$

A, B, C and D are constants for a particular network and called *parameters*. For equation [1] when $\mathbf{I}_2 = 0$ then $A = \mathbf{V}_1/\mathbf{V}_2$ and when $\mathbf{V}_2 = 0$ then $B = -\mathbf{V}_1/\mathbf{I}_2$. For equation [2] when $\mathbf{I}_2 = 0$ then $C = \mathbf{I}_1/\mathbf{V}_2$ and when $\mathbf{V}_2 = 0$ then $D = -\mathbf{I}_1/\mathbf{I}_2$. Thus since A and D are ratios they are pure numbers, while B is an impedance and C an admittance. These parameters are sometimes called the *general circuit* or *transmission parameters* and are generally used for the analysis of two-port networks which are heavy current circuits or power frequency transmission lines.

2 *z parameters* The input and output voltages are expressed in terms of the input and output currents.

$$\mathbf{V}_1 = z_{11}\mathbf{I}_1 + z_{12}\mathbf{I}_2 \quad [3]$$

$$\mathbf{V}_2 = z_{21}\mathbf{I}_1 + z_{22}\mathbf{I}_2 \quad [4]$$

For equation [3] when $\mathbf{I}_2 = 0$ then $z_{11} = \mathbf{V}_1/\mathbf{I}_1$ and when $\mathbf{I}_1 = 0$ then $z_{12} = \mathbf{V}_1/\mathbf{I}_2$. For equation [4] when $\mathbf{I}_2 = 0$ then $z_{21} = \mathbf{V}_2/\mathbf{I}_1$ and when $\mathbf{I}_1 = 0$ then $z_{22} = \mathbf{V}_2/\mathbf{I}_2$. The subscripts used with a z relate to the subscripts of the two variables which specify it. The z terms are called *parameters* and, since all are impedances specified under open-circuit conditions, i.e. a current equal to zero, they are often called the *open-circuit impedance parameters*.

3 *y parameters* The input and output currents are expressed in terms of the input and output voltages.

$$\mathbf{I}_1 = y_{11}\mathbf{V}_1 + y_{12}\mathbf{V}_2 \quad [5]$$

$$\mathbf{I}_2 = y_{21}\mathbf{V}_1 + y_{22}\mathbf{V}_2 \quad [6]$$

With equation [5] when $\mathbf{V}_2 = 0$ then $y_{11} = \mathbf{I}_1/\mathbf{V}_1$ and when $\mathbf{V}_1 = 0$ then $y_{12} = \mathbf{I}_1/\mathbf{V}_2$. With equation [6] when $\mathbf{V}_2 = 0$ then $y_{21} = \mathbf{I}_2/\mathbf{V}_1$ and when $\mathbf{V}_1 = 0$ then $y_{22} = \mathbf{I}_2/\mathbf{V}_2$. The y parameters are all admittances and since all are admittances specified under short-circuit conditions, i.e. a voltage equal to zero, they are often called the *short-circuit admittance parameters*.

4 *h parameters* The input voltage and output current are

expressed in terms of the input current and output voltage.

$$V_1 = h_{11}I_1 + h_{12}V_2 \qquad [7]$$

$$I_2 = h_{21}I_1 + h_{22}V_2 \qquad [8]$$

With equation [7] when $V_2 = 0$ then $h_{11} = V_1/I_1$ and when $I_1 = 0$ then $h_{12} = V_1/V_2$. With equation [8] when $V_2 = 0$ then $h_{21} = I_2/I_1$ and when $I_1 = 0$ then $h_{22} = I_2/V_2$. These h parameters are called *hybrid parameters* because they may be an impedance, admittance or a pure number. These parameters are widely used for the representation of two-port networks which involve bipolar transistors.

5 *g parameters* The input current and output voltage are expressed in terms of the input voltage and output current.

$$I_1 = g_{11}V_1 + g_{12}I_2 \qquad [9]$$

$$V_2 = g_{21}V_1 + g_{22}I_2 \qquad [10]$$

With equation [9] when $I_2 = 0$ then $g_{11} = I_1/V_1$ and when $V_1 = 0$ then $g_{12} = I_1/I_2$. With equation [10] when $I_2 = 0$ then $g_{21} = V_2/V_1$ and when $V_1 = 0$ then $g_{22} = V_2/I_2$. The g parameters are sometimes called the *inverse h parameters* because they express the inverse of the h parameter relationships.

The sixth way of writing the parameters is to give the output voltage and current in terms of the input voltage and current. This is not generally used in practice. The six sets of parameters are obviously related. It is thus possible to express one set of parameters in terms of another set. Thus, for example, consider the z parameters (equations [3] and [4]).

$$V_1 = z_{11}I_1 + z_{12}I_2$$

$$V_2 = z_{21}I_1 + z_{22}I_2$$

Multiplying the first equation by z_{22} and the second equation by z_{12} gives

$$z_{22}V_1 = z_{22}z_{11}I_1 + z_{22}z_{12}I_2$$

$$z_{12}V_2 = z_{12}z_{21}I_1 + z_{22}z_{12}I_2$$

Subtracting then gives

$$z_{22}V_1 - z_{12}V_2 = z_{22}z_{11}I_1 - z_{12}z_{21}I_1$$

and hence

$$V_1 = \left(\frac{z_{22}z_{11} - z_{12}z_{21}}{z_{22}}\right)I_1 + \frac{z_{12}}{z_{22}}V_2$$

Comparison of this equation with equation [7]

$$\mathbf{V}_1 = h_{11}\mathbf{I}_1 + h_{12}\mathbf{V}_2$$

indicates that

$$h_{11} = \frac{z_{22}z_{11} - z_{12}z_{21}}{z_{22}}$$

$$h_{12} = \frac{z_{12}}{z_{22}}$$

Relationships between other parameters can be obtained in a similar way.

Example 1

Determine the y parameters for the two-port network shown in Fig. 8.2.

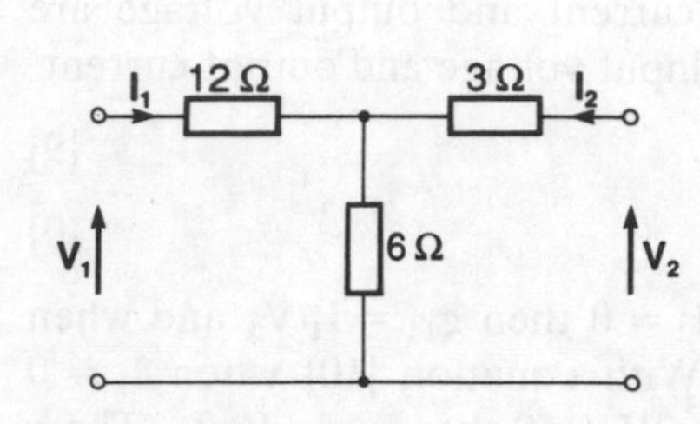

Fig. 8.2 Example 1

Answer

For the y parameters we have (equations [5] and [6]),

$$\mathbf{I}_1 = y_{11}\mathbf{V}_1 + y_{12}\mathbf{V}_2$$

$$\mathbf{I}_2 = y_{21}\mathbf{V}_1 + y_{22}\mathbf{V}_2$$

When $\mathbf{V}_1$ is 0, i.e. the input port is short-circuited (Fig. 8.3(*a*)), then the resulting parallel arrangement of 12 Ω and 6 Ω is (12 × 6)/(12 + 6) = 4 Ω. Thus, since the total resistance is now (3 + 4) Ω, $\mathbf{I}_2 = \mathbf{V}_2/7$. But with $\mathbf{V}_1 = 0$

$$\mathbf{I}_2 = y_{22}\mathbf{V}_2$$

Hence $y_{22} = 1/7$ S. The potential difference across the parallel arrangement of the 6 Ω and 12 Ω resistors is $4\mathbf{I}_2 = 4\mathbf{V}_2/7$. Thus the current $\mathbf{I}_1$ through the 12 Ω resistor is $\mathbf{I}_1 = -(4\mathbf{V}_2/7)/12 = \mathbf{V}_2/21$. The minus sign is because the direction of the current is opposite to that used to define the direction in Fig. 8.1. But with $\mathbf{V}_1 = 0$

$$\mathbf{I}_1 = y_{12}\mathbf{V}_2$$

Hence $y_{12} = -1/21$ S

When $\mathbf{V}_2 = 0$, i.e. the output port is short-circuited (Fig. 8.3(*b*)), then the resulting parallel arrangement of 3 Ω and 6 Ω is (3 × 6)/(3 + 6) = 2 Ω. Thus, since the total resistance is now (2 + 12) = 14 Ω, $\mathbf{I}_1 = \mathbf{V}_1/14$. But with $\mathbf{V}_2 = 0$

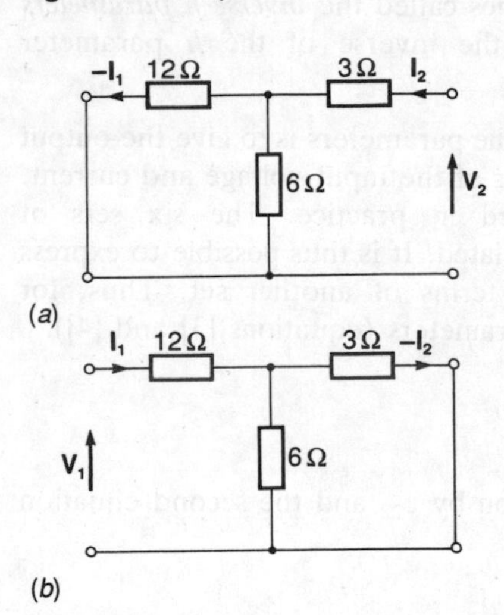

Fig. 8.3 Example 1

$$\mathbf{I}_1 = y_{11}\mathbf{V}_1$$

Hence $y_{11} = 1/14$ S. The potential difference across the parallel arrangement of the 6 Ω and 3 Ω resistors is $\mathbf{V} = 2\mathbf{I}_1 = 2\mathbf{V}_1/14 = \mathbf{V}_1/7$. Thus the current $\mathbf{I}_2$ through the 3 Ω resistor is $-\mathbf{I}_2 = (\mathbf{V}_1/7)/3 = -\mathbf{V}_1/21$. But with $\mathbf{V}_2 = 0$

$$\mathbf{I}_2 = y_{21}\mathbf{I}_1$$

Hence $y_{21} = -1/21$ S.

Thus the y parameter equations describing the two-port network can be written as

$$\mathbf{I}_1 = (1/14)\mathbf{V}_1 - (1/21)\mathbf{V}_2$$

$$\mathbf{I}_2 = -(1/21)\mathbf{V}_1 + (1/7)\mathbf{V}_2$$

Example 2

A two-port network has the parameters $A = 1 + j1$, $B = 2\,\Omega$, $C = 1 + j1.5\,\text{S}$, and $D = 3$. What is the input current and voltage when the output is a current of 100 mA through a resistive load of 10 Ω?

Answer

Equation [1] gives

$$\mathbf{V}_1 = A\mathbf{V}_2 - B\mathbf{I}_2 = (1 + j1)(10 \times 0.100) - 2 \times 0.100$$
$$= 0.8 + j1 = 1.28 \quad \underline{/51.3^\circ\,\text{V}}$$

Equation [2] gives

$$\mathbf{I}_1 = C\mathbf{V}_2 - D\mathbf{I}_2 = (1 + j1.5)(10 \times 0.100) - 3 \times 0.100$$
$$= 0.7 + j1.5 = 1.66 \quad \underline{/65.0^\circ\,\text{A}}$$

Example 3

A two-port network has the following transmission parameters: $A = 2$, $B = 50\,\Omega$, $C = 0.02\,\text{S}$, $D = 1$. What are the values of the input impedance when the output is (*a*) open-circuit, (*b*) short-circuited?

Answer

(*a*) Equation [1] gives when the output is open circuit, i.e. $\mathbf{I}_2 = 0$,

$$\mathbf{V}_1 = A\mathbf{V}_2 - B\mathbf{I}_2 = A\mathbf{V}_2 - 0$$

Equation [2] gives

$$\mathbf{I}_1 = C\mathbf{V}_2 - D\mathbf{I}_2 = C\mathbf{V}_2 - 0$$

The input impedance Z is thus

$$Z = \frac{\mathbf{V}_1}{\mathbf{I}_1} = \frac{A\mathbf{V}_2}{C\mathbf{V}_2} = \frac{A}{C} = \frac{2}{0.02} = 100\,\Omega$$

(*b*) Equation [1] gives when the output is short-circuited, i.e. $\mathbf{V}_2 = 0$,

$$\mathbf{V}_1 = A\mathbf{V}_2 - B\mathbf{I}_2 = 0 - B\mathbf{I}_2$$

Equation [2] gives

$$\mathbf{I}_1 = C\mathbf{V}_2 - D\mathbf{I}_2 = 0 - D\mathbf{I}_2$$

The input impedance Z is thus

$$Z = \frac{\mathbf{V}_1}{\mathbf{I}_1} = \frac{-B\mathbf{I}_2}{-D\mathbf{I}_2} = \frac{B}{D} = \frac{50}{1} = 50\,\Omega$$

Matrices

All the above descriptions of the behaviour of two-port networks in terms of parameters involve two simultaneous equations. Such equations can be described, and solved, by matrices. Thus, for example, equations [1] and [2]

$$\mathbf{V}_1 = A\mathbf{V}_2 - B\mathbf{I}_2$$

$$\mathbf{I}_1 = C\mathbf{V}_2 - D\mathbf{I}_2$$

can be represented in matrix notation by

$$\begin{bmatrix} \mathbf{V}_1 \\ \mathbf{I}_1 \end{bmatrix} = \begin{bmatrix} A & B \\ C & D \end{bmatrix} \begin{bmatrix} \mathbf{V}_2 \\ -\mathbf{I}_2 \end{bmatrix} \qquad [11]$$

The matrix involving A, B, C and D is often referred to as the $ABCD$ or transmission matrix. The following is a brief indication of how matrices can be used to solve simultaneous equations.

A matrix which has two rows and two columns is called a 2 by 2 matrix, one which has just one column a column matrix. Thus the two equations are represented by a 2 by 1 column matrix which is equal to the product of a 2 by 2 matrix and a 2 by 1 column matrix.

When a matrix is multiplied by a number then each element in that matrix is multiplied by the number. Thus, for example,

$$5\begin{bmatrix} 1 & 2 \\ 3 & 4 \end{bmatrix} = \begin{bmatrix} 5 & 10 \\ 15 & 20 \end{bmatrix}$$

When two matrices are multiplied, each row of the first matrix is combined with each column of the second matrix. Thus, for example,

$$\begin{bmatrix} 1 & 2 \\ 3 & 4 \end{bmatrix} \begin{bmatrix} 5 \\ 6 \end{bmatrix}$$

is

$$\begin{bmatrix} 1 & 2 \end{bmatrix} \begin{bmatrix} 5 \\ 6 \end{bmatrix} = 1 \times 5 + 2 \times 6 = 17$$

$$\begin{bmatrix} 3 & 4 \end{bmatrix} \begin{bmatrix} 5 \\ 6 \end{bmatrix} = 3 \times 5 + 4 \times 6 = 39$$

Thus the product of the two matrices is

$$\begin{bmatrix} 1 & 2 \\ 3 & 4 \end{bmatrix} \begin{bmatrix} 5 \\ 6 \end{bmatrix} = \begin{bmatrix} 17 \\ 39 \end{bmatrix}$$

The procedure for solving the two simultaneous equations by matrices is to:

1 Write the equations as a matrix equation, as in equation [11].

2 Determine the inverse of the square matrix. The inverse

matrix is defined as that matrix which multiplied by the original matrix gives the unit matrix. A unit matrix is one for which all the elements in the diagonal going from top left to bottom right have the value 1 and all the other elements have the value 0. This can be shown to give the same result for the inverse matrix as interchanging the positions of A and D, changing the signs of B and C and multiplying this new matrix by the reciprocal of the determinant of the matrix. The determinant is defined as $AD - BC$ and is written between vertical lines as shown:

$$\text{The inverse of } \begin{bmatrix} A & B \\ C & D \end{bmatrix} = \frac{1}{\begin{vmatrix} A & B \\ C & D \end{vmatrix}} \begin{bmatrix} D & -B \\ -C & A \end{bmatrix}$$

$$= \frac{1}{AD - BC} \begin{bmatrix} D & -B \\ -C & A \end{bmatrix}$$

3 Multiply each side of the matrix equation by the inverse matrix. The product of the $ABCD$ matrix and its inverse has the value 1. Thus the matrix equation then becomes

$$\begin{bmatrix} \mathbf{V}_2 \\ -\mathbf{I}_2 \end{bmatrix} = \frac{1}{AD - BC} \begin{bmatrix} D & -B \\ -C & A \end{bmatrix} \begin{bmatrix} \mathbf{V}_1 \\ \mathbf{I}_1 \end{bmatrix}$$

4 Solve for the unknowns by equating corresponding elements.

Example 4

For a two-port network

$$\mathbf{V}_1 = 6 = 3\,\mathbf{V}_2 - 2\,\mathbf{I}_2$$

$$\mathbf{I}_1 = 2 = 4\,\mathbf{V}_2 - 5\,\mathbf{I}_2$$

The voltage is in volts and the current in amps. What is (*a*) the transmission matrix and (*b*) the values of $\mathbf{V}_2$ and $\mathbf{I}_2$?

Answer

(*a*) The transmission matrix is

$$\begin{bmatrix} 3 & 2 \\ 4 & 5 \end{bmatrix}$$

(*b*) The two simultaneous equations can be written as

$$\begin{bmatrix} 6 \\ 2 \end{bmatrix} = \begin{bmatrix} 3 & 2 \\ 4 & 5 \end{bmatrix} \begin{bmatrix} \mathbf{V}_2 \\ -\mathbf{I}_2 \end{bmatrix}$$

The inverse of the transmission matrix is

$$\frac{1}{15 - 8} \begin{bmatrix} 5 & -2 \\ -4 & 3 \end{bmatrix} = \begin{bmatrix} 5/7 & -2/7 \\ -4/7 & 3/7 \end{bmatrix}$$

Thus

$$\begin{bmatrix} \mathbf{V}_2 \\ -\mathbf{I}_2 \end{bmatrix} = \begin{bmatrix} 6 \\ 2 \end{bmatrix} \begin{bmatrix} 5/7 & -2/7 \\ -4/7 & 3/7 \end{bmatrix} = \begin{bmatrix} 26/7 \\ -18/7 \end{bmatrix}$$

Thus $\mathbf{V}_2 = 26/7$ V and $\mathbf{I}_2 = 18/7$ A.

Cascaded networks

Consider two networks in cascade and their *ABCD* parameters (Fig. 8.4). For network 1 we have

$$\begin{bmatrix} \mathbf{V}_1 \\ \mathbf{I}_1 \end{bmatrix} = \begin{bmatrix} A_1 & B_1 \\ C_1 & D_1 \end{bmatrix} \begin{bmatrix} \mathbf{V}_2 \\ -\mathbf{I}_2 \end{bmatrix}$$

and for network 2

$$\begin{bmatrix} \mathbf{V}_2 \\ -\mathbf{I}_2 \end{bmatrix} = \begin{bmatrix} A_2 & B_2 \\ C_2 & D_2 \end{bmatrix} \begin{bmatrix} \mathbf{V}_3 \\ -\mathbf{I}_3 \end{bmatrix}$$

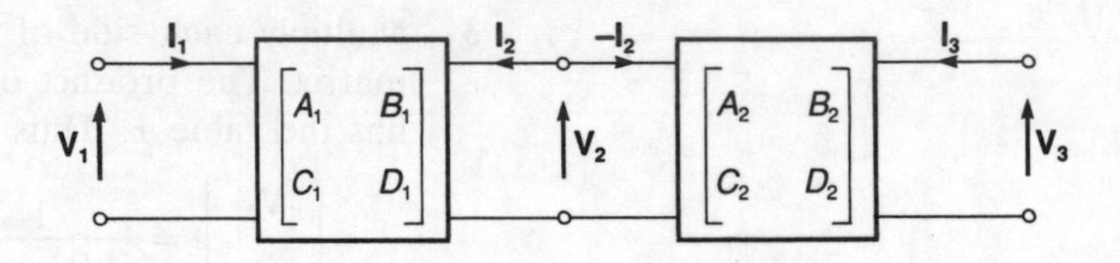

Fig. 8.4 Networks in cascade

Thus we can write

$$\begin{bmatrix} \mathbf{V}_1 \\ \mathbf{I}_2 \end{bmatrix} = \begin{bmatrix} A_1 & B_1 \\ C_1 & D_1 \end{bmatrix} \begin{bmatrix} A_2 & B_2 \\ C_2 & D_2 \end{bmatrix} \begin{bmatrix} \mathbf{V}_3 \\ -\mathbf{I}_3 \end{bmatrix}$$

$$= \begin{bmatrix} A_1A_2 + B_1C_2 & A_1B_2 + B_1D_2 \\ C_1A_2 + D_1C_2 & C_1B_2 + D_1D_2 \end{bmatrix} \begin{bmatrix} \mathbf{V}_3 \\ -\mathbf{I}_3 \end{bmatrix} \quad [12]$$

Thus the cascaded network behaves as a network with the parameters

$$A = A_1A_2 + B_1C_2$$

$$B = A_1B_2 + B_1D_2$$

$$C = C_1A_2 + D_1C_2$$

$$D = C_1B_2 + D_1D_2$$

Example 5

Derive the matrix for the *y* parameters of two networks in parallel (Fig. 8.5).

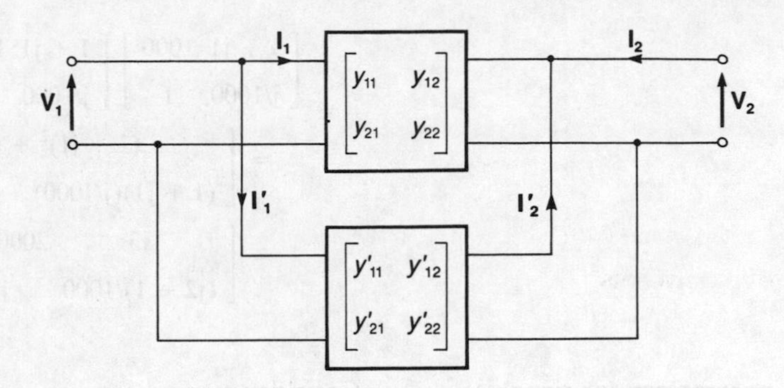

Fig. 8.5 Example 5

Answer

Equations [5] and [6] give

$$\mathbf{I}_1 = y_{11}\mathbf{V}_1 + y_{12}\mathbf{V}_2$$

$$\mathbf{I}_2 = y_{21}\mathbf{V}_1 + y_{22}\mathbf{V}_2$$

and hence for element 1

$$\begin{bmatrix} \mathbf{I}_1 \\ \mathbf{I}_2 \end{bmatrix} = \begin{bmatrix} y_{11} & y_{12} \\ y_{21} & y_{22} \end{bmatrix} \begin{bmatrix} \mathbf{V}_1 \\ \mathbf{V}_2 \end{bmatrix}$$

Similarly for element 2, since the voltages are the same because they are in parallel,

$$\begin{bmatrix} \mathbf{I}'_1 \\ \mathbf{I}'_2 \end{bmatrix} = \begin{bmatrix} y'_{11} & y'_{12} \\ y'_{21} & y'_{22} \end{bmatrix} \begin{bmatrix} \mathbf{V}_1 \\ \mathbf{V}_2 \end{bmatrix}$$

The total current input to the networks is $\mathbf{I}_1 + \mathbf{I}'_1$ and the total output is $\mathbf{I}_2 + \mathbf{I}'_2$. The corresponding elements in two matrices may be added to form a single matrix. Thus adding the matrix equations for the two parallel elements gives

$$\begin{bmatrix} \mathbf{I}_1 + \mathbf{I}'_1 \\ \mathbf{I}_2 + \mathbf{I}'_2 \end{bmatrix} = \begin{bmatrix} y_{11} + y'_{11} & y_{12} + y'_{12} \\ y_{21} + y'_{21} & y_{22} + y'_{22} \end{bmatrix} \begin{bmatrix} \mathbf{V}_1 \\ \mathbf{V}_2 \end{bmatrix}$$

Thus the network as a whole has y parameters which are the sums of the corresponding y parameters of the two networks.

Example 6

A network has the transmission parameters of

$$\begin{bmatrix} 1 + j1 & 1000 \\ j/1000 & 1 \end{bmatrix}$$

What will be the parameters for two such networks in cascade?

Answer

The parameters for two such networks in cascade will be

$$\begin{bmatrix} 1 + j1 & 1000 \\ j/1000 & 1 \end{bmatrix} \begin{bmatrix} 1 + j1 & 1000 \\ j/1000 & 1 \end{bmatrix}$$

$$= \begin{bmatrix} (1 + j1)^2 + j1 & 1000(1 + j1) + 1000 \\ (1 + j1)(j/1000) + (j/1000) & j1 + 1 \end{bmatrix}$$

$$= \begin{bmatrix} j3 & 2000 + j1000 \\ (j2 - 1)/1000 & j1 + 1 \end{bmatrix}$$

ABCD parameters for passive networks

Consider a two-port network and the *ABCD* parameters when one pair of terminals is short-circuited with an input **V** to the other terminals (Fig. 8.6(*a*)). Equations [1] and [2] give

$$\mathbf{V} = A\mathbf{V}_2 - B\mathbf{I}_2 = 0 - B\mathbf{I}_2 \qquad [13]$$

$$\mathbf{I}_1 = C\mathbf{V}_2 - D\mathbf{I}_2 = 0 - D\mathbf{I}_2 \qquad [14]$$

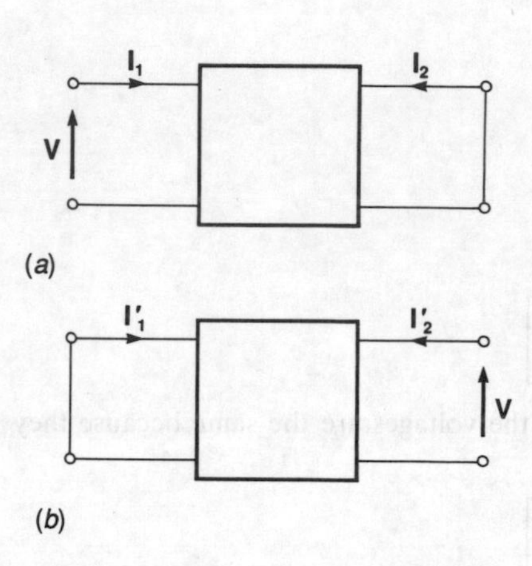

Fig. 8.6 *ABCD* parameters with a passive network

Now consider the situation when the other port is short-circuited with an input **V** at the port that was initially short-circuited (Fig. 8.6(*b*)). Equations [1] and [2] now give

$$\mathbf{V}_1 = 0 = A\mathbf{V} - B\mathbf{I}'_2 \qquad [15]$$

$$\mathbf{I}'_1 = C\mathbf{V} - D\mathbf{I}'_2 \qquad [16]$$

Equation [15] gives $\mathbf{I}'_2 = (A/B)\mathbf{V}$ and so equation [16] can be written as

$$\mathbf{I}'_1 = C\mathbf{V} - D(A/B)\mathbf{V} = \frac{(BC - AD)\mathbf{V}}{B}$$

Substituting for **V** by using equation [13] now gives

$$\mathbf{I}'_1 = (AD - BC)\mathbf{I}_2$$

There is a theorem called the *reciprocity theorem* which states that if a voltage is applied to a linear passive network at one point and produces a current at another, then the same voltage applied to the second point will produce the same current at the first point. Thus, since we have applied the same voltage at the two points, we must have $\mathbf{I}'_1 = \mathbf{I}_2$. Thus

$$AD - BC = 1 \qquad [17]$$

This condition must be satisfied by any linear passive network.

Example 7

Tests on a passive two-port network gave the following results:

Output open-circuit: $\mathbf{V}_1 = 10\,\text{V}$, $\mathbf{V}_2 = 5\,\text{V}$, $\mathbf{I}_1 = 0.1\,\text{A}$

Output short-circuited: $\mathbf{V}_1 = 10\,\text{V}$, $\mathbf{I}_1 = 0.2\,\text{A}$, $\mathbf{I}_2 = 0.2\,\text{A}$

Determine the values of the *ABCD* parameters and confirm that the network is passive.

Answer

Equation [1] gives, when the output is open-circuit,

$$\mathbf{V}_1 = A\mathbf{V}_2 - B\mathbf{I}_2 = A\mathbf{V}_2 - 0$$

$$10 = 5A$$

Hence $A = 2$. Equation [2] gives, when the output is open-circuit,

$$\mathbf{I}_1 = C\mathbf{V}_2 - D\mathbf{I}_2 = C\mathbf{V}_2 - 0$$

$$0.1 = 5C$$

Hence $C = 0.02\,\text{S}$. Equation [1] gives, when the output is short-circuited,

$$\mathbf{V}_1 = A\mathbf{V}_2 - B\mathbf{I}_2 = 0 - B\mathbf{I}_2$$

$$10 = -(-0.2)B$$

Hence B is $50\,\Omega$ if we assume that $\mathbf{I}_2$ is in the same direction as $\mathbf{I}_1$. Equation [2] gives, when the output is short-circuited,

$$\mathbf{I}_1 = C\mathbf{V}_2 - D\mathbf{I}_2 = 0 - D\mathbf{I}_2$$

$$0.2 = -(-0.2)D$$

Hence $D = 1$.

For a passive network equation [17] must be true, i.e.

$$AD - BC = 1$$

Thus substituting the values of the parameters

$$2 \times 1 - 50 \times 0.02 = 1$$

Thus the network is passive

ABCD parameters for networks

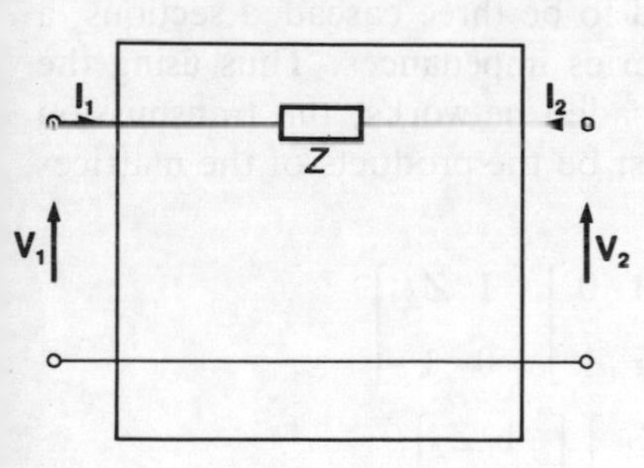

Fig. 8.7 Series impedance

Consider a two-port network which is just a *series impedance* Z (Fig. 8.7). Kirchoff's voltage law applied to the circuit gives

$$\mathbf{V}_1 = \mathbf{V}_2 - Z_2\mathbf{I}_2$$

The input current $\mathbf{I}_1$ must equal $-\mathbf{I}_2$ and thus we can write

$$\mathbf{I}_1 = 0 - \mathbf{I}_2$$

Comparing these equations with equations [1] and [2]

$$\mathbf{V}_1 = A\mathbf{V}_2 - B\mathbf{I}_2$$

$$\mathbf{I}_1 = C\mathbf{V}_2 - D\mathbf{I}_2$$

then we must have $A = 1$, $B = \text{Z}$, $C = 0$ and $D = 1$ and the transmission matrix is

$$\begin{bmatrix} 1 & Z \\ 0 & 1 \end{bmatrix} \qquad [18]$$

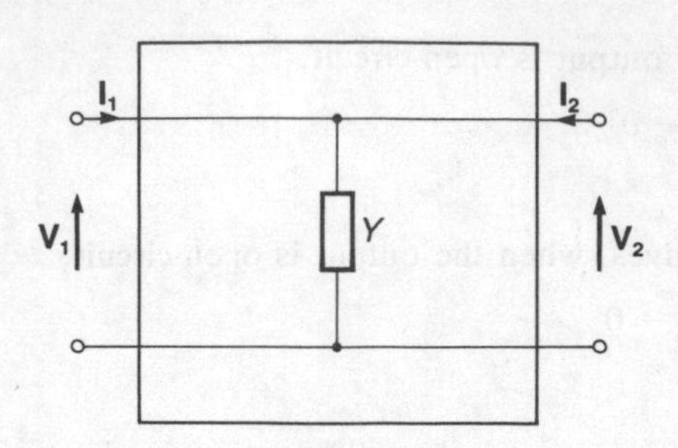

Fig. 8.8 Shunt admittance

Consider a two-port network which is just a *shunt admittance* Y (Fig. 8.8). We must have

$$V_1 = V_2 + 0$$

and, since the current through the admittance Y is $(I_1 + I_2)$ then $(I_1 + I_2) = YV_2$ and so

$$I_1 = YV_2 - I_2$$

Comparing these equations with equations [1] and [2]

$$V_1 = AV_2 - BI_2$$

$$I_1 = CV_2 - DI_2$$

then we must have $A = 1$, $B = 0$, $C = Y$, $D = 1$ and the transmission matrix is

$$\begin{bmatrix} 1 & 0 \\ Y & 1 \end{bmatrix} \quad [19]$$

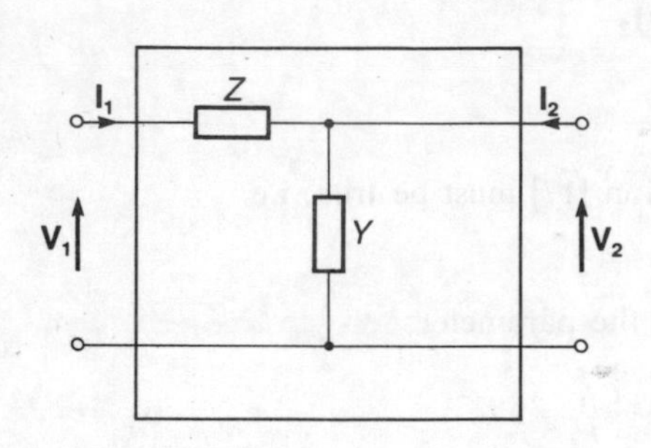

Fig. 8.9 L-network

Consider a two-port network which is an *L-network* (Fig. 8.9). This can be considered to be the cascade connection of a single series impedance and a single shunt admittance. Thus using the matrix equation [12] for cascaded networks, the transmission matrix for the L-network must be the product of the matrices [18] and [19], i.e.

$$\begin{bmatrix} A & B \\ C & D \end{bmatrix} = \begin{bmatrix} 1 & Z \\ 0 & 1 \end{bmatrix}\begin{bmatrix} 1 & 0 \\ Y & 1 \end{bmatrix} = \begin{bmatrix} 1 + YZ & Z \\ Y & 1 \end{bmatrix} \quad [20]$$

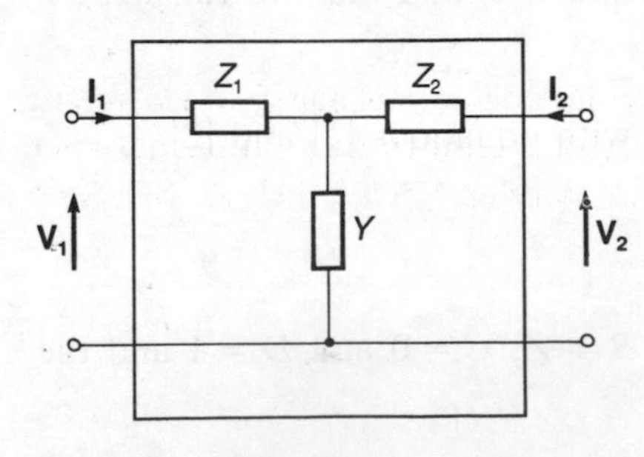

Fig. 8.10 T-network

Consider a two-port network which is a *T-network* (Fig. 8.10). This can be considered to be three cascaded sections, a shunt admittance and two series impedances. Thus using the matrix equation [12] for cascaded networks, the transmission matrix for the T-network must be the products of the matrices [18], [19] and [18], i.e.

$$\begin{bmatrix} A & B \\ C & D \end{bmatrix} = \begin{bmatrix} 1 & Z_1 \\ 0 & 1 \end{bmatrix}\begin{bmatrix} 1 & 0 \\ Y & 1 \end{bmatrix}\begin{bmatrix} 1 & Z_2 \\ 0 & 1 \end{bmatrix}$$

$$= \begin{bmatrix} 1 + YZ_1 & Z_1 \\ Y & 1 \end{bmatrix}\begin{bmatrix} 1 & Z_2 \\ 0 & 1 \end{bmatrix}$$

$$= \begin{bmatrix} 1 + YZ_1 & Z_1 + Z_2 + YZ_1Z_2 \\ Y & 1 + YZ_2 \end{bmatrix} \quad [21]$$

Consider a two-port network which is a π-network (Fig. 8.11). This can be considered to be three cascaded sections, two shunt admittances and a series impedance. Thus using the

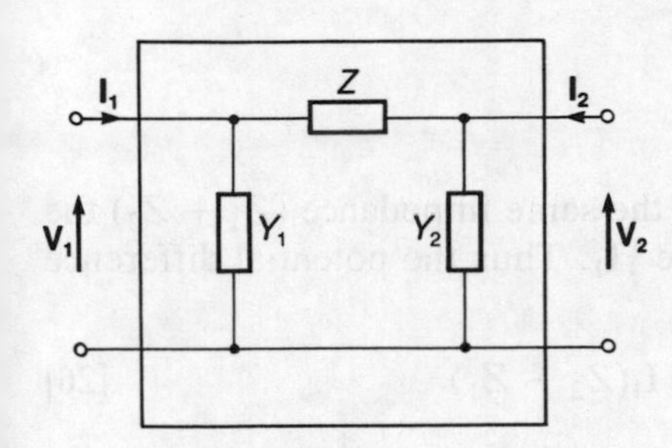

Fig. 8.11 π-network

matrix equation [12] for cascaded networks, the transmission matrix for the π-network must be the products of the matrices [19], [18] and [19], i.e.

$$\begin{bmatrix} A & B \\ C & D \end{bmatrix} = \begin{bmatrix} 1 & 0 \\ Y_1 & 1 \end{bmatrix} \begin{bmatrix} 1 & Z \\ 0 & 1 \end{bmatrix} \begin{bmatrix} 1 & 0 \\ Y_2 & 1 \end{bmatrix}$$

$$= \begin{bmatrix} 1 & Z \\ Y_1 & 1 + Y_1 Z \end{bmatrix} \begin{bmatrix} 1 & 0 \\ Y_2 & 1 \end{bmatrix}$$

$$= \begin{bmatrix} 1 + Y_2 Z & Z \\ Y_1 + Y_2 + Y_1 Y_2 Z & 1 + Y_1 Z \end{bmatrix} \qquad [22]$$

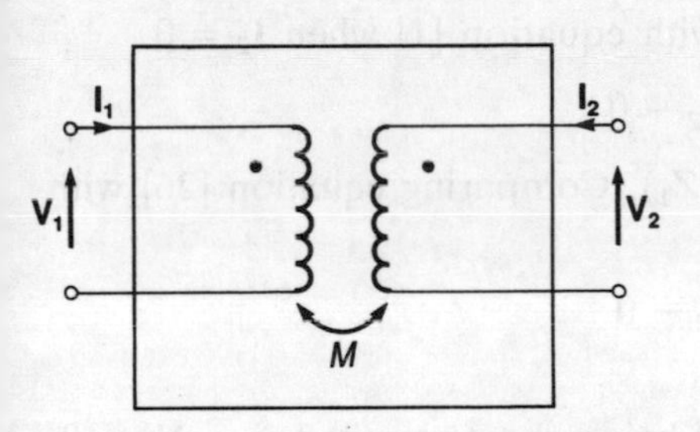

Fig. 8.12 Pure mutual inductance

Consider a two-port network which is a pure *mutual inductance* (Fig. 8.12). Applying Kirchoff's voltage law to the primary circuit (see the discussion in Chapter 5 and the equivalent circuit in Fig. 5.3)

$$\mathbf{V}_1 + j\omega M \mathbf{I}_2 = 0$$

$$\mathbf{V}_1 = -j\omega M \mathbf{I}_2 \qquad [23]$$

and for the secondary circuit

$$\mathbf{V}_2 = j\omega M \mathbf{I}_1$$

$$\mathbf{I}_1 = (1/j\omega M)\mathbf{V}_2 \qquad [24]$$

Comparing these equations [23] and [24] with [1] and [2]

$$\mathbf{V}_1 = A\mathbf{V}_2 - B\mathbf{I}_2$$

$$\mathbf{I}_1 = C\mathbf{V}_2 - D\mathbf{I}_2$$

then we must have $A = 0$, $B = j\omega M$, $C = 1/j\omega M$ and $D = 0$. Thus the transmission matrix is

$$\begin{bmatrix} 0 & j\omega M \\ 1/j\omega M & 0 \end{bmatrix} \qquad [25]$$

Consider a two-port network which is a *symmetrical lattice* (Fig. 8.13(*a*)). This circuit can be redrawn as a bridge network (Fig. 8.13(*b*)). With the output open-circuit, i.e. $\mathbf{I}_2 = 0$, then Fig. 8.13(*b*) indicates that

$$\mathbf{I}_1 = \frac{\mathbf{V}_1}{Z}$$

where

$$\frac{1}{Z} = \frac{1}{Z_1 + Z_2} + \frac{1}{Z_1 + Z_2}$$

$$Z = \tfrac{1}{2}(Z_1 + Z_2)$$

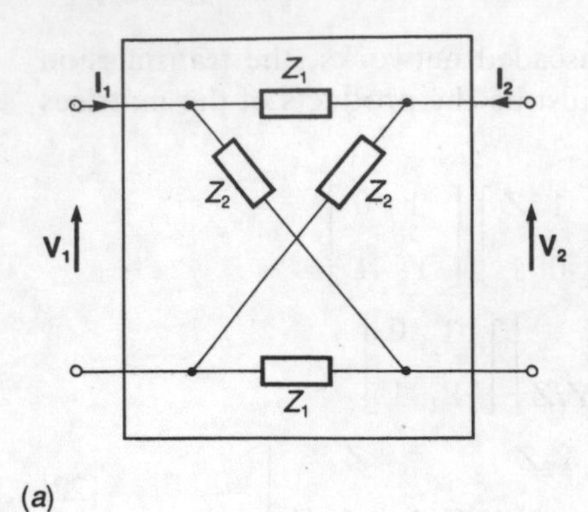

(a)

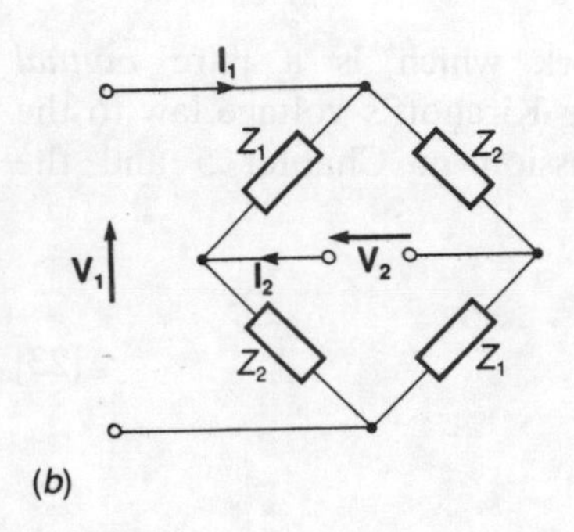

(b)

Fig. 8.13 The symmetrical lattice

Thus

$$\mathbf{I}_1 = \frac{\mathbf{V}_1}{\frac{1}{2}(Z_1 + Z_2)}$$

Because the two arms have the same impedance $(Z_1 + Z_2)$ the current through each will be $\frac{1}{2}\mathbf{I}_1$. Thus the potential difference $\mathbf{V}_2$ is

$$\mathbf{V}_2 = \tfrac{1}{2}\mathbf{I}_1 Z_2 - \tfrac{1}{2}\mathbf{I}_1 Z_1 = \tfrac{1}{2}\mathbf{I}_1(Z_2 - Z_1) \quad [26]$$

$$= \frac{\mathbf{V}_1(Z_2 - Z_1)}{Z_1 + Z_2}$$

$$\mathbf{V}_1 = \left(\frac{Z_1 + Z_2}{Z_2 - Z_1}\right)\mathbf{V}_2 \quad [27]$$

Comparing equation [27] with equation [1] when $\mathbf{I}_2 = 0$

$$\mathbf{V}_1 = A\mathbf{V}_2 - B\mathbf{I}_2 = A\mathbf{V}_2 - 0$$

then $A = (Z_1 + Z_2)/(Z_2 - Z_1)$. Comparing equation [26] with equation [2] when $\mathbf{I}_2 = 0$

$$\mathbf{I}_1 = C\mathbf{V}_2 - D\mathbf{I}_2 = C\mathbf{V}_2 - 0$$

then $C = 2/(Z_2 - Z_1)$.

Now consider the output being short-circuited, i.e. $\mathbf{V}_2 = 0$. The current $\mathbf{I}_2$ will then be

$$\mathbf{I}_2 = \frac{\mathbf{V}}{Z_2} - \frac{\mathbf{V}}{Z_1}$$

where $\mathbf{V}$ is the potential difference across Z_1 and also across Z_2 (this must be the same potential difference in order that $\mathbf{V}_2 = 0$). But because the network is symmetrical $\mathbf{V} = \frac{1}{2}\mathbf{V}_1$. Hence

$$\mathbf{I}_2 = \frac{\mathbf{V}_1}{2}\left(\frac{1}{Z_2} - \frac{1}{Z_1}\right)$$

$$\mathbf{V}_1 = \left(\frac{2Z_1Z_2}{Z_1 - Z_2}\right)\mathbf{I}_2 \quad [28]$$

Comparing equation [28] with equation [1] when $\mathbf{V}_2 = 0$

$$\mathbf{V}_1 = A\mathbf{V}_2 - B\mathbf{I}_2 = 0 - B\mathbf{I}_2$$

then $B = 2Z_1Z_2/(Z_2 - Z_1)$. Since

$$\mathbf{I}_1 = \frac{\mathbf{V}}{Z_1} + \frac{\mathbf{V}}{Z_2} = \frac{\mathbf{V}_1}{2}\left(\frac{1}{Z_1} + \frac{1}{Z_2}\right)$$

$$= \frac{\mathbf{I}_2}{2}\left(\frac{2Z_1Z_2}{Z_1 - Z_2}\right)\left(\frac{Z_1 + Z_2}{Z_1Z_2}\right)$$

$$= \mathbf{I}_2\left(\frac{Z_1 + Z_2}{Z_1 - Z_2}\right) \quad [29]$$

then, on comparing equation [29] with equation [2] when $V_2 = 0$

$$I_1 = CV_2 - DI_2 = 0 - DI_2$$

we have $D = (Z_1 + Z_2)/(Z_2 - Z_1)$. Thus the transmission matrix for the symmetrical lattice is

$$\begin{bmatrix} (Z_1 + Z_2)/(Z_2 - Z_1) & 2Z_1Z_2/(Z_2 - Z_1) \\ 2/(Z_2 - Z_1) & (Z_1 + Z_2)/(Z_2 - Z_1) \end{bmatrix} \quad [30]$$

Example 8

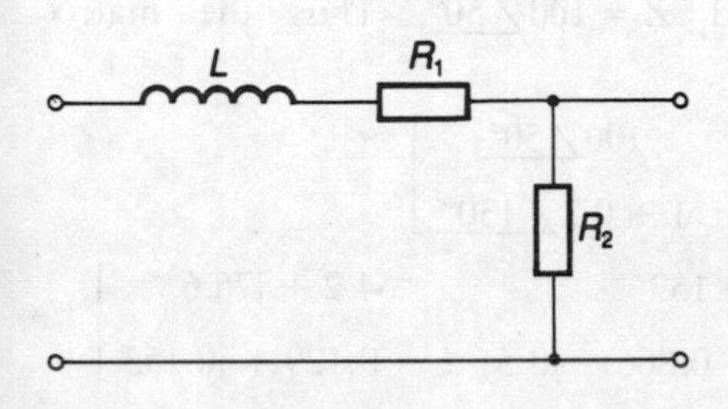

Fig. 8.14 Example 8

What are the transmission parameters for the L circuit in Fig. 8.14?

Answer

As derived above in equation [20], an L-network has the transmission parameters

$$\begin{bmatrix} A & B \\ C & D \end{bmatrix} = \begin{bmatrix} 1 + YZ & Z \\ Y & 1 \end{bmatrix}$$

where Y is the admittance of the shunt element, in this case $1/R_2$, and Z is the impedance of the series element, in this case $R_1 + j\omega L$. Thus

$$\begin{bmatrix} A & B \\ C & D \end{bmatrix} = \begin{bmatrix} 1 + (R_1 + j\omega L)/R_2 & R_1 + j\omega L \\ 1/R_2 & 1 \end{bmatrix}$$

Example 9

What are the transmission parameters of a symmetrical T-network which has series arms of $(100 + j50)\,\Omega$ and a shunt arm of $-j100\,\Omega$.

Answer

As derived above in equation [21] the transmission parameters for a T-network are

$$\begin{bmatrix} 1 + YZ_1 & Z_1 + Z_2 + YZ_1Z_2 \\ Y & 1 + YZ_2 \end{bmatrix}$$

and when symmetrical with the series arms the same, i.e. $Z_1 = Z_2 = 100 + j50$, and the shunt arm with an admittance of $1/(-j100) = j/100$

$$\begin{bmatrix} 1 - 0.5 + j1 & (100 + j50)(2 - 0.5 + j1) \\ j/100 & 1 + 0.5 + j1 \end{bmatrix}$$

$$= \begin{bmatrix} 0.5 + j1 & 50(2 + j3.5) \\ j/100 & 0.5 + j1 \end{bmatrix}$$

Example 10

A symmetrical π-network has a series impedance of $100\angle 50°\,\Omega$ and shunt admittances of $0.002\angle 80°\,S$. What is (*a*) the transmission matrix and (*b*) the input voltage and current when there is a load resistance of 50 Ω across the output and the input voltage produces a current of 2 mA through it?

Answer

(*a*) The transmission matrix for a π-network is given by equation [22] as

$$\begin{bmatrix} 1 + Y_2Z & Z \\ Y_1 + Y_2 + Y_1Y_2Z & 1 + Y_1Z \end{bmatrix}$$

$Y_1 = Y_2 = 0.002\angle 80°$ and $Z = 100\angle 50°$. Thus the matrix becomes

$$\begin{bmatrix} 1 + 0.2\angle 130° & 100\angle 50° \\ 0.004\angle 80° + 0.0004\angle 210° & 1 + 0.2\angle 130° \end{bmatrix}$$

$$= \begin{bmatrix} 1 - 0.129 + j0.153 & 64.2 + j76.6 \\ 0.0004(1.736 + j9.848 - 0.866 - j0.5) & 1 - 0.129 + j0.153 \end{bmatrix}$$

$$= \begin{bmatrix} 0.871 + j0.153 & 64.2 + j76.6 \\ 3.48 \times 10^{-4} + j3.739 \times 10^{-3} & 0.871 + j0.153 \end{bmatrix}$$

(*b*) The potential difference across the output $\mathbf{V}_2$ is $\mathbf{I}_2R = 0.002 \times 50 = 0.1$ V. The current $\mathbf{I}_2$ will be -0.002 A. Equation [1] gives

$$\mathbf{V}_1 = A\mathbf{V}_2 - B\mathbf{I}_2$$

and so

$$\mathbf{V}_1 = (0.871 + j0.153) \times 0.1 + (64.2 + j76.6) \times 0.002$$
$$= 0.216 + j0.169$$
$$= 0.274\angle 38°\ \text{V}$$

Equation [2] gives

$$\mathbf{I}_1 = C\mathbf{V}_2 - D\mathbf{I}_2$$

and so

$$\mathbf{I}_1 = (3.48 \times 10^{-4} + j3.739 \times 10^{-3}) \times 0.1 + (0.871 + j0.153) \times 0.002$$
$$= 1.777 \times 10^{-3} + j6.799 \times 10^{-3}$$
$$= 1.90\angle 20.9°\ \text{mA}$$

Equivalent circuits

Circuits that have the same terminal behaviour are termed *equivalent*. Thus if we consider a T-network and a π-network, the two are equivalent if there are the same relationships between their input voltages, input currents, output voltages and output currents. This must mean that the two networks have the same parameters.

In terms of the z parameters, a two-port network has the terminal characteristics represented by the following equations (equations [3] and [4])

$$\mathbf{V}_1 = z_{11}\mathbf{I}_1 + z_{12}\mathbf{I}_2$$

$$\mathbf{V}_2 = z_{21}\mathbf{I}_1 + z_{22}\mathbf{I}_2$$

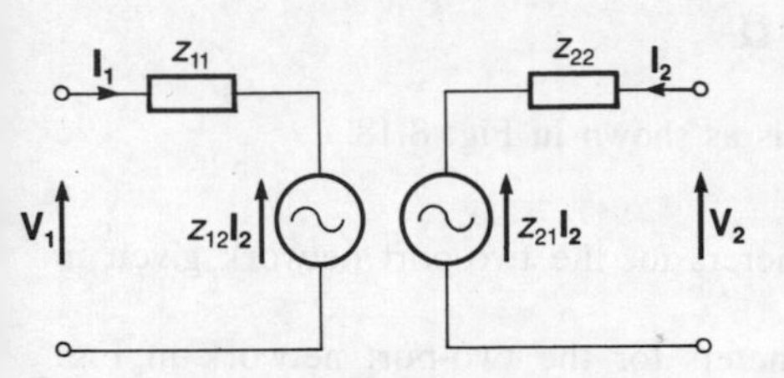

Fig. 8.15 Equivalent z parameter circuit

The first equation could however be describing an input circuit which with an input voltage of $\mathbf{V}_1$ has a series impedance of z_{11} and a voltage source of $z_{12}\mathbf{I}_2$. The second equation could describe an output circuit which with an input voltage of $\mathbf{V}_2$ has a series impedance of z_{22} and a voltage source of $z_{21}\mathbf{I}_1$. Figure 8.15 thus shows the equivalent z parameter circuit.

In terms of the h parameters, a two-port network has the terminal characteristics represented by the following equations (equations [7] and [8])

$$\mathbf{V}_1 = h_{11}\mathbf{I}_1 + h_{12}\mathbf{V}_2$$

$$\mathbf{I}_2 = h_{21}\mathbf{I}_1 + h_{22}\mathbf{V}_2$$

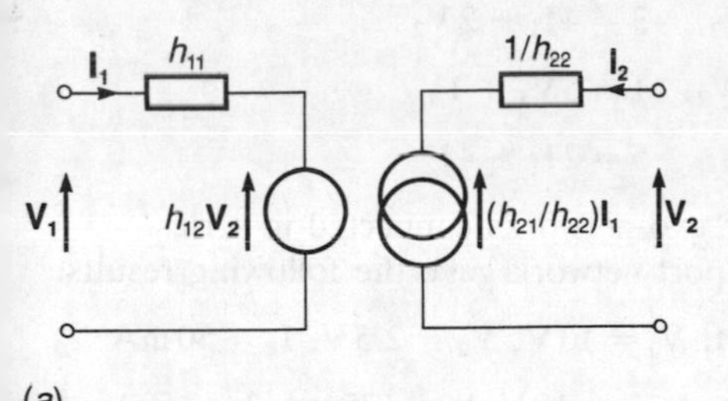

(*a*)

The first equation could however be describing an input circuit which with an input voltage of $\mathbf{V}_1$ has a series impedance of h_{11} and a voltage source of $h_{12}\mathbf{V}_2$. The second equation can be written as

$$\mathbf{V}_2 = (1/h_{22})\mathbf{I}_2 - (h_{21}/h_{22})\mathbf{I}_1$$

and thus could describe an output circuit which with an input voltage of $\mathbf{V}_2$ has a series impedance of $(1/h_{22})$ and a current source of $(h_{21}/h_{22})\mathbf{I}_1$. Figure 8.16($a$) thus shows the equivalent h parameter circuit. A more usual representation of this is, however, the circuit in Fig. 8.16(b).

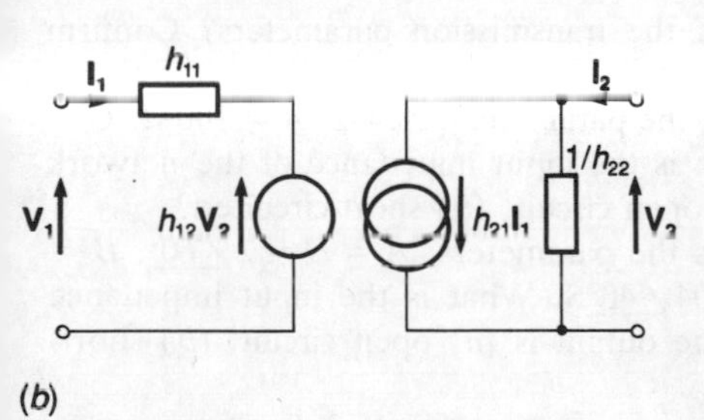

(*b*)

Fig. 8.16 Equivalent h parameter circuit

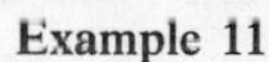

Example 11

What is the equivalent T-network for the π-network in Fig. 8.17?

Answer

For the circuits to be equivalent they must have the same transmission parameters. Since the π-circuit is symmetrical we must also have a symmetrical T-section. Thus for the π-circuit we have

$$\begin{bmatrix} 1 + YZ & Z \\ 2Y + Y^2Z & 1 + YZ \end{bmatrix} = \begin{bmatrix} 1 + 0.1 \times 20 & 20 \\ 2 \times 0.1 + 0.1^2 \times 20 & 1 + 0.1 \times 20 \end{bmatrix}$$

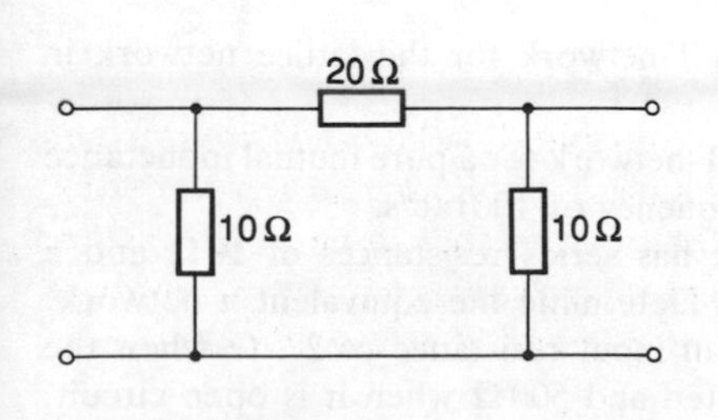

Fig. 8.17 Example 11

For a symmetrical T-network we have

$$\begin{bmatrix} 1 + YZ & 2Z + YZ^2 \\ Y & 1 + YZ \end{bmatrix}$$

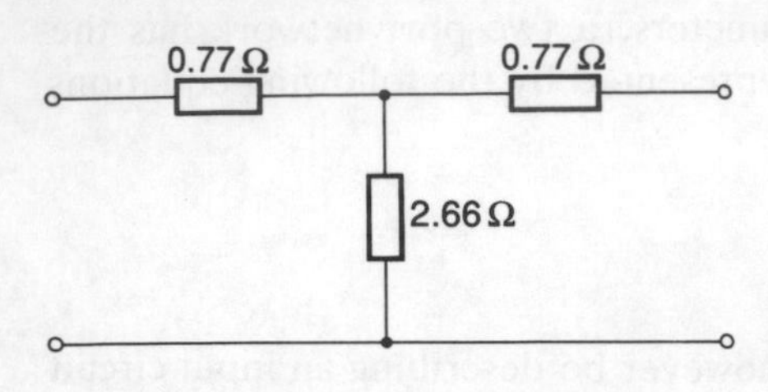

Fig. 8.18 Example 11

Thus we must have

$$Y = 2 \times 0.1 + 0.1^2 \times 20 = 0.4\,\text{S}$$

and

$$1 + YZ = 1 + 0.1 \times 20$$

$$Z = \frac{0.1 \times 20}{0.4} = 5\,\Omega$$

Thus the equivalent circuit is as shown in Fig. 8.18.

Problems

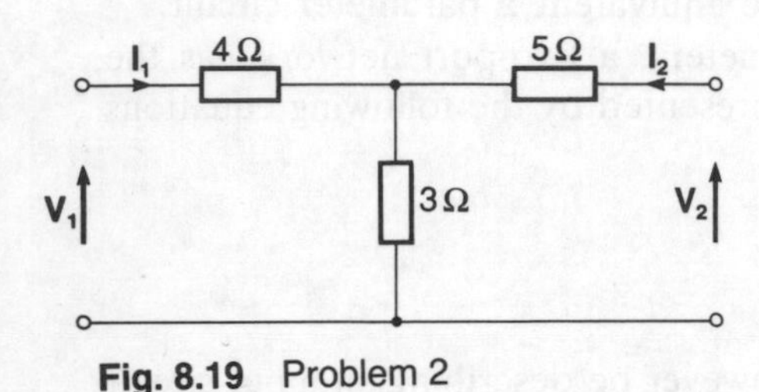

Fig. 8.19 Problem 2

1. Determine the h parameters for the two-port network given in Fig. 8.2.
2. Determine the z parameters for the two-port network in Fig. 8.19.
3. Solve, using matrices, the following sets of simultaneous equations:

   (*a*) $2 = 2\mathbf{I}_1 + 3\mathbf{V}_2, \quad 3 = 4\mathbf{I}_1 + 2\mathbf{V}_2$

   (*b*) $5 = 2\mathbf{V}_1 + 2\mathbf{V}_2, \quad 1 = 5\mathbf{V}_1 + 1\mathbf{V}_2$

   (*c*) $3 = 1\mathbf{I}_1 + 3\mathbf{I}_2, \quad 5 = 2\mathbf{I}_1 + 4\mathbf{I}_2$
4. Derive the z matrix for two networks connected in series.
5. Tests on a passive two-port network gave the following results:

   Output open circuit: $\mathbf{V}_1 = 10\,\text{V}$, $\mathbf{V}_2 = 2.5\,\text{V}$, $\mathbf{I}_1 = 50\,\text{mA}$

   Output short-circuited: $\mathbf{V}_1 = 10\,\text{V}$, $\mathbf{I}_1 = 175\,\text{mA}$, $\mathbf{I}_2 = 500\,\text{mA}$

   What are the values of the transmission parameters? Confirm that the network is passive.
6. A two-port network has the parameters: $A = 2$, $B = 100\,\Omega$, $C = 0.01\,\text{S}$ and $D = 1$. What is the input impedance of the network when the output is (*a*) open circuit, (*b*) short-circuited?
7. A two-port network has the parameters: $A = D = 2\angle 60°$, $B = 100\angle 20°\,\Omega$ and $C = 0.04\angle 40°\,\text{S}$. What is the input impedance of the network when the output is (*a*) open circuit, (*b*) short-circuited?
8. What are the transmission parameters for the following circuits:

   (*a*) A shunt of $200\,\Omega$,

   (*b*) a symmetrical T-network with series arms of $200\,\Omega$ and a shunt arm of $j100\,\Omega$,

   (*c*) a π-network with a series arm of $j100\,\Omega$ and shunt arms of $20\,\Omega$.
9. Determine the equivalent T-network for the lattice network in Fig. 8.20.
10. Determine the equivalent T-network for a pure mutual inductance of 0.1 H at an angular frequency of 100 rad/s.
11. A symmetrical T-network has series resistances of $10\,\Omega$ and a shunt admittance of 0.1 S. Determine the equivalent π-network.
12. A two-port network has an input resistance of $275\,\Omega$ when the output port is short-circuited and $500\,\Omega$ when it is open circuit. When the input port is open circuit the resistance at the output

Fig. 8.20 Problem 9

port is 400 Ω. Determine the equivalent T-network.

13 Determine the characteristic impedance of a two-port network in terms of the *ABCD* parameters.

9 Transmission lines

Introduction

Transmission lines are conductors along which electrical signals and energy can be sent. Telephone lines and power distribution lines are typical examples. Transmission lines can be visualized as a pair of parallel conductors, separated by a dielectric, to give 'go' and 'return' paths. A common form is a pair of parallel wires kept a constant distance apart by suitable spacers (Fig. 9.1(*a*)). Another common form is a coaxial cable, this essentially being a central conducting 'go' wire insulated from a coaxial outer 'return' conductor (Fig. 9.1(*b*)). On a printed circuit board the transmission lines might be just a metallic strip separated by a layer of dielectric from a ground metallic sheet, such an arrangement being known as a microstrip (Fig. 9.1(*c*)). Transmission lines can vary in length from centimetres, where for example the line is used as an integral part of high frequency circuits, to thousands of kilometres when it is used for the transmission of electrical power. The frequencies involved can be as low as 50 Hz or 60 Hz for lines used to transmit electrical power, or as high as a few GHz for electrical circuits used in the reception and

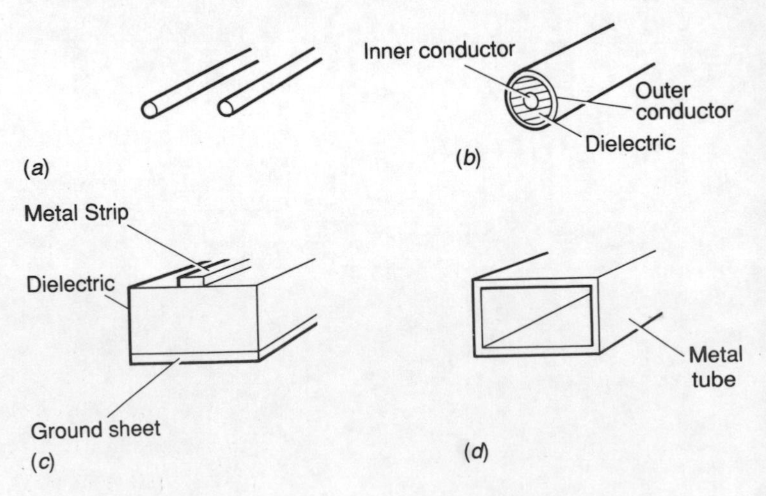

Fig. 9.1 (*a*) Parallel conductors, (*b*) coaxial cable, (*c*) microstrip, (*d*) rectangular waveguide

amplification of radio waves. The term high frequency tends to be used for frequencies above a few tens of MHz and up to a few tens of GHz.

At very high frequencies the transmission system used may be waveguides. These can be in the form of rectangular or circular metal tubes with the electrical energy being transmitted as a wave down the inside of the tube (Fig. 9.1(*d*)). Waveguides are just a one-conductor form of transmission line. The basic theory of transmission lines can still, however, be applied to them.

Transmission line primary constants

There are four *primary constants* or *coefficients* which affect the transmission of signals along transmission lines. These depend on the physical dimensions of the conductors and the nature of the dielectric used. The constants are:

1 *Resistance* The resistance of a transmission line is the sum of the resistances of the two conductors comprising the 'go' and 'return' paths. The resistance R is given by

$$R = \frac{\rho l}{A} \tag{1}$$

where ρ is the resistivity of the conductor material, l the total length of conductor and A its cross-sectional area. The resistance is usually stated as the resistance per unit loop length of a line, the term loop often being used to indicate that there are two conductors.

2 *Inductance* The magnetic flux produced by a current in a conductor results in inductance due to the interaction of this flux with the current in the other conductor and also with the current within itself. This internal inductance element can generally be neglected at high frequencies, and the inductance per unit length of transmission line is, for parallel conductors of radius a and centres a distance D apart,

$$\text{inductance/unit length} = \frac{\mu}{\pi} \ln\left(\frac{D}{a}\right) \tag{2}$$

and for coaxial cable, inner conductor radius a and outer conductor with an internal radius b,

$$\text{inductance/unit length} = \frac{\mu}{2\pi} \ln\left(\frac{b}{a}\right) \tag{3}$$

μ is the absolute permeability of the material between the conductors and is often written as $\mu = \mu_r\mu_0$ where μ_r is the relative permeability and μ_0 is the permeability of free space and has the value $4\pi \times 10^{-7}$ H/m. Generally with transmission lines the relative permeability is assumed to be 1.

3 *Capacitance* The capacitance per unit length of transmission line is due to the electric field produced between the conductors. For a pair of parallel conductors, each radius a and centres D apart,

$$\text{capacitance/unit length} = \frac{\pi\varepsilon}{\ln(D/a)} \quad [4]$$

and for coaxial cable, inner conductor radius a and outer conductor internal radius b,

$$\text{capacitance per unit length} = \frac{2\pi\varepsilon}{\ln(b/a)} \quad [5]$$

ε is the absolute permittivity of the dielectric between the conductors. It is usually expressed as $\varepsilon = \varepsilon_r\varepsilon_0$, where ε_r is the relative permittivity and ε_0 is the permittivity of free space and has the value 8.85×10^{-12} F/m. Usually with transmission lines ε_r is taken as having the value 1.

4 *Conductance* This is sometimes referred to as the *leakance* and represents the imperfection of the insulation between the conductors allowing some current to leak from one to the other. The conductance G is expressed in units of siemens per unit length. The conductance depends on the frequency, increasing with frequency and at high frequencies becoming proportional to the frequency.

Travelling waves

Consider a length of uniform rope lying on the ground. If you suddenly start to wag one end of the rope back-and-forth then a wave motion can be seen to travel along the rope (Fig. 9.2). Such a wave is called a *travelling* or *progressive wave* because the wave disturbance travels along the rope with a finite speed. If we concentrate on watching any one point on the rope then it undergoes the same back-and-forth movements that were made at the end of the rope, but not necessarily in phase with those movements.

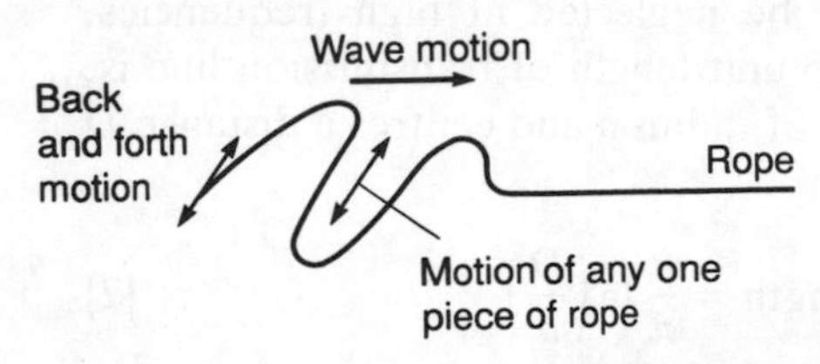

Fig. 9.2 A travelling wave on a rope

Transmission lines have resistance, inductance, capacitance and conductance. If we consider the lines broken up into segments then each section will have resistance, inductance, capacitance and conductance. A reasonable model for each segment is a symmetrical T-section which has series resistance and inductance and shunt capacitance and conductance. The transmission lines then become a cascaded sequence of such T-sections (Fig. 9.3). The series resistance and the shunt conductance represent losses. If we ignore them then we can

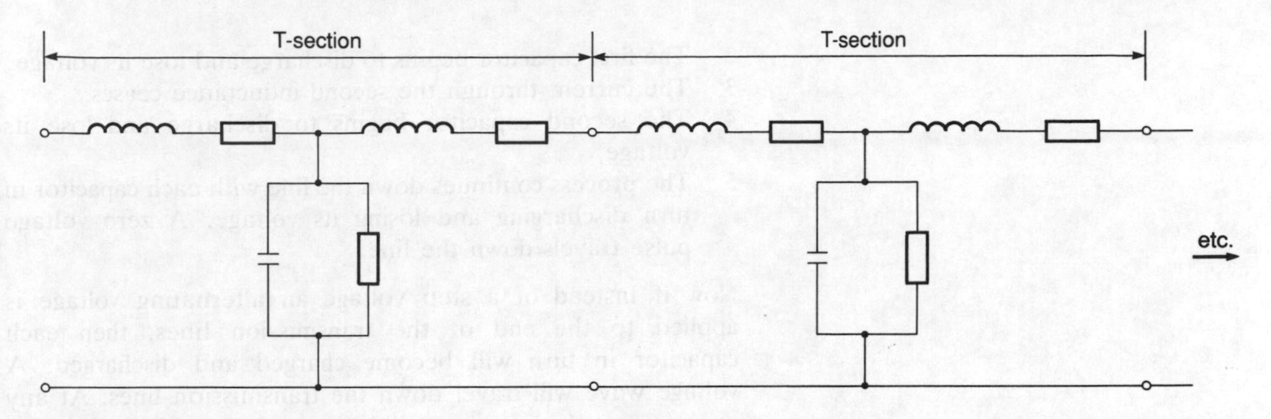

Fig. 9.3 Representation of transmission lines by cascaded T-sections

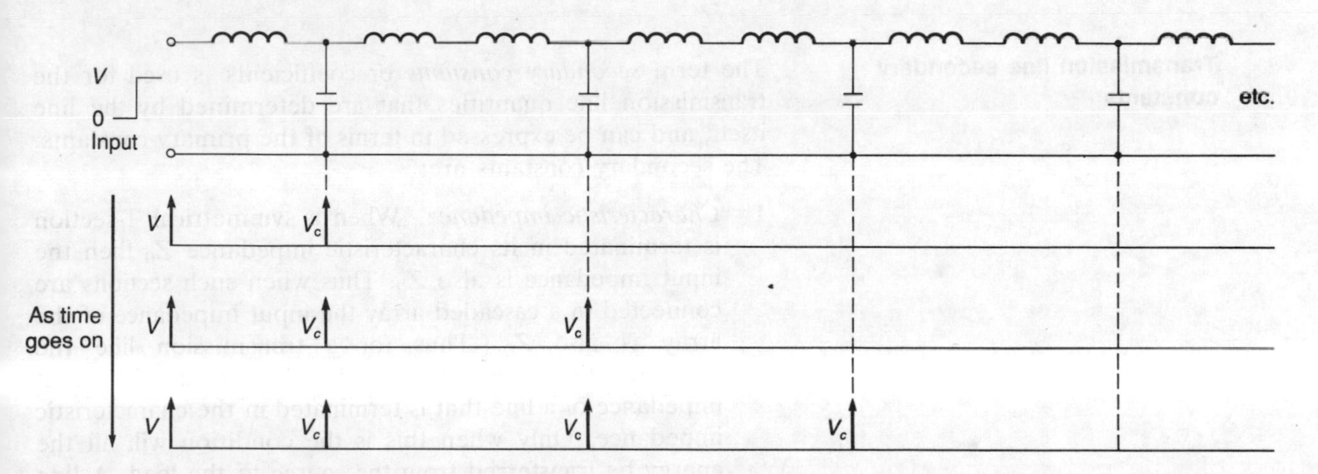

Fig. 9.4 Step voltage applied to a transmission line

see that when a voltage (a so-called step voltage, in that the voltage is suddenly increased from zero to some constant value) is first connected to one end of the transmission lines (Fig. 9.4) that the sequence becomes:

1 A current begins to flow through the first inductance.
2 The first capacitor begins to charge up and a voltage develops across it.
3 A current begins to flow through the second inductance.
4 The second capacitor begins to charge up and a voltage develops across it.
5 The process continues down the line with each capacitor in turn becoming charged up to the voltage. A voltage pulse travels down the line.

Now if the step voltage ceases, then

1 The current through the first inductance ceases.

2 The first capacitor begins to discharge and lose its voltage.
3 The current through the second inductance ceases.
4 The second capacitor begins to discharge and lose its voltage.
5 The process continues down the line with each capacitor in turn discharging and losing its voltage. A zero voltage pulse travels down the line.

Now if instead of a step voltage an alternating voltage is applied to the end of the transmission lines, then each capacitor in turn will become charged and discharged. A voltage wave will travel down the transmission lines. At any one segment the capacitor will have a voltage which undergoes the same alternations as that of the alternating voltage applied to the end of the line.

Transmission line secondary constants

The term *secondary constants* or coefficients is used for the transmission line quantities that are determined by the line itself, and can be expressed in terms of the primary constants. The secondary constants are:

1 *Characteristic impedance* When a symmetrical T-section is terminated in its characteristic impedance Z_0 then the input impedance is also Z_0. Thus when such sections are connected in a cascaded array the input impedance of the array is also Z_0. Thus for a transmission line the characteristic impedance can be defined as the input impedance of a line that is terminated in the characteristic impedance. Only when this is the condition will all the energy be transferred from the source to the load. A line terminated in its characteristic impedance is said to be *correctly terminated* or *matched*.

2 *Propagation coefficient* The propagation coefficient γ is defined as (see Chapter 7, equation [9])

$$\gamma = \ln\left(\frac{I_1}{I_2}\right) \quad [6]$$

where I_2 is the current leaving a section and I_1 is the current entering it. Since, if the line is correctly terminated, each section has the same characteristic impedance then we can also write

$$\gamma = \ln\left(\frac{V_1}{V_2}\right) \quad [7]$$

where V_2 is the voltage across the load of a section and V_1 that across the input. If we consider γ to be the propagation coefficient for unit length of a transmission line, e.g. 1 km, then if two such lengths are in cascade, i.e. the line length is doubled, and if $\mathbf{I_1}$ is the input current to the first length, $\mathbf{I_2}$ the output current from the first length and the input current to the second length, and $\mathbf{I_3}$ the output current from the second length,

$$\frac{\mathbf{I_1}}{\mathbf{I_3}} = \frac{\mathbf{I_1}}{\mathbf{I_2}} \times \frac{\mathbf{I_2}}{\mathbf{I_3}}$$

$$\ln\left(\frac{\mathbf{I_1}}{\mathbf{I_3}}\right) = \ln\left(\frac{\mathbf{I_1}}{\mathbf{I_2}}\right) + \ln\left(\frac{\mathbf{I_2}}{\mathbf{I_3}}\right)$$

Since both the lengths have the same attenuation coefficient γ then

$$\ln\left(\frac{\mathbf{I_1}}{\mathbf{I_3}}\right) = 2\gamma$$

$$\mathbf{I_1} = \mathbf{I_3}\,e^{2\gamma}$$

Thus for a line made up of l such lengths, i.e. a length l, the relationship between the current at the source end $\mathbf{I_S}$ and the current at the receiving end $\mathbf{I_R}$ is

$$\mathbf{I_S} = \mathbf{I_R}\,e^{l\gamma}$$

or

$$\mathbf{I_R} = \mathbf{I_S}\,e^{-l\gamma} \qquad [8]$$

Similarly we can write

$$\mathbf{V_R} = \mathbf{V_S}\,e^{-l\gamma} \qquad [9]$$

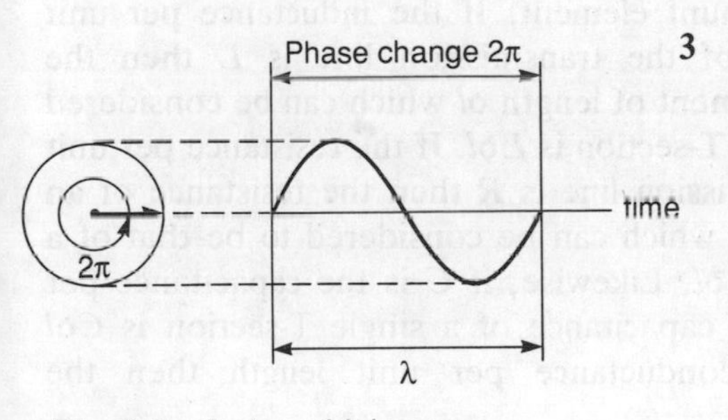

Fig. 9.5 A sinusoidal wave

3 *Phase velocity* The term phase velocity is used for the velocity with which a wave travels. For a wave of frequency f and wavelength λ the phase velocity v is $v = f\lambda$. One wavelength is the distance between two points on a wave for which there is a phase change of 2π rad (Fig. 9.5). Thus in some distance d there will be d/λ wavelengths and so a phase change of $2\pi d/\lambda$. The phase change per unit length (often 1 km) of transmission line β is thus

$$\beta = \frac{2\pi}{\lambda} \qquad [10]$$

Hence

$$v = f\lambda = \frac{2\pi f}{\beta} = \frac{\omega}{\beta} \qquad [11]$$

where ω is the angular frequency. Because the wave travels with a finite velocity down the line there will be a time delay between the start of the wave and its reception at the other end of the line. The time delay will be l/v, where l is the length of the line.

Example 1

What is (*a*) the wavelength, (*b*) the phase velocity of transmission of a 1000 rad/s signal along transmission lines which give a phase shift of 0.05 rad/km, and (*c*) the time delay with a line of length 10 km?

Answer

(*a*) Equation [10] gives $\beta = 2\pi/\lambda$ and so

$$\lambda = \frac{2\pi}{0.05} = 126\,\text{km}$$

(*b*) Equation [11] gives

$$v = \frac{\omega}{\beta} = \frac{1000}{0.05} = 2.0 \times 10^4\,\text{km/s}$$

(*c*) The time delay is l/v and so

$$\text{time delay} = \frac{10}{2.0 \times 10^4} = 5 \times 10^{-4}\,\text{s}$$

Characteristic impedance

The characteristic impedance Z_0 of a single symmetrical T-section is given by (Chapter 7, equation [14])

$$Z_0 = \sqrt{(\tfrac{1}{4}Z_1^2 + Z_1Z_2)}$$

where $\frac{1}{2}Z_1$ is the impedance of the series element and Z_2 is the impedance of the shunt element. If the inductance per unit length, e.g. 1 km, of the transmission line is L then the inductance of an element of length δl which can be considered to be that of a single T-section is $L\delta l$. If the resistance per unit length of the transmission line is R then the resistance of an element of length δl which can be considered to be that of a single T-section is $R\delta l$. Likewise, if C is the capacitance per unit length then the capacitance of a single T-section is $C\delta l$ and if G is the conductance per unit length then the

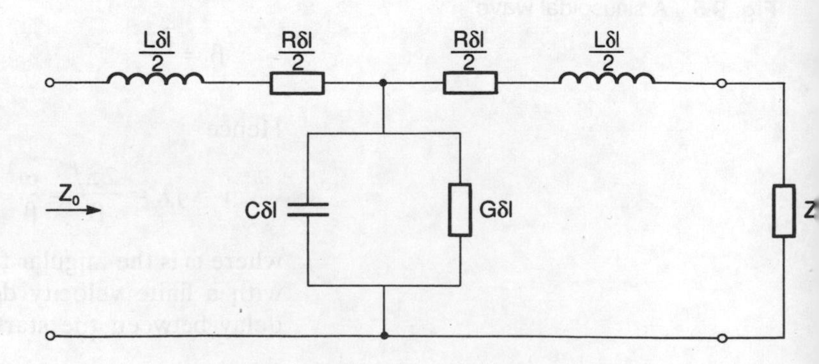

Fig. 9.6 A single T-section

conductance of a single T-section is $G\delta l$. Figure 9.6 shows such a section. Thus

$$Z_1 = R\delta l + j\omega L\delta l$$

$$Z_2 = \frac{1}{G\delta l + j\omega C\delta l}$$

Therefore

$$Z_0 = \sqrt{\left[\frac{(R + j\omega L)^2 \delta l^2}{4} + \frac{R + j\omega L}{G + j\omega C}\right]}$$

δl^2 is very small. Thus the expression approximates to

$$Z_0 \approx \sqrt{\left(\frac{R + j\omega L}{G + j\omega C}\right)} \qquad [12]$$

At very low frequencies $R \gg \omega L$ and $G \gg \omega C$ and so $Z_0 \approx \sqrt{(R/G)}$. At high frequencies when $\omega L \gg R$ and $\omega C \gg G$ then $Z_0 \approx \sqrt{(L/C)}$. This approximation is invariably used for radio frequency lines.

Example 2

A transmission line has a resistance of 50 Ω/km, an inductance of 1 mH/km, a conductance of 1 μS/km and a capacitance of 0.03 μF/km. What is the characteristic impedance of the line at an angular frequency of 1000 rad/s?

Answer

Using equation [12]

$$Z_0 = \sqrt{\left(\frac{R + j\omega L}{G + j\omega C}\right)} = \sqrt{\left(\frac{50 + j1}{10^{-6} + j0.03 \times 10^{-3}}\right)}$$

$$= \sqrt{\left(\frac{50\angle 1.1^\circ}{30 \times 10^{-6}\angle 88.1^\circ}\right)} = 1291\angle -43.5^\circ\,\Omega$$

Note that the $(\angle 43.5^\circ)(\angle 43.5^\circ) = \angle 87.0^\circ$ and so the square root of $\angle 87.0^\circ$ is $\angle 43.5^\circ$.

The propagation coefficient

For a symmetrical T-section the propagation coefficient γ is (Chapter 7, equation [51])

$$e^\gamma = \frac{Z_1}{2Z_2} + 1 + \frac{Z_0}{Z_2}$$

But for the sections in a transmission line (as in Fig. 9.6)

$$Z_1 = R\delta l + j\omega L\delta l$$

$$Z_2 = \frac{1}{G\delta l + j\omega C\delta l}$$

and Z_0 is given by equation [12] as

$$Z_0 \approx \sqrt{\left(\frac{R + j\omega L}{G + j\omega C}\right)}$$

Thus

$$e^{\gamma} = \frac{(R + j\omega L)(G + j\omega C)\delta l^2}{2} + 1$$
$$+ (G + j\omega C)\,\delta l \sqrt{\left(\frac{R + j\omega L}{G + j\omega C}\right)}$$
$$= 1 + \delta l \sqrt{[(R + j\omega L)(G + j\omega C)]}$$
$$+ \delta l^2 \tfrac{1}{2}(R + j\omega L)(G + j\omega C)$$

This can be compared with the expression for e^x as a series, i.e.

$$e^x = 1 + x + \frac{x^2}{2} +$$

and thus

$$\gamma = \sqrt{[(R + j\omega L)(G + j\omega C)]} \qquad [13]$$

At very low frequencies when $R \gg \omega L$ and $G \gg \omega C$ then we have $\gamma \approx \sqrt{(RG)}$. At high frequencies when $\omega L \gg R$ and $\omega C \gg G$ we have

$$\gamma = \sqrt{\left[j\omega L\left(1 + \frac{R}{j\omega L}\right) j\omega C\left(1 + \frac{G}{j\omega C}\right)\right]}$$

and using the binomial theorem

$$\gamma \approx j\omega\sqrt{(LC)}\left[1 + \frac{R}{2j\omega L}\right]\left[1 + \frac{G}{2j\omega C}\right]$$
$$\approx j\omega\sqrt{(LC)}\left[1 + \frac{R}{2j\omega L} + \frac{G}{2j\omega C}\right]$$
$$= \left[\frac{R}{2}\sqrt{\left(\frac{C}{L}\right)} + \frac{G}{2}\sqrt{\left(\frac{L}{C}\right)}\right] + j\omega\sqrt{(LC)}$$

and since at high frequencies $Z_0 = \sqrt{(L/C)}$

$$\gamma = \left[\frac{R}{2Z_0} + \frac{GZ_0}{2}\right] + j\omega\sqrt{(LC)} \qquad [14]$$

Hence since $\gamma = \alpha + j\beta$, equating the real terms gives

$$\alpha = \frac{R}{2Z_0} + \frac{GZ_0}{2} \qquad [15]$$

In most practical lines G is very small and equation [15] becomes

$$\alpha \approx \frac{R}{2Z_0} \qquad [16]$$

Equating the j terms gives

$$\beta = \omega\sqrt{(LC)} \qquad [17]$$

The phase velocity, given by equation [11], can thus be written as

$$v = \frac{\omega}{\beta} = \frac{1}{\sqrt{LC}} \qquad [18]$$

Example 3

What is the propagation coefficient of a transmission line at an angular frequency of 1000 rad/s if the resistance is 50 Ω/km, the inductance 30 mH/km, the capacitance 0.05 μF/km and the conductance 2.0 μS/km

Answer

Using equation [13]

$$\gamma = \sqrt{[(R + j\omega L)(G + j\omega C)]}$$

$$= \sqrt{[(50 + j30)(2.0 \times 10^{-6} + j0.05 \times 10^{-3})]}$$

$$= \sqrt{[(58.3 \angle 31.0^\circ)(0.05 \times 10^{-3} \angle 87.7^\circ)]}$$

$$= 0.054 \angle 59.4^\circ/\text{km}$$

Note that $(\angle 59.4^\circ)(\angle 59.4^\circ) = \angle(59.4 + 59.4^\circ)$ and so the square root of $\angle 118.8^\circ$ is $\angle 59.4^\circ$.

Example 4

A transmission line has an inductance of 0.6 mH/km and a capacitance of 0.03 μF/km. What will be (*a*) the phase velocity of a signal along and (*b*) the phase delay per km if the signal has an angular frequency of 1000 rad/s?

Answer

(*a*) Using equation [18],

$$v = \frac{1}{\sqrt{LC}} = \frac{1}{\sqrt{(0.6 \times 10^{-3} \times 0.03 \times 10^{-6})}}$$

$$= 2.4 \times 10^5 \text{ km/S}$$

(*b*) Using equation [17],

$$\beta = \omega\sqrt{LC} = 1000\sqrt{(0.6 \times 10^{-3} \times 0.03 \times 10^{-6})}$$

$$= 0.0042 \text{ rad/km}$$

Example 5

What is the characteristic impedance and the attenuation for a transmission line at 1 MHz if $R = 0.5\ \Omega/\text{m}$, $L = 0.2\ \mu\text{H/m}$, $C = 0.09\ \text{nF/m}$ and G is negligible?

Answer

At high frequencies equation [12] becomes

$$Z_0 = \sqrt{\frac{L}{C}} = \sqrt{\left(\frac{0.2 \times 10^{-6}}{0.09 \times 10^{-9}}\right)} = 47\,\Omega$$

At high frequencies the attenuation is given by equation [16] as

$$\alpha = \frac{R}{2Z_0} = \frac{0.5}{2 \times 47} = 5.3 \times 10^{-3}\,\text{Np/m}$$

Voltage and current relationships

The propagation coefficient is a complex quantity and can be expressed as $\gamma = \alpha + j\beta$, where α is the attenuation coefficient and β the phase change coefficient. Thus for a correctly terminated line we can write for the voltage $\mathbf{V_x}$ at some point a distance x from the source end, using equation [9],

$$\mathbf{V_x} = \mathbf{V_S}e^{-x\gamma}$$
$$= \mathbf{V_S}e^{-x(\alpha+j\beta)}$$
$$= \mathbf{V_S}e^{-\alpha x}\,(e^{-j\beta x}) = \mathbf{V_S}e^{-\alpha x}\,\underline{\angle -\beta x} \qquad [19]$$

We can similarly write for the current $\mathbf{I_x}$ at the distance x from the source end

$$\mathbf{I_x} = \mathbf{I_S}e^{-\alpha x}\,(e^{-j\beta x}) = \mathbf{I_S}e^{-\alpha x}\,\underline{\angle -\beta x} \qquad [20]$$

At any point along the line

$$\frac{\mathbf{V_x}}{\mathbf{I_x}} = \frac{\mathbf{V_S}}{\mathbf{I_S}} = Z_0 \qquad [21]$$

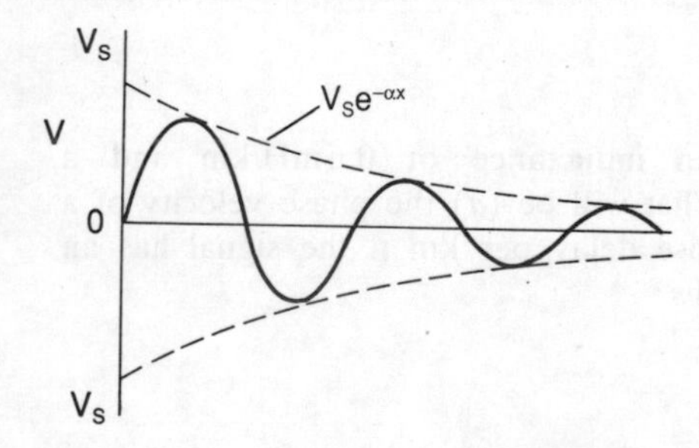

Fig. 9.7 Voltage attenuation along the line

The voltage at all points along the transmission line will have the same frequency as that of the source but will be progressively attenuated from its initial value of magnitude $\mathbf{V_s}$ at $x = 0$ to $\mathbf{V_s}e^{-\alpha x}$ at distance x. Figure 9.7 shows a snapshot of the wave at some instant of time. The current is likewise attenuated.

An unchanging sinusoidal voltage, or current, can be represented by a rotating phasor of constant length, the length representing the maximum magnitude of the voltage or current. At any instant the angle of the phasor relative to some reference axis gives the phase angle (see Chapter 3). For each point along the transmission line we can draw such a diagram, the length of the rotating phasor being the magnitude of $\mathbf{V_x}$, i.e. $\mathbf{V_S}e^{-\alpha x}$, and the phase angle relative to that of the source being $-\beta x$ (Fig. 9.8). As x increases so the length of the rotating phasor decreases and its phase angle increases relative to the voltage at the start of the line. If we draw all the phasors on the same diagram, then the tip of the phasor traces out a spiral form of path as x increases.

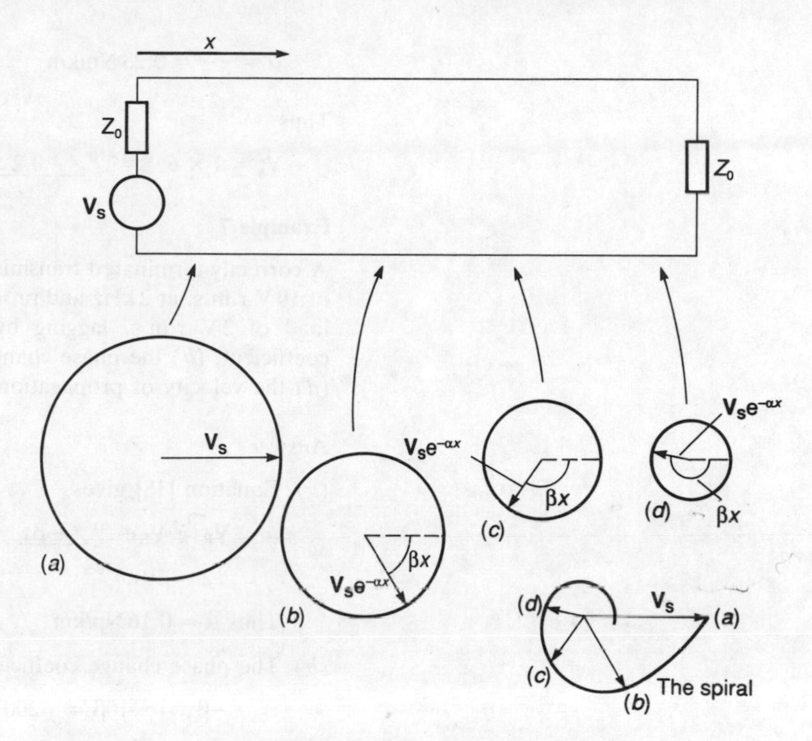

Fig. 9.8 Phasor diagrams for a transmission line

At the receiving end of a correctly terminated line the voltage $\mathbf{V_R}$ across the load is equal to $\mathbf{V_S}\,e^{-\gamma l}$, where l is the length of the line from source to receiver, and the current $\mathbf{I_R}$ is $\mathbf{I_S}\,e^{-\gamma l}$. The ratio $\mathbf{V_R}/\mathbf{I_R}$ is the characteristic impedance Z_0. The power dissipated in the load P_R is

$$P_R = |V_R||I_R|\cos\theta_R \qquad [22]$$

where θ_R is the phase angle between the received current and voltage. The input power to the line is

$$P_S = |V_S||I_S|\cos\theta_S \qquad [23]$$

where θ_S is the phase angle between the input current and voltage.

Example 6

A correctly terminated transmission line has an attenuation coefficient of 2 dB/km and a phase change coefficient of 0.2 rad/km. What is the voltage at a point 10 km along the line if there is an input of 1 V r.m.s.?

Answer

Using equation [19]

$$\mathbf{V_x} = \mathbf{V_S}\,e^{-\alpha x}\,(e^{-j\beta x}) = \mathbf{V_S}\,e^{-\alpha x}\,\underline{\angle -\beta x}$$

where α is in Np/km and β in rad/km. Using equation [8], Chapter 7, the attenuation in dB/km can be converted to Np/km.

$$\alpha = \frac{2}{8.7} = 0.23\,\text{Np/km}$$

Thus

$$\mathbf{V_x} = 1 \times e^{-0.23\times 10}\underline{/-0.2 \times 10} = 0.10\underline{/-2} = 0.10\underline{/-115^\circ}\,\text{V}$$

Example 7

A correctly terminated transmission line of length 10 km has an input of 10 V r.m.s. at 2 kHz and produces an output across the terminating load of 2 V r.m.s. lagging by 200°. What is (*a*) the attenuation coefficient, (*b*) the phase change coefficient, (*c*) the wavelength and (*d*) the velocity of propagation?

Answer

(*a*) Equation [18] gives

$$\mathbf{V_R} = \mathbf{V_S}\,e^{-\alpha x}\underline{/-\beta x}$$

$$2 = 10\,e^{-10\alpha}$$

Thus $\alpha = 0.16\,\text{Np/km}$

(*b*) The phase change coefficient β is given

$$-\beta x = -10\beta = -200^\circ$$

$$\beta = 20^\circ/\text{km} = 0.35\,\text{rad/km}$$

(*c*) The wavelength is given by equation [10] as $\beta = 2\pi/l$. Thus

$$\lambda = \frac{2\pi}{0.35} = 18.0\,\text{km}$$

(*d*) Equation [11] gives

$$v = \frac{\omega}{\beta} = \frac{2\pi \times 2000}{0.35} = 3.6 \times 10^4\,\text{km/s}$$

Example 8

Plot the current phasor spiral for a correctly terminated transmission line of length 4 km when subject to an input of 10 V r.m.s.. The line has an attenuation coefficient of 0.10 Np/km, a phase change coefficient of 0.15 rad/km and a characteristic resistance of $200\underline{/-30^\circ}\,\Omega$.

Answer

The input current $\mathbf{I_S}$ is $\mathbf{V_S}/Z_0$ and so is

$$\mathbf{I_S} = \frac{10}{200\underline{/-30^\circ}} = 50\underline{/30^\circ}\,\text{mA}$$

The current $\mathbf{I_x}$ at a distance x from the source is given by equation [20] as

$$\mathbf{I_x} = \mathbf{I_S}\,e^{-\alpha x}\underline{/-\beta x}$$

Thus the current at $x = 1\,\text{km}$ is, since $0.15\,\text{rad} = 8.6^\circ$,

$$(50\underline{/30^\circ})\,e^{-0.10}\underline{/-8.6^\circ} = 45.2\underline{/21.4^\circ}\,\text{mA}$$

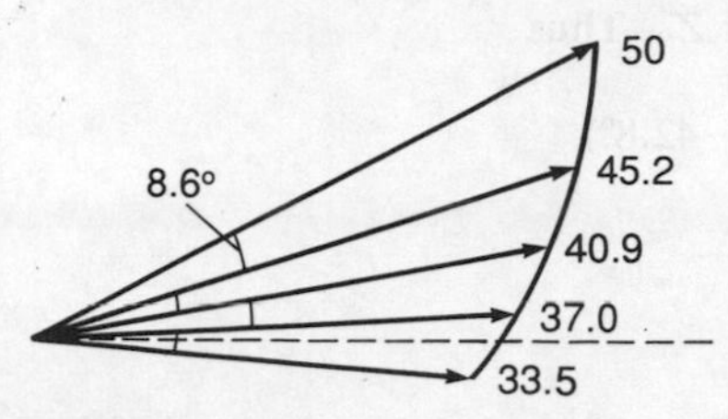

Fig. 9.9 Example 8

At $x = 2$ km the current is

$$(50\angle 30°)\,\mathrm{e}^{-0.20}\ \angle -17.2° = 40.9\angle 12.8°\,\mathrm{mA}$$

At $x = 3$ km the current is

$$(50\angle 30°)\,\mathrm{e}^{-0.30}\ \angle -25.8° = 37.0\angle 4.2°\,\mathrm{mA}$$

At $x = 4$ km the current is

$$(50\angle 30°)\,\mathrm{e}^{-0.40}\ \angle -34.4° = 33.5\angle -4.4°\,\mathrm{mA}$$

Figure 9.9 shows the above data plotted as the spiral.

Example 9

A correctly terminated transmission line of length 10 km has a resistance of 50 Ω/km, an inductance of 0.6 mH/km, a capacitance of 0.04 μF/km and conductance of 1 μS/km. If the power input to the line is 1.0 W at a frequency of 1 kHz, what is (*a*) the characteristic impedance of the line, (*b*) the propagation coefficient, (*c*) the attenuation coefficient, (*d*) the phase change coefficient, (*e*) the magnitudes of the voltage and current at the source and receiving ends of the line, and (*f*) the receiving-end power?

Answer

(*a*) The characteristic impedance is given by equation [12] as

$$Z_0 = \sqrt{\left(\frac{R + \mathrm{j}\omega L}{G + \mathrm{j}\omega C}\right)}$$

$$= \sqrt{\left(\frac{50 + \mathrm{j}2\pi \times 1000 \times 0.6 \times 10^{-3}}{10^{-6} + \mathrm{j}2\pi \times 1000 \times 0.04 \times 10^{-6}}\right)}$$

$$= \sqrt{\left(\frac{50 + \mathrm{j}3.77}{10^{-6} + \mathrm{j}2.51 \times 10^{-4}}\right)} = \sqrt{\left(\frac{50.1\angle 4.3°}{2.51 \times 10^{-4}\angle 89.8°}\right)}$$

$$= 446.8\angle -42.8°\,\Omega$$

(*b*) Using equation [13]

$$\gamma = \sqrt{[(R + \mathrm{j}\omega L)(G + \mathrm{j}\omega C)]}$$

$$= \sqrt{[(50 + \mathrm{j}2\pi \times 1000 \times 0.6 \times 10^{-3})(10^{-6} + \mathrm{j}2\pi \times 1000 \times 0.04 \times 10^{-6})]}$$

$$= \sqrt{[(50.1\angle 4.3°)(2.51 \times 10^{-4}\angle 89.8°)]}$$

$$= 0.112\angle 47.1° = 0.0762 + \mathrm{j}0.0820$$

(*c*) The real part of the propagation coefficient is 0.0762 and thus the attenuation coefficient is 0.0762 Np/km.

(*d*) The imaginary part of the propagation coefficient is 0.0820 and thus the phase change coefficient is 0.0820 rad/km.

(*e*) Since the line is correctly terminated then $|Z_0| = |V_S|/|I_S|$ and so the input power (equation [23]) is

$$P_S = |V_S||I_S|\cos\theta_S = \left|\frac{V_S^2}{Z_0}\right|\cos\theta_S$$

But θ_S will be the angle of Z_0. Thus

$$P_S = 1 = \frac{|V_S^2|}{446.8}\cos(-42.8°)$$

$$|V_S| = 24.7\,\text{V}$$

Since $|Z_0| = |V_S|/|I_S|$ then

$$|I_S| = \frac{24.7}{446.8} = 0.0553\,\text{A}$$

The voltage at the receiving end is given by equation [19] as

$$\mathbf{V_R} = \mathbf{V_S}\,e^{-\alpha l}\angle{-\beta l}$$

and thus

$$|V_R| = |V_S|\,e^{-\alpha l} = 24.7\,e^{-0.762} = 11.5\,\text{V}$$

The current at the receiving end is given by equation [20] as

$$\mathbf{I_R} = \mathbf{I_S}\,e^{-\alpha l}\angle{-\beta l}$$

and thus

$$|I_R| = |I_S|\,e^{-\alpha l} = 0.0553\,e^{-0.762} = 0.0258\,\text{A}$$

(*f*) The power at the receiving end of the line is

$$P_R = |I_R||V_R|\cos\theta_R$$

The phase angle θ_R between the receiving end current and voltage will be that of Z_0 since the line is correctly terminated. Thus

$$P_R = 0.0258 \times 11.5 \times \cos(-42.8°) = 0.218\,\text{W}$$

Conditions for distortion

Distortion is said to occur if the waveform at the receiving end of a transmission line is not the same shape as the waveform at the sending end. There are three main causes of distortion:

1 The *characteristic impedance* Z_0 depends on the operating frequency, equation [12] indicating the relationship, i.e.

$$Z_0 = \sqrt{\left(\frac{R + j\omega L}{G + j\omega C}\right)}$$

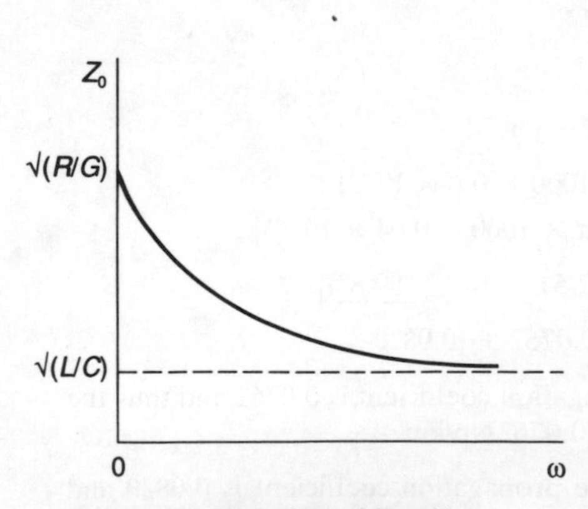

Fig. 9.10 Variation of Z_0 with frequency

Figure 9.10 shows a graph of how Z_0 varies with ω. When ω is very low then Z_0 becomes $\sqrt{(R/G)}$ and when ω is very high Z_0 becomes $\sqrt{(L/C)}$. The terminating impedance of the line may not, however, vary with frequency in the same way. Thus the line will not be correctly terminated at all frequencies.

2 The *attenuation* of the line varies with frequency (equation [13]),

$$\gamma = \sqrt{[(R + j\omega L)(G + j\omega C)]}$$

Thus waves of different frequency and the component

waves of complex waves are attenuated by different amounts.

3 The *phase velocity* depends on the frequency (equation [11]),

$$v = \frac{\omega}{\beta}$$

Thus the delay time, i.e. the time taken for a wave to move from source to receiver, depends on the frequency and so different frequencies arrive at the receiver at different times.

The distortionless line

Consider the distortion, outlined above, which results from the variation of the characteristic impedance with frequency. For Z_0 to be independent of the frequency and present a constant value then we must have, using the equation [12] given above,

$$\frac{R + j\omega L}{G + j\omega C} = K$$

where K is a constant. Thus

$$R + j\omega L = KG + jK\omega C$$

This means we must have $R = KG$ and $\omega L = K\omega C$, i.e.

$$\frac{R}{G} = \frac{L}{C}$$

or

$$LG = CR \qquad [24]$$

For this condition

$$Z_0 = \sqrt{\left(\frac{R + j\omega L}{G + j\omega C}\right)} = \sqrt{\left[\frac{(GL/C) + j\omega L}{G + j\omega C}\right]}$$

$$= \sqrt{\left[\frac{L(G + j\omega C)}{C(G + j\omega C)}\right]}$$

$$= \sqrt{\left(\frac{L}{C}\right)} \qquad [25]$$

or, using equation [23],

$$Z_0 = \sqrt{\left(\frac{R}{G}\right)} \qquad [26]$$

Since there is no imaginary term the characteristic impedance is $\sqrt{(L/C)}\angle 0°$.

Now consider the problem of the attenuation varying with frequency. Using the equation [13] given above,

$$\gamma = \sqrt{[(R + j\omega L)(G + j\omega C)]}$$

$$\gamma^2 = (R + j\omega L)(G + j\omega C) = RG + j\omega(CR + LG) - \omega^2 LC$$

The distortionless condition for the characteristic impedance of $LG = CR$ (equation [24]) thus gives for γ^2

$$\gamma^2 = RG + j\omega(2LG) - \omega^2 LC$$

$$= RG + j\omega 2\sqrt{(LC)}\sqrt{(RG)} - \omega^2 LC$$

$$= [\sqrt{(RG)} + j\omega\sqrt{(LC)}]^2$$

Thus

$$\gamma = \sqrt{(RG)} + j\omega\sqrt{(LC)}$$

Hence, since $\gamma = \alpha + j\beta$, the attenuation coefficient α is

$$\alpha = \sqrt{(RG)} \qquad [27]$$

and is independent of frequency. The phase change coefficient β is

$$\beta = \omega\sqrt{(LC)} \qquad [28]$$

Now consider the velocity of propagation of the wave down a transmission line. Since $v = \omega/\beta$ (equation [11]), then using equation [28]

$$v = \frac{\omega}{\beta} = \frac{\omega}{\omega\sqrt{LC}} = \frac{1}{\sqrt{LC}} \qquad [29]$$

Thus the velocity is independent of the frequency.

Thus the condition for no distortion as a result of the frequency dependence of the characteristic impedance, the attenuation and the velocity of propagation is that $LG = CR$. However, in the practical forms that transmission lines can take, LG does not equal CR but is much smaller than CR. The inductance is too low and the capacitance too high. The capacitance is not easily reduced and an increase in G would not be desirable since it would increase the losses and the attenuation. Thus the only way to move towards a distortion-free line is to increase L. Such an increase is referred to as *loading* the line. The most common way of doing this is by adding inductance coils in series with the line at regularly spaced intervals along the line. Such a method of adding loading in lumps is called *lumped loading*. It is, however, not generally practical to add sufficient loading to obtain completely distortionless conditions since the amount of inductance required would be rather large. Also, an increase in inductance decreases the phase velocity and may produce unacceptable time delays.

Example 10

An underground transmission line has a resistance of 50 Ω/km, an inductance of 1.0 mH/km, a capacitance of 0.05 μF/km and a conductance of 2.0 μS/km. What amount of inductance should be added to give a distortionless line?

Answer

Equation [24] gives the condition for a line free of distortion as $LG = CR$. Thus the total inductance required per km is

$$L = \frac{CR}{G} = \frac{0.05 \times 10^{-6} \times 50}{2.0 \times 10^{-6}} = 1.25\,\text{H}$$

Thus the inductance to be added is $1.25 - 0.001 = 1.249$ H/km.

Mismatch on transmission lines

A correctly terminated transmission line has a terminating impedance which is the characteristic impedance of the line. In practice this often does not occur. Under such conditions the transmission line is said to have a *mismatched load*.

When the line is correctly terminated all the energy delivered along the line by the wave motion is absorbed by the load. We can think of the transmission line as being represented by a rope with the wave produced by one end being wagged back-and-forth. The wave travels down the rope and all its energy is absorbed at the far end (Fig. 9.11(*a*)). However, if all the energy is not absorbed at the far end of the

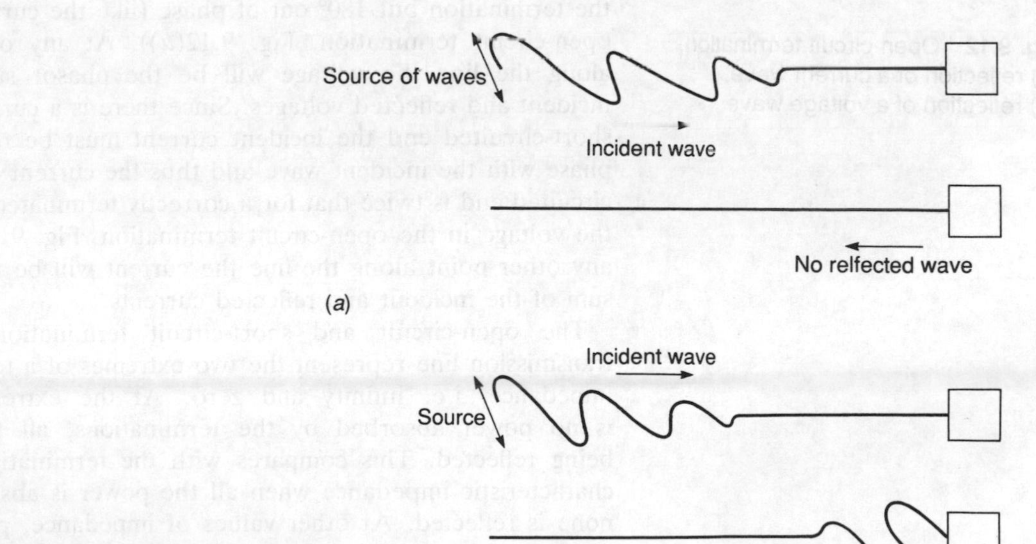

Fig. 9.11 Wave moving along a rope (*a*) with no reflection, (*b*) with reflection at the far end

rope then a reflected wave travels back down the rope (Fig. 9.11(*b*)). This is what happens with the transmission line with the mismatched load: part of the energy of the incident wave is absorbed at the far end and part is reflected back down the line. With a correctly terminated line there is no reflected wave.

Suppose the transmission line is open circuit at its termination. No power can be absorbed by this termination as the current through the termination must be zero. We can consider this no current situation to be one where we send an alternating current down the line and at the termination there is a reflected alternating current which at the termination is exactly equal in amplitude to the incident current but with a phase difference of 180° (Fig. 9.12(*a*)). At the termination the two cancel each other out and hence, since the phasor sum is zero, the current there is zero. At any other point along the line the current will be the phasor sum of the incident and reflected currents. Since there is a voltage at the open-circuit end the incident voltage wave must be reflected in phase with the incident wave and thus the voltage at an open-circuit end is twice that for a correctly terminated line (Fig. 9.12(*b*)). At any other point along the line the current will be the phasor sum of the incident and reflected currents.

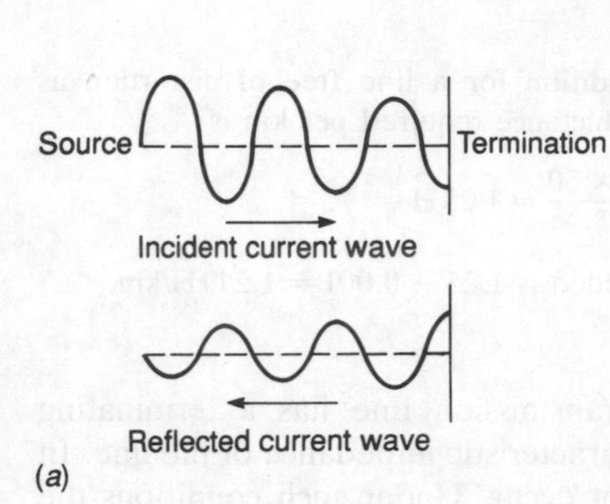

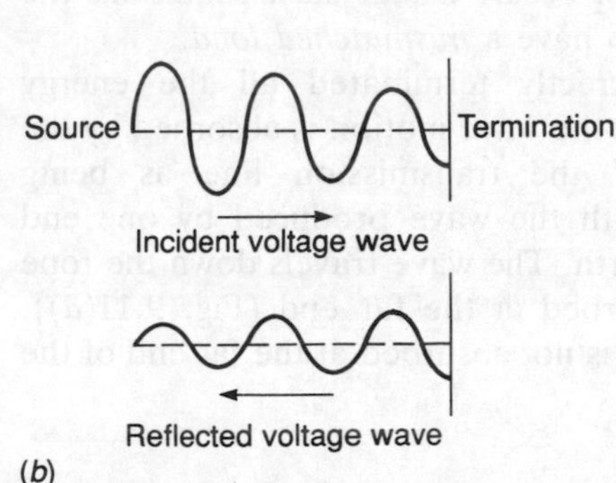

Fig. 9.12 Open-circuit termination, (*a*) reflection of a current wave, (*b*) reflection of a voltage wave

Now consider the case of a line with the termination short-circuited. Because the impedance will be zero the voltage must be zero. Thus no power will be absorbed by the termination. In order for the voltage to be zero the incident voltage wave and the reflected voltage wave must be equal in amplitude at the termination but 180° out of phase (like the current in the open-circuit termination, Fig. 9.12(*a*)). At any other point along the line the voltage will be the phasor sum of the incident and reflected voltages. Since there is a current at the short-circuited end the incident current must be reflected in phase with the incident wave and thus the current at a short-circuited end is twice that for a correctly terminated line (like the voltage in the open-circuit termination, Fig. 9.12(*b*)). At any other point along the line the current will be the phasor sum of the incident and reflected currents.

The open-circuit and short-circuit terminations of the transmission line represent the two extremes of a termination impedance, i.e. infinity and zero. At the extremes there is no power absorbed by the terminations, all the power being reflected. This compares with the termination by the characteristic impedance when all the power is absorbed and none is reflected. At other values of impedance, part of the power is absorbed by the termination and the remainder reflected.

Current and voltage on mismatched lines

The current and voltage at any point along a transmission line will be the phasor sum of the incident and reflected waves at that point (Fig. 9.13). The figure indicates that the distances are measured from the load end of the line; this is because it is generally more convenient to measure distances from the load end rather than the source end of a transmission line. The incident wave current $\mathbf{I}_{ix}$ at some distance $(l - x)$ from the source $\mathbf{I}_S$ is

$$\mathbf{I}_{ix} = \mathbf{I}_S e^{-\gamma(l-x)} = \mathbf{I}_S e^{-\gamma l} e^{\gamma x}$$

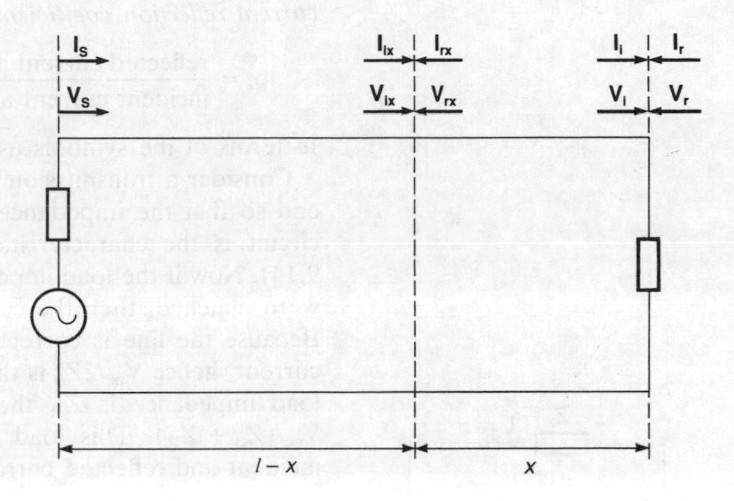

Fig. 9.13 Current and voltage at a point along a line

But $\mathbf{I}_S e^{-\gamma l}$ is the incident current at the load end. If we denote this by $\mathbf{I}_i$ then

$$\mathbf{I}_{ix} = \mathbf{I}_i e^{\gamma x}$$

Likewise

$$\mathbf{V}_{ix} = \mathbf{V}_i e^{\gamma x}$$

where $\mathbf{V}_i$ is the incident voltage at the load end.

We can thus think of the reflected wave as being due to a source of current $\mathbf{I}_r$ and voltage $\mathbf{V}_r$ at the receiving end and becoming attenuated as the distance x from that end increases. Thus the reflected current $\mathbf{I}_{rx}$ at some distance x is

$$\mathbf{I}_{rx} = \mathbf{I}_r e^{-\gamma x}$$

and the reflected voltage $\mathbf{V}_{rx}$ at distance x is

$$\mathbf{V}_{rx} = \mathbf{V}_r e^{-\gamma x}$$

Thus the total current at some distance x from the load end is

$$\mathbf{I}_x = \mathbf{I}_{ix} + \mathbf{I}_{rx} = \mathbf{I}_i e^{\gamma x} + \mathbf{I}_r e^{-\gamma x} \qquad [30]$$

and the total voltage as

$$\mathbf{V}_x = \mathbf{V}_{ix} + \mathbf{V}_{rx} = \mathbf{V}_i e^{\gamma x} + \mathbf{V}_r e^{-\gamma x} \qquad [31]$$

Reflection coefficients

The *voltage reflection coefficient* ϱ_v of a transmission line is defined as

$$\varrho_v = \frac{\text{reflected voltage at the load}}{\text{incident voltage at the load}} \qquad [32]$$

In terms of the symbols used in Fig. 9.13, $\varrho_v = \mathbf{V}_r/\mathbf{V}_i$. The *current reflection coefficient* ϱ_i is defined as

$$\varrho_i = \frac{\text{reflected current at the load}}{\text{incident current at the load}} \qquad [33]$$

In terms of the symbols used in Fig. 9.13, $\varrho_i = \mathbf{I}_r/\mathbf{I}_i$.

Consider a transmission line which is matched at its source end so that the impedance measured when the load is open-circuit is the characteristic impedance Z_0 of the line (Fig. 9.14). Now if the load impedance were to be Z_0, i.e. if the line were matched, then the current through it would be $\mathbf{V}_{oc}/2Z_0$. Because the line is correctly terminated there is no reflected current, hence $\mathbf{V}_{oc}/2Z_0$ is the incident current. However, if the load impedance is Z_R then the current through it of $\mathbf{I}_R$ is $\mathbf{V}_{oc}/(Z_0 + Z_R)$. This load current is the phasor sum of the incident and reflected currents, i.e.

$$\mathbf{I}_R = \mathbf{I}_r + \mathbf{I}_i$$

Hence

$$\mathbf{I}_r = \mathbf{I}_R - \mathbf{I}_i = \frac{\mathbf{V}_{oc}}{Z_0 + Z_R} - \frac{\mathbf{V}_{oc}}{2Z_0} = \frac{\mathbf{V}_{oc}(Z_0 - Z_R)}{2Z_0(Z_0 + Z_R)}$$

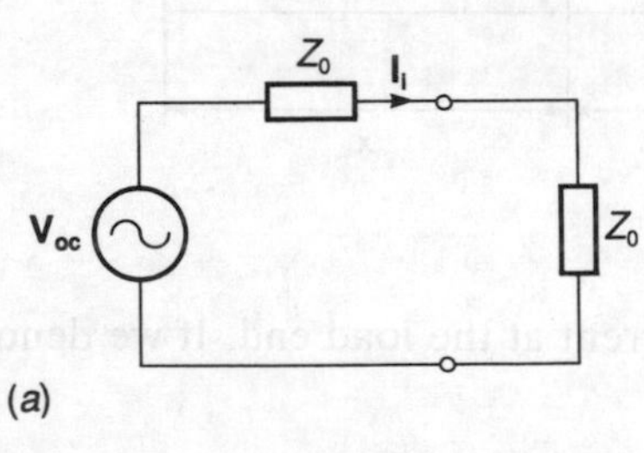

(a)

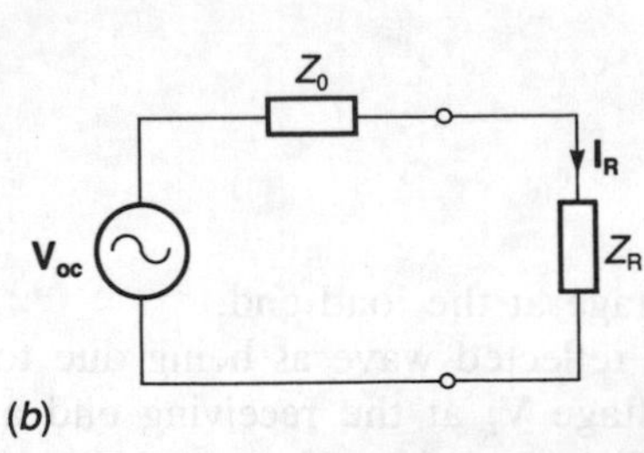

(b)

Fig. 9.14 Transmission line, (*a*) correctly terminated, (*b*) with load Z_R

Therefore the current reflection coefficient is, since $\mathbf{I}_i = \mathbf{V}_{oc}/2Z_0$,

$$\varrho_i = \frac{Z_0 - Z_R}{Z_0 + Z_R} \qquad [34]$$

The voltage across the correctly terminated load (Fig. 9.14(*a*)) will be $\mathbf{I}_i Z_0$ and that across the load Z_R in Fig. 9.14(*b*) $\mathbf{I}_R Z_R$. The load voltage is the phasor sum of the incident and reflected voltages, i.e.

$$\mathbf{V}_R = \mathbf{V}_r + \mathbf{V}_i$$

Thus

$$\mathbf{V}_r = \mathbf{V}_R - \mathbf{V}_i = \mathbf{I}_R Z_R - \mathbf{I}_i Z_0 = \frac{\mathbf{V}_{oc} Z_R}{Z_0 + Z_R} - \frac{\mathbf{V}_{oc} Z_0}{2Z_0}$$

$$= \frac{V_{oc}(Z_R - Z_0)}{2(Z_0 + Z_R)}$$

Therefore the voltage reflection coefficient is, since $V_i = V_{oc}/2$,

$$\varrho_v = \frac{Z_R - Z_0}{Z_0 + Z_R} \qquad [35]$$

Notice that $\varrho_i = -\varrho_r$, i.e. the current and the voltage reflection currents have the same magnitude but different signs. Notice also that when $Z_R = Z_0$ both coefficients are zero and there is no reflection.

The load voltage V_R is $(V_r + V_i)$, with $V_r = \varrho_v V_i$. Thus

$$V_R = \varrho_v V_i + V_i$$

But $V_i = V_S e^{-\gamma l}$ when l is the length of the line from source to load and V_S the source voltage. Thus

$$V_R = (1 + \varrho_v)V_S e^{-\gamma l} \qquad [36]$$

Similarly the load current I_R is

$$I_R = (1 + \varrho_i)I_S e^{-\gamma l} \qquad [37]$$

Example 11

A transmission line with a characteristic impedance of 100 Ω is terminated in a resistive load of 300 Ω. If the voltage measured across the load is 10 V r.m.s., what is (*a*) the voltage reflection coefficient for the line, (*b*) the incident current and voltage at the load, (*c*) the reflected current and voltage at the load?

Answer

(*a*) The voltage reflection coefficient is given by equation [35] as

$$\varrho_v = \frac{Z_R - Z_0}{Z_0 + Z_R} = \frac{300 - 100}{100 + 300} = 0.5$$

(*b*) As the voltage across the load is 10 V and the load has a resistance of 300 Ω, then the load current I_R is 10/300 = 0.033 A. But

$$I_R = I_i + I_r$$

and $I_r = \varrho_i I_i$. Thus

$$I_R = I_i(1 + \varrho_i)$$

and so, since $\varrho_i = -\varrho_v$,

$$I_i = \frac{I_R}{1 + \varrho_i} = \frac{0.033}{1 - 0.5} = 0.066\ \text{A}$$

The incident voltage V_i is

$$V_i = \frac{V_R}{1 + \varrho_v} = \frac{10}{1 + 0.5} = 6.7\ \text{V}$$

(*c*) Since $\mathbf{I_R} = \mathbf{I_i} + \mathbf{I_r}$, then $\mathbf{I_r} = 0.033 - 0.066 = -0.033$ A. The minus sign indicates that the reflected signal suffers a phase change of 180°. Since $\mathbf{V_R} = \mathbf{V_i} + \mathbf{V_r}$, then $\mathbf{V_r} = 10 - 6.7 = 3.3$ V.

Example 12

What is the voltage reflection coefficient for a line with a characteristic impedance of 50 Ω and a load of (*a*) 50 Ω, (*b*) 0 Ω, (*c*) infinity, (*d*) 150 + j100 Ω?

Answer

(*a*) The voltage reflection coefficient is given by equation [35] as

$$\varrho_v = \frac{Z_R - Z_0}{Z_0 + Z_R}$$

Thus when terminated by 50 Ω,

$$\varrho_v = \frac{50 - 50}{50 + 50} = 0$$

There is no reflection since the line is terminated by the characteristic impedance.

(*b*) When terminated by 0 Ω, i.e. short circuit,

$$\varrho_v = \frac{0 - 50}{50 + 0} = -1$$

There is complete reflection, i.e. no absorption, and a phase change of 180°.

(*c*) When terminated by infinity, i.e. open circuit,

$$\varrho_v = \frac{Z_R - Z_0}{Z_0 + Z_R} = \frac{1 - (Z_0/Z_R)}{(Z_0/Z_R) + 1} = \frac{1 - 0}{0 + 1} = 1$$

There is complete reflection with no phase change.

(*d*) When terminated by 150 + j100 Ω, then

$$\varrho_v = \frac{150 + j100 - 50}{50 + 150 + j100} = \frac{100 + j100}{200 + j100} = \frac{141\angle 45°}{224\angle 26.6°}$$

$$= 0.63\angle 18.4°$$

Input impedance

The impedance at any point on a transmission line is the ratio of the total voltage, i.e. the phasor sum of the incident and reflected voltages, to the total current, i.e. the phasor sum of the incident and reflected currents, at the point concerned. Thus since the total current a distance x from the load is (equation [30])

$$\mathbf{I}_x = \mathbf{I}_{ix} + \mathbf{I}_{rx} = \mathbf{I}_i e^{\gamma x} + \mathbf{I}_r e^{-\gamma x}$$

and the total voltage (equation [31])

$$\mathbf{V}_x = \mathbf{V}_{ix} + \mathbf{V}_{rx} = \mathbf{V}_i e^{\gamma x} + \mathbf{V}_r e^{-\gamma x}$$

then the impedance Z_x is

$$Z_x = \frac{\mathbf{V_i}\,e^{\gamma x} + \mathbf{V_r}\,e^{-\gamma x}}{\mathbf{I_i}\,e^{\gamma x} + \mathbf{I_r}\,e^{-\gamma x}}$$

But equation [32] gives $\mathbf{V_r} = \varrho_v \mathbf{V_i}$ and equation [33] $\mathbf{I_r} = \varrho_i \mathbf{I_i}$. Thus

$$Z_x = \frac{\mathbf{V_i}(e^{\gamma x} + \varrho_v\,e^{-\gamma x})}{\mathbf{I_i}(e^{\gamma x} + \varrho_i\,e^{-\gamma x})}$$

But $Z_0 = \mathbf{V_i}/\mathbf{I_i}$ and thus

$$Z_x = \frac{Z_0(e^{\gamma x} + \varrho_v\,e^{-\gamma x})}{e^{\gamma x} + \varrho_i\,e^{-\gamma x}} \qquad [38]$$

Since ϱ_v is $(Z_R - Z_0)/(Z_0 + Z_R)$ (equation [35]) and ϱ_i is $(Z_0 - Z_R)/(Z_0 + Z_R)$ (equation [34]). Z_R is the load impedance. Thus

$$Z_x = \frac{Z_0[(Z_0 + Z_R)\,e^{\gamma x} + (Z_R - Z_0)\,e^{-\gamma x}]}{(Z_0 + Z_R)\,e^{\gamma x} + (Z_0 - Z_R)\,e^{-\gamma x}}$$

$$= \frac{Z_0[Z_0(e^{\gamma x} - e^{-\gamma x}) + Z_R(e^{\gamma x} + e^{-\gamma x})]}{Z_0(e^{\gamma x} + e^{-\gamma x}) + Z_R(e^{\gamma x} - e^{-\gamma x})}$$

Since $\cosh \gamma x = \frac{1}{2}(e^{\gamma x} + e^{-\gamma x})$ and $\sinh \gamma x = \frac{1}{2}(e^{\gamma x} - e^{-\gamma x})$ then

$$Z_x = \frac{Z_0[Z_0 \sinh \gamma x + Z_R \cosh \gamma x]}{Z_0 \cosh \gamma x + Z_R \sinh \gamma x} \qquad [39]$$

The input impedance is the impedance at a distance l from the load, where l is the length of the transmission line. Thus the input impedance Z_S is

$$Z_s = \frac{Z_0[Z_0 \sinh \gamma l + Z_R \cosh \gamma l]}{Z_0 \cosh \gamma l + Z_R \sinh \gamma l} \qquad [40]$$

Another way of expressing the impedance at any point along the line and at the input is by rearranging equation [38]. Since $\varrho_i = -\varrho_v$ then equation [38] becomes

$$Z_x = \frac{Z_0(e^{\gamma x} + \varrho_v\,e^{-\gamma x})}{e^{\gamma x} - \varrho_v\,e^{-\gamma x}}$$

$$= \frac{Z_0\,e^{\gamma x}(1 + \varrho_v\,e^{-2\gamma x})}{e^{\gamma x}(1 - \varrho_v\,e^{-2\gamma x})}$$

$$= \frac{Z_0(1 + \varrho_v\,e^{-2\gamma x})}{1 - \varrho_v\,e^{-2\gamma x}} \qquad [41]$$

The input impedance is thus

$$Z_S = \frac{Z_0(1 + \varrho_v\,e^{-2\gamma l})}{1 - \varrho_v\,e^{-2\gamma l}} \qquad [42]$$

Example 13

What is the input impedance of a line which has a characteristic impedance of 500 Ω, has an attenuation of 6 dB and a phase shift of 180° in its length, and is terminated by a load of 1000 Ω?

Answer

The voltage reflection coefficient is given by equation [35] as

$$\varrho_v = \frac{Z_R - Z_0}{Z_0 + Z_R} = \frac{1000 - 500}{500 + 1000} = 0.333$$

The current reflection coefficient is thus −0.333, i.e. $0.333\underline{/180^\circ}$.

An attenuation of 6 dB means that the voltage, or current, is reduced by the factor given by

$$6 = 20\lg(V_1/V_2)$$
$$10^{0.3} = 2.0 = V_1/V_2$$

i.e. the voltage, or current, is halved in traversing the length of the line. In addition there is a phase change of −180°. Thus, if V_i is the magnitude of the incident voltage at the source end of the line, it will have become $V_i/2\underline{/-180^\circ}$ by the time it reaches the load end. There it is reflected, to become

$$(0.333\underline{/0^\circ})(V_i/2\underline{/-180^\circ}) = 0.167V_i\underline{/-180^\circ}$$

It then travels back down the line to the source end, being attenuated to half its value and suffering a phase change of −180°, to become $0.167V_i/2\underline{/-360^\circ}$. The total voltage at the source end is thus

$$V_S = V_i/0^\circ + 0.084V_i\underline{/-360^\circ} = 1.084V_i\underline{/0^\circ}$$

If I_i is the incident current at the source end of the line, it will have become $I_i/2\underline{/-180^\circ}$ by the time it reaches the load end. There it is reflected, to become

$$(0.333\underline{/180^\circ})(I_i/2\underline{/-180^\circ}) = 0.167I_i\underline{/0^\circ}$$

It then travels back down the line to the source end, being attenuated to half its value and suffering a phase change of −180°, to become $0.167I_i/2\underline{/-180^\circ}$. The total current at the source end is thus

$$I_S = I_i\underline{/0^\circ} + 0.084I_i\underline{/-180^\circ} = I_i - 0.084I_i = 0.916I_i\underline{/0^\circ}$$

Thus the input impedance, which is V_S/I_S is

$$Z_S = \frac{1.084V_i\underline{/0^\circ}}{0.916I_i\underline{/0^\circ}}$$

and since $V_i/I_i = 500\,\Omega$, then $Z_S = 592\,\Omega$. Equation [40] or [42] could have been used to obtain the same answer.

Low loss lines

The term *low loss line* is used for a line for which the resistance R and the conductance G are very small. The propagation coefficient γ is given by equation [13] as

$$\gamma = \sqrt{[(R + j\omega L)(G + j\omega C)]}$$
$$= j\omega\sqrt{(LC)}\left(1 + \frac{R}{j\omega L}\right)^{\frac{1}{2}}\left(1 + \frac{G}{j\omega C}\right)^{\frac{1}{2}}$$

This can be expanded by the binomial series to give

$$\gamma = j\omega\sqrt{(LC)}\left(1 + \frac{R}{2j\omega L} - \ldots\right)\left(1 + \frac{G}{2j\omega C} - \ldots\right)$$

For a low loss line we can neglect all terms in R^2 and G^2. Thus

$$\gamma \approx j\omega\sqrt{(LC)}\left(1 - \frac{RG}{4\omega^2 LC} - \frac{jR}{2\omega L} - \frac{jG}{2\omega C}\right)$$
$$\approx \frac{R\sqrt{(LC)}}{2L} + \frac{G\sqrt{(LC)}}{2C} + j\omega\left[\sqrt{(LC)} - \frac{RG\sqrt{(LC)}}{4\omega^2 LC}\right]$$

Since $\gamma = \alpha + j\beta$, then equating real terms

$$\alpha \approx R\sqrt{\left(\frac{C}{L}\right)} + G\sqrt{\left(\frac{L}{C}\right)} \qquad [43]$$

and equating imaginary terms

$$\beta \approx \omega\sqrt{(LC)} - \frac{RG}{4\omega\sqrt{(LC)}}$$

The second term is usually negligible and so

$$\beta \approx \omega\sqrt{(LC)} \qquad [44]$$

The characteristic impedance is given by equation [12] as

$$Z_0 = \sqrt{\left(\frac{R + j\omega L}{G + j\omega C}\right)}$$
$$= \sqrt{\left(\frac{j\omega L}{j\omega C}\right)}\left(1 + \frac{R}{j\omega L}\right)^{\frac{1}{2}}\left(1 + \frac{G}{j\omega C}\right)^{-\frac{1}{2}}$$

Expansion by means of the binomial theorem gives

$$Z_0 = \sqrt{\left(\frac{L}{C}\right)}\left(1 + \frac{R}{2j\omega L} - \ldots\right)\left(1 - \frac{G}{2j\omega C} - \ldots\right)$$

For a low loss line we can neglect terms involving R^2, G^2 and RG, thus

$$Z_0 \approx \sqrt{\left(\frac{L}{C}\right)}\left(1 - \frac{jR}{2\omega L} - \frac{jG}{2\omega C}\right)$$

Generally $R/\omega L$ and $G/\omega C$ will be small enough to be neglected. Then

$$Z_0 = \sqrt{\left(\frac{L}{C}\right)} \qquad [45]$$

Loss-free lines

For lines used at radio frequencies the losses with transmission lines are generally small enough to be ignored, such lines then being referred to as *loss-free lines*. For such lines the resistance R, the conductance G and the attenuation coefficient α are effectively zero. Thus the characteristic impedance Z_0 is $\sqrt{(L/C)}$, as given by equation [45], and the propagation coefficient is effectively just $j\beta$ and so β is $\omega\sqrt{(LC)}$, as given by equation [44]. With $\gamma = j\beta$ then the input impedance (equation [40])

$$Z_S = \frac{Z_0[Z_0 \sinh \gamma l + Z_R \cosh \gamma l]}{Z_0 \cosh \gamma l + Z_R \sinh \gamma l}$$

becomes

$$Z_S = \frac{Z_0[Z_0 \sinh j\beta l + Z_R \cosh j\beta l]}{Z_0 \cosh j\beta l + Z_R \sinh j\beta l}$$

But

$$\sinh j\beta l = \tfrac{1}{2}(e^{j\beta l} - e^{-j\beta l})$$

and since

$$\sin \beta l = \frac{1}{2j}(e^{j\beta l} - e^{-j\beta l})$$

then

$$\sinh j\beta l = j \sin \beta l$$

Also

$$\cosh j\beta l = \tfrac{1}{2}(e^{j\beta l} + e^{j\beta l})$$

and since

$$\cos \beta l = \tfrac{1}{2}(e^{j\beta l} + e^{-j\beta l})$$

then

$$\cosh j\beta l = \cos \beta l$$

Thus

$$Z_S = \frac{Z_0[Z_0 j \sin \beta l + Z_R \cos \beta l]}{Z_0 \cos \beta l + Z_R j \sin \beta l}$$

$$= \frac{Z_0 \cos \beta l\,[Z_R + jZ_0 \tan \beta l]}{\cos \beta l\,[Z_0 + jZ_R \tan \beta l]}$$

$$= \frac{Z_0[Z_R + jZ_0 \tan \beta l]}{Z_0 + jZ_R \tan \beta l} \qquad [46]$$

The above is the general equation for the input impedance of a loss-free line. Consider the following special situations:

1 *Short-circuited line* When $Z_R = 0$ then equation [46] gives

$$Z_S = \frac{Z_0[0 + jZ_0 \tan \beta l]}{Z_0 + 0} = jZ_0 \tan \beta l \qquad [47]$$

This means that the input impedance of the loss-free line when short-circuited is a pure reactance whose magnitude and sign, i.e. whether capacitive or inductive, depend on the length l of the line.

2 *Open-circuited line* When $Z_R = \infty$ then equation [46] gives

$$Z_S = \frac{Z_0 Z_R[1 + j(Z_0/Z_R) \tan \beta l]}{Z_R[(Z_0/Z_R) + j \tan \beta l]} = \frac{Z_0[1 + 0]}{[0 + j \tan \beta l]}$$

$$Z_S = \frac{Z_0}{j \tan \beta} = -jZ_0 \cot \beta l \qquad [48]$$

This means that the input impedance of the loss-free line when open-circuited is a pure reactance whose magnitude and sign depend on the length l of the line.

3 *Quarter wavelength line* When $l = \lambda/4$ then, since $\beta = 2\pi/\lambda$, we have $\beta l = \pi/2$. Since $\tan \pi/2 = \infty$, then equation [46] gives

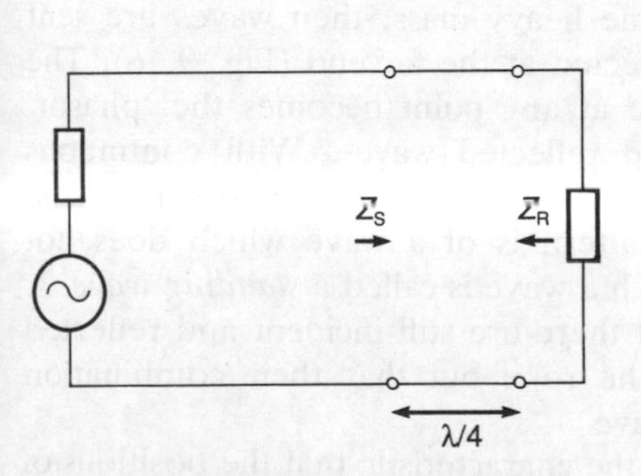

Fig. 9.15 Matching using a quarter wave section

$$Z_S = \frac{Z_0 \tan \beta l[(Z_R/\tan \beta l) + jZ_0]}{\tan \beta l[(Z_0/\tan \beta l) + jZ_R]} = \frac{Z_0[0 + jZ_0]}{0 + jZ_R}$$

$$= \frac{Z_0^2}{Z_R} \qquad [49]$$

This gives a means of matching a load Z_R to an input Z_S by placing between them a quarter wavelength section of loss-free line (Fig. 9.15) with the characteristic impedance Z_0 indicated by equation [49]. Such a section of line is known as a *quarter wave impedance transformer*.

4 *Half wavelength line* When $l = \lambda/2$ then, since $\beta = 2\pi/\lambda$, we have $\beta l = \pi$. Since $\tan \pi = 0$, then equation [46] gives

$$Z_S = \frac{Z_0(Z_R + 0)}{Z_0 + 0}$$

$$Z_S = Z_R \qquad [50]$$

Thus the input impedance of a half wavelength section of loss-free line is equal to the load impedance.

Example 14

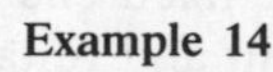

Two correctly terminated loss-free transmission lines have characteristic resistances of 200 Ω and 400 Ω. What should be the characteristic resistance of the quarter wavelength section used to connect them if there is to be perfect matching?

Answer

The input impedance for the quarter wavelength section is 200 Ω and the load is 400 Ω, thus using equation [49]

$$Z_0^2 = Z_S Z_R = 200 \times 400$$

Hence $Z_0 = 283\ \Omega$.

Example 15

What is the characteristic resistance of a loss-free line having an inductance of 1.2 mH/km and a capacitance of 0.02 μF/km?

Answer

Using equation [45],

$$Z_0 = \sqrt{\left(\frac{L}{C}\right)} = \sqrt{\left(\frac{1.2 \times 10^{-3}}{0.02 \times 10^{-6}}\right)} = 245\ \Omega$$

Standing waves

Fig. 9.16 Standing waves on a rope

If you wag back-and-forth one end of a rope, with the other end perhaps fixed to some heavy mass, then waves are sent down the rope to be reflected at the far end (Fig. 9.16). The displacement of the rope at any point becomes the 'phasor' sum of the incident and reflected waves. With continuous waves of constant frequency being sent down the rope the resulting displacement pattern is of a wave which does not move along the rope. Such a wave is called a *standing wave*. It needs to be realized that there are still incident and reflected waves travelling along the rope but that their combination results in a stationary wave.

A stationary wave has the characteristic that the positions of no displacement are fixed and do not move, such no-displacement positions being known as *nodes*. Also the positions of maximum displacement are fixed though the size of the maximum displacement can vary, these maximum displacement positions being known as *antinodes*. With the rope, the tethered end must be a position of no displacement. This means that the incident and reflected waves must always cancel out at that point. The end that is wagged back-and-forth must be a position of maximum displacement. Thus the standing wave must have a wavelength that fits the length of the rope so that the wagged end is a maximum displacement point and the fixed end a zero displacement point.

With current and voltage waves moving along a transmission line, the current and voltage at any point along the line is the phasor sum of the incident and reflected waves. Consider a loss-free line with an open-circuit termination. At the load end there must be zero current through the load, and so a current node, but a maximum voltage, and so a voltage antinode. The standing wave is thus of the form shown in Fig. 9.17(a). If the line had a short-circuit termination then the voltage at the end must be zero and so a node while the current is a maximum and so an antinode (Fig. 9.17(b)).

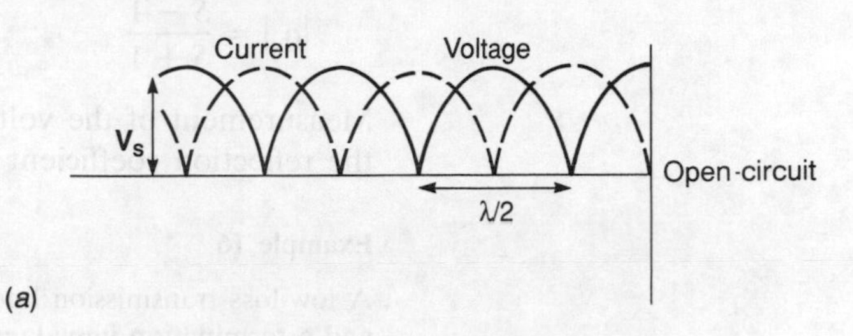

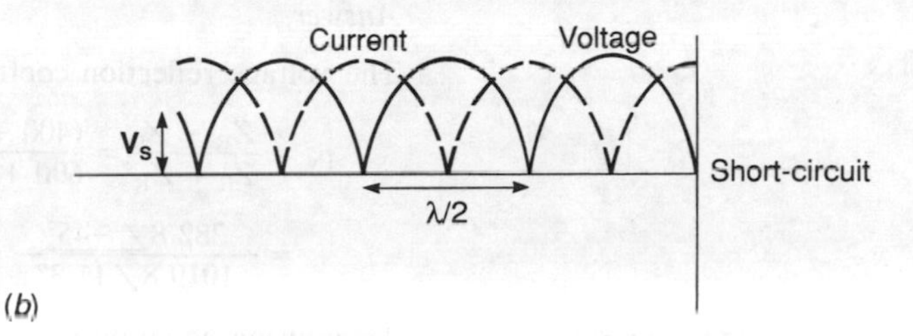

Fig. 9.17 Standing waves on a transmission line

With standing waves, adjacent minima are a distance of $\lambda/2$ apart. Thus, since $\beta = 2\pi/\lambda$, the adjacent minima are a distance of π/β apart.

Voltage standing wave ratio

The *voltage standing wave ratio* (VSWR) is the ratio of the maximum voltage to the minimum voltage of the standing wave along a transmission line.

$$\text{VSWR} = \left|\frac{V_{\max}}{V_{\min}}\right| \qquad [51]$$

But the maximum voltage occurs when the incident and reflected voltage waves are in phase, the minimum voltage when they are 180° out-of-phase. Thus

$$|V_{\max}| = |V_i| + |V_r|$$

$$|V_{\min}| = |V_i| - |V_r|$$

But $|V_r| = |\varrho_v||V_i|$, where ϱ_v is the reflection coefficient. Thus

$$|V_{\text{max}}| = |V_{\text{i}}|(1 + |\varrho_{\text{v}}|)$$
$$|V_{\text{min}}| = |V_{\text{i}}|(1 - |\varrho_{\text{v}}|)$$

Thus, if S is the VSWR,

$$S = \frac{1 + |\varrho_{\text{v}}|}{1 - |\varrho_{\text{v}}|} \quad [52]$$

This can be rearranged to give

$$|\varrho_{\text{v}}| = \frac{S - 1}{S + 1} \quad [53]$$

Measurement of the voltage standing wave ratio thus enables the reflection coefficient to be determined.

Example 16

A low-loss transmission line has a characteristic impedance of 600 Ω and a termination impedance of (400 + j200) Ω. What is the voltage standing wave ratio?

Answer

The voltage reflection coefficient is given by equation [35] as

$$\varrho_{\text{v}} = \frac{Z_{\text{R}} - Z_0}{Z_0 + Z_{\text{R}}} = \frac{(400 + \text{j}200) - 600}{600 + (400 + \text{j}200)} = \frac{-200 + \text{j}200}{1000 + \text{j}200}$$

$$= \frac{282.8\angle -45^\circ}{1019.8\angle 11.3^\circ} = 0.28\angle -56.3^\circ$$

Thus, using equation [52],

$$S = \frac{1 + |\varrho_{\text{v}}|}{1 - |\varrho_{\text{v}}|} = \frac{1 + 0.28}{1 - 0.28} = 1.8$$

Single stub matching

The term *matching* is used when all the power transmitted along a line is completely absorbed by the load and there is no reflection, or when it passes along one line and into another without any loss and hence reflection. Thus the load does not set up any standing waves. A quarter wavelength section of transmission line can be used to connect two transmission lines and match them and this has been discussed above and illustrated by Example 14. There are, however, other methods that can be used to match transmission lines. The term *stub* is applied to a short length of transmission line which is used to provide an adjustable impedance for use in matching. Such a short length may be inserted in the line (Fig. 9.18(*a*)) or connected across it (Fig. 9.18(*b*)), i.e. in series or parallel. Stubs are loss-free lengths of line and are either open-circuited or short-circuited. The two factors then to be determined for a stub are its length l and its distance x from the load.

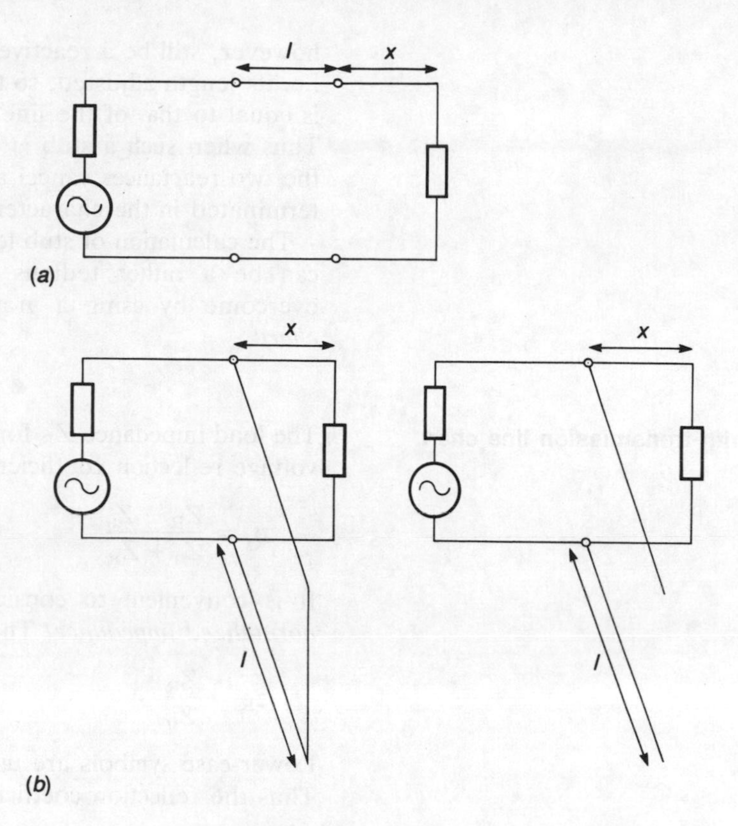

Fig. 9.18 Matching using a stub

For a short-circuited length of loss-free line, the input impedance Z_S is given by equation [47] as

$$Z_S = jZ_0 \tan \beta l$$

where l is the length of the line and Z_0 its characteristic impedance. The input impedance is a pure reactance whose magnitude and sign, i.e. whether capacitive or inductive, depend on the length of the line. For an open-circuited length of loss-free line, the input impedance is given by equation [48] as

$$Z_S = -jZ_0 \cot \beta l$$

The input impedance is a pure reactance whose magnitude and sign depend on the length l of the line.

The impedance Z_x at a point a distance x from the load is given by equation [39] as

$$Z_x = \frac{Z_0[Z_0 \sinh \gamma x + Z_R \cosh \gamma x]}{Z_0 \cosh \gamma x + Z_R \sinh \gamma x}$$

As x changes then so does the impedance of the line. At some distance from the load the impedance Z_x will have a real component equal to the characteristic impedance. There will, however, still be a reactive element. A stub can be designed, i.e. its length adjusted, so that it has a reactive element which is equal to that of the line at that point but of opposite sign. Thus when such a stub is connected to the line at that point the two reactances cancel and the line appears to be perfectly terminated in the characteristic impedance.

The calculation of stub length l and distance x from the load can be a rather tedious calculation. Most of this can be overcome by using a graphical technique called the *Smith chart*.

Smith transmission line chart

The load impedance Z_R for a transmission line is related to the voltage reflection coefficient ϱ_v by (equation [35])

$$\varrho_v = \frac{Z_R - Z_0}{Z_0 + Z_R}$$

It is convenient to consider this equation in terms of the *normalized impedance*. The normalized load impedance z_R is

$$z_R = \frac{Z_R}{Z_0} \qquad [54]$$

Lower-case symbols are used for the normalized impedances. Thus the reflection coefficient is

$$\varrho_v = \frac{Z_0[(Z_R/Z_0) - 1]}{Z_0[1 + (Z_R/Z_0)]} = \frac{z_R - 1}{z_R + 1}$$

This can be rearranged to give

$$z_R = \frac{\varrho_v + 1}{1 - \varrho_v} \qquad [55]$$

In general, both z_R and ϱ_v are complex quantities. Thus if we let

$$z_R = r + \mathrm{j}x$$

with r being the normalized resistance and x the normalized reactance, and

$$\varrho_v = u + \mathrm{j}v$$

then equation [55] becomes

$$r + \mathrm{j}x = \frac{(u + \mathrm{j}v) + 1}{1 - (u + \mathrm{j}v)} = \frac{(u + 1) + \mathrm{j}v}{(1 - u) - \mathrm{j}v}$$

$$= \frac{[(u + 1) + \mathrm{j}v][(1 - u) + \mathrm{j}v]}{[(1 - u) - \mathrm{j}v][(1 - u) + \mathrm{j}v]}$$

$$= \frac{1 - u^2 - v^2 + 2\mathrm{j}v}{(1 - u)^2 + v^2} \qquad [56]$$

Equating the real components of equation [56], then

$$r = \frac{1 - u^2 - v^2}{(1 - u)^2 + v^2}$$

$$r(1 - u)^2 + rv^2 = 1 - u^2 - v^2$$

$$r - 2ur + u^2r + rv^2 = 1 - u^2 - v^2$$

$$u^2(1 + r) - 2ur + v^2(1 + r) = 1 - r$$

$$u^2 - \frac{2ur}{1 + r} + v^2 = \frac{1 - r}{1 + r}$$

Adding $r^2/(1 + r)^2$ to each side of the equation,

$$u^2 - \frac{2ur}{1 + r} + \frac{r^2}{(1 + r)^2} + v^2 = \frac{1 - r}{1 + r} + \frac{r^2}{(1 + r)^2}$$

$$\left(u - \frac{r}{1 + r}\right)^2 + v^2 = \frac{1}{(1 + r)^2} \qquad [57]$$

This equation is of the form

$$(x - x_1)^2 + (y - y_1)^2 = R^2$$

which is the equation for a circle of radius R with centre at the coordinates x_1, y_1. Thus if r is constant, i.e. the real part of the load impedance is constant, then the graph of u against v for equation [57] is a graph of a circle with radius $1/(1 + r)$ and centre at $u = r/(1 + r)$ and $v = 0$. Figure 9.19 shows a group of such circles for different values of r.

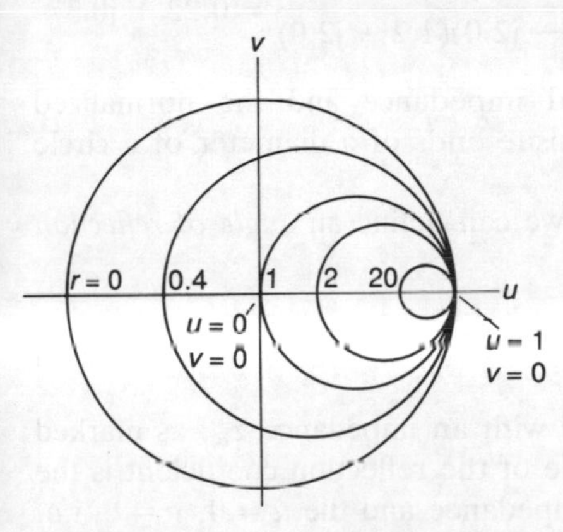

Fig. 9.19 Normalized resistance circles

Equating the imaginary components of equation [56] gives

$$x = \frac{2v}{(1 - u)^2 + v^2}$$

$$(1 - u)^2 + v^2 - \frac{2v}{x} = 0$$

Adding $1/x^2$ to both sides of the equation,

$$(1 - u)^2 + v^2 - \frac{2v}{x} + \frac{1}{x^2} = \frac{1}{x^2}$$

$$(1 - u)^2 + \left(v - \frac{1}{x}\right)^2 = \frac{1}{x^2} \qquad [58]$$

This equation is of the form of that of a circle. Thus if x is constant, i.e. the imaginary part of the load impedance is constant, then the graph of u against v for equation [58] is a graph of a circle with radius $1/x$ and centre at $u = 1$ and $v = 1/x$. Figure 9.20 shows a group of such circles for different values of x. Only those segments of circles for which r is 0 or greater have been plotted. The arcs above the $x = 0$ axis are

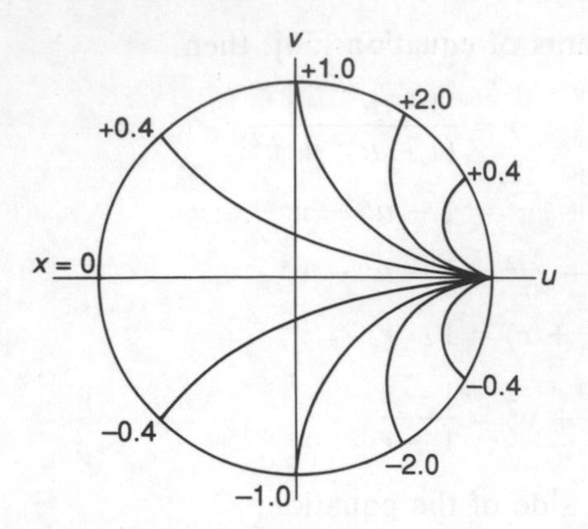

Fig. 9.20 Normalized resistance circles

for positive imaginary components, while those below the axis are for negative imaginary components.

The *Smith chart* consists of plots of the circles given by equations [57] and [58] for various values of r and x. The centre of the chart has the coordinates $u = 0$, $v = 0$, which is the same as $r = 1$ and $x = 0$. It is thus the load $z_r = 1 + j0$ and so is a line terminated by the characteristic impedance. When stubs are used to produce matching the aim is to change the impedance at the matching point to the value represented by the coordinates $r = 1$, $x = 0$.

Consider a normalized load impedance of, say, $z_R = 1.2 - j2.0$. Figure 9.21 shows this plotted on the Smith chart. The point is one which has a radius for r of 1.2 and a radius for x of 2.0. If a circle is drawn through this point with its centre at $r = 1$, $x = 0$, then a diameter drawn from the z_R point through $r = 1$, $x = 0$ intersects the other side of the circle at $0.22 + j0.37$. This in fact is $1/z_R$, i.e. the admittance y_R.

$$\frac{1}{z_R} = \frac{1}{1.2 - j2.0} = \frac{1.2 + j2.0}{(1.2 - j2.0)(1.2 + j2.0)} = 0.22 + j0.37$$

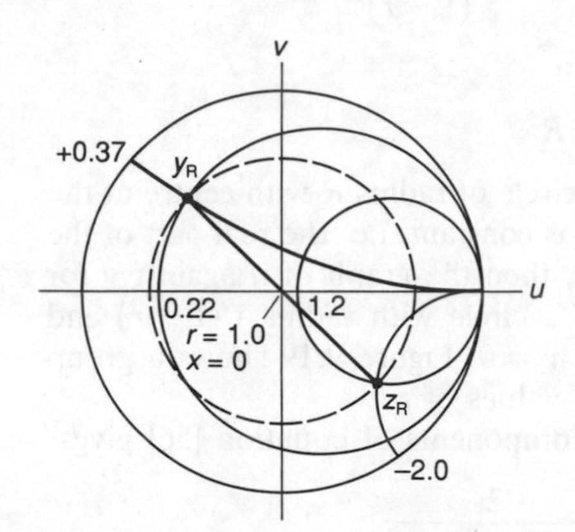

Fig. 9.21 Impedance and admittance

Thus the normalized load impedance and the normalized admittance are at the opposite ends of a diameter of a circle with centre (1,0).

Since $\varrho_v = u + jv$, then we can define an *angle of reflection coefficient* ψ as

$$\tan \psi = \frac{v}{u} \qquad [59]$$

Thus if we consider a load with an impedance z_R, as marked on Fig. 9.22, then the angle of the reflection coefficient is the angle a line through the impedance and the $u = 0$, $v = 0$, i.e. $r = 1$, $v = 0$, point makes with the $v = 0$ axis. Hence a scale in degrees can be marked round the circumference of the circle which will enable the angle of the reflection coefficient to be directly read.

The magnitude of the reflection coefficient is given by

$$|\varrho_v|^2 = u^2 + v^2$$

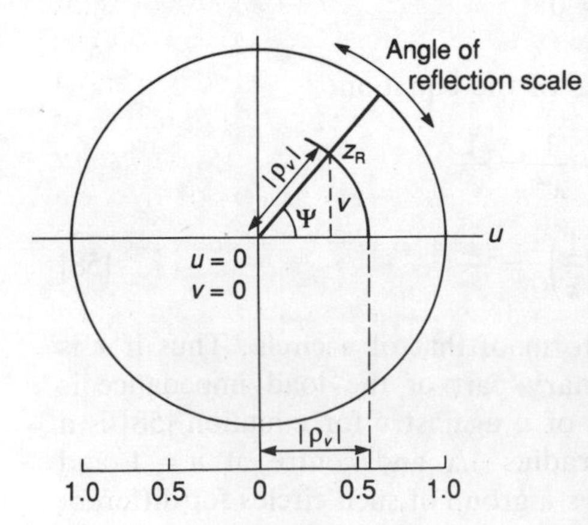

Fig. 9.22 Angle of reflection coefficient

This just represents Pythagoras' theorem, with $|\varrho_v|$ being the length of the line joining the centre, i.e. $u = 0$, $v = 0$, to the point concerned (Fig. 9.22). If a circle is drawn with the radius $|\varrho_v|$ then the magnitude of the reflection coefficient can be directly read off the chart from a radial scale.

For a circle of radius $|\varrho_v|$ and centre $r = 1$, $x = 0$, the maximum and minimum values of impedance will be where the circle cuts the $v = 0$ axis (Fig. 9.23). The maximum value of the impedance thus occurs when the angle of the reflection

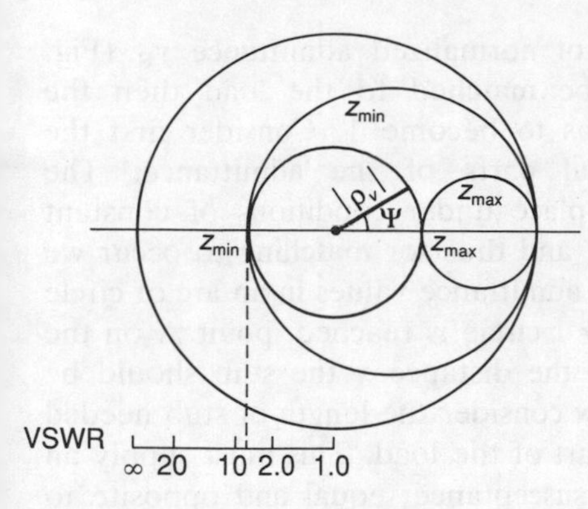

Fig. 9.23 Minimum and maximum voltages

coefficient ψ is 0° and the minimum value when it is 180°. These maximum and minimum impedances will give the minimum and maximum values of the voltage along the standing wave.

Equation [55] gives for the normalized impedance

$$z_R = \frac{\rho_v + 1}{1 - \rho_v}$$

Thus for $\psi = 0°$ we have

$$z_{max} = \frac{|\rho_v| + 1}{1 - |p_v|}$$

But the VSWR is given by (equation [52])

$$S = \frac{1 + |\rho_v|}{1 - |\rho_v|}$$

Thus $S = z_{max}$. For $\psi = 180°$, then $\rho_v = |\rho_v| \quad \angle 180° = -|\rho_v|$. Thus

$$z_{min} = \frac{-|\rho_v| + 1}{1 - (-|\rho_v|)} = \frac{1}{S}$$

Thus the VSWR ratio can be read off from a radial scale (Fig. 9.23).

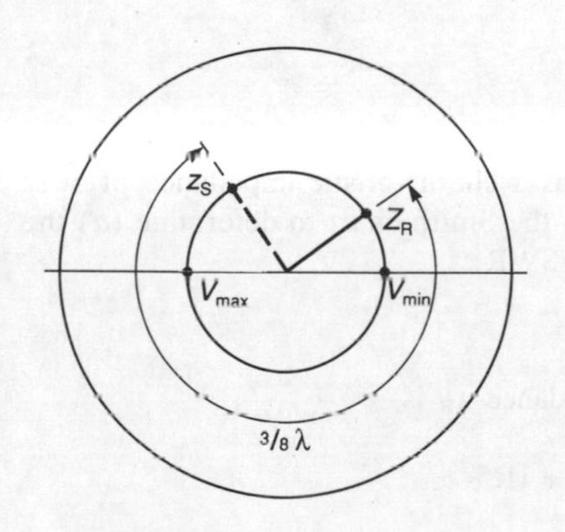

Fig. 9.24 Input impedance z_S

Consider a normalized load impedance z_R at the end of a transmission line of length l. A circle of radius $|\rho_v|$ which passes through the load impedance and has its centre at $u = 0$, $v = 0$, has one point of minimum and one of maximum impedance. Moving round the circumference of the circle is thus a movement from one minimum to the next one and thus is a movement of λ/2 (Fig. 9.24). Thus if we start at the load end, point z_R on Fig. 9.24, and have a transmission line of length $l = (3/8)\lambda$ then the line joining z_R to the centre must sweep out an arc of length $(3/8)\lambda$ to arrive at the source. The value of the impedance at that point is the normalized input impedance z_S. The distances from the load along the line at which the minima and maxima occur can be found in terms of the arc lengths swept out to reach the minima and maxima positions. If the length of the line is greater than λ/2 then the line sweeps round the circle more than once: where it ends up is still the input impedance. It is an important point to note that clockwise movement from any position on the chart gives a shift in position towards the source, while counter-clockwise movement is a shift towards the load.

Consider the problem of determining the position and length of a stub in order to match a load to a line. For series stubs it is necessary to consider impedances, for shunt stubs admittances. Consider the problem of a shunt stub to be used

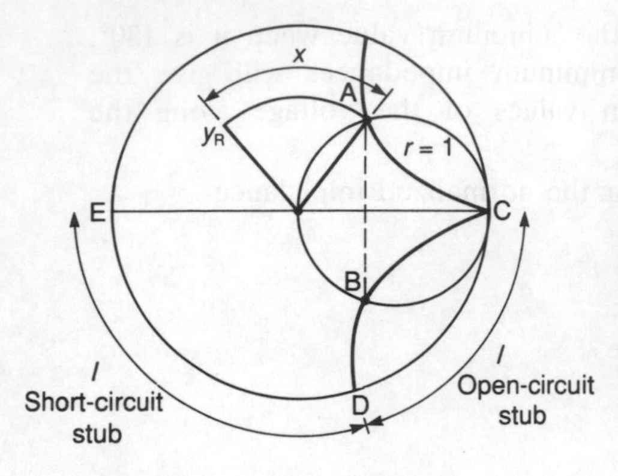

Fig. 9.25 Stub matching

with a line with a load of normalized admittance y_R (Fig. 9.25). For the line to be matched to the load then the normalized admittance has to become 1. Consider first the matching up of the real parts of the admittance. The transformation will take place under conditions of constant VSWR, i.e. constant $|\rho_v|$, and thus for matching to occur we must move between these admittance values in an arc of circle of radius $|\rho_v|$ until the $r = 1$ circle is reached, point A on the figure. This arc length is the distance x the stub should be placed from the load. Now consider the length of stub needed to match the imaginary part of the load. This must supply an amount of reactance, or susceptance, equal and opposite to that at A. This is point B. If the stub is to be open-circuited then the length l is the arc length from C, which corresponds to open-circuit, to D. If the stub is to be short-circuited then l is the arc length from E, which corresponds to short-circuit, to D.

The above discussion has outlined, on the basis of simplified sketches of the chart, how the Smith chart can be used for some basic measurements. The chart itself is of the form shown in Fig. 9.26, with all its scales to simplify the extraction of data from the chart. For simplicity not all the r circles and the x arcs have been included; also, just some of the scale readings are included.

Example 17

A loss-less transmission line has a characteristic impedance of 50 Ω and a load of 150 + j75 Ω. Use the Smith chart to determine (*a*) the reflection coefficient, (*b*) the VSWR.

Answer

(*a*) The normalized load impedance z_R is

$$z_R = \frac{150 + j75}{50} = 3.0 + j1.5$$

Figure 9.27 shows this plotted on a simplified version of a Smith chart. By drawing a line from $r = 1$, $x = 0$ through the z_R point the angle ψ can be determined. It is about 16°. The value of $|\rho|$ can be determined from the appropriate radial scale, being the length of the line joining the centre to z_R. The value is about 0.59. Thus

$$\rho = 0.59/16° = 0.57 + j0.16$$

(*b*) Where the circle drawn with radius $|\rho_v|$ crosses the $x = 0$ axis then the value of r at that point is the VSWR. In this case it is about 3.8.

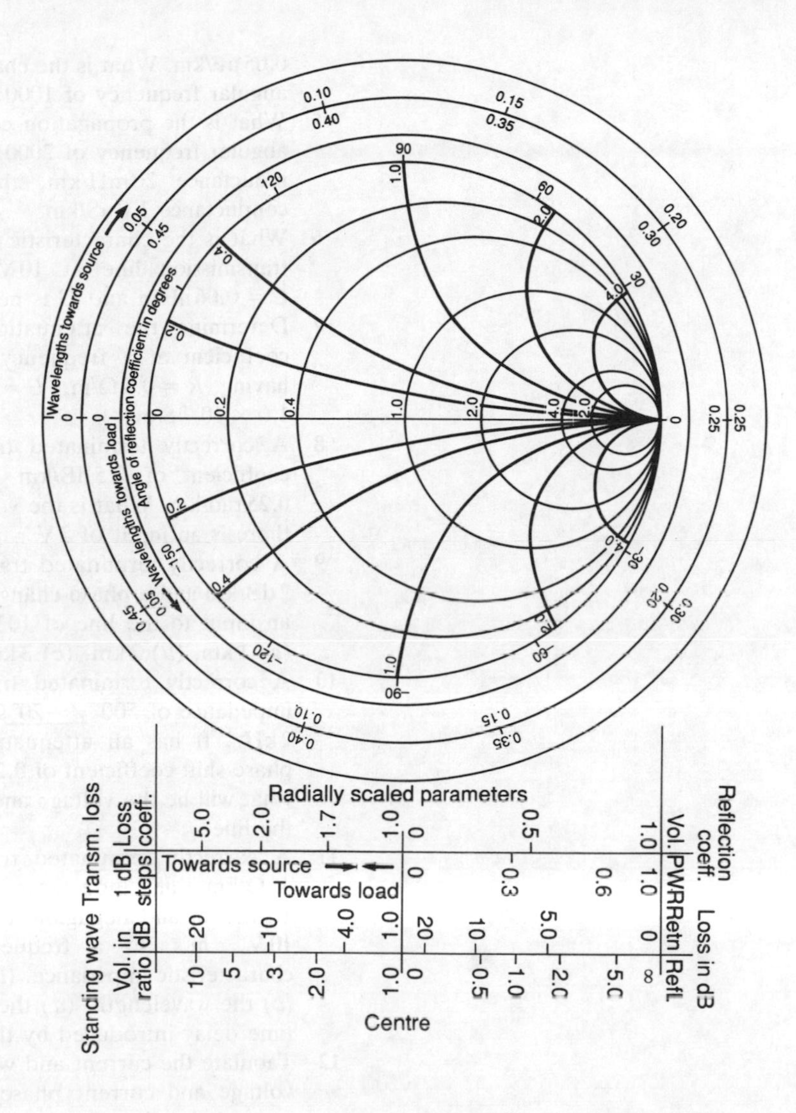

Fig. 9.26 The Smith chart

Problems

1 What is (*a*) the wavelength and (*b*) the phase velocity of transmission of a 1000 rad/s signal along transmission lines which give a phase shift of 0.04 rad/km?

2 A coaxial cable has an inductance of 0.3 mH/km and a capacitance of 0.05 μF/km. What will be (*a*) the phase velocity of a signal along them and (*b*) the phase delay per km if the signal has a frequency of 10 kHz?

3 A transmission line has a capacitance of 4 nF/km and an inductance of 5 mH/km. When used at a frequency of 5 kHz, what will be (*a*) the phase delay per km, (*b*) the wavelength on the line, (*c*) the phase velocity along the line, and (*d*) the time delay introduced by a 10 km length of the line?

4 A transmission line has a resistance of 60 Ω/km, an inductance of 10 mH/km, a conductance of 4 μS/km and a capacitance of

0.05 μF/km. What is the characteristic impedance of the line at an angular frequency of 1000 rad/s?

5 What is the propagation coefficient of a transmission line at an angular frequency of 2000 rad/s if the resistance is 60 Ω/km, the inductance 20 mH/km, the capacitance 0.04 μF/km and the conductance 1.0 μS/km.

6 What is the characteristic impedance and the attenuation for a transmission line at 10 MHz if $R = 0.6\,\Omega/\text{m}$, $L = 0.2\,\mu\text{H/m}$, $C = 0.06\,\text{nF/m}$ and G is negligible?

7 Determine the attenuation coefficient and the phase shift coefficient at a frequency of 10 MHz for a transmission line having $R = 0.5\,\Omega/\text{m}$, $L = 250\,\text{nH/m}$, $C = 100\,\text{pF/m}$ and $G = 1.0 \times 10^{-6}\,\text{S/m}$.

8 A correctly terminated transmission line has an attenuation coefficient of 2.5 dB/km and a phase change coefficient of 0.25 rad/km. What is the voltage at a point 5 km along the line if there is an input of 2 V r.m.s.?

9 A correctly terminated transmission line has an attenuation of 2 dB/km and a phase change coefficient of π/4 rad/km. If there is an input to the line of 10 V r.m.s., what are the voltages after (*a*) 1 km, (*b*) 2 km, (*c*) 3 km?

10 A correctly terminated transmission line has a characteristic impedance of $500\,\angle{-20^\circ}\,\Omega$ and is operating at a frequency of 2 kHz. It has an attenuation coefficient of 0.25 Np/km and a phase shift coefficient of 0.20 rad/km. For an input of 10 V r.m.s. what will be the voltage and current at a distance of 10 km down the line?

11 A correctly terminated transmission line has a resistance of 45 Ω/km, an inductance of 3.5 mH/km, a capacitance of 1.3 nF/km and negligible conductance. The input to the line is 10 V r.m.s. at a frequency of 10 kHz. What is (*a*) the characteristic impedance, (*b*) the voltage and current after 8 km, (*c*) the wavelength, (*d*) the velocity of propagation, and (*e*) the time delay introduced by the line?

12 Tabulate the current and voltage data necessary for plotting the voltage and current phasor spirals for a correctly terminated transmission line of length 12 km with an input of 10 V r.m.s. and a characteristic impedance of $1000\,\angle{-\pi/4}\,\Omega$. The attenuation coefficient is 0.05 Np/km and the phase change coefficient π/6 rad/km.

13 If the input impedance of a transmission line is $2.0\,\angle{30^\circ}\,\text{k}\Omega$ when the output end is open-circuit and $0.2\,\angle{-40^\circ}\,\text{k}\Omega$ when it is short-circuited, what is the characteristic resistance of the line?

14 A correctly terminated transmission line of length 10 km has a resistance of 55 Ω/km, an inductance of 0.60 mH/km, a capacitance of 0.033 μF/km and conductance of 1 μS/km. If the power input to the line is 1.0 W at a frequency of 1 kHz, what is (a) the characteristic impedance of the line, (*b*) the propagation coefficient, (*c*) the attenuation coefficient, (*d*) the phase change coefficient, (*e*) the wavelength, (*f*) the velocity of propagation, (*g*) the time delay introduced by the line, (*h*) the magnitudes of the voltage and current at the source and receiving ends of the

line, and (*i*) the receiving-end power?

15 A transmission line has a resistance of 20 Ω/km, an inductance of 2.0 mH/km, a capacitance of 0.06 μF/km and a conductance of 1.0 μS/km. What amount of inductance should be added to give a distortionless line?

16 A transmission line with a characteristic impedance of 80 Ω is terminated in a resistive load of 240 Ω. If the voltage measured across the load is 10 V r.m.s., what is (*a*) the voltage reflection coefficient for the line, (*b*) the incident current and voltage at the load, (*c*) the reflected current and voltage at the load?

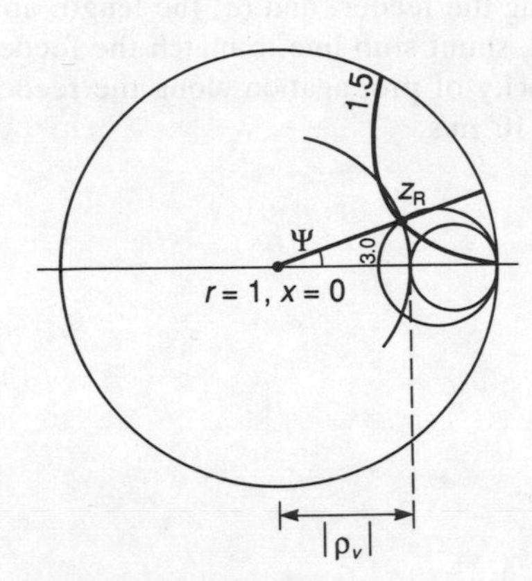

Fig. 9.27 Example 17

17 What are the voltage reflection coefficient and current reflection coefficient for a line of characteristic impedance 50 Ω and terminated by an impedance of (*a*) 50 Ω, (*b*) 25 Ω, (*c*) 75 Ω?

18 What is the voltage reflection coefficient for a line of characteristic impedance 100 $\angle -20°$ Ω and a load of 50 $\angle 0°$ Ω?

19 What is the input impedance of a line which has a characteristic impedance of 600 Ω, has an attenuation of 6 dB and a phase shift of 180° in its length, and is terminated by a load of 2000 Ω?

20 Two correctly terminated loss-free transmission lines have characteristic resistances of 80 Ω and 100 Ω. What should be the characteristic resistance of the quarter wavelength section used to connect them if there is to be perfect matching?

21 A loss-free transmission line with an inductance of 1.3 mH/km and capacitance 0.010 μF/km is to be matched to a length of loss-free line having an inductance of 2.0 mH/km and capacitance 0.10 μF/km by a quarter wavelength section of line. What should be the characteristic resistance of the section?

22 What should be the characteristic impedance of a loss-free quarter wavelength line used to match an aerial of impedance 100 Ω to a 600 Ω transmission line?

23 A low loss transmission line has a characteristic impedance of 600 Ω and a load of (500 + j200) Ω. What is the voltage standing wave ratio?

24 A low loss transmission line has a voltage reflection coefficient of 0.2 $\angle 30°$. If the line has a characteristic impedance of 80 Ω what is (*a*) the voltage standing wave ratio, and (*b*) the load impedance?

25 Explain how the Smith chart can be used to determine (*a*) impedance at any point along a line, (*b*) the input impedance, (*c*) the distance from the load to voltage maxima and minima, and (*d*) the position and length of a shunt stub to produce matching.

26 Explain how the Smith chart can be used to determine the voltage reflection coefficient and VSWR for a line, given the line characteristic impedance and the load impedance.

27 Use the Smith chart to determine the input impedance and the VSWR for a transmission line of length 0.8Λ, characteristic impedance 50 Ω and load (50 − j100) Ω.

28 Use the Smith chart to determine the length and position of a shunt, short-circuited, stub if a loss-free transmission line with a characteristic admittance of 0.02 S is to be matched to a load of (0.04 + j0.01) S.

29 A loss-free transmission line of characteristic impedance 600 Ω is used to connect a 100 MHz transmiter to an aerial. If the aerial has an impedance of 75 Ω and the power delivered to it is 1 kW, what is (*a*) the VSWR, (*b*) the maximum and minimum values of the voltage and current along the feeder, and (*c*) the length and position of a short-circuited, shunt stub line to match the feeder line to the aerial? The velocity of propagation along the feeder may be assumed to be 3×10^8 m/s.

10 Three-phase circuits

Introduction

The earliest generation, transmission and distribution of power was by direct current, with the transmission being restricted to very localized areas. This was superseded by single phase alternating current, transformers enabling the voltage to be raised or lowered and so give more efficient distribution over greater distances. The power loss in passing a current along a power cable is proportional to the square of the current, and since for a fixed power input the current is inversely proportional to the voltage then the power loss is inversely proportional to the square of the voltage. Thus stepping-up the voltage enables transmission to occur with less loss. Nowadays electrical power is generated, transmitted and distributed by means of three-phase circuits. Three-phase has the advantages that the windings in rotating machines more efficiently use the space available for them, for the same overall cross-sectional area of conductors and power the transmission efficiency is greater, rectification to d.c. gives less ripple than with a single-phase supply, and it is easier to produce with rotating machines a rotating magnetic field.

Three-phase supply

A single-phase generator can be considered to be effectively a single loop or coil of wire rotating in a magnetic field (Fig. 10.1(*a*)). A three-phase generator can be considered to be three loops or coils of wire rotating in the magnetic field, the coils being at such angles that the induced e.m.f.s in the three coils are displaced one from another by 120° (Fig. 10.1(*b*)). The three induced e.m.f.s are usually distinguished by colour coding the wires from the three coils and thus are frequently referred to as the red, yellow and blue phases. With another convention they are referred to as the a-phase, b-phase and c-phase. The order in which the three phases attain their peak e.m.f.s is known as the *phase sequence* for Fig. 10.1(*b*), this is red–yellow–blue.

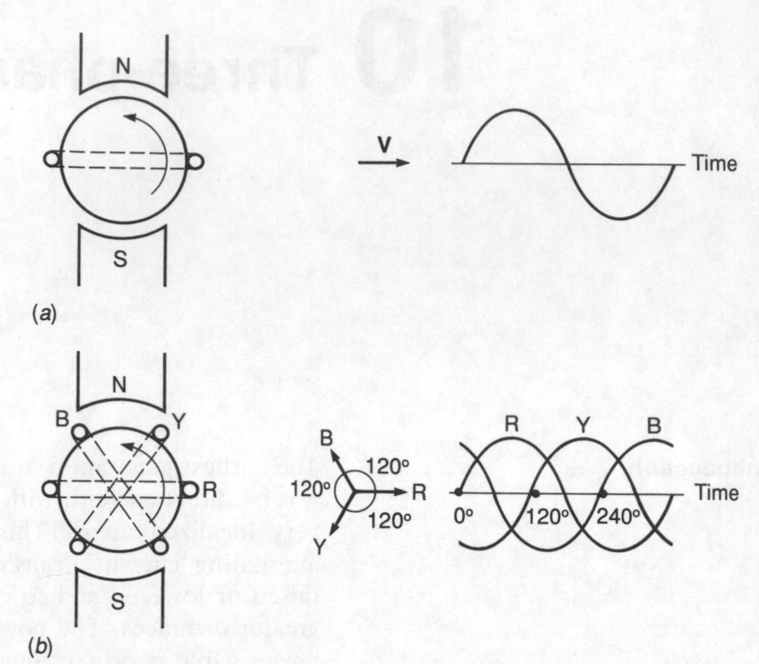

Fig. 10.1 (*a*) Single-phase generation, (*b*) three-phase generation

The term *balanced* is used for the three-phase voltages if they have equal amplitudes and frequency but are out of phase with each other by exactly 120°. In phasor notation the three balanced phase voltages are

$$\mathbf{V_R} = V_m\angle 0° \quad [1]$$

$$\mathbf{V_Y} = V_m\angle -120° \quad [2]$$

$$\mathbf{V_B} = V_m\angle -240° \quad [3]$$

where V_m is the maximum value of the voltage and is the same for the three-phase voltages.

Connection of phases

Each of the three outputs from a three-phase generator may be independently connected to its own load and thus three completely independent circuits obtained (Fig. 10.2). Such a form of connection requires six lines between the generator and the loads.

It is however possible to reduce the number of lines, and so effect a saving in cost. There are two methods of interconnection between the three generator outputs which result in only three or four lines being necessary. These are called *star*, or *Y*, and *delta* or *mesh*. Figure 10.3 shows these two forms of connection.

With the *star* connection the three generator coils are connected to a common point, called the *neutral* or *star point*.

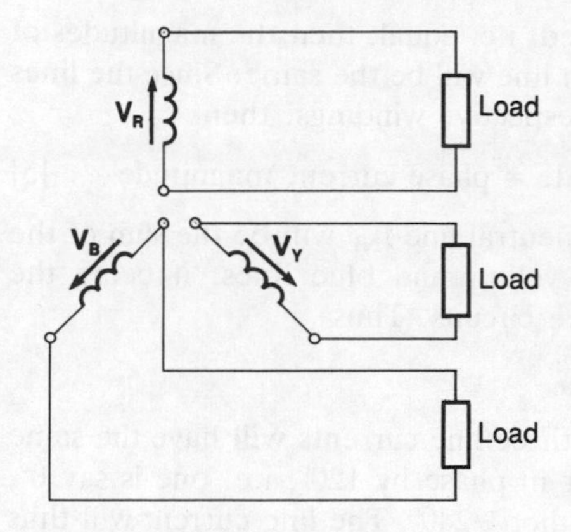

Fig. 10.2 Independent connection of phases

The line from this point is called the *neutral line*. The voltages of each of the terminals R, Y, B in relation to the neutral point are the so-called *phase voltages*. The star connection is such that the voltage in each winding acts outwards from the star point when positive. It should however be realized that not all the voltages will be positive at the same instant of time. The term *line-to-line voltages* (or *line voltage*) is used for the voltages between lines, e.g. between the red and yellow lines. Thus in Fig. 10.3(*a*), the phase voltage between the red line and the neutral line is $\mathbf{V_{RN}}$, for the yellow line $\mathbf{V_{YN}}$, and for the blue line $\mathbf{V_{BN}}$. If V_p is the magnitude of the phase voltage, this being the same for each of them since we are considering a balanced system, then

$$\mathbf{V_{RN}} = V_p\angle 0° = V_p(1 + j0)$$

$$\mathbf{V_{YN}} = V_p\angle -120° = V_p(-0.5 - j0.87)$$

$$\mathbf{V_{BN}} = V_p\angle -240° = V_p(-0.5 + j0.87)$$

The line voltage between the red and yellow lines $\mathbf{V_{RY}}$ is the phasor difference between $\mathbf{V_{RN}}$ and $\mathbf{V_{YN}}$. Figure 10.4 shows the phasor diagram. Thus

$$\mathbf{V_{RY}} = \mathbf{V_{RN}} - \mathbf{V_{YN}} = V_p(1 + j0) - V_p(-0.5 - j0.87)$$

$$= V_p(1.5 + j0.87) = \sqrt{3}V_p\angle 30°$$

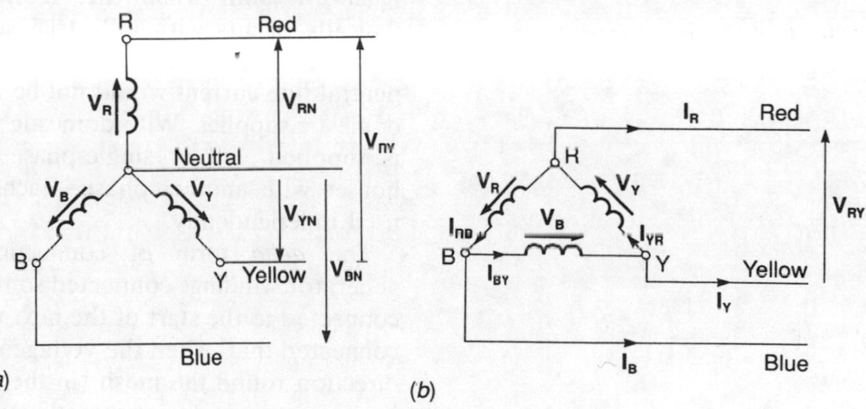

Fig. 10.3 (*a*) Star connection, (*b*) delta connection

A similar calculation can be carried out for each of the other line voltages $\mathbf{V_{BR}}$ and $\mathbf{V_{YB}}$. Figure 10.5 shows the phasor diagram for all the line voltages. For each

$$\text{line voltage magnitude} = \sqrt{3} \times \text{phase voltage magnitude} \qquad [4]$$

Thus, for example, a three-phase star-connected supply with a phase voltage of 240 V has line-to-line voltages of 416 V.

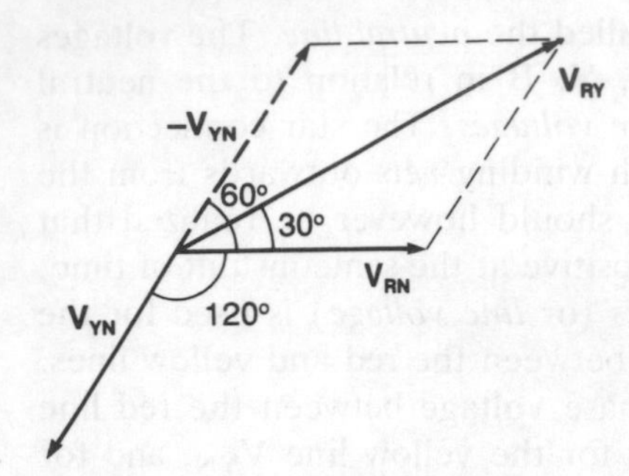

Fig. 10.4 Line voltage phasor diagram

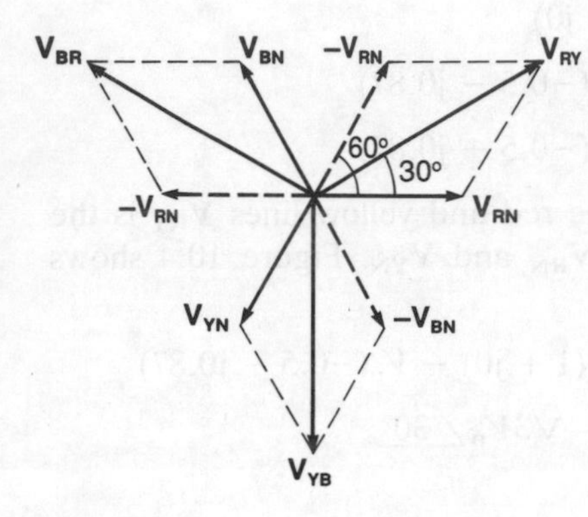

Fig. 10.5 Line voltages phasor diagram

If the loads are balanced, i.e. equal, then the magnitudes of the currents through each line will be the same. Since the lines are in series with their respective windings, then

$$\text{line current magnitude} = \text{phase current magnitude} \qquad [5]$$

The current along the neutral line $\mathbf{I}_{NL}$ will be the sum of the currents along the red, yellow and blue lines, it being the 'return' wire for the three circuits. Thus

$$\mathbf{I}_{NL} = \mathbf{I}_{RL} + \mathbf{I}_{YL} + \mathbf{I}_{BL}$$

For a balanced load the three line currents will have the same magnitude but will differ in phase by 120°, i.e. one is say 0°, another −120° and the other −240°. The line current will thus be

$$\mathbf{I}_{NL} = I_L\angle 0° + I_L\angle -120° + I_L\angle -240°$$

where I_L is the magnitude of the line currents. Thus

$$\mathbf{I}_{NL} = I_L(1 + \text{j}0) + I_L(-0.5 - \text{j}0.87) + I_L(-0.5 + \text{j}0.87)$$
$$= 0$$

The neutral line current with a balanced system is zero.

With the star form of connection we can have a three-wire or a four-wire system of lines, depending on whether the neutral line is used. A three-wire system is commonly used for balanced loads when the neutral line current is zero, as typically occurs with industrial supplies, while the four-wire system is used when the load is likely to be unbalanced and the neutral line current would not be zero, as typically occurs with domestic supplies. With domestic supplies one group of houses is supplied with a single-phase voltage, another group of houses with another phase. Each of the three phases is then used independently.

The *delta* form of connection (Fig. 10.3(*b*)) has the generator windings connected so that the end of one winding is connected to the start of the next winding. The windings are so connected that, when the voltage is positive, it acts in the same direction round the mesh (in the figure this is shown as being in a counter-clockwise direction). It needs to be realized that not all the voltages will be positive at the same instant of time. If the magnitude of each of the phase voltages is the same, namely V_p, then

$$\mathbf{V_R} = V_p\angle 0° = V_p(1 + \text{j}0)$$
$$\mathbf{V_Y} = V_p\angle -120° = V_p(-0.5 - \text{j}0.87)$$
$$\mathbf{V_B} = V_p\angle -240° = V_p(-0.5 + \text{j}0.87)$$

The sum of the voltages round the mesh is

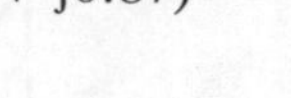

$$\mathbf{V_R} + \mathbf{V_Y} + \mathbf{V_B} = V_p(1 + j0 - 0.5 - j0.87 - 0.5 + j0.87)$$
$$= 0$$

There is thus no circulating current in the mesh. Hence the line voltages are equal to the phase voltages, e.g. $\mathbf{V_{RY}} = \mathbf{V_R}$.

$$\text{Line voltage magnitude} = \text{phase voltage magnitude} \qquad [6]$$

The above relationship is only valid if the directions of the voltages when positive are all in the same direction round the mesh. If this is not the case then there is a current circulating in the mesh. Such a mesh is said to be *incorrectly closed*. Suppose one of the windings connections to be reversed, say the blue winding. Then we have

$$\mathbf{V_R} = V_p\angle 0° = V_p(1 + j0)$$
$$\mathbf{V_Y} = V_p\angle -120° = V_p(-0.5 - j0.87)$$
$$\mathbf{V_B} = -V_p\angle -240° = -V_p(-0.5 + j0.87)$$

The sum of the voltages round the mesh is

$$\mathbf{V_R} + \mathbf{V_Y} + \mathbf{V_B} = V_p(1 + j0 - 0.5 - j0.87 + 0.5 - j0.87)$$
$$= V_p(1 - j1.74) = 2V_p\angle -60°$$

The resultant loop voltage has thus a magnitude which is twice the value of a phase voltage.

For a correctly closed mesh, applying Kirchoff's current law to node R of the mesh gives

$$\mathbf{I_R} = \mathbf{I_{YR}} - \mathbf{I_{RB}}$$

Fig. 10.6 Line current phasor diagram

This equation represents the phasor diagram given in Fig. 10.6. If the loads are balanced and each has unity power factor $\mathbf{I_{YR}}$ is in phase with $\mathbf{V_{YR}}$, $\mathbf{I_{RB}}$ is in phase with $\mathbf{V_{RB}}$ and each of the currents has the same magnitude. If this magnitude is denoted by I_p, then

$$\mathbf{I_{BR}} = I_p\angle 0° = I_p(1 + j0)$$
$$\mathbf{I_{RY}} = I_p\angle -240° = I_p(-0.5 - j0.87)$$

Thus

$$\mathbf{I_R} = I_p(1 + j0 + 0.5 + j0.87) = I_p(1.5 + j0.87)$$
$$= \sqrt{3}I_p\angle 150°$$

The same current magnitude is obtained for each of the other lines if the above analysis is repeated for the nodes B and Y. Figure 10.7 shows the phasor diagram for all the line currents. Thus the

$$\text{line current magnitude}$$
$$= \sqrt{3} \times \text{phase current magnitude} \qquad [7]$$

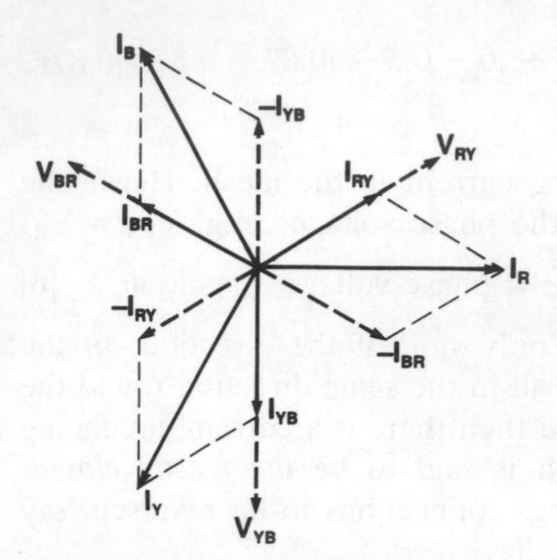

Fig. 10.7 Line currents phasor diagram

The loads can be connected to the star or delta connected windings in two ways, as a star or delta. Figure 10.8 shows these two forms of connection. If each of the loads is the same then the load is said to be balanced. A star-connected generator may be connected to a star-connected or a delta-connected load, and similarly a delta-connected generator may be connected to a star-connected or delta-connected load.

With star-connected balanced loads the voltage across one will be the same as the phase e.m.f. that would occur with a star-connected generator and thus (line voltage/$\sqrt{3}$). The current through each load will be the same as the line current. With delta-connected loads the voltage across each one will be the same as the line voltage, and hence the phase voltage since the line voltage equals the phase voltage. The current through each load will be the same as the phase current that would occur in a delta-connected generator and so (line current/$\sqrt{3}$).

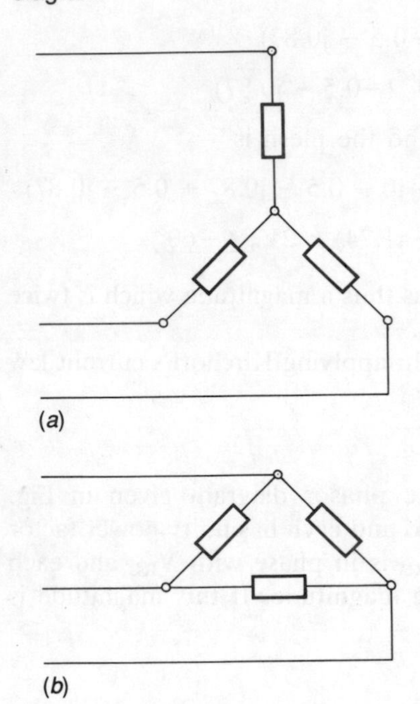

Fig. 10.8 Loads, (*a*) star connection, (*b*) delta connection

Example 1

For a balanced star-connected three-phase generator and star-connected load system with a line voltage of 440 V and three equal resistive loads of 100 Ω, what will be the magnitudes of (*a*) the phase voltage, (*b*) the phase current, (*c*) the line current?

Answer

Figure 10.9 shows the circuit.

(*a*) The phase voltage will be, according to equation [4], 440/$\sqrt{3}$ = 254 V.

(*b*) The phase current will (phase voltage/load) = 254/100 = 2.54 A.

(*c*) For star connection the line current equals the phase current, thus the line current = 2.54 A.

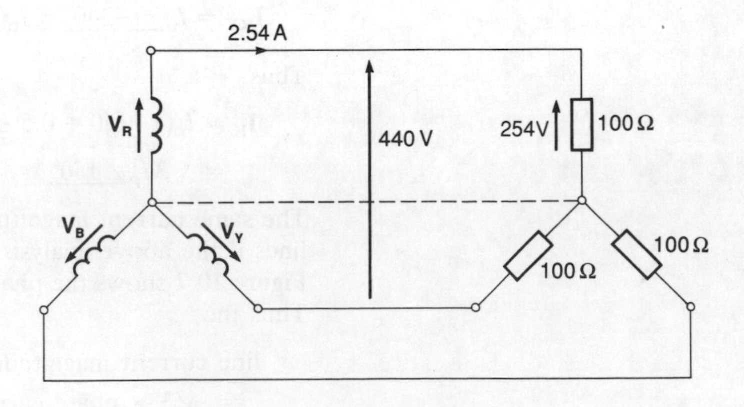

Fig. 10.9 Example 1

Example 2

For a balanced delta-connected three-phase generator and delta-connected load system with a line voltage of 440 V and three equal resistive loads of 100 Ω, what will be the magnitudes of (*a*) the phase voltage, (*b*) the phase current, (*c*) the line current?

Answer

Figure 10.10 shows the circuit.

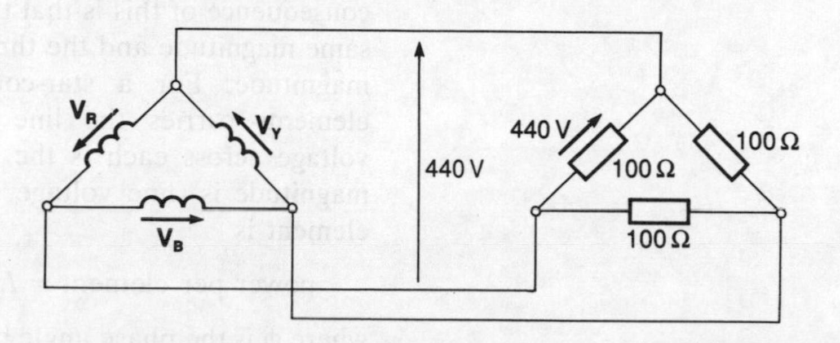

Fig. 10.10 Example 2

(*a*) For a delta-connected system the phase voltage equals the line voltage, thus the phase voltage = 440 V.

(*b*) The phase current will be the same as the current through a 100 Ω resistor. Thus the phase current = (phase voltage/100) = 440/100 = 4.40 A.

(*c*) The line current is, according to equation [7], $\sqrt{3}$ × phase current and is thus $\sqrt{3} \times 4.40 = 7.62$ A.

Example 3

A balanced three-phase generator is star-connected and has a phase e.m.f. of 80 V. The generator is connected to a delta-connected balanced load consisting of three 120 Ω resistors. What is (*a*) the line voltage, (*b*) the voltage across a load resistor, (*c*) the current through a load resistor?

Answer

Figure 10.11 shows the circuit.

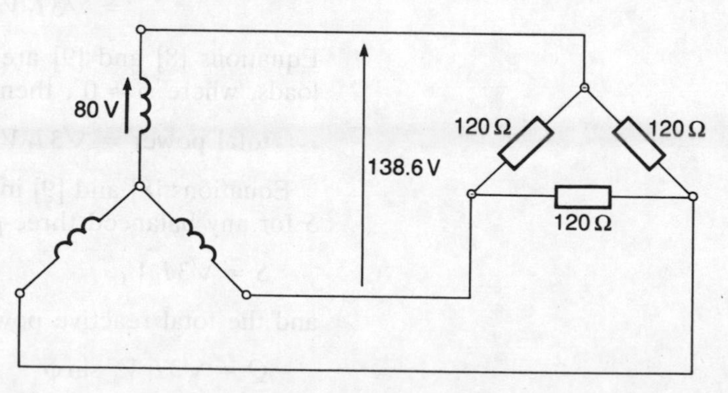

Fig. 10.11 Example 3

(*a*) The line voltage produced by a star-connected generator is given by equation [4] as ($\sqrt{3}$ × phase voltage) = $\sqrt{3} \times 80 = 138.6$ V.

(*b*) The voltage across a delta-connected load resistor is the line voltage and hence 138.6 V.

(*c*) The current through load resistor is 138.6/120 = 1.16 A.

Power in a balanced system

Consider a balanced generator and a balanced load. A consequence of this is that the three-line voltages will have the same magnitude and the three-line currents will have the same magnitude. For a star-connected load, each of the load elements carries the line current, magnitude I_L, and the voltage across each is the phase voltage. The phase voltage magnitude is (line voltage $V_L/\sqrt{3}$). Thus the power per load element is

$$\text{power per element} = I_L \times (V_L/\sqrt{3}) \cos\phi$$

where ϕ is the phase angle between the voltage and the current for the load impedance, cos ϕ thus being the power factor for it. The total power, for the three-load elements, is thus

$$\begin{aligned}\text{total power} &= 3I_L \times (V_L/\sqrt{3}) \cos\phi \\ &= \sqrt{3} I_L V_L \cos\phi \qquad [8]\end{aligned}$$

For a delta-connected load, each of the load elements carries the phase current and the voltage across it is the line voltage, magnitude V_L. The phase current magnitude is (line current $I_L/\sqrt{3}$). Thus the power per load element is

$$\text{power per element} = (I_L/\sqrt{3}) \times V_L \cos\phi$$

where ϕ is the phase angle between the voltage and the current for the load impedance, cos ϕ thus being the power factor for it. The total power, for the three-load elements, is thus

$$\begin{aligned}\text{total power} &= 3(I_L/\sqrt{3}) \times V_L \cos\phi \\ &= \sqrt{3} I_L V_L \cos\phi \qquad [9]\end{aligned}$$

Equations [8] and [9] are identical. Thus for purely resistive loads, where $\phi = 0°$, then

$$\text{total power} = \sqrt{3} I_L V_L \qquad [10]$$

Equations [8] and [9] indicate that the total apparent power S for any balanced three-phase system is

$$S = \sqrt{3} I_L V_L \qquad [11]$$

and the total reactive power Q is

$$Q = \sqrt{3} I_L V_L \sin\phi \qquad [12]$$

Example 4

What is the power consumed when a three-phase supply with a line voltage of 440 V is connected to a star-connected load composed of three identical 100 Ω resistors?

Answer

For a star-connected load the voltage across a load element is the phase voltage. This is $440/\sqrt{3} = 254$ V. The current through a load element is thus $254/100 = 2.54$ A. Thus the power consumed per load element is

$$\text{power per element} = 254 \times 2.54 = 645.2\,\text{W}$$

The total power is thus $3 \times 645.2 = 1936$ W.

The above answer could have been obtained by using equation [8]. The line current is the current through a load element and because the load is purely resistive the power factor is 1. Thus

$$\text{total power} = \sqrt{3} I_L V_L \cos\phi = \sqrt{3} \times 2.54 \times 440 \times 1$$

$$= 1936\,\text{W}.$$

Example 5

What is the power consumed when a three-phase supply with a line voltage of 440 V supplies a balanced delta-connected load, each of the load elements having an impedance of 50 Ω and a power factor of 0.8?

Answer

For a delta-connected load, the current through a load element is the phase current and the voltage across it is the line voltage. The magnitude of the current through the load is $440/50 = 8.8$ A. Thus the power consumed per load element is

$$\text{power per element} = 8.8 \times 440 \times 0.8 = 3097.6\,\text{W}$$

Thus the total power consumed is $3 \times 3097.6 = 9293$ W.

Alternatively, equation [9] can be used. Since the line current is $\sqrt{3} \times 8.8$ A then

$$\text{total power} = \sqrt{3} I_L V_L \cos\phi = \sqrt{3} \times \sqrt{3} \times 8.8 \times 440 \times 0.8$$

$$= 9293\,\text{W}$$

Example 6

A three-phase motor is connected to a three-phase supply and provides a balanced load at a power factor of 0.8. What is the power input to the motor if the line voltage is 415 V and the line current 3.0 A?

Answer

It does not matter whether the motor windings are star- or delta-connected. The power is given by equations [8] or [9] as

$$\text{power} = \sqrt{3} I_L V_L \cos\phi = \sqrt{3} \times 3.0 \times 415 \times 0.8 = 1725\,\text{W}$$

Power measurement in a balanced system

When there is a balanced load with a three-phase system, it is possible to measure the total power consumed by using just a single wattmeter. With a balanced load the total power is the sum of the power consumed per load element. Figure 10.12 shows how a wattmeter can be used to make such a measurement with a star-connected balanced load. The total power consumed by the load is thus three times the instrument reading.

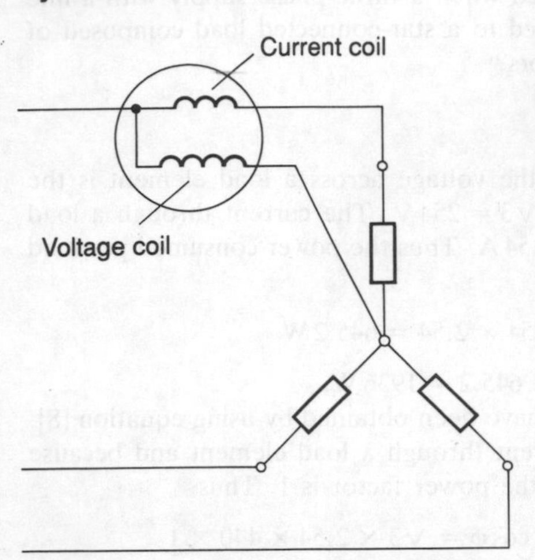

Fig. 10.12 Measurement of power with a balanced load

The power in any three-phase system, whether with a balanced or unbalanced load, can be measured using two or three wattmeters. For a star-connected balanced, or unbalanced, load a wattmeter may be connected to each load element and the total power thus becomes the sum of the three wattmeter readings (Fig. 10.13).

Figure 10.14 shows how two wattmeters may be used to determine the power in a three-phase balanced, or unbalanced, three-wire circuit. For that circuit when balanced, the power reading indicated by wattmeter 1 is

$$P_1 = V_{RB} I_R \cos(\text{angle between } \mathbf{I_R} \text{ and } \mathbf{V_{RB}})$$

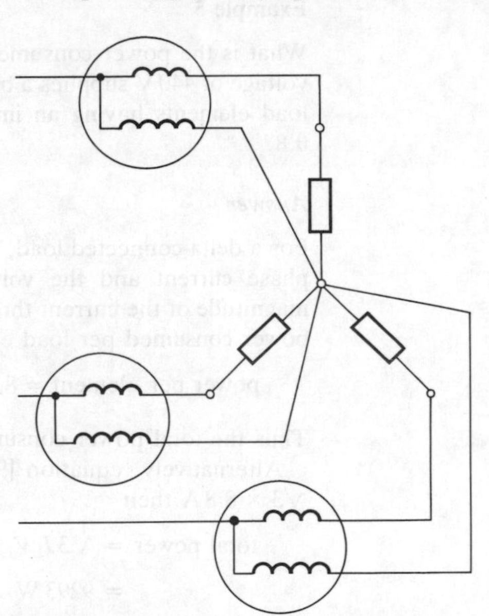

Fig. 10.13 Three-wattmeter method

The voltage $\mathbf{V_{RB}}$ is $\mathbf{V_{RS}} - \mathbf{V_{BS}}$ and is indicated by the phasor diagram in Fig. 10.15. For a balanced system the magnitudes of $\mathbf{V_{RS}}$ and $\mathbf{V_{BS}}$ are the same and so the phase angle between $\mathbf{V_{RB}}$ and $\mathbf{V_{RS}}$ is 30°. If the load has a lagging phase angle ϕ, i.e. the phase angle between $\mathbf{I_R}$ and $\mathbf{V_{RS}}$ is ϕ, then the phase angle between $\mathbf{I_R}$ and $\mathbf{V_{RB}}$ is $(30° - \phi)$. Hence

$$P_1 = V_L I_L \cos(30° - \phi) = V_L I_L (0.87 \cos\phi + 0.5 \sin\phi) \quad [13]$$

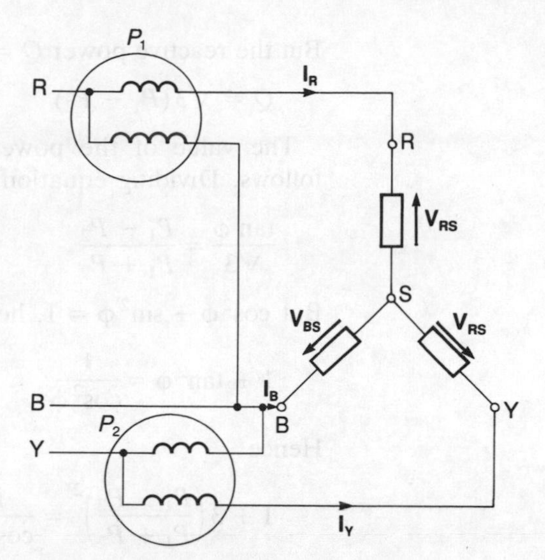

Fig. 10.14 Two-wattmeter method

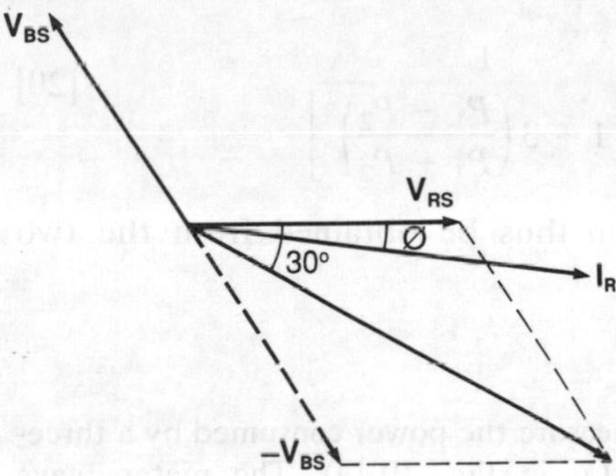

Fig. 10.15 Phasor diagram with a lagging power factor

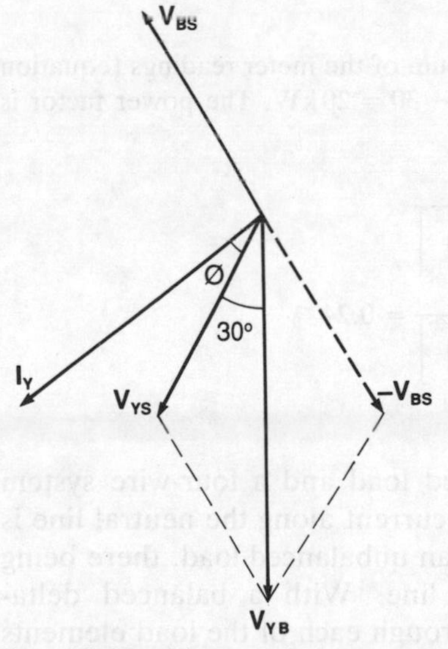

Fig. 10.16 Phasor diagram with a lagging power factor

where V_L is the magnitude of the line voltage and I_L that of the line current.

The power reading indicated by wattmeter 2 is

$$P_2 = V_{YB} I_Y \cos(\text{angle between } \mathbf{I_Y} \text{ and } \mathbf{V_{YB}})$$

The voltage $\mathbf{V_{YB}}$ is $\mathbf{V_{YS}} - \mathbf{V_{BS}}$ and is indicated by Fig. 10.16. For a balanced system we thus have a phase angle of 30° between $\mathbf{V_{YB}}$ and $\mathbf{V_{YS}}$ and since the phase angle between $\mathbf{V_{YS}}$ and $\mathbf{I_Y}$ is ϕ then

$$P_2 = V_L I_L \cos(30° + \phi) = V_L I_L (0.87 \cos\phi - 0.5 \sin\phi) \quad [14]$$

Thus the sum of the wattmeter readings is, using equations [13] and [14],

$$P_1 + P_2 = V_L I_L (0.87 \cos\phi + 0.5 \sin\phi + 0.87 \cos\phi - 0.5 \sin\phi)$$

$$= \sqrt{3}\, V_L I_L \cos\phi \quad [15]$$

But this is the total power consumed by the balanced load (equation [9]). Thus

$$\text{total power} = P_1 + P_2 \quad [16]$$

i.e. the sum of the two meter readings is the total power consumed.

The difference between the two wattmeter readings is, using equations [13] and [14],

$$P_1 - P_2 = V_L I_L (0.87 \cos\phi + 0.5 \sin\phi - 0.87 \cos\phi + 0.5 \sin\phi)$$

$$= V_L I_L \sin\phi \quad [17]$$

But the reactive power $Q = \sqrt{3}\, I_L V_L \sin\phi$, equation [12], Thus

$$Q = \sqrt{3}(P_1 - P_2) \qquad [18]$$

The value of the power factor $\cos\phi$ can be obtained as follows. Dividing equation [17] by equation [15] gives

$$\frac{\tan\phi}{\sqrt{3}} = \frac{P_1 - P_2}{P_1 + P_2} \qquad [19]$$

But $\cos^2\phi + \sin^2\phi = 1$, hence when this is divided by $\cos^2\phi$

$$1 + \tan^2\phi = \frac{1}{\cos^2\phi}$$

Hence

$$1 + 3\left(\frac{P_1 - P_2}{P_1 + P_2}\right)^2 = \frac{1}{\cos^2\phi}$$

$$\cos\phi = \frac{1}{\sqrt{\left[1 + 3\left(\frac{P_1 - P_2}{P_1 + P_2}\right)^2\right]}} \qquad [20]$$

The power factor $\cos\phi$ can thus be obtained from the two wattmeter readings.

Example 7

Two wattmeters are used to measure the power consumed by a three-wire, balanced load system (as in Fig. 10.14). The meters gave readings of 50 kW and −30 kW. What is the total power consumed and the power factor?

Answer

The total power consumed is the sum of the meter readings (equation [16]). Thus the total power is 50 − 30 = 20 kW. The power factor is given by equation [20] as

$$\cos\phi = \frac{1}{\sqrt{\left[1 + 3\left(\frac{P_1 - P_2}{P_1 + P_2}\right)^2\right]}}$$

$$= \frac{1}{\sqrt{\left[1 + 3\left(\frac{50 + 30}{50 - 30}\right)^2\right]}} = 0.24$$

Unbalanced loads

With a balanced star-connected load and a four-wire system the instantaneous value of the current along the neutral line is zero. This is not the case with an unbalanced load, there being a current along the neutral line. With a balanced delta-connected load the currents through each of the load elements have the same magnitude and their phase angles differ by 120°.

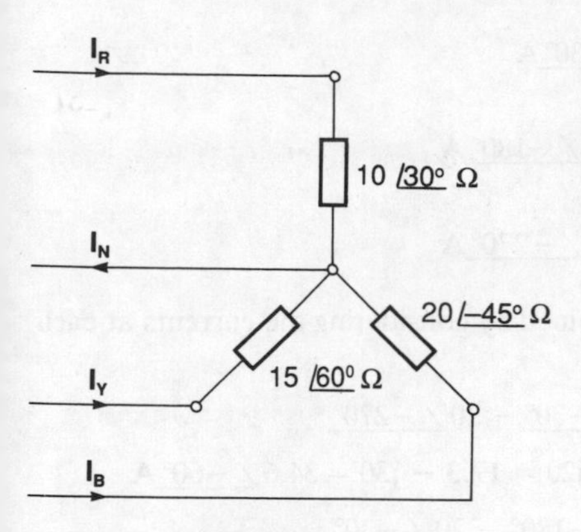

Fig. 10.17 Example 8

With an unbalanced delta-connected load the magnitudes may differ and the phase angles differ from the balanced values.

Example 8

Determine the current in the neutral wire for the star-connected load shown in Fig. 10.17 and the total power consumed. The line voltage is 415 V.

Answer

The voltage across a load element will be $415/\sqrt{3} = 240$ V. Hence, if we take $\mathbf{V_R}$ to be $240\angle 0°$ V, then

$$\mathbf{I_R} = \frac{240\angle 0°}{10\angle 30°} = 24\angle -30° \text{ A}$$

For the yellow line we have $\mathbf{V_Y} = 240\angle -120°$ V and so

$$\mathbf{I_Y} = \frac{240\angle -120°}{15\angle 60°} = 16\angle -180° \text{ A}$$

For the blue line we have $\mathbf{V_B} = 240\angle -240°$ V and so

$$\mathbf{I_B} = \frac{240\angle -240°}{20\angle -45°} = 12\angle -195° \text{ A}$$

Thus the neutral line current is

$$\begin{aligned}\mathbf{I_N} = \mathbf{I_R} + \mathbf{I_Y} + \mathbf{I_B} &= 24\angle -30° + 16\angle -180° + 12\angle -195° \\ &= 20.8 - j12 - 16 + j0 - 11.6 + j3.1 \\ &= -6.8 - j8.9 = 11.2\angle -127° \text{ A}\end{aligned}$$

The total power consumed will be the sum of the powers consumed by each of the load elements. Thus

$$\text{power for red phase} = 240 \times 24 \cos 30° = 4988 \text{ W}$$

$$\text{power for yellow phase} = 240 \times 16 \cos 60° = 1920 \text{ W}$$

$$\text{power for blue phase} = 240 \times 12 \cos 45° = 2036 \text{ W}$$

and so the total power is 8944 W.

Example 9

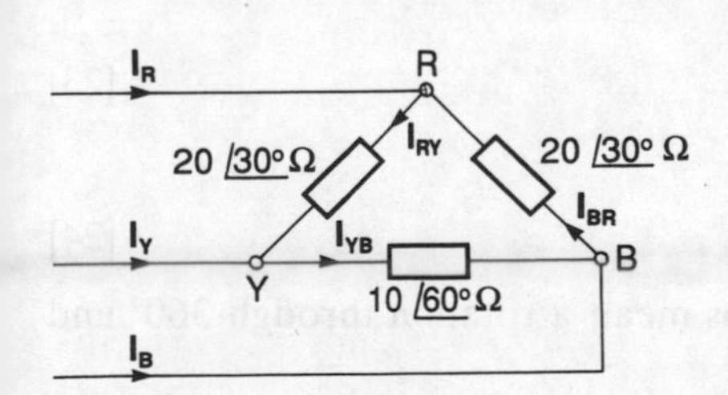

Fig. 10.18 Example 9

Determine the line currents and the currents through the load elements when a three-phase supply is connected to a delta-connected load having elements with impedances of $20\angle 30°\,\Omega$, $10\angle 60°\,\Omega$ and $20\angle 30°\,\Omega$, as shown in Fig. 10.18. The line voltage is 400 V.

Answer

The potential difference across each of the resistors in the mesh is 400 V. Thus if we take $\mathbf{V_{RY}} = 400\angle 0°$ V, $\mathbf{V_{YB}} = 400\angle -120°$ V and $\mathbf{V_{BR}} = 400\angle -240°$ V, then the currents through the load elements are

$$\mathbf{I_{RY}} = \frac{400\angle 0^\circ}{20\angle 30^\circ} = 20\angle -30^\circ \text{ A}$$

$$\mathbf{I_{YB}} = \frac{400\angle -120^\circ}{10\angle 60^\circ} = 40\angle -180^\circ \text{ A}$$

$$\mathbf{I_{BR}} = \frac{400\angle -240^\circ}{20\angle 30^\circ} = 20\angle -270^\circ \text{ A}$$

The lines currents can be obtained by considering the currents at each of the nodes R, Y and B.

$$\mathbf{I_R} = \mathbf{I_{RY}} - \mathbf{I_{BR}} = 20\angle -30^\circ - 20\angle -270^\circ$$

$$= 17.3 - \text{j}10 + 0 - \text{j}20 = 17.3 - \text{j}30 = 34.6\angle -60^\circ \text{ A}$$

$$\mathbf{I_Y} = \mathbf{I_{YB}} - \mathbf{I_{RY}} = 40\angle -180^\circ - 20\angle -30^\circ$$

$$= -40 + 0 - 17.3 + \text{j}10 = -57.3 + \text{j}10 = 58.2\angle -170^\circ \text{ A}$$

$$\mathbf{I_B} = \mathbf{I_{BR}} - \mathbf{I_{YB}} = 20\angle -270^\circ - 40\angle -180^\circ$$

$$= 0 + \text{j}20 + 40 + \text{j}0 = 44.7\angle 26.6^\circ \text{ A}$$

As a check, the line currents should sum to give zero. Thus

$$\mathbf{I_R} + \mathbf{I_Y} + \mathbf{I_B} = 17.3 - \text{j}30 - 57.3 + \text{j}10 + 40 + \text{j}20 = 0.$$

The 120° operator

Because the phase displacement of a balanced set of phasors is 120° it is convenient to have a method of indicating the rotation of a phasor through 120°. The letter a is used to specify an operator that causes a phasor to rotate through 120° in the anticlockwise direction, i.e. add 120°. Any phasor operated on by this operator remains unchanged in magnitude but is just rotated by 120° in the anticlockwise direction. Thus

$$a = 1\angle 120^\circ \qquad [21]$$

or

$$a = -0.5 + \text{j}0.866 \qquad [22]$$

Two successive operations mean a rotation through 240° and is thus represented by

$$a^2 = 1\angle 240^\circ \qquad [23]$$

or

$$a^2 = -0.5 - \text{j}0.866 \qquad [24]$$

Three successive operations mean a rotation through 360° and is thus represented by

$$a^3 = 1\angle 360^\circ = 1\angle 0^\circ$$

Four successive operations mean a rotation through 480°, thus

$$a^4 = 1\angle 480^\circ = 1\angle 120^\circ = a$$

When a phasor is operated on by $-a$ then it is rotated by 60° in a clockwise direction, since

$$-a = a \times (-1) = 1\angle 120° \times 1\angle 180° = 1\angle 300°$$
$$= 1\angle -60°$$

Consider a balanced three-phase system of phase voltages,

$$\mathbf{V_R} = V_p\angle 0° = V_p$$
$$\mathbf{V_Y} = V_p\angle -120° = V_p\angle +240° = a^2V_p$$
$$\mathbf{V_B} = V_p\angle -240° = V_p\angle 120° = aV_p$$

There are situations where we can end up summing a number of phasors where each has the same magnitude but differ in phase. Thus we might have

$$V_p\angle 0° + V_p\angle 120° + V_p\angle 240° = V_p + aV_p + a^2V_p$$
$$= V_p(1 + a + a^2)$$

Using equations [22] and [24] then the sum is

$$V_p(1 - 0.5 + j0.866 - 0.5 - j0.866) = 0$$

Thus

$$1 + a + a^2 = 0 \qquad [25]$$

Since $a^4 = a$ then we can also write

$$1 + a^4 + a^2 = 0 \qquad [26]$$

Example 10

If $\mathbf{I_R} = 20\angle 30°$ A, what is $a\mathbf{I_R}$ and $a^2\mathbf{I_R}$?

Answer

$$a\mathbf{I_R} = 1\angle 120° \times 20\angle 30° = 20\angle 150° \text{ A}$$
$$a^2\mathbf{I_R} = 1\angle 120° \times 1\angle 120° \times 20\angle 30° = 20\angle 270° \text{ A}$$

Symmetrical components

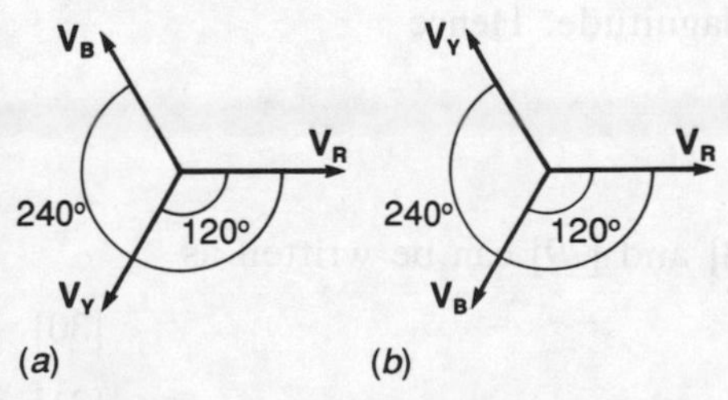

Fig. 10.19 Phase sequence (*a*) RYB, (*b*) RBY

Figure 10.19(*a*) shows what can be termed *normal phase rotation* or *positive phase rotation*. This is, assuming the normal anticlockwise rotation of phasors, the sequence red, yellow, blue for the maximum values to be reached (see also Fig. 10.1). If the sequence were red, blue, yellow (Fig. 10.19(*b*)) then the sequence is said to be *reverse phase rotation* or *negative phase rotation*, since the same situation as with the normal phase rotation could be produced by the red, yellow, blue phasors rotating in the reverse direction, i.e. clockwise.

Figure 10.20(*a*) shows a typical phasor diagram for the line currents of an unbalanced three-phase system with normal

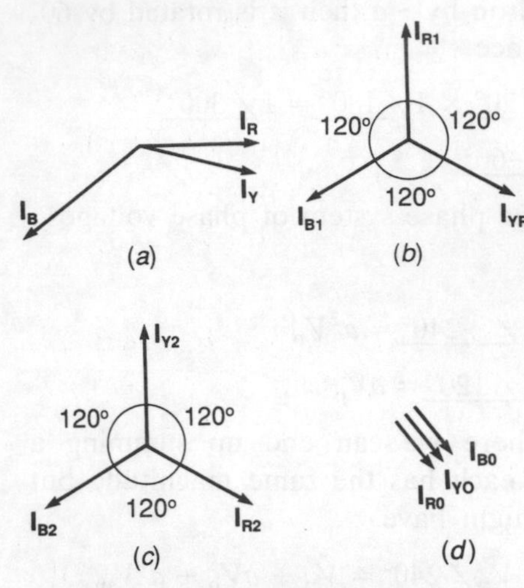

Fig. 10.20 (*a*) The unbalanced set of phasors, and their equivalent, (*b*) the positive-sequence set, (*c*) the negative-sequence set, (*d*) the zero-sequence set

rotation. Such a set of phasors can be represented by the sum of three balanced systems of phasors, namely:

1. Three phasors equal in magnitude, displaced from each other by 120° in phase, and having the same sequence as the original phasors, i.e. a positive-sequence balanced phasor system (Fig. 10.20(*b*)).
2. Three phasors equal in magnitude, displaced from each other by 120° in phase, and having the phase sequence opposite to that of the original phasors, i.e. a negative-sequence balanced phasor system (Fig. 10.20(*c*)).
3. Three phasors equal in magnitude and with zero phase displacement from each other, i.e. a zero-sequence balanced phasor system (Fig. 10.20(*d*)).

These three sets of balanced phasors are called the *symmetrical components*.

Since the unbalanced set of phasors is the sum of its symmetrical components, then

$$\mathbf{I_R} = \mathbf{I_{R1}} + \mathbf{I_{R2}} + \mathbf{I_{R0}} \quad [27]$$

$$\mathbf{I_Y} = \mathbf{I_{Y1}} + \mathbf{I_{Y2}} + \mathbf{I_{Y0}} \quad [28]$$

$$\mathbf{I_B} = \mathbf{I_{B1}} + \mathbf{I_{B2}} + \mathbf{I_{B0}} \quad [29]$$

But $\mathbf{I_{R1}}$, $\mathbf{I_{Y1}}$ and $\mathbf{I_{B1}}$ are a set of balanced, positive-sequence, phasors. We can regard the $\mathbf{I_{Y1}}$ as just being the $\mathbf{I_{R1}}$ phasor rotated through +240° and the $\mathbf{I_{B1}}$ as it rotated through +120°. Thus we can represent them by means of the 120° operator as

$$\mathbf{I_{Y1}} = a^2\mathbf{I_{R1}}$$

$$\mathbf{I_{B1}} = a\mathbf{I_{R1}}$$

Similarly, since $\mathbf{I_{R2}}$, $\mathbf{I_{Y2}}$ and $\mathbf{I_{B2}}$ are a set of balanced, negative-sequence, phasors we can represent them as

$$\mathbf{I_{Y2}} = a\mathbf{I_{R2}}$$

$$\mathbf{I_{B2}} = a^2\mathbf{I_{R2}}$$

The zero sequence set of phasors are $\mathbf{I_{R0}}$, $\mathbf{I_{Y0}}$ and $\mathbf{I_{B0}}$ and are all the same phase and magnitude. Hence

$$\mathbf{I_{Y0}} = \mathbf{I_{R0}}$$

$$\mathbf{I_{B0}} = \mathbf{I_{R0}}$$

Hence equations [27], [28] and [29] can be written as

$$\mathbf{I_R} = \mathbf{I_{R1}} + \mathbf{I_{R2}} + \mathbf{I_{R0}} \quad [30]$$

$$\mathbf{I_Y} = a^2\mathbf{I_{R1}} + a\mathbf{I_{R2}} + \mathbf{I_{R0}} \quad [31]$$

$$\mathbf{I_B} = a\mathbf{I_{R1}} + a^2\mathbf{I_{R2}} + \mathbf{I_{R0}} \quad [32]$$

We can obtain a value for $\mathbf{I_{R1}}$ by adding equation [30], equation [31] multiplied by a and equation [32] multiplied by a^2.

$$\mathbf{I_R} + a\mathbf{I_Y} + a^2\mathbf{I_B} = \mathbf{I_{R1}} + \mathbf{I_{R2}} + \mathbf{I_{R0}} + a^3\mathbf{I_{R1}} + a^2\mathbf{I_{R2}} + a\mathbf{I_{R0}} + a^3\mathbf{I_{R1}} + a^4\mathbf{I_{R2}} + a^2\mathbf{I_{R0}}$$

$$= \mathbf{I_{R1}}(1 + a^3 + a^3) + \mathbf{I_{R2}}(1 + a^2 + a^4) + \mathbf{I_{R0}}(1 + a + a^2)$$

Since $a^3 = 1 + \mathrm{j}0$, then $1 + a^3 + a^3 = 3$. Since $a^2 = -0.5 - \mathrm{j}0.87$ and $a^4 = 1\angle 480° = 1\angle 120° = a$ then $1 + a^2 + a^4 = 1 - 0.5 - \mathrm{j}0.87 - 0.5 + \mathrm{j}0.87 = 0$. Also $1 + a + a^2 = 1 - 0.5 + \mathrm{j}0.87 - 0.5 - \mathrm{j}0.87 = 0$. Thus

$$\mathbf{I_R} + a\mathbf{I_Y} + a^2\mathbf{I_B} = 3\mathbf{I_{R1}}$$

$$\mathbf{I_{R1}} = \tfrac{1}{3}(\mathbf{I_R} + a\mathbf{I_Y} + a^2\mathbf{I_B}) \qquad [33]$$

Similarly, we can derive

$$\mathbf{I_{R2}} = \tfrac{1}{3}(\mathbf{I_R} + a^2\mathbf{I_Y} + a\mathbf{I_B}) \qquad [34]$$

$$\mathbf{I_{R0}} = \tfrac{1}{3}(\mathbf{I_R} + \mathbf{I_Y} + \mathbf{I_B}) \qquad [35]$$

All the above equations have been derived for current; they could equally well have been derived for voltages. The forms of the resulting equations are identical.

With a three-phase, four-wire system the neutral line current is $\mathbf{I_R} + \mathbf{I_Y} + \mathbf{I_B}$ and is thus $3\mathbf{I_{R0}}$. With a three-phase, three-wire, star or delta, system the sum of the three-line currents is zero and so $\mathbf{I_{R0}}$ is zero and thus there are no zero-sequence components.

Example 11

What are the symmetrical red line components of a system of unbalanced three-phase currents where the line currents are $\mathbf{I_R} = 20\angle 0°$ A, $\mathbf{I_Y} = 10\angle -90°$ A, $\mathbf{I_B} = 5\angle 45°$ A?

Answer

We can write $\mathbf{I_R}$ as $20 + \mathrm{j}0$, $\mathbf{I_Y}$ as $0 - \mathrm{j}10$ and $\mathbf{I_B}$ as $3.5 + \mathrm{j}3.5$. Thus, since $a\mathbf{I_Y} = 10\angle +30° = 8.7 + \mathrm{j}5$ and $a^2\mathbf{I_B} = 5\angle 285° = 1.3 - \mathrm{j}4.8$, then using equation [33]

$$\mathbf{I_{R1}} = \tfrac{1}{3}(\mathbf{I_R} + a\mathbf{I_Y} + a^2\mathbf{I_B})$$

$$= \tfrac{1}{3}(20 + \mathrm{j}0 + 8.7 + \mathrm{j}5 + 1.3 - \mathrm{j}4.8)$$

$$= 10 + \mathrm{j}0.07 = 10.6\angle 0.4°\ \mathrm{A}$$

Since $a^2\mathbf{I_Y} = 10\angle 150° = -8.7 + \mathrm{j}5$ and $a\mathbf{I_B} = 5\angle 165° = -4.8 + \mathrm{j}1.3$, then using equation [34]

$$\mathbf{I_{R2}} = \tfrac{1}{3}(\mathbf{I_R} + a^2\mathbf{I_Y} + a\mathbf{I_B})$$

$$= \tfrac{1}{3}(20 + j0 - 8.7 + j5 - 4.8 + j1.3)$$
$$= 2.2 + j2.1 = 3.0\angle 43.7^\circ \text{ A}$$

Using equation [35]

$$\mathbf{I_{R0}} = \tfrac{1}{3}(\mathbf{I_R} + \mathbf{I_Y} + \mathbf{I_B}) = \tfrac{1}{3}(20 + j0 + 0 - j10 + 3.5 + j3.5)$$
$$= 7.8 - j6.5 = 10.2\angle -39.8^\circ \text{ A}$$

Unbalanced star load and symmetrical components

Consider a star load with elements having impedances of Z_R, Z_Y and Z_B (Fig. 10.21) and supplied from a three-phase, three-line, supply. If the star point of the load is S then the zero phase sequence component is, using the voltage equivalent equation to equation [35],

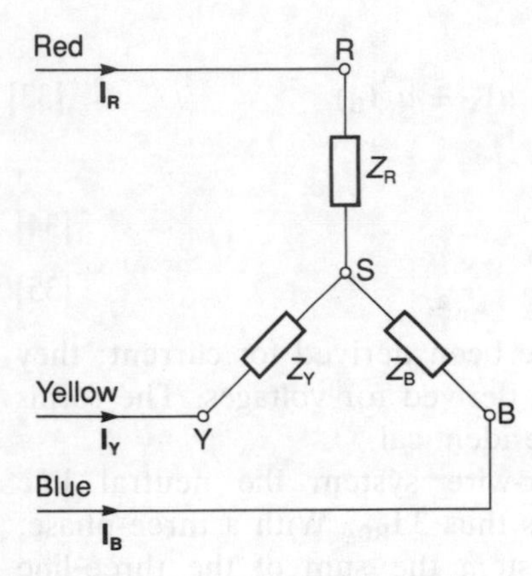

Fig. 10.21 Unbalanced star load

$$\mathbf{V_{R0}} = \tfrac{1}{3}(\mathbf{V_{RS}} + \mathbf{V_{YS}} + \mathbf{V_{BS}})$$

But $\mathbf{V_{RS}} = \mathbf{I_R}Z_R$, $\mathbf{V_{YS}} = \mathbf{I_Y}Z_Y$ and $\mathbf{V_{BS}} = \mathbf{I_B}Z_B$. Thus

$$\mathbf{V_{R0}} = \tfrac{1}{3}(\mathbf{I_R}Z_R + \mathbf{I_Y}Z_Y + \mathbf{I_B}Z_B)$$

Since the system is three-wire there is no zero-phase sequence line current component, thus equation [30] gives

$$\mathbf{I_R} = \mathbf{I_{R1}} + \mathbf{I_{R2}} + \mathbf{I_{R0}} = \mathbf{I_{R1}} + \mathbf{I_{R2}}$$

and equation [31] gives

$$\mathbf{I_Y} = a^2\mathbf{I_{R1}} + a\mathbf{I_{R2}} + \mathbf{I_{R0}} = a^2\mathbf{I_{R1}} + a\mathbf{I_{R2}}$$

and equation [32]

$$\mathbf{I_B} = a\mathbf{I_{R1}} + a^2\mathbf{I_{R2}} + \mathbf{I_{R0}} = a\mathbf{I_{R1}} + a^2\mathbf{I_{R2}}$$

Thus

$$\mathbf{V_{R0}} = \tfrac{1}{3}[Z_R(\mathbf{I_{R1}} + \mathbf{I_{R2}}) + Z_Y(a^2\mathbf{I_{R1}} + a\mathbf{I_{R2}}) + Z_B(a\mathbf{I_{R1}} + a^2\mathbf{I_{R2}})]$$
$$= \tfrac{1}{3}[\mathbf{I_{R1}}(Z_R + a^2Z_Y + aZ_B) + \mathbf{I_{R2}}(Z_R + aZ_Y + a^2Z_B)] \qquad [36]$$

Similarly, the positive phase sequence component is, using the voltage equivalent equation to equation [33],

$$\mathbf{V_{R1}} = \tfrac{1}{3}(\mathbf{V_R} + a\mathbf{V_Y} + a^2\mathbf{V_B}) = \tfrac{1}{3}(\mathbf{I_R}Z_R + a\mathbf{I_Y}Z_Y + a^2\mathbf{I_B}Z_B)$$
$$= \tfrac{1}{3}[Z_R(\mathbf{I_{R1}} + \mathbf{I_{R2}}) + aZ_Y(a^2\mathbf{I_{R1}} + a\mathbf{I_{R2}}) + a^2Z_B(a\mathbf{I_{R1}} + a^2\mathbf{I_{R2}})]$$
$$= \tfrac{1}{3}[\mathbf{I_{R1}}(Z_R + a^3Z_Y + a^3Z_B) + \mathbf{I_{R2}}(Z_R + a^2Z_Y + a^4Z_B)]$$

and since $a^3 = 1$ and $a^4 = a$

$$\mathbf{V_{R1}} = \tfrac{1}{3}[\mathbf{I_{R1}}(Z_R + Z_Y + Z_B)$$

$$+ \mathbf{I_{R2}}(Z_R + a^2 Z_Y + a Z_B)] \quad [37]$$

Similarly, the negative phase sequence component is, using the voltage equivalent equation to equation [34],

$$\mathbf{V_{R2}} = \tfrac{1}{3}(\mathbf{V_R} + a^2\mathbf{V_Y} + a\mathbf{V_B})$$
$$= \tfrac{1}{3}(\mathbf{I_R}Z_R + a^2\mathbf{I_Y}Z_Y + a\mathbf{I_B}Z_B)$$
$$= \tfrac{1}{3}[\mathbf{I_{R1}}(Z_R + aZ_Y + a^2Z_B)$$
$$+ \mathbf{I_{R2}}(Z_R + Z_Y + Z_B)] \quad [38]$$

The symmetrical components can be also expressed in terms of the line voltages using the voltage equivalent equations of equations [30], [31] and [32]. Thus the line-to-line voltage $\mathbf{V_{RY}}$ is

$$\mathbf{V_{RY}} = \mathbf{V_{RS}} - \mathbf{V_{YS}}$$

and since equations [30] and [31] give

$$\mathbf{V_{RS}} = \mathbf{V_{R1}} + \mathbf{V_{R2}} + \mathbf{V_{R0}}$$

$$\mathbf{V_{YS}} = a^2\mathbf{V_{R1}} + a\mathbf{V_{R2}} + \mathbf{V_{R0}}$$

then

$$\mathbf{V_{RY}} = \mathbf{V_{R1}} + \mathbf{V_{R2}} + \mathbf{V_{R0}} - a^2\mathbf{V_{R1}} - a\mathbf{V_{R2}} - \mathbf{V_{R0}}$$
$$= \mathbf{V_{R1}}(1 - a^2) + \mathbf{V_{R2}}(1 - a) \quad [39]$$

The line-to-line voltage $\mathbf{V_{YB}}$ is

$$\mathbf{V_{YB}} = \mathbf{V_{YS}} - \mathbf{V_{BS}}$$

and since equations [31] and [32] give

$$\mathbf{V_{YS}} = a^2\mathbf{V_{R1}} + a\mathbf{V_{R2}} + \mathbf{V_{R0}}$$

$$\mathbf{V_{BS}} = a\mathbf{V_{R1}} + a^2\mathbf{V_{R2}} + \mathbf{V_{R0}}$$

then

$$\mathbf{V_{YB}} = a^2\mathbf{V_{R1}} + a\mathbf{V_{R2}} + \mathbf{V_{R0}} - a\mathbf{V_{R1}} - a^2\mathbf{V_{R2}} - \mathbf{V_{R0}}$$
$$= \mathbf{V_{R1}}(a^2 - a) + \mathbf{V_{R2}}(a - a^2) \quad [40]$$

The two equations [39] and [40] can be solved to give $\mathbf{V_{R1}}$ and $\mathbf{V_{R2}}$. Thus multiplying equation [39] by $(a^2 - a)$ and subtracting from it equation [40] multiplied by $(1 - a^2)$ gives

$$(a^2 - a)\mathbf{V_{RY}} - (1 - a^2)\mathbf{V_{YB}} = (1 - a)(a^2 - a)\mathbf{V_{R2}}$$
$$- (a - a^2)(1 - a^2)\mathbf{V_{R2}}$$
$$= 3(a^2 - a)\mathbf{V_{R2}}$$

Thus

$$\mathbf{V_{R2}} = \frac{1}{3}\left[\mathbf{V_{RY}} - \left(\frac{1 - a^2}{a^2 - a}\right)\mathbf{V_{YB}}\right]$$

$$= \tfrac{1}{3}[\mathbf{V_{RY}} + (0.5 - j0.866)\mathbf{V_{YB}}] \quad [41]$$

Similarly

$$\mathbf{V_{R1}} = \tfrac{1}{3}[\mathbf{V_{RY}} + (0.5 + j0.866)\mathbf{V_{YB}}] \quad [42]$$

Example 12

An unsymmetrical three-phase, three-wire, system supplies line voltages of $\mathbf{V_{RY}} = 30\angle 0°\,\text{V}$, $\mathbf{V_{YB}} = 40\angle -90°\,\text{V}$ and $\mathbf{V_{BR}} = 50\angle 127°\,\text{V}$ to a star-connected load consisting of three identical $10\,\Omega$ resistors. What is the current in the red line?

Answer

The voltage symmetrical components can be determined using equations [41] and [42]

$$\mathbf{V_{R2}} = \tfrac{1}{3}[\mathbf{V_{RY}} + (0.5 - j0.866)\mathbf{V_{YB}}]$$

$$= \tfrac{1}{3}[30 + (0.5 - j0.866)(-j40)] = -1.55 - j6.67$$

$$\mathbf{V_{R1}} = \tfrac{1}{3}[\mathbf{V_{RY}} + (0.5 + j0.866)\mathbf{V_{YB}}]$$

$$= \tfrac{1}{3}[30 + (0.5 + j0.866)(-j40)] = 21.55 - j6.67$$

Because the system is three wire, $\mathbf{V_{R0}} = 0$.

The symmetrical current components can be determined by means of equations [36], [37] and [38].

$$\mathbf{I_{R0}} = \tfrac{1}{3}[\mathbf{I_{R1}}(Z_R + a^2Z_Y + aZ_B) + \mathbf{I_{R2}}(Z_R + aZ_Y + a^2Z_B)]$$

$$= \tfrac{1}{3}[\mathbf{I_{R1}}(10 + 10a^2 + 10a) + \mathbf{I_{R2}}(10 + 10a + 10a^2)]$$

But $(1 + a + a^2) = 0$ and so $\mathbf{I_{R0}} = 0$.

$$\mathbf{V_{R1}} = \tfrac{1}{3}[\mathbf{I_{R1}}(Z_R + Z_Y + Z_B) + \mathbf{I_{R2}}(Z_R + a^2Z_Y + aZ_B)]$$

$$= \tfrac{1}{3}[\mathbf{I_{R1}}(10 + 10 + 10) + \mathbf{I_{R2}}(10 + 10a^2 + 10a)]$$

Thus, since $(1 + a + a^2) = 0$ then

$$\mathbf{V_{R1}} = 10\,\mathbf{I_{R1}}$$

and hence, using the value of $\mathbf{V_{R1}}$ derived earlier,

$$\mathbf{I_{R1}} = \frac{21.55 - j6.67}{10} = 2.155 - j0.667$$

The $\mathbf{I_{R2}}$ symmetrical current component is given by equation [38] as

$$\mathbf{V_{R2}} = \tfrac{1}{3}[\mathbf{I_{R1}}(Z_R + aZ_Y + a^2Z_B) + \mathbf{I_{R2}}(Z_R + Z_Y + Z_B)]$$

$$= \tfrac{1}{3}[\mathbf{I_{R1}}(10 + 10a + 10a^2) + \mathbf{I_{R2}}(10 + 10 + 10)]$$

$$= 10\,\mathbf{I_{R2}}$$

Thus

$$-1.55 - j6.67 = 10\,\mathbf{I_{R2}}$$

$$\mathbf{I_{R2}} = -0.155 - j0.667$$

The current through the red line $\mathbf{I_R}$ is given by equation [30] as

$$\mathbf{I_R} = \mathbf{I_{R1}} + \mathbf{I_{R2}} + \mathbf{I_{R0}}$$

$$= 2.155 - j0.667 - 0.155 - j0.667 + 0$$

$$= 2 - j1.334 = 2.40\angle -33.7° \text{ A}$$

Unsymmetrical faults

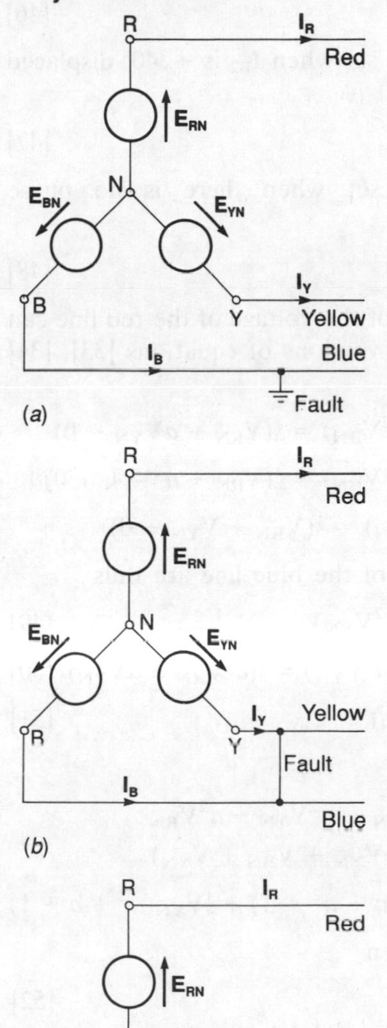

Fig. 10.22 (*a*) Single line-to-earth fault, (*b*) line-to-line fault, (*c*) double line-to-earth fault

Generally the faults that occur with power systems are unsymmetrical faults. These may, for example, be unsymmetrical short circuits or open circuits. Unsymmetrical faults occur as single line-to-earth faults (Fig. 10.22(*a*)), line-to-line faults (Fig. 10.22(*b*)), or double line-to-line faults (Fig. 10.22(*c*)). In the following analysis we will assume that the impedance of the fault is negligible, the network impedances up to the fault are balanced, the generator supply is positive sequence and the generator is unloaded.

A consequence of considering an unloaded balanced generator is that the open-circuit voltages, i.e. e.m.f.s, of $\mathbf{E_{RN}}$, $\mathbf{E_{YN}}$ and $\mathbf{E_{BN}}$, give symmetrical component e.m.f.s for the red line, using the voltage version of equations [33], [34] and [35], of

$$\mathbf{E_{R1}} = \tfrac{1}{3}(\mathbf{E_{RN}} + a\mathbf{E_{YN}} + a^2\mathbf{E_{BN}})$$

$$\mathbf{E_{R2}} = \tfrac{1}{3}(\mathbf{E_{RN}} + a^2\mathbf{E_{YN}} + a\mathbf{E_{BN}})$$

$$\mathbf{E_{R0}} = \tfrac{1}{3}(\mathbf{E_{RN}} + \mathbf{E_{YN}} + \mathbf{E_{BN}})$$

But, because the system is balanced, $\mathbf{E_{YN}} = a^2\mathbf{E_{RN}}$ and $\mathbf{E_{BN}} = a\mathbf{E_{RN}}$. Thus

$$\mathbf{E_{R1}} = \tfrac{1}{3}(\mathbf{E_{RN}} + a^3\mathbf{E_{RN}} + a^3\mathbf{E_{RN}}) = \tfrac{1}{3}\mathbf{E_{RN}}(1 + 1 + 1)$$

$$= \mathbf{E_{RN}} \qquad [43]$$

$$\mathbf{E_{R2}} = \tfrac{1}{3}(\mathbf{E_{RN}} + a^4\mathbf{E_{RN}} + a^2\mathbf{E_{RN}}) = \tfrac{1}{3}\mathbf{E_{RN}}(1 + a + a^2)$$

$$= 0 \qquad [44]$$

$$\mathbf{E_{R0}} = \tfrac{1}{3}(\mathbf{E_{RN}} + a^2\mathbf{E_{RN}} + a\mathbf{E_{RN}}) = \tfrac{1}{3}\mathbf{E_{RN}}(1 + a^2 + a)$$

$$= 0 \qquad [45]$$

since $(1 + a + a^2) = 0$. Thus the unloaded balanced generator gives only positive phase sequence voltages.

Consider the *line-to-earth fault* (Fig. 10.22(*a*)) with the fault of the three-wire system being an earthing of the blue line. Since the star point N is earthed then we must have the potential difference between the star point and the fault as zero potential, i.e. $\mathbf{V_{BN}} = 0$. Because the generator is unloaded then $\mathbf{I_R} = 0$ and $\mathbf{I_Y} = 0$. Hence, using equations [33], [34], and [35], the symmetrical current components in the red line are

$$\mathbf{I_{R1}} = \tfrac{1}{3}(\mathbf{I_R} + a\mathbf{I_Y} + a^2\mathbf{I_B}) = \tfrac{1}{3}(0 + 0 + a^2\mathbf{I_B}) = \tfrac{1}{3}a^2\mathbf{I_B}$$

$$\mathbf{I_{R2}} = \tfrac{1}{3}(\mathbf{I_R} + a^2\mathbf{I_Y} + a\mathbf{I_B}) = \tfrac{1}{3}(0 + 0 + a\mathbf{I_B}) = \tfrac{1}{3}a\mathbf{I_B}$$

$$\mathbf{I_{R0}} = \tfrac{1}{3}(\mathbf{I_R} + \mathbf{I_Y} + \mathbf{I_B}) = \tfrac{1}{3}(0 + 0 + \mathbf{I_B}) = \tfrac{1}{3}\mathbf{I_B}$$

The symmetrical components in the blue line are, for the positive-sequence set when $\mathbf{I_{B1}}$ is +120° displaced from $\mathbf{I_{R1}}$ (see Fig. 10.20(*b*))

$$\mathbf{I_{B1}} = a\mathbf{I_{R1}} = \tfrac{1}{3}a^3\mathbf{I_B} = \tfrac{1}{3}\mathbf{I_B} \qquad [46]$$

and for the negative-sequence set when $\mathbf{I_{B2}}$ is +240° displaced from $\mathbf{I_{R2}}$ (see Fig. 10.20(*c*))

$$\mathbf{I_{B2}} = a^2\mathbf{I_{R2}} = \tfrac{1}{3}a^3\mathbf{I_B} = \tfrac{1}{3}\mathbf{I_B} \qquad [47]$$

and for the zero-sequence set when there is no phase difference (see Fig. 10.20(*d*))

$$\mathbf{I_{B0}} = \mathbf{I_{R0}} = \tfrac{1}{3}\mathbf{I_B} \qquad [48]$$

The symmetrical components of the voltage of the red line can be obtained from the voltage versions of equations [33], [34] and [35].

$$\mathbf{V_{R1}} = \tfrac{1}{3}(\mathbf{V_{RN}} + a\mathbf{V_{YN}} + a^2\mathbf{V_{BN}}) = \tfrac{1}{3}(\mathbf{V_{RN}} + a\mathbf{V_{YN}} + 0)$$
$$\mathbf{V_{R2}} = \tfrac{1}{3}(\mathbf{V_{RN}} + a^2\mathbf{V_{YN}} + a\mathbf{V_{BN}}) = \tfrac{1}{3}(\mathbf{V_{RN}} + a^2\mathbf{V_{YN}} + 0)$$
$$\mathbf{V_{R0}} = \tfrac{1}{3}(\mathbf{V_{RN}} + \mathbf{V_{YN}} + \mathbf{V_{BN}}) = \tfrac{1}{3}(\mathbf{V_{RN}} + \mathbf{V_{YN}} + 0)$$

The symmetrical components of the blue line are thus

$$\mathbf{V_{B1}} = a\mathbf{V_{R1}} = \tfrac{1}{3}(a\mathbf{V_{RN}} + a^2\mathbf{V_{YN}}) \qquad [49]$$

$$\mathbf{V_{B2}} = a^2\mathbf{V_{R2}} = \tfrac{1}{3}(a^2\mathbf{V_{RN}} + a^4\mathbf{V_{YN}}) = \tfrac{1}{3}(a^2\mathbf{V_{RN}} + a\mathbf{V_{YN}}) \qquad [50]$$

$$\mathbf{V_{B0}} = \mathbf{V_{R0}} = \tfrac{1}{3}(\mathbf{V_{RN}} + \mathbf{V_{YN}}) \qquad [51]$$

Thus

$$\mathbf{V_{B1}} + \mathbf{V_{B2}} + \mathbf{V_{B0}} = \tfrac{1}{3}(a\mathbf{V_{RN}} + a^2\mathbf{V_{YN}} + a^2\mathbf{V_{RN}} + a\mathbf{V_{YN}} + \mathbf{V_{RN}} + \mathbf{V_{YN}})$$
$$= \tfrac{1}{3}\mathbf{V_{RN}}(a + a^2 + 1) + \tfrac{1}{3}\mathbf{V_{YN}}(a^2 + a + 1)$$

and since $(1 + a + a^2) = 0$ then

$$\mathbf{V_{B1}} + \mathbf{V_{B2}} + \mathbf{V_{B0}} = 0 \qquad [52]$$

If, for the blue line, the impedance up to the fault for the positive-sequence symmetrical component is Z_1 then the generated phase e.m.f. $\mathbf{E_{B1}}$ must supply the potential difference $\mathbf{V_{B1}}$ and the potential drop across Z_1, namely $\mathbf{I_{B1}}Z_1$. Thus

$$\mathbf{E_{B1}} = \mathbf{V_{B1}} + \mathbf{I_{B1}}Z_1 \qquad [53]$$

If the impedance up to the fault for the negative-sequence symmetrical component is Z_2 then

$$\mathbf{E_{B2}} = \mathbf{V_{B2}} + \mathbf{I_{B2}}Z_2 \qquad [54]$$

If the impedance up to the fault for the zero-sequence symmetrical component is Z_0 then

$$\mathbf{E_{B0}} = \mathbf{V_{B0}} + \mathbf{I_{B0}}Z_0 \quad [55]$$

But $\mathbf{E_{B1}} = a\mathbf{E_{R1}}$, $\mathbf{E_{B2}} = a^2\mathbf{E_{R2}}$ and $\mathbf{E_{B0}} = \mathbf{E_{R0}}$. However, as equations [44] and [45] give $\mathbf{E_{R2}} = 0$ and $\mathbf{E_{R0}} = 0$, then equations [53], [54] and [55] become

$$a\mathbf{E_{R1}} = \mathbf{V_{B1}} + \mathbf{I_{B1}}Z_1$$

$$0 = \mathbf{V_{B2}} + \mathbf{I_{B2}}Z_2$$

$$0 = \mathbf{V_{B0}} + \mathbf{I_{B0}}Z_0$$

Adding these three equations gives

$$a\mathbf{E_{R1}} = \mathbf{V_{B1}} + \mathbf{V_{B2}} + \mathbf{V_{B0}} + \mathbf{I_{B1}}Z_1 + \mathbf{I_{B2}}Z_2 + \mathbf{I_{B0}}Z_0$$

and substituting for the current components using equations [46], [47] and [48], gives

$$a\mathbf{E_{R1}} = \mathbf{V_{B1}} + \mathbf{V_{B2}} + \mathbf{V_{B0}} + \tfrac{1}{3}\mathbf{I_B}(Z_1 + Z_2 + Z_0)$$

Using equation [52] then

$$a\mathbf{E_{R1}} = \tfrac{1}{3}\mathbf{I_B}(Z_1 + Z_2 + Z_3)$$

and thus the fault current of $\mathbf{I_B}$ is

$$\mathbf{I_B} = \frac{3a\mathbf{E_{R1}}}{Z_1 + Z_2 + Z_3} \quad [56]$$

The above is the equation for the fault current when the fault is on the blue line. If the fault is on the red or yellow line then the current has the same magnitude as that given by equation [56] but the phase angle is different. Thus for the red line

$$\mathbf{I_R} = \frac{3\mathbf{E_{R1}}}{Z_1 + Z_2 + Z_3} \quad [57]$$

Now consider a *line-to-line fault*, a short circuit between two lines as illustrated in Fig. 10.22(*b*). Because the generator is unloaded we must have $\mathbf{I_R} = 0$ and $\mathbf{I_Y} = -\mathbf{I_B}$. The symmetrical current components are thus, using equations [33], [34] and [35]

$$\mathbf{I_{R1}} = \tfrac{1}{3}(\mathbf{I_R} + a\mathbf{I_Y} + a^2\mathbf{I_B}) = \tfrac{1}{3}(0 + a\mathbf{I_Y} - a^2\mathbf{I_Y})$$
$$= \tfrac{1}{3}\mathbf{I_Y}(a - a^2) \quad [58]$$

$$\mathbf{I_{R2}} = \tfrac{1}{3}(\mathbf{I_R} + a^2\mathbf{I_Y} + a\mathbf{I_B}) = \tfrac{1}{3}(0 + a^2\mathbf{I_Y} - a\mathbf{I_Y})$$
$$= \tfrac{1}{3}\mathbf{I_Y}(a^2 - a) \quad [59]$$

$$\mathbf{I_{R0}} = \tfrac{1}{3}(\mathbf{I_R} + \mathbf{I_Y} + \mathbf{I_B}) = \tfrac{1}{3}(0 + \mathbf{I_Y} - \mathbf{I_Y}) = 0 \quad [60]$$

The symmetrical components of the voltage of the red line can be obtained from the voltage versions of equations [33], [34] and [35], with $\mathbf{V_{BN}} = \mathbf{V_{YN}}$.

$$\mathbf{V_{R1}} = \tfrac{1}{3}(\mathbf{V_{RN}} + a\mathbf{V_{YN}} + a^2\mathbf{V_{BN}})$$

$$= \tfrac{1}{3}[\mathbf{V_{RN}} + (a + a^2)\mathbf{V_{YN}}] \quad [61]$$

$$\mathbf{V_{R2}} = \tfrac{1}{3}(\mathbf{V_{RN}} + a^2\mathbf{V_{YN}} + a\mathbf{V_{BN}})$$

$$= \tfrac{1}{3}[\mathbf{V_{RN}} + (a + a^2)\mathbf{V_{YN}}] \quad [62]$$

$$\mathbf{V_{R0}} = \tfrac{1}{3}(\mathbf{V_{RN}} + \mathbf{V_{YN}} + \mathbf{V_{BN}})$$

$$= \tfrac{1}{3}(\mathbf{V_{RN}} + 2\mathbf{V_{YN}}) \quad [63]$$

If, for the red line, the impedance up to the fault for the positive-sequence symmetrical component is Z_1 then the generated phase e.m.f. $\mathbf{E_{R1}}$ must supply the potential difference $\mathbf{V_{R1}}$ and the potential drop across Z_1, namely $\mathbf{I_{R1}}Z_1$. Thus

$$\mathbf{E_{R1}} = \mathbf{V_{R1}} + \mathbf{I_{R1}}Z_1 \quad [64]$$

If the impedance up to the fault for the negative sequence symmetrical component is Z_2 then

$$\mathbf{E_{R2}} = \mathbf{V_{R2}} + \mathbf{I_{R2}}Z_2 \quad [65]$$

If the impedance up to the fault for the zero sequence symmetrical component is Z_0 then

$$\mathbf{E_{R0}} = \mathbf{V_{R0}} + \mathbf{I_{R0}}Z_0 \quad [66]$$

But equations [44] and [45] give $\mathbf{E_{R2}} = 0$ and $\mathbf{E_{R0}} = 0$, so equations [64], [65] and [66] become

$$\mathbf{E_{R1}} = \mathbf{V_{R1}} + \mathbf{I_{R1}}Z_1$$

$$0 = \mathbf{V_{R2}} + \mathbf{I_{R2}}Z_2$$

$$0 = \mathbf{V_{R0}} + \mathbf{I_{R0}}Z_0$$

Subtracting the first two equations gives

$$\mathbf{E_{R1}} = \mathbf{V_{R1}} - \mathbf{V_{R2}} + \mathbf{I_{R1}}Z_1 - \mathbf{I_{R2}}Z_2$$

But equations [61] and [62] give $\mathbf{V_{R1}} = \mathbf{V_{R2}}$, thus

$$\mathbf{E_{R1}} = \mathbf{I_{R1}}Z_1 - \mathbf{I_{R2}}Z_2$$

Equations [58] and [59] give $\mathbf{I_{R2}} = -\mathbf{I_{R1}}$, thus

$$\mathbf{E_{R1}} = \mathbf{I_{R1}}(Z_1 + Z_2)$$

and so

$$\mathbf{I_{R1}} = \frac{\mathbf{E_{R1}}}{Z_1 + Z_2}$$

The fault current is $\mathbf{I_Y}$ and this is the sum of the symmetrical components of the current, i.e. $\mathbf{I_{Y1}} + \mathbf{I_{Y2}} + \mathbf{I_{Y0}}$. Since $\mathbf{I_{Y1}} =$

$a^2\mathbf{I}_{\mathbf{R1}}$ and $\mathbf{I}_{\mathbf{Y2}} = a\mathbf{I}_{\mathbf{R2}} = -a\mathbf{I}_{\mathbf{R1}}$ (equations [58] and [59]), with $\mathbf{I}_{\mathbf{Y0}} = \mathbf{I}_{\mathbf{R0}} = 0$ (equation [60]), then

$$\mathbf{I}_{\mathbf{Y}} = a^2\mathbf{I}_{\mathbf{R1}} - a\mathbf{I}_{\mathbf{R1}} = \frac{(a^2 - a)\mathbf{E}_{\mathbf{R1}}}{Z_1 + Z_2} \quad [67]$$

Now consider a *double line-to-earth fault*, as illustrated in Fig. 10.22(c). Because the generator is unloaded we must have $\mathbf{I}_{\mathbf{R}} = 0$ and because it is balanced $\mathbf{V}_{\mathbf{YN}} = \mathbf{V}_{\mathbf{BN}} = 0$. The symmetrical components of the voltage for the red line at the fault are

$$\mathbf{V}_{\mathbf{R1}} = \tfrac{1}{3}(\mathbf{V}_{\mathbf{RN}} + a\mathbf{V}_{\mathbf{YN}} + a^2\mathbf{V}_{\mathbf{BN}}) = \tfrac{1}{3}\mathbf{V}_{\mathbf{RN}} \quad [68]$$

$$\mathbf{V}_{\mathbf{R2}} = \tfrac{1}{3}(\mathbf{V}_{\mathbf{RN}} + a^2\mathbf{V}_{\mathbf{YN}} + a\mathbf{V}_{\mathbf{BN}}) = \tfrac{1}{3}\mathbf{V}_{\mathbf{RN}} \quad [69]$$

$$\mathbf{V}_{\mathbf{R0}} = \tfrac{1}{3}(\mathbf{V}_{\mathbf{RN}} + \mathbf{V}_{\mathbf{YN}} + \mathbf{V}_{\mathbf{BN}}) = \tfrac{1}{3}\mathbf{V}_{\mathbf{RN}} \quad [70]$$

If, for the red line, the impedance up to the fault for the positive-sequence symmetrical component is Z_1 then the generated phase e.m.f. $\mathbf{E}_{\mathbf{R1}}$ must supply the potential difference $\mathbf{V}_{\mathbf{R1}}$ and the potential drop across Z_1, namely $\mathbf{I}_{\mathbf{R1}}Z_1$. Thus

$$\mathbf{E}_{\mathbf{R1}} = \mathbf{V}_{\mathbf{R1}} + \mathbf{I}_{\mathbf{R1}}Z_1 \quad [71]$$

If the impedance up to the fault for the negative-sequence symmetrical component is Z_2 then

$$\mathbf{E}_{\mathbf{R2}} = \mathbf{V}_{\mathbf{R2}} + \mathbf{I}_{\mathbf{R2}}Z_2 \quad [72]$$

If the impedance up to the fault for the zero-sequence symmetrical component is Z_0 then

$$\mathbf{E}_{\mathbf{R0}} = \mathbf{V}_{\mathbf{R0}} + \mathbf{I}_{\mathbf{R0}}Z_0 \quad [73]$$

But equations [43],[44] and [45] give $\mathbf{E}_{\mathbf{R1}} = \mathbf{E}_{\mathbf{RN}}$, $\mathbf{E}_{\mathbf{R2}} = 0$ and $\mathbf{E}_{\mathbf{R0}} = 0$; then equations [71], [72] and [73] become

$$\mathbf{E}_{\mathbf{RN}} = \mathbf{V}_{\mathbf{R1}} + \mathbf{I}_{\mathbf{R1}}Z_1$$

$$0 = \mathbf{V}_{\mathbf{R2}} + \mathbf{I}_{\mathbf{R2}}Z_2$$

$$0 = \mathbf{V}_{\mathbf{R0}} + \mathbf{I}_{\mathbf{R0}}Z_0$$

Thus, rearranging these equations gives

$$\mathbf{I}_{\mathbf{R1}} = \frac{\mathbf{E}_{\mathbf{RN}}}{Z_1} - \frac{\mathbf{V}_{\mathbf{R1}}}{Z_1} \quad [74]$$

$$\mathbf{I}_{\mathbf{R2}} = -\frac{\mathbf{V}_{\mathbf{R2}}}{Z_2} \quad [75]$$

$$\mathbf{I}_{\mathbf{R0}} = -\frac{\mathbf{V}_{\mathbf{R0}}}{Z_0} \quad [76]$$

Since the current in the red line is zero then the sum of the red line components must be zero, i.e.

$$\mathbf{I_{R1}} + \mathbf{I_{R2}} + \mathbf{I_{R0}} = 0$$

Thus

$$\frac{\mathbf{E_{RN}}}{Z_1} - \frac{\mathbf{V_{R1}}}{Z_1} - \frac{\mathbf{V_{R2}}}{Z_2} - \frac{\mathbf{V_{R0}}}{Z_0} = 0$$

Since equations [68], [69] and [70] indicate that $\mathbf{V_{R0}} = \mathbf{V_{R1}} = \mathbf{V_{R2}}$, then

$$\frac{\mathbf{E_{RN}}}{Z_1} - \frac{\mathbf{V_{R1}}}{Z_1} - \frac{\mathbf{V_{R1}}}{Z_2} - \frac{\mathbf{V_{R1}}}{Z_0} = 0$$

$$\mathbf{V_{R1}}\left(\frac{1}{Z_1} + \frac{1}{Z_2} + \frac{1}{Z_0}\right) = \frac{\mathbf{E_{RN}}}{Z_1}$$

$$\mathbf{V_{R1}} = \frac{\mathbf{E_{RN}}Z_2Z_0}{Z_1Z_2 + Z_1Z_0 + Z_2Z_0} \qquad [77]$$

This voltage value can be substituted into equations [74], [75] and [76] to give the red line current components.

$$\mathbf{I_{R1}} = \frac{\mathbf{E_{RN}}}{Z_1} - \frac{\mathbf{V_{R1}}}{Z_1} = \frac{\mathbf{E_{RN}}}{Z_1}\left(1 - \frac{Z_2Z_0}{Z_1Z_2 + Z_1Z_0 + Z_2Z_0}\right)$$

$$= \frac{\mathbf{E_{RN}}(Z_2 + Z_0)}{Z_1Z_2 + Z_1Z_0 + Z_2Z_0} \qquad [78]$$

Since $\mathbf{V_{R0}} = \mathbf{V_{R1}} = \mathbf{V_{R2}}$, then

$$\mathbf{I_{R2}} = -\frac{\mathbf{V_{R1}}}{Z_2} = \frac{-\mathbf{E_{RN}}Z_0}{Z_1Z_2 + Z_1Z_0 + Z_2Z_0} \qquad [79]$$

$$\mathbf{I_{R0}} = -\frac{\mathbf{V_{R1}}}{Z_0} = \frac{-\mathbf{E_{RN}}Z_2}{Z_1Z_2 + Z_1Z_0 + Z_2Z_0} \qquad [80]$$

The fault current is $\mathbf{I_Y} + \mathbf{I_B}$ and can be found using the above symmetrical component values.

Example 13

A three-phase, balanced, star-connected generator has its star point earthed and gives a phase voltage of 3 kV. It has a positive-sequence reactance of 2.0 Ω, a negative-sequence reactance of 0.5 Ω and a zero-sequence reactance of 0.3 Ω. When operating unloaded it sustains a short-circuit to earth of the red line. What is the fault current?

Answer

The phase e.m.f. is 3000/√3 = 1732 V. The fault current is, using equation [57],

$$\mathbf{I_R} = \frac{3\,\mathbf{E_{R1}}}{Z_1 + Z_2 + Z_3} = \frac{3 \times 1732}{\mathrm{j}2.0 + \mathrm{j}0.5 + \mathrm{j}0.3} = \frac{1856}{\mathrm{j}} = -\mathrm{j}1836\ \mathrm{A}$$

$$= 1836\angle -90^\circ\ \mathrm{A}$$

Power transmission lines

A three-phase power distribution system has at the sending end a three-phase generator or transformer and uses a wire for each phase to transmit power to a load at the receiving end. The transmission lines will have resistance, inductance, capacitance and shunt conductance (see Chapter 9 and the definition of transmission line primary constants). Normally the generator and load are both balanced.

For a *short transmission line*, less than about 80 km, in air with frequencies of 50 Hz or 60 Hz the effects of the capacitance and conductance are small enough to be ignored and the line can be considered to have only resistance and inductance. These can be considered to be concentrated in 'lumps' along resistanceless and inductionless lines, as in Fig. 10.23. Calculations can be carried out per phase, considering the line voltages, line currents and phase voltages to have equal magnitudes in each phase. One phase of such a three-phase supply can be considered to be a simple series a.c. circuit (Fig. 10.24). Thus if $\mathbf{V_S}$ is the sending-end line-to-neutral voltage, $\mathbf{V_R}$ the receiving-end line-to-neutral voltage, $\mathbf{I_S}$ the sending-end current and $\mathbf{I_R}$ the receiving-end current then because it is a simple series circuit

$$\mathbf{I_S} = \mathbf{I_R} \qquad [81]$$

and

$$\mathbf{V_S} = \mathbf{V_R} + \mathbf{I_R}(R + j\omega L) \qquad [82]$$

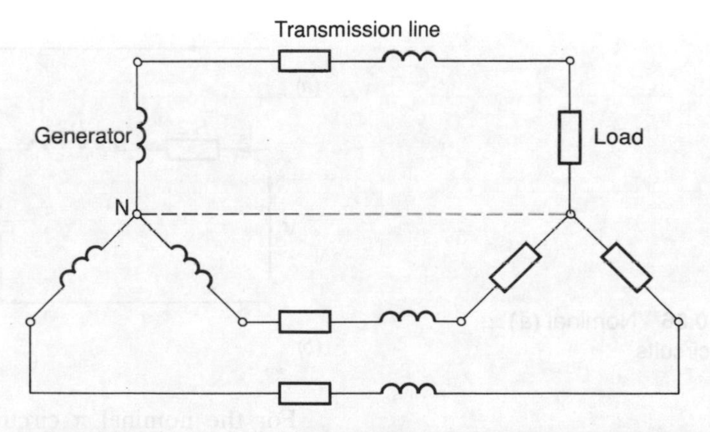

Fig. 10.23 A short transmission line

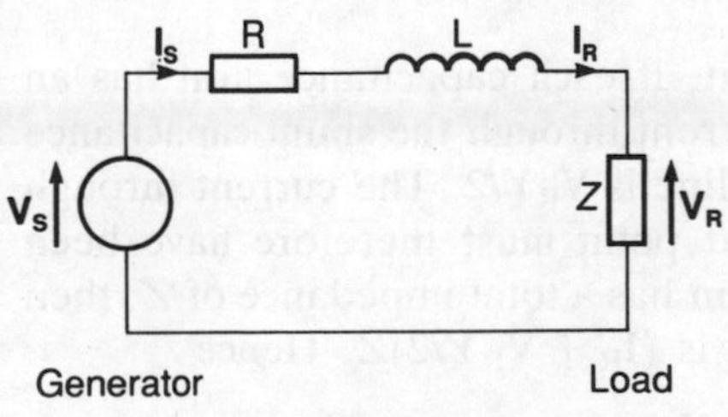

Fig. 10.24 Equivalent circuit for one phase of a short transmission line

Figure 10.25 shows the phasor diagram for this equation, the form of the diagram depending on the power factor of the load, i.e. the value of the phase difference ϕ between $\mathbf{V_R}$ and $\mathbf{I_R}$. The effect of the power factor can be described by what is termed the *voltage regulation* of the line. This is defined as

$$\text{percent regulation} = \frac{|V_{\text{R.NL}}| - |V_{\text{R.FL}}|}{|V_{\text{R.FL}}|} \times 100\% \qquad [83]$$

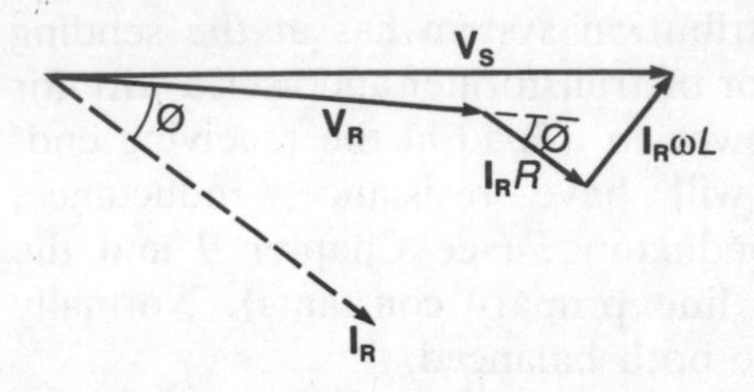

Fig. 10.25 Phasor diagram for short transmission line

where $|V_{R.NL}|$ is the magnitude of the receiving-end voltage at no load and $|V_{R.FL}|$ is the magnitude at full load. With no load the voltage at the receiving end of a short transmission line is equal to the voltage at the sending end $|V_S|$. With the full load the voltage at the receiving end is, as designated in Fig. 10.24, $|V_R|$. Thus

$$\text{percent regulation} = \frac{|V_S| - |V_R|}{|V_R|} \times 100\% \qquad [84]$$

The higher the value of the voltage regulation the greater the voltage drop along the line.

For a *medium-length transmission line*, between about 80 km and 240 km, in air the capacitance, which increases with the length of line, is no longer small enough to be ignored. Reasonable accuracy in calculations is possible if the resistance, inductance and capacitance are considered to be *lumped*, the circuit thus being a nominal π or nominal T circuit (Fig. 10.26).

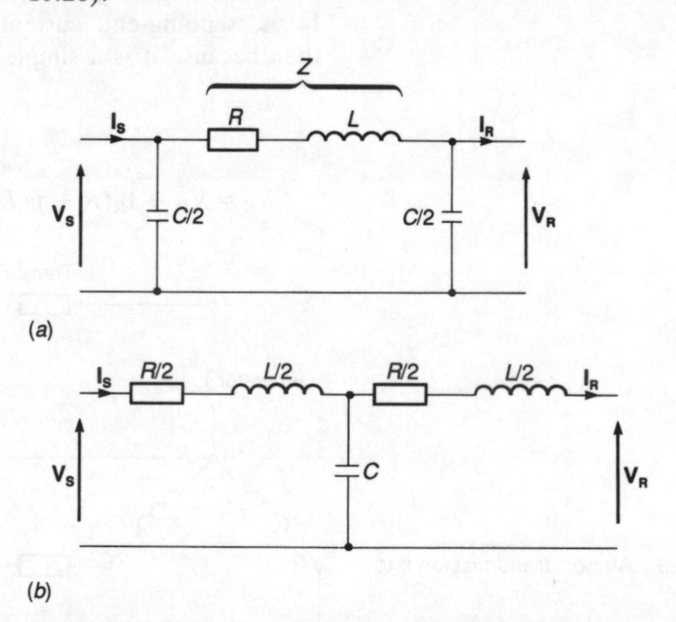

Fig. 10.26 Nominal (*a*) π, (*b*) T circuits

For the nominal π circuit, if each capacitance arm has an admittance $Y/2$ then the current through the shunt capacitance at the receiving end of the line is $\mathbf{V_R}Y/2$. The current through the series arm prior to that point must therefore have been $\mathbf{I_R} + \mathbf{V_R}Y/2$. If the series arm has a total impedance of Z, then the potential drop across it is $(\mathbf{I_R} + \mathbf{V_R}Y/2)Z$. Hence

$$\mathbf{V_S} = (\mathbf{I_R} + \mathbf{V_R}Y/2)Z + \mathbf{V_R}$$

$$= \mathbf{V_R}(1 + ZY/2) + Z\mathbf{I_R} \qquad [85]$$

The current through the shunt capacitance at the sending end is $\mathbf{V_S}Y/2$. Thus, since the current in the series arm is $\mathbf{I_R} + \mathbf{V_R}Y/2$, then

$$\mathbf{I_S} = \mathbf{V_S}Y/2 + \mathbf{I_R} + \mathbf{V_R}Y/2$$

Substituting for $\mathbf{V_S}$, using equation [85] gives

$$\mathbf{I_S} = [\mathbf{V_R}(1 + ZY/2) + Z\mathbf{I_R}]Y/2 + \mathbf{I_R} + \mathbf{V_R}Y/2$$

$$= \mathbf{V_R}Y(1 + ZY/4) + \mathbf{I_R}(1 + ZY/2) \qquad [86]$$

For a *long transmission line*, i.e. over about 240 km, in air or underground cables of any length, it is necessary to take account of the resistance, inductance, capacitance and shunt conductance and also of the fact that they cannot be considered to be lumped at just one place along a line. They have to be considered to be distributed along the length of a line. In Chapter 9 transmission lines were discussed in terms of a cascaded sequence of π or T sections and it is this approach and the equations developed in Chapter 9 which are valid for a long transmission line (see in particular the section on current and voltage on mismatched lines). Thus, at some distance x from the load (Fig. 10.27) the current $\mathbf{I}_x$ will be the sum of the incident $\mathbf{I_i}$ and reflected current waves $\mathbf{I_r}$ and the voltage $\mathbf{V}_x$ the sum of the incident $\mathbf{V_i}$ and reflected voltage waves $\mathbf{V_r}$. Considering first the current,

$$\mathbf{I}_x = \mathbf{I_i} + \mathbf{I_r}$$

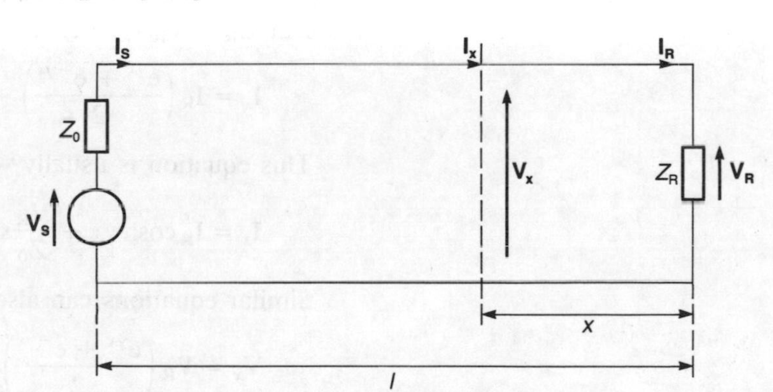

Fig. 10.27 Long transmission line

But $\mathbf{I_i} = \mathbf{I_S}\,e^{-\gamma(l-x)}$ and $\mathbf{I_r}$ can be considered to be due to the attenuation of a current wave from a source of $\varrho_i\mathbf{I_S}\,e^{-\gamma l}$ located at the load end, where ϱ_i is the current reflection coefficient (see Chapter 9), and so equal to $\varrho_i\mathbf{I_S}\,e^{-\gamma l}e^{-\gamma x}$. Thus

$$\mathbf{I}_x = \mathbf{I_S}\,e^{-\gamma(l-x)} + \varrho_i\mathbf{I_S}\,e^{-\gamma l}e^{-\gamma x}$$

$$= \mathbf{I_S}\,e^{-\gamma l}(e^{\gamma x} + \varrho_i\,e^{-\gamma x})$$

But (see equation [34], Chapter 9),

$$\varrho_i = \frac{Z_0 - Z_R}{Z_0 + Z_R}$$

where Z_0 is the characteristic impedance of the line. With power lines Z_0 is sometimes referred to as the *surge impedance*. Thus

$$\mathbf{I}_x = \mathbf{I}_S e^{-\gamma l}\left[e^{\gamma x} + \left(\frac{Z_0 - Z_R}{Z_0 + Z_R}\right)e^{-\gamma x}\right]$$

$$= \frac{\mathbf{I}_S e^{-\gamma l}}{Z_0 + Z_R}[e^{\gamma x}(Z_0 + Z_R) + (Z_0 - Z_R)e^{-\gamma x}]$$

$$= \frac{\mathbf{I}_S e^{-\gamma l}}{Z_0 + Z_R}[Z_0(e^{\gamma x} + e^{-\gamma x}) + Z_R(e^{\gamma x} - e^{-\gamma x})] \quad [87]$$

When $x = 0$ then $\mathbf{I}_x = \mathbf{I}_R$. Thus for this condition

$$\mathbf{I}_R = \frac{\mathbf{I}_S e^{-\gamma l}}{Z_0 + Z_R}[Z_0(1 + 1) + Z_R(1 - 1)]$$

$$\frac{\mathbf{I}_R}{2Z_0} = \frac{\mathbf{I}_S e^{-\gamma l}}{Z_0 + Z_R}$$

Substituting this in equation [87] gives

$$\mathbf{I}_x = \frac{\mathbf{I}_R}{2Z_0}[Z_0(e^{\gamma x} + e^{-\gamma x}) + Z_R(e^{\gamma x} - e^{-\gamma x})]$$

But $Z_R = \mathbf{V}_R/\mathbf{I}_R$, hence

$$\mathbf{I}_x = \mathbf{I}_R\left(\frac{e^{\gamma x} + e^{-\gamma x}}{2}\right) + \frac{\mathbf{V}_R}{Z_0}\left(\frac{e^{\gamma x} - e^{-\gamma x}}{2}\right) \quad [88]$$

This equation is usually written as

$$\mathbf{I}_x = \mathbf{I}_R \cosh \gamma x + \frac{\mathbf{V}_R}{Z_0}\sinh \gamma x \quad [89]$$

Similar equations can also be derived for the voltage

$$\mathbf{V}_x = \mathbf{V}_R\left(\frac{e^{\gamma x} + e^{-\gamma x}}{2}\right) + \mathbf{I}_R Z_0\left(\frac{e^{\gamma x} - e^{-\gamma x}}{2}\right) \quad [90]$$

$$\mathbf{V}_x = \mathbf{V}_R \cosh \gamma x + \mathbf{I}_R Z_0 \sinh \gamma x \quad [91]$$

When $x = l$ then $\mathbf{I}_x = \mathbf{I}_S$ and $\mathbf{V}_x = \mathbf{V}_S$, thus equations [89] and [90] become

$$\mathbf{I}_S = \mathbf{I}_R \cosh \gamma l + \frac{\mathbf{V}_R}{Z_0}\sinh \gamma l \quad [92]$$

$$\mathbf{V}_S = \mathbf{V}_R \cosh \gamma l + \mathbf{I}_R Z_0 \sinh \gamma l \quad [93]$$

Example 14

A three-phase 50 Hz transmission line of length 10 km has a resistance of 0.14 Ω/km/phase and an inductance of 1.2 mH/km/phase. What is the sending-end voltage if the line supplies a load of 10 MW at 30 kV and 0.80 power factor lagging?

Answer

The voltage across a load element at the receiving end is $30\,000/\sqrt{3} = 17\,321$ V. Since the power in a three-phase system is $\sqrt{3}\,V_L I_L \cos\phi$ then at the receiving end

$$10 \times 10^6 = \sqrt{3} \times 30\,000 \times I_L \times 0.80$$

Hence I_L, which is the receiving-end current, is 241 A. If the receiving-end voltage is taken as the reference then $\mathbf{I_R}$ lags $\mathbf{V_R}$ by ϕ, where $\cos\phi = 0.80$. This is $-36.9°$. Hence $\mathbf{I_R} = 241\angle{-36.9°}$ A.

Since the line is short equation [82] can be used. Thus

$$\begin{aligned}\mathbf{V_S} &= \mathbf{V_R} + \mathbf{I_R}(R + j\omega L)\\ &= 17\,321\angle 0° + 241\angle{-36.9°}(1.4 + j2\pi \times 50 \times 12 \times 10^{-3})\\ &= 17\,321\angle 0° + 241\angle{-36.9°}(4.02\angle 69.6°)\\ &= 17\,321\angle 0° + 969\angle 32.7°\\ &= 17\,321 + j0 + 815 + j523 = 18\,136 + j523\\ &= 18\,144\angle 1.65° \text{ V}\end{aligned}$$

Two-port network parameters

The *ABCD* parameters for a two-port network (see Chapter 8) can be used to simplify the consideration of three-phase transmission lines. The two-port network equations can be written as (see equations [1] and [2] in Chapter 8)

$$\mathbf{V_S} = \mathbf{A}\mathbf{V_R} - B\mathbf{I_R}$$

$$\mathbf{I_S} = C\mathbf{V_R} - D\mathbf{I_R}$$

Note that the above equations adopt the current direction convention used in Chapter 8, both $\mathbf{I_S}$ and $\mathbf{I_R}$ being inwards to the network. Adopting the current directions used in this chapter, e.g. in Fig. 10.24, then

$$\mathbf{V_S} = \mathbf{A}\mathbf{V_R} + B\mathbf{I_R} \qquad [94]$$

$$\mathbf{I_S} = C\mathbf{V_R} + D\mathbf{I_R} \qquad [95]$$

For a *short transmission line* where Fig. 10.24 is valid, the two-port network has only series impedance of Z. For a network with only series impedance $A = 1$, $B = Z$, $C = 0$ and $D = 1$ (see equation [18] Chapter 8). For the short transmission line $Z = R + j\omega L$.

For the *medium-length transmission* line where Fig. 10.26 is valid, with a π-section the parameters are $A = 1 + Y_2 Z$,

$B = Z$, $C = Y_1 + Y_2 + Y_1Y_2Z$ and $D = 1 + Y_1Z$, with $Y_1/2$ and $Y_2/2$ being the admittances of the two shunt capacitors and Z being the impedance of the series element. Since $Y_1 = Y_2$ then $A = 1 + YZ$, $B = Z$, $C = 2Y + Y^2Z$ and $D = 1 + YZ$, with Y being the shunt admittance of $j/\omega C$ and Z being $R + j\omega L$. If the T-section were used then $A = 1 + YZ$, $B = 2Z + YZ^2$, $C = Y$ and $D = 1 + YZ$ (see equation 21, Chapter 8), with Y being $j/\omega C$ and Z being $R + j\omega L$.

For a transmission line with uniformly distributed resistance, inductance, capacitance and conductance, e.g. a *long transmission line*, then equations [92] and [93], namely

$$\mathbf{I_S} = \mathbf{I_R} \cosh \gamma l + \frac{\mathbf{V_R}}{Z_0} \sinh \gamma l$$

$$\mathbf{V_S} = \mathbf{V_R} \cosh \gamma l + \mathbf{I_R} Z_0 \sinh \gamma l$$

when compared with the two-port equations [94] and [95] indicate that $A = \cosh \gamma l$, $B = Z_0 \sinh \gamma l$, $C = (1/Z_0) \sinh \gamma l$ and $D = \cosh \gamma l$.

Example 15

A three-phase 50 Hz transmission line of length 10 km has a resistance of 0.14 Ω/km/phase and an inductance of 1.2 mH/km/phase. What is the sending-end voltage and the voltage regulation if the line supplies a current of $241\angle{-36.9°}$ A at 30 kV to the load?

Answer

This is a repeat of Example 14 but using the *ABCD* parameters for its solution. For a short transmission line $A = 1$, $B = Z$, $C = 0$, $D = 1$. Thus

$$B = R + j\omega L = 1.4 + j2\pi \times 50 \times 12 \times 10^{-3} = 4.02\angle 69.6°\ \Omega$$

The voltage across a load element at the receiving end is $30\,000/\sqrt{3} = 17\,321$ V. Hence, equation [94] gives

$$\begin{aligned}\mathbf{V_S} = A\mathbf{V_R} + B\mathbf{I_R} &= 17\,321 + 4.02\angle 69.6°(241\angle{-36.9°}) \\ &= 17\,321\angle 0° + 969\angle 32.7° \\ &= 17\,321 + j0 + 815 + j523 = 18\,136 + j523 \\ &= 18\,144\angle 1.65°\ \text{V}\end{aligned}$$

The voltage regulation for a short transmission line is given by equation [84] as

$$\begin{aligned}\text{percent regulation} &= \frac{|V_S| - |V_R|}{|V_R|} \times 100\% \\ &= \frac{18\,144 - 17\,321}{17\,321} \times 100\% \\ &= 4.8\%\end{aligned}$$

Example 16

A three-phase 50 Hz transmission line of length 100 km has a resistance of 16 Ω/phase, an inductive reactance of 40 Ω/phase and a shunt capacitive susceptance of $4 \times 10^{-4} \angle 90°$ S/phase. Consider each phase as a nominal T-network and hence determine the sending-end voltage and current, the voltage regulation and the transmission efficiency if the line supplies a load at the receiving end with 125 MW at 215 kV and 0.85 power factor lagging.

Answer

The voltage across a load element is $215/\sqrt{3} = 124$ kV. Since the receiving-end power is $\sqrt{3} V_L I_L \cos\phi$ then since $I_R = I_L$

$$125 \times 10^6 = \sqrt{3} \times 215 \times 10^3 \times I_R \times 0.85$$

Hence $I_R = 395$ A. Using $\mathbf{V_R}$ as the reference, since $\cos\phi = 0.85$ and so $\phi = 31.8°$, then $\mathbf{I_R} = 395 \angle -31.8°$ A. Since the line is being considered as a medium-length line in the form of a T-network, the the *ABCD* parameters are

$$A = D = 1 + YZ = 1 + 4 \times 10^{-4} \angle 90° \times (8 + j20)$$
$$= 1 + 4 \times 10^{-4} \angle 90° \times 21.5 \angle 68.2°$$
$$= 1 + 8.6 \times 10^{-3} \angle 158.2°$$
$$= 1 - 8.0 \times 10^{-3} + j3.2 \times 10^{-3}$$
$$= 0.99 + j3.2 \times 10^{-3}$$
$$= 0.992 \angle 0.19°$$

$$B = 2Z + Z^2Y = 2(8 + j20) + (8 + j20)^2 \times 4 \times 10^{-4} \angle 90°$$
$$= 16 + j40 + (21.5 \angle 68.2°)^2 \times 4 \times 10^{-4} \angle 90°$$
$$= 16 + j40 + 0.18 \angle 226.4° = 16 + j40 - 0.12 - j0.13$$
$$= 15.9 + j39.9 = 43.0 \angle 68.3° \, \Omega$$

$$C = Y = 4 \times 10^{-4} \text{ S}$$

Thus, using these values in equation [94] gives

$$\mathbf{V_S} = A\mathbf{V_R} + B\mathbf{I_R}$$
$$= (0.992 \angle 0.19°) 124 \times 10^3 \angle 0° + (43.0 \angle 68.3°) 395 \angle -31.8°$$
$$= 123 \times 10^3 \angle 0.19° + 17.0 \times 10^3 \angle 36.5°$$
$$= 123 \times 10^3 + j408 + 13.7 \times 10^3 + j10.1 \times 10^3$$
$$= 137 \angle 4.4° \text{ kV}$$

This is the phase voltage; the line voltage is $\sqrt{3} \times 137 = 237$ kV.
Using equation [95],

$$\mathbf{I_S} = C\mathbf{V_R} + D\mathbf{I_R}$$
$$= (4 \times 10^{-4}) 124 \times 10^3 \angle 0° + (0.99 \angle 0.19°) 395 \angle -31.8°$$
$$= 49.6 \angle 0° + 391 \angle -31.6°$$

$$= 49.6 + j0 + 333.0 - j204.9$$

$$= 434\angle{-28.2°}\text{ A}$$

The voltage regulation is given by equation [83] as

$$\text{percent regulation} = \frac{|V_{R.NL}| - |V_{R.FL}|}{|V_{R.FL}|} \times 100\%$$

At no load $\mathbf{I_R} = 0$ and so

$$\mathbf{V_S} = \mathbf{A}\mathbf{V_R} + B\mathbf{I_R} = \mathbf{A}\mathbf{V_{R.NL}} + 0$$

Thus

$$\mathbf{V_{R.NL}} = \frac{\mathbf{V_S}}{A} = \frac{137\angle 4.4°}{0.99\angle 0.19°} = 138\angle 4.2°\text{ kV}$$

Thus

$$\text{percent regulation} = \frac{138 - 124}{124} \times 100\% = 11.3\%$$

The transmission efficiency is the ratio output power to input power. Since the sending-end power is $\sqrt{3}\, V_L I_L \cos\phi$ then

$$\text{efficiency} = \frac{125 \times 10^6}{\sqrt{3} \times 237 \times 10^3 \times 434 \times 0.85} \times 100\%$$

$$= 82.5\%$$

Example 17

A 400 km three-phase 50 Hz transmission line has a distributed series impedance of $0.5\angle 80°\ \Omega$/km and a shunt admittance of $3.0 \times 10^{-6}\angle 90°$ S/km and delivers power to a load of 125 MW at 215 kV and 0.85 power factor lagging. What are the voltage, current and power at the sending end of the line?

Answer

The load voltage $V_R = 215/\sqrt{3} = 124$ kV. Since the receiving-end power is $\sqrt{3}\, V_R I_R \cos\phi$ then

$$125 \times 10^6 = \sqrt{3} \times 124 \times 10^3 \times I_R \times 0.85$$

and so $I_R = 685$ A. Taking $\mathbf{V_R}$ as the reference, since $\cos\phi = 0.85$ and so $\phi = 31.8°$, then $\mathbf{I_R} = 685\angle{-31.8°}$ A. The propagation coefficient γ is (equation [13], Chapter 9)

$$\gamma = \sqrt{[(R + j\omega L)(G + j\omega C)]}$$

$$= \sqrt{[(0.5\angle 80°)(3.0 \times 10^{-6}\angle 90°)]}$$

$$= 1.22 \times 10^{-3}\angle 85° = 1.06 \times 10^{-4} + j1.22 \times 10^{-3}$$

Thus

$$\gamma l = 0.488\angle 85° = 0.0424 + j0.488$$

The characteristic impedance Z_o is (equation [12], Chapter 9)

$$Z_0 = \sqrt{\left(\frac{R + j\omega L}{G + j\omega C}\right)}$$

$$= \sqrt{\left(\frac{0.5\angle 80^\circ}{3.0 \times 10^{-6}\angle 90^\circ}\right)}$$

$$= 408\angle -5^\circ\,\Omega$$

Since the line is a long line the *ABCD* parameters are

$$A = D = \cosh \gamma l$$

There are a number of ways of evaluating $\cosh \gamma l$. Since $\gamma = \alpha + j\beta$, then as

$$\cosh(\alpha l + j\beta l) = \cosh \alpha l \cos \beta l + j \sinh \alpha l \sin \beta l$$

with

$$\cosh \alpha l = \frac{e^{\alpha l} + e^{-\alpha l}}{2} = \frac{e^{0.0424} + e^{-0.0424}}{2} = 1.001$$

$$\sinh \alpha l = \frac{e^{\alpha l} - e^{-\alpha l}}{2} = \frac{e^{0.0424} - e^{-0.0424}}{2} = 0.042$$

$$\cos \beta l = \cos 0.488 = \cos 28.0^\circ = 0.853$$

$$\sin \beta l = \sin 0.488 = \sin 28.0^\circ = 0.469$$

Thus

$$A = D = 1.001 \times 0.883 + j0.042 \times 0.469 = 0.864 + j0.020$$

$$= 0.884\angle 1.3^\circ$$

$B = Z_0 \sinh \gamma l$ and since

$$\sinh \gamma l = \sinh(\alpha + j\beta)$$

$$= \sinh \alpha l \cos \beta l + j \cosh \alpha l \sin \beta l$$

$$= 0.042 \times 0.883 + j1.001 \times 0.469 = 0.0371 + j0.469$$

$$= 0.470\angle 85.5^\circ$$

Thus

$$B = 408\angle -5^\circ(0.470\angle 85.5^\circ) = 191\angle 80.5^\circ\,\Omega$$

Since $C = (1/Z_0)\sinh \gamma l$ then

$$C = \frac{0.470\angle 85.5^\circ}{408\angle -5^\circ} = 1.15 \times 10^{-3}\angle 90.5^\circ\,\text{S}$$

Equations [94] and [95] can then be used to determine the sending voltage and current.

$$\mathbf{V_S} = A\mathbf{V_R} + B\mathbf{I_R}$$

$$= 0.884\angle 1.3^\circ(124 \times 10^3\angle 0^\circ) + 191\angle 80.5^\circ(685\angle -31.8^\circ)$$

$$= 110 \times 10^3\angle 1.3^\circ + 131 \times 10^3\angle 48.7^\circ$$

$$= 110 \times 10^3 + j2.52 \times 10^3 + 86.5 \times 10^3 + j98.4 \times 10^3$$

$$= 197 \times 10^3 + j101 \times 10^3 = 220 \times 10^3\angle 27.1^\circ\,\text{V}$$

This is the sending-end phase voltage; the sending-end line voltage is $\sqrt{3} \times 220 \times 10^3 = 381\,\text{kV}$.

$$\begin{aligned}\mathbf{I_S} &= C\mathbf{V_R} + D\mathbf{I_R}\\ &= 1.15 \times 10^{-3}\angle 90.5°(124 \times 10^3 \angle 0°)\\ &\quad + 0.884\angle 1.3°(685\angle -31.8°)\\ &= 143\angle 90.5° + 606\angle -30.5°\\ &= -1.248 + \text{j}143 + 522 - \text{j}308\\ &= 522 - \text{j}165 = 547\angle -17.5°\,\text{A}\end{aligned}$$

The power at the sending end is

$$\begin{aligned}\text{power} &= \sqrt{3}\,V_L I_L \cos\phi\\ &= \sqrt{3} \times 381 \times 10^3 \times 547 \times 0.71 = 257\,\text{MW}\end{aligned}$$

Circuit elements in series

With two two-port networks cascaded in series the *ABCD* parameters for the combination (see Chapter 8, equation [12])

$$A = A_1A_2 + B_1C_2$$

$$B = A_1B_2 + B_1D_2$$

$$C = C_1A_2 + D_1C_2$$

$$D = C_1B_2 + D_1D_2$$

where A_1, B_1, C_1 and D_1 are the parameters of one network and A_2, B_2, C_2 and D_2 those of the second network. Thus, suppose we have a transmission line represented by a nominal π circuit and add a capacitor in series. The parameters for the nominal π circuit are $A_1 = 1 + YZ$, $B_1 = Z$, $C_1 = 2Y + Y^2Z$ and $D_1 = 1 + YZ$ (see previous section in this chapter). The parameters for a series impedance are (equation [18], Chapter 8) $A_2 = 1$, $B_2 =$ capacitor impedance, $C_2 = 0$ and $D_2 = 1$. Thus since the capacitor impedance is $1/(\text{j}\omega C)$ then the parameters for the cascaded π circuit become

$$A = (1 + YZ)1 + Z \times 0$$

$$B = (1 + YZ)(1/\text{j}\omega C) + Z \times 1$$

$$C = (2Y + Y^2Z) \times 1 + (1 + YZ) \times 0$$

$$D = (2Y + Y^2Z)(1/\text{j}\omega C) + (1 + YZ) \times 1$$

The capacitor thus changes the B and D parameters. Such a capacitor may be introduced to improve the power flow through the line.

Inductors are sometimes connected between the line and neutral for a transmission line. A shunt admittance Y_2, in this case $1/\text{j}\omega L$, has *ABCD* parameters $A_2 = 1$, $B_2 = 0$, $C_2 = 1/\text{j}\omega L$, $D_2 = 1$. An effect of such an introduction is a reduction in the no-load voltage and hence a reduction in the voltage regulation.

Example 18

What is the effect on the voltage regulation of the transmission line described in Example 16 of connecting an inductor with inductive susceptance 3×10^{-4} S between the line and the neutral for each phase?

Answer

The percentage voltage regulation is given by

$$\text{percent regulation} = \frac{|V_{R.NL}| - |V_{R.FL}|}{|V_{R.FL}|} \times 100\%$$

At no load $\mathbf{I_R} = 0$ and so

$$\mathbf{V_S} = A\mathbf{V_R} + B\mathbf{I_R} = A\mathbf{V_{R.NL}} + 0$$

Thus

$$\mathbf{V_{R.NL}} = \frac{\mathbf{V_S}}{A}$$

But

$$A = A_1A_2 + B_1C_2$$

and since $A_1 = 0.99\angle 0.19°$, $B_1 = 43\angle 68.3°$, $A_2 = 1$ and $C_2 = 3 \times 10^{-4}$ S, then

$$\begin{aligned} A &= (0.99\angle 0.19°) \times 1 + (43\angle 68.3°)(3 \times 10^{-4}\angle -90°) \\ &= 0.99 + j3.2 \times 10^{-3} + 12 \times 10^{-3} - j4.8 \times 10^{-3} \\ &= 1.00 - j1.6 \times 10^{-3} = 1.00\angle -0.092° \end{aligned}$$

Thus

$$\mathbf{V_{R.NL}} = \frac{137\angle 4.4°}{1.004\angle -0.092°} = 137\angle 4.5° \text{ V}$$

Hence

$$\text{percent regulation} = \frac{137 - 124}{124} \times 100\% = 10.5\%$$

This is an improvement on the original 11.3%.

Problems

1 A balanced star-connected three-phase generator has a phase voltage of 100 V. What will be the magnitude of the line voltages?

2 A balanced delta-connected three-phase generator is connected to balanced resistive loads. If the magnitude of the phase current is 10 A, what will be the magnitude of the line currents?

3 A three-phase supply has a balanced star-connected load composed of three 10 Ω resistors. If the line voltage is 200 V what

is (*a*) the current in each resistor, (*b*) the voltage across each resistor?

4 A three-phase supply has a balanced delta-connected load composed of three 10 Ω resistors. If the line voltage is 200 V what is (*a*) the current in each resistor, (*b*) the voltage across each resistor?

5 A balanced star-connected three-phase generator has a phase voltage of 80 V and is connected to a balanced star-connected load consisting of three 100 Ω resistors. What is (*a*) the phase current for a generator winding, (*b*) the current through a load resistor, (*c*) the voltage across a load resistor, (*d*) the line voltage, (*e*) the line current?

6 A balanced delta-connected three-phase generator has a phase voltage of 100 V and is connected to a balanced delta-connected load consisting of three 150 Ω resistors. What is (*a*) the phase current for a generator winding, (*b*) the current through a load resistor, (*c*) the voltage across a load resistor, (*d*) the line voltage, (*e*) the line current?

7 A balanced delta-connected three-phase generator has a phase voltage of 80 V and is connected to a balanced star-connected load consisting of three 120 Ω resistors. What is (*a*) the line current, (*b*) the voltage across a load resistor, (*c*) the current through a load resistor?

8 A balanced three-phase supply with a line voltage of 500 V has a star-connected load composed of three 100 Ω resistors. What is (*a*) the voltage across a load element, (*b*) the current through a load element, (*c*) the total power dissipated?

9 A balanced three-phase supply with a line voltage of 415 V has a delta-connected load composed of three 100 Ω resistors. What is (*a*) the voltage across a load element, (*b*) the current through a load element, (*c*) the total power dissipated?

10 A three-phase motor with an output of 1.2 kW is connected to a three-phase supply and provides a balanced load at a power factor of 0.8. What is the efficiency of the motor if the line voltage is 415 V and the line current 2.5 A?

11 A three-phase motor with an output of 3 kW and an efficiency of 95% is a balanced delta-connected load with a power factor of 0.9 for a three-phase supply. If the supply has a line voltage of 415 V, what is the current in each winding of the motor?

12 A balanced star-connected load is supplied by a three-phase supply with 20 kW at a power factor of 0.8 lagging. If the line voltage is 500 V, what is the current through a load element?

13 Two wattmeters are used to measure the power consumed by a three-wire, balanced load system (as in Fig. 10.14). The meters gave readings of 7 kW and 3 kW. What is the total power consumed and the power factor?

14 Determine the current in the neutral wire for a star-connected load with load elements having impedances of $10\angle 30^\circ\,\Omega$, $15\angle -45^\circ\,\Omega$ and $20\angle 60^\circ\,\Omega$ when connected to a balanced three-phase system having a line voltage of 440 V.

15 Determine the line currents for a mesh having load elements of $Z_{RY} = 10\angle -45^\circ\,\Omega$, $Z_{YB} = 15\angle 0^\circ\,\Omega$ and $Z_{BR} = 12\angle 30^\circ\,\Omega$

when there is a line voltage of 300 V.

16 If $I_Y = 10\angle -40°$, what is aI_Y and a^2I_Y?

17 For the 120° operator, determine the values of (a) $a - a^2$, (b) $2 - a - a^2$, (c) $1 + a + a^2$.

18 What are the symmetrical red line components of a system of unbalanced three-phase currents where the line currents are $I_R = 10\angle 0°$ A, $I_Y = 0$, $I_B = 10\angle 180°$ A?

19 A three-phase, four-wire system has line currents of $I_R = 10\angle 30°$ A, $I_Y = 5\angle -90°$ A and $I_B = 4\angle 180°$ A. What are the symmetrical current components in the red line and the current through the neutral line?

20 A three-phase, four-wire system has line currents of $I_R = 100\angle 30°$ A, $I_Y = 50\angle 300°$ A and $I_B = 30\angle 180°$ A. What are the symmetrical current components in the red line and the current through the neutral line?

21 An unsymmetrical three-phase, three-wire system supplies line voltages of $V_{RY} = 100\angle 0°$ V, $V_{YB} = 200\angle -90°$ V and $V_{BR} = 224\angle 117°$ V to a star-connected load consisting of three identical 100 Ω resistors. What is the current in the red line?

22 A three-phase, balanced, star-connected generator has its star point earthed and gives a phase voltage of 10 kV. It has a positive sequence reactance of 3.0 Ω, a reactance sequence impedance of 1.5 Ω and a zero-sequence reactance of 0.5 Ω. When operating unloaded it sustains a short-circuit to earth of the blue line. What is the fault current?

23 For the three-phase generator specified in problem 22, what would be the fault current if the fault was a line-to-line short-circuit?

24 For the three-phase generator specified in problem 22, what would be the red line voltage if the fault was the yellow and blue lines both short-circuiting to earth?

25 A three-phase 50 Hz transmission line of length 10 km has a resistance of 0.18 Ω/km/phase and an inductance of 1.2 mH/km/phase. What is the sending-end voltage if the line supplies a load of 2 MW at 11 kV and 0.75 power factor lagging?

26 A three-phase 50 Hz transmission line of length 100 km has a resistance of 0.20 Ω/km/phase, an inductance of 1.6 mH/km/phase and a capacitance of 13 nF/km/phase. What is the sending-end voltage if the line supplies a load of 60 MW at 132 kV and power factor 0.85?

27 A three-phase 50 Hz transmission line in air has a length of 10 km, a resistance of 2 Ω/phase and an inductive reactance of 6 Ω/phase. What are the *ABCD* parameters for the line?

28 A three-phase 50 Hz transmission line in air has a length of 100 km, a resistance of 20 Ω/phase, an inductive reactance of 60 Ω/phase and a shunt capacitive susceptance of 4×10^{-4} S. What are the *ABCD* parameters for the line if it is considered to be a lump π-network?

29 A long three-phase transmission line has $\gamma l = 0.48\angle 85°$ and a characteristic impedance of $400\angle -5.5°$ Ω. If the line delivers 125 MW at 215 kV with power factor 1, what are the voltage, current and power at the sending end and the voltage regulation?

30 A three-phase 50 Hz transmission line in air of length 160 km has a resistance of 0.18 Ω/km/phase, an inductance of 1.3 mH/km/phase and a shunt capacitance of 9 nF/km/phase. Consider the line to be represented by a lumped T or π-network and hence determine the sending-end voltage and current if the power delivered to the receiving end at a line voltage of 132 kV is 40 MW at 0.8 power factor lagging. What is the transmission efficiency?

31 A three-phase 50 Hz transmission line in air of length 100 km has a resistance of 15 Ω/phase, an inductance of 180 mH/phase and a shunt capacitance of 1.6 μF/phase. Determine the *ABCD* parameters for the nominal T circuit which can be used to represent the line.

32 What is the effect on the *ABCD* parameters for the line described in problem 31 if a series capacitance with an impedance of $50\angle -90°$ Ω is included in the line?

33 What is the effect on the voltage regulation for the line described in problem 29 if a shunt inductor is included and has an admittance which cancels out 70% of the shunt admittance of the line?

11 The Laplace transform

Introduction

In Chapter 2 it was seen how the analysis of circuits involving capacitors and inductors and a step input involves differential equations. This chapter is concerned with a more powerful approach to such problems which will allow the response to be determined to inputs in general and which will allow networks to be tackled and not just a few components in series or parallel. This powerful approach is called the *Laplace transform*.

A simple example of a mathematical transform is when the problem of multiplication is changed into the simpler operation of addition by means of the *logarithm transform* (Fig. 11.1). Thus the multiplication of B by C to give A, i.e. $A = BC$, can be transformed by using logarithms to

$$\log A = \log BC = \log B + \log C$$

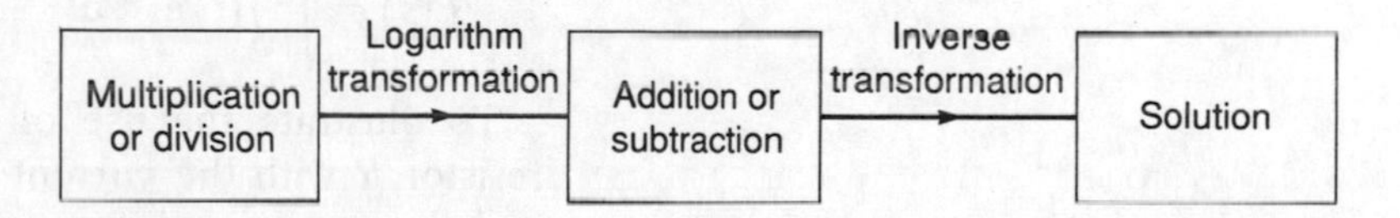

Fig. 11.1 The logarithm transform

We can then add $\log B$ and $\log C$ to give the number D. Then

$$\log A = D$$

To find A we have to carry out the inverse logarithm, or antilogarithm, operation.

$$A = \text{antilog}\, D$$

The *Laplace transform* is a similar type of mathematical operation (Fig. 11.2). Differential equations which describe how a circuit behaves with time are transformed into simple algebraic relationships, not involving time, where we can carry out normal algebraic manipulations of the quantities. We talk of the circuit behaviour in the *time domain* being transformed to the *s-domain*. This transformation is rather like the taking of logarithms. Then algebraic manipulations can be carried out. Finally an inverse transform, like the antilogarithm, is used in order to obtain the solution describing how a signal varies with time, i.e. transform from the *s*-domain back to the time domain.

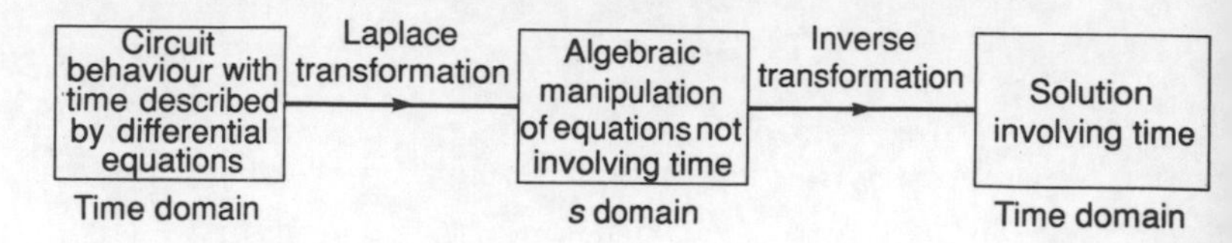

Fig. 11.2 The Laplace transform

The Laplace transformation

The French mathematician P. S. de Laplace (1749–1827) discovered a means of solving differential equations: multiply each term in the equation by e^{-st} and then integrate each such term with respect to time from zero to infinity; s is a constant with the unit of 1/time. The result is what we now call the *Laplace transform*. Thus the Laplace transform of some term which is a function of time is

$$\int_0^\infty (\text{term})\,e^{-st}\,dt$$

A term which is a function of time is usually written as $f(t)$ with the Laplace transform, since it is a function of s, written as $F(s)$. It is usual to use a capital letter F for the Laplace transform of the time varying function $f(t)$. Thus

$$F(s) = \int_0^\infty f(t)\,e^{-st}\,dt \qquad [1]$$

To illustrate the use of the function notation, consider a resistor R with the current through it at some instant being i and the potential difference across it v. Generally we would write

$$v = Ri$$

But since both v and i are functions of time, we should ideally indicate this by writing the equation to indicate this, i.e.

$$v(t) = Ri(t)$$

The (t) does not indicate that the preceding term should be multiplied by it but just that the preceding term is a function of time, i.e. its value depends on what time we are considering.

If we take the Laplace transforms of i and v the equation becomes

$$V(s) = RI(s)$$

$V(s)$ indicates that the term is the Laplace transform of v, similarly $I(s)$ indicates that the term is the Laplace transform of i. The (s) does not indicate that the preceding term should be multiplied by s.

The Laplace transform for a step function

As an illustration of how a Laplace transform can be developed from first principles, consider a step function. Figure 11.3 shows the form that would be taken by a step input when the abrupt change in input takes place at time $t = 0$ and the size of the step is 1 unit. Such a form of signal is very common in electrical circuits, occurring when a voltage is abruptly applied to a circuit (see Chapter 2 for a discussion of the potential difference and current changes in circuits involving resistors with capacitors and/or inductors when step voltage signals are applied). The equation for this function is

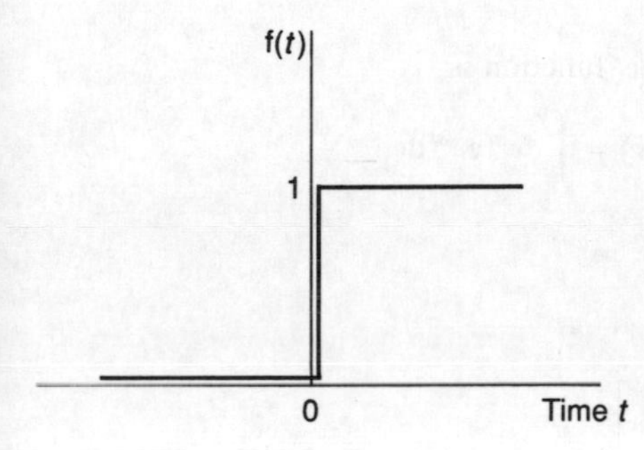

Fig. 11.3 A step function of height 1

$$f(t) = 1$$

for all values of t greater than 0. For values of t less than 0 the equation is

$$f(t) = 0.$$

The Laplace transform of this step function, for values greater than 0, is thus

$$F(s) = \int_0^\infty 1\,e^{-st}$$

$$F(s) = -\frac{1}{s}\left[e^{-st}\right]_0^\infty$$

Since when $t = \infty$ the value of e^∞ is 0 and when $t = 0$ the value of e^{-0} is -1, then

$$F(s) = \frac{1}{s} \qquad [2]$$

Suppose now instead of a step input signal of height 1 unit we have one of height a units. Then, for all values of t greater than 0 we have

$$f(t) = a$$

The Laplace transform of this function is

$$F(s) = \int_0^\infty a\,e^{-st}\,dt$$

$$= a\int_0^\infty e^{-st}\,dt$$

But this is just a multiplied by the transform of the unit step. Thus

$$F(s) = \frac{a}{s} \quad [3]$$

The multiplication of some function of time by a constant a gives a Laplace transform which is just the multiplication of the Laplace transform of that function by the constant.

Example 1

Determine, from first principles, the Laplace transform of the function e^{at}, where a is a constant.

Answer

The Laplace transform of this function is

$$\text{Laplace transform} = F(s) = \int_0^\infty e^{at}\, e^{-st}\, dt$$

This can be simplified to

$$F(s) = \int_0^\infty e^{-(s-a)t}\, dt$$

$$F(s) = -\frac{1}{s-a}\left[e^{-(s-a)t}\right]_0^\infty$$

When $t = \infty$ then the term in the brackets becomes 0 and when $t = 0$ it becomes -1. Thus

$$F(s) = \frac{1}{s-a}$$

Using Laplace transforms

Fortunately it is not usually necessary to evaluate the integrals obtained in carrying out the Laplace transformation since tables are available which give the Laplace transforms of all the commonly occurring electrical signals and these combined with some basic rules for handling such transforms enable most electrical circuit problems to be solved. Table 11.1 gives some of the more common Laplace transforms and their corresponding time functions.

The basic rules are:

1 The addition of two functions becomes the addition of their two Laplace transforms.

$$f_1(t) + f_2(t) \quad \text{becomes} \quad F_1(s) + F_2(s)$$

2 The subtraction of two functions becomes the subtraction of their two Laplace transforms.

$$f_1(t) - f_2(t) \quad \text{becomes} \quad F_1(s) - F_2(s)$$

3 The multiplication of some function by a constant becomes the multiplication of the Laplace transform of the function by the same constant.

Table 11.1 Laplace transforms

Laplace transform	*Time function*	*Description of time function*
1		A unit impulse
$\frac{1}{s}$		A unit step function
$\frac{e^{-st}}{s}$		A delayed unit step function
$\frac{1 - e^{-st}}{s}$		A rectangular pulse of duration T
$\frac{1}{s^2}$	t	A unit slope ramp function
$\frac{1}{s^3}$	$\frac{t^2}{2}$	
$\frac{1}{s + a}$	e^{-at}	Exponential decay
$\frac{1}{(s + a)^2}$	$t e^{-at}$	
$\frac{2}{(s + a)^3}$	$t^2 e^{-at}$	
$\frac{a}{s(s + a)}$	$1 - e^{-at}$	Exponential growth
$\frac{a}{s^2(s + a)}$	$t - \frac{(1 - e^{-at})}{a}$	
$\frac{s}{(s + a)^2}$	$(1 - at)e^{-at}$	
$\frac{\omega}{s^2 + \omega^2}$	$\sin \omega t$	Sine wave
$\frac{s}{s^2 + \omega^2}$	$\cos \omega t$	Cosine wave
$\frac{\omega}{(s + a)^2 + \omega^2}$	$e^{-at} \sin \omega t$	Damped sine wave
$\frac{\omega s}{(s + a)^2 + \omega^2}$	$\sqrt{(a^2 + \omega^2)}\, e^{-at} \sin(\omega t + \phi)$ where $\phi = \tan^{-1}(\omega/-a)$	
$\frac{s + a}{(s + a)^2 + \omega^2}$	$e^{-at} \cos \omega t$	Damped cosine wave
$\frac{\omega^2}{s(s^2 + \omega^2)}$	$1 - \cos \omega t$	
$\frac{\omega^2}{s^2 + 2\zeta\omega s + \omega^2}$	$\frac{\omega}{\sqrt{(1 - \zeta^2)}} e^{-\zeta\omega t} \sin[\omega\sqrt{(1 - \zeta^2)}t]$	
$\frac{\omega}{s^2 - \omega^2}$	$\sinh \omega t$	
$\frac{s}{s^2 - \omega^2}$	$\cosh \omega t$	

$$af(t) \quad \text{becomes} \quad aF(s)$$

4 A function which is delayed by a time T, i.e. $f(t - T)$, becomes $e^{-Ts}F(s)$ for values of T greater than or equal to zero.

5 The first derivative of some function becomes s times the Laplace transform of the function minus the value of $f(t)$ at $t = 0$.

$$\frac{d}{dt}f(t) \quad \text{becomes} \quad sF(s) - f(0)$$

where $f(0)$ is the value of the function at $t = 0$.

6 The second derivative of some function becomes s^2 times the Laplace transform of the function minus s times the value of the function at $t = 0$ minus the value of the first derivative of $f(t)$ at $t = 0$.

$$\frac{d^2}{dt^2}f(t) \quad \text{becomes} \quad s^2F(s) - sf(0) - \frac{df(0)}{dt}$$

where $sf(0)$ is s multiplied by the value of the function at $t = 0$ and $df(0)/dt$ is the first derivative of the function at $t = 0$.

7 The nth derivative of some function becomes s^n times the Laplace transform of the function minus terms involving the values of $f(t)$ and its derivatives at $t = 0$.

$$\frac{d^n}{dt^n}f(t) \quad \text{becomes} \quad s^nF(s) - s^{n-1}f(0) - \ldots - \frac{d^{n-1}f(0)}{dt^{n-1}}$$

8 The first integral of some function, between zero time and time t, becomes $(1/s)$ times the Laplace transform of the function.

$$\int_0^t f(t) \quad \text{becomes} \quad \frac{1}{s}F(s).$$

Example 2

Determine, using Table 11.1, the Laplace transforms for:

(*a*) a step voltage of size 4 V which starts at $t = 0$,
(*b*) a step voltage of size 4 V which starts at $t = 2$ s,
(*c*) a ramp voltage which starts at $t = 0$ and increases at the rate of 3 V/s,
(*d*) a ramp voltage which starts at $t = 2$ s and increases at the rate of 3 V/s,
(*e*) an impulse voltage of size 4 V which starts at $t = 3$ s,
(*f*) a sinusoidal voltage of amplitude 2 V and angular frequency 10 Hz.

Answer

Figure 11.4 shows the form of the six functions, which represent common forms of input signals to systems.

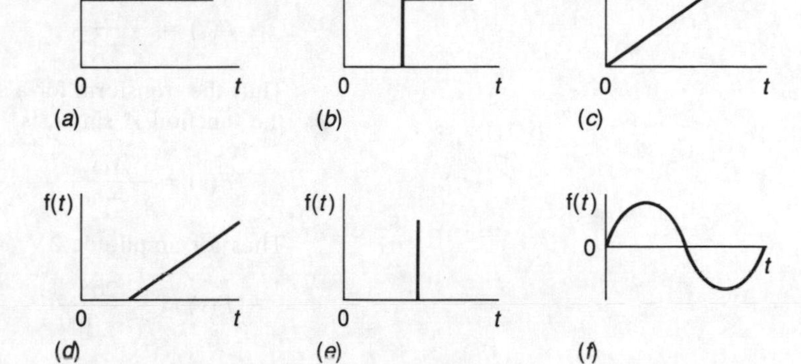

Fig. 11.4 (*a*) Step function, (*b*) delayed step function, (*c*) ramp function, (*d*) delayed ramp function, (*e*) delayed impulse, (*f*) sinusoidal function

(*a*) The step voltage is a function of the form $f(t) = a$, where a is 4 V. The Laplace transform of the step function of size 1 is $1/s$ and thus the step function of size a has the Laplace transform of

$$F(s) = a \times \frac{1}{s}$$

Hence

$$F(s) = \frac{4}{s}$$

(*b*) The step function in (*a*) is delayed by 2 s. For a delayed function the Laplace transform is that of the undelayed function, i.e. the function starting at $t = 0$, multiplied by e^{-sT}. Thus the Laplacc transform is

$$F(s) = \frac{a}{s} e^{-sT} = \frac{4}{s} e^{-2s}$$

(*c*) The ramp function is of the form $f(t) = at$ with a having the value 3 V/s. Because a is a constant then the Laplace transform of the function will be a multiplied by the transform of t which is $1/s^2$. Thus

$$F(s) = \frac{a}{s^2} = \frac{3}{s^2}$$

(*d*) The ramp voltage is delayed by a time T, where $T = 2$ s. For a delayed function the Laplace transform is that of the undelayed function, i.e. the function starting at $t = 0$, multiplied by e^{-sT}. Thus the Laplace transform is

$$F(s) = \frac{a\,e^{-T/s}}{s^2} = \frac{3\,e^{-2s}}{s^2}$$

(*e*) The Laplace transform of a unit impulse occurring at $t = 0$ is 1.

For an impulse of 4 V the transform will be 4. Delaying the impulse means the undelayed function is multiplied by $e^{-T/s}$. Thus the Laplace transform with $T = 3$ s is

$$F(s) = 4\,e^{-3/s}$$

(*f*) The Laplace transform of a sinusoidal function $\sin \omega t$ is

$$F(s) = \frac{\omega}{s^2 + \omega^2}$$

Thus the transform for a sinusoidal function of amplitude A, i.e. the function $A \sin \omega t$, is

$$F(s) = \frac{A\omega}{s^2 + \omega^2}$$

Thus for amplitude 2 V and angular frequency 10 Hz,

$$F(s) = \frac{20}{s^2 + 100}$$

Example 3

Determine, using Table 11.1, the Laplace transforms for (*a*) t^2, (*b*) $t^2 e^{-at}$.

Answer

(*a*) The table gives the Laplace transform of $\frac{1}{2}t^2$ as $1/s^3$. Thus to obtain the Laplace transform of t^2 we have to multiply the function in the table by 2. Since this is a constant, the Laplace transform of t^2 will be

$$F(s) = \frac{2}{s^3}$$

(*b*) Using the table, the transform is

$$F(s) = \frac{2}{(s + a)^3}$$

Example 4

Determine, using Table 11.1, the inverse transformations of

$$(a)\ \frac{2}{s},\quad (b)\ \frac{3}{2s + 1},\quad (c)\ \frac{2}{s - 5}.$$

Answer

(*a*) The table includes a Laplace transform of $1/s$ and thus since this is just multiplied by the constant 2 the inverse transformation will be the function which gives $1/s$, i.e. 1, multiplied by the same constant. The inverse transformation is thus 2.

(*b*) This transform can be rearranged to give

$$\frac{(3/2)}{s + (1/2)}$$

The table contains the transform $1/(s + a)$, the inverse of which is e^{-at}. Thus the inverse transformation is just this multiplied by the constant (3/2) with $a = (1/2)$, i.e. $(3/2)\,e^{-t/2}$.

(*c*) This transform is of the same form as in (*b*) with $a = -5$. Thus the inverse transformation is $2\,e^{5t}$.

Partial fractions

The process of converting an algebraic expression into simple fraction terms is called resolving into *partial fractions*. For example, it is the conversion of

$$\frac{3x + 4}{x^2 + 3x + 2} \quad \text{into} \quad \frac{1}{x + 1} + \frac{2}{x + 2}$$

Such a procedure is often necessary in order that Laplace transformed equations can be put into a form that can be recognized in tables of standard Laplace transform pairs and the inverse transformation procedure carried out.

In order to resolve an algebraic expression into partial fractions the denominator must factorize, e.g. in the above expression the $(x^2 + 3x + 2)$ yields the factors $(x + 1)$ and $(x + 2)$, and the numerator must be at least one degree less than the denominator, e.g. in the above example the numerator only contains x to the power 1 while the denominator has x to the power 2. When the degree of the numerator is equal to or higher than the degree of the denominator, the numerator must be divided by denominator to give terms which each have the numerator at least one degree less than the denominator.

There are basically three types of partial fractions. The form of the partial fractions for each of these types is as follows:

1 *Linear factors in the denominator*

Expression $$\frac{F(s)}{(s + a)(s + b)(s + c)}$$

Partial fraction $$\frac{A}{s + a} + \frac{B}{s + b} + \frac{C}{s + c}$$

2 *Repeated linear factors in the denominator*

Expression $$\frac{F(s)}{(s + a)^n}$$

Partial fraction $$\frac{A}{s + a} + \frac{B}{(s + a)^2} + \frac{C}{(s + a)^3} \dots + \frac{N}{(s + a)^n}$$

3 *Quadratic factors in the denominator, when the quadratic does not factorize without imaginary terms*

$$\text{Expression} \quad \frac{F(s)}{as^2 + bs + c}$$

$$\text{Partial fraction} \quad \frac{As + B}{as^2 + bs + c}$$

or if there is also a linear factor in the denominator,

$$\text{Expression} \quad \frac{F(s)}{(as^2 + bs + c)(s + d)}$$

$$\text{Partial fraction} \quad \frac{As + B}{as^2 + bs + c} + \frac{C}{s + d}$$

Example 5

Determine the partial fractions of $\dfrac{s + 5}{s^2 + 3s + 2}$.

Answer

The expression can be rearranged to give

$$\frac{s + 5}{(s + 1)(s + 2)}$$

This can be resolved into the partial fractions

$$\frac{A}{s + 1} + \frac{B}{s + 2}$$

which on bringing to a common denominator is

$$\frac{A(s + 2) + B(s + 1)}{(s + 1)(s + 2)}$$

Since this expression has the same denominator as the initial expression then we must have, for the expressions to be equal,

$$A(s + 1) + B(s + 2) = s + 5$$

To determine the values of the constants A and B, values of s are chosen to make the terms in A or B equal to zero. Thus if we choose $s = -1$, then

$$A(-1 + 1) + B(-1 + 2) = -1 + 5$$

and B equals 4. When $s = -2$, then

$$A(-2 + 1) + B(-2 + 2) = -2 + 5$$

and A equals -3. Thus the expression in partial fractions is

$$-\frac{3}{s + 1} + \frac{4}{s + 2}$$

The initial and final value theorems

If a Laplace transform is multiplied by s, the value of the product as s tends to infinity is the value of the inverse transform as the time t tends to zero.

$$\lim_{s\to\infty} sF(s) = \lim_{t\to 0} f(t) \qquad [4]$$

This is known as the *initial value theorem*.

If a Laplace transform is multiplied by s, the value of the product as s tends to zero is the value of the inverse transform as t tends to infinity.

$$\lim_{s\to 0} sF(s) = \lim_{t\to\infty} f(t) \qquad [5]$$

This is known as the *final value theorem*.

The initial and the final value theorems are useful when there is a need to determine from the Laplace transform the behaviour of the function $f(t)$ at 0 and ∞

Example 6

Without evaluating the Laplace transforms, what are the initial and final values of the functions giving the following transforms?

$$(a)\ F(s) = \frac{s+a}{s^2}, \quad (b)\ V_C(s) = \frac{V(1/RC)}{[s + (1/RC)]s}$$

Answer

(*a*) If the expression is multiplied by s it becomes

$$sF(s) = \frac{s+a}{s} = 1 + \frac{a}{s}$$

Using the initial value theorem, when $s \to \infty$ then the expression tends to the value 1. So the initial value of the function is 1. Using the final value theorem, when $s \to 0$ then the expression tends to the value ∞. So the final value of the function is ∞.

(*b*) If the expression is multiplied by s it becomes

$$sV_C(s) = \frac{V(1/RC)}{[s + (1/RC)]}$$

Using the initial value theorem, when $s \to \infty$ then the expression tends to the value 0. So the initial value of v_C is 0. Using the final value theorem, when $s \to 0$ then the expression tends to the value $V(1/RC)/(1/RC)$ or V.

Circuit elements in the s-domain

For a *resistor* the relationship between the potential difference v across it at some instant of time and the current i through it is

$$v = Ri$$

where R is the resistance. The Laplace transform of the equation is

$$V(s) = RI(s)$$

where $V(s)$ is the transform of the potential difference and $I(s)$ the transform of the current. So in the s-domain if we define resistance as $V(s)/I(s)$,

$$R(s) = \frac{V(s)}{I(s)} = R \qquad [6]$$

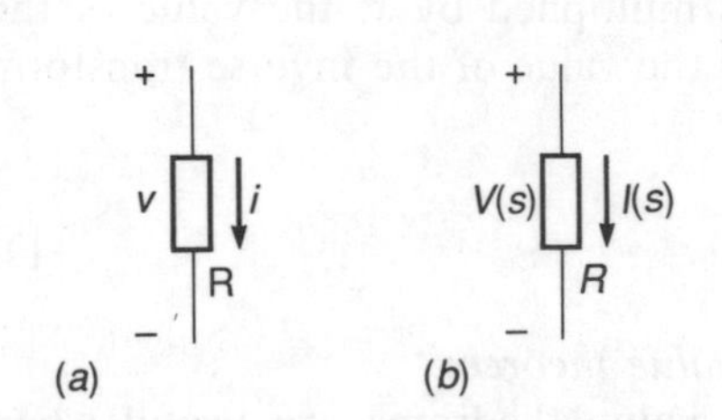

Fig. 11.5 The resistor element (*a*) in the time domain, (*b*) in the *s*-domain

Thus the s-domain equivalent circuit of a resistor is simply a resistance of R that has a current $I(s)$ and a potential difference $V(s)$ (Fig. 11.5).

For an *inductor* with no initial current, i.e. $i = 0$ at $t = 0$, the equation relating the potential difference v across it with the rate of change of current is

$$v = L\frac{\mathrm{d}i}{\mathrm{d}t}$$

where L is the inductance. The Laplace transform of the equation is

$$V(s) = L[sI(s) - i(0)]$$

and as $i(0) = 0$ then

$$V(s) = sLI(s)$$

The impedance of the inductor in the s-domain of the inductor can be considered to be

$$Z(s) = \frac{V(s)}{I(s)} = sL \qquad [7]$$

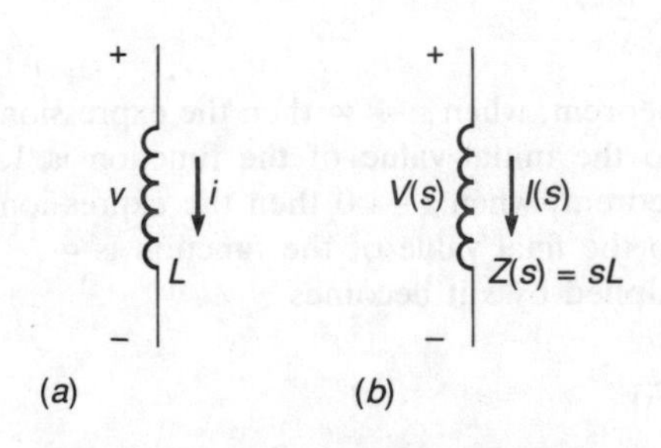

Fig. 11.6 The inductor element (*a*) in the time domain, (*b*) in the *s*-domain

Thus the s-domain equivalent circuit of an inductor is an impedance of sL (Fig. 11.6).

If, for the inductor, instead of the current being zero at $t = 0$ but some value I_0, then the transform would be

$$V(s) = L[sI(s) - i(0)] = sLI(s) - LI_0$$

Since sL is the impedance in the s-domain then the potential difference in the s-domain across the inductor is $sLI(s)$. Thus the equation can be considered to describe two series elements with

$$V(s) = \text{p.d. across } L + \text{voltage generator of } -LI_0 \qquad [8]$$

We can thus consider such an inductor to be an impedance of sL in series with an independent voltage source of $-LI_0$.

Alternatively we can consider it to be an impedance of sL in parallel with an independent current source of I_0/s (Fig. 11.7). If the equation is rearranged, then

$$I(s) = \frac{V(s) + LI_0}{sL} = \frac{V(s)}{sL} + \frac{I_0}{s}$$

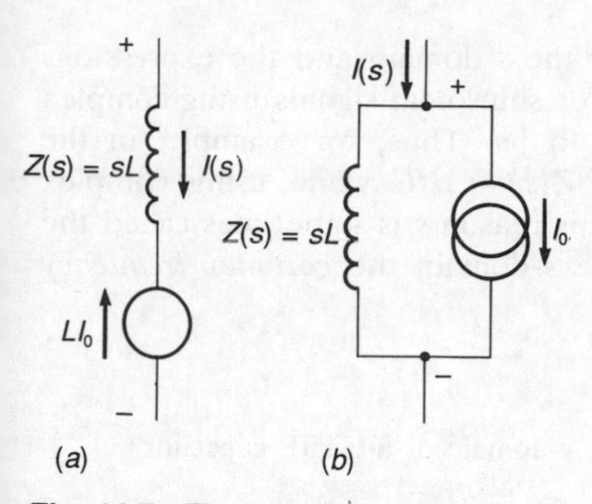

Fig. 11.7 The inductor element in the s-domain with current I_0 at $t = 0$, (a) series equivalent circuit, (b) parallel equivalent circuit

and so for a parallel arrangement

$$I(s) = \text{current through } L \text{ plus current source } I_0/s \qquad [9]$$

Both these arrangements satisfy equation [7].

For a *capacitor* with no initial potential difference

$$i = C\frac{dv}{dt}$$

The Laplace transform of this equation is

$$I(s) = C[sV(s) - v(0)]$$

Since $v(0) = 0$ then the impedance of the capacitor in the s-domain is (Fig. 11.8)

$$Z(s) = \frac{V(s)}{I(s)} = \frac{1}{sC} \qquad [10]$$

Fig. 11.8 The capacitor element (a) in the time domain, (b) in the s-domain

If the capacitor has an initial potential difference V_0 at $t = 0$, then

$$I(s) = C[sV(s) - V_0] = CsV(s) - CV_0 \qquad [11]$$

An equivalent circuit in the s-domain for a capacitor with an initial potential difference is a capacitor with impedance $1/sC$ in series with an independent voltage source of V_0/s (Fig. 11.9(a)). Rearranging equation [11] gives

$$V(s) = \left(\frac{1}{sC}\right)I(s) + \frac{V_0}{s}$$

and thus for a series arrangement of capacitor and a voltage source

$$V(s) = \text{p.d. across } C + \text{voltage source } V_0/s \qquad [12]$$

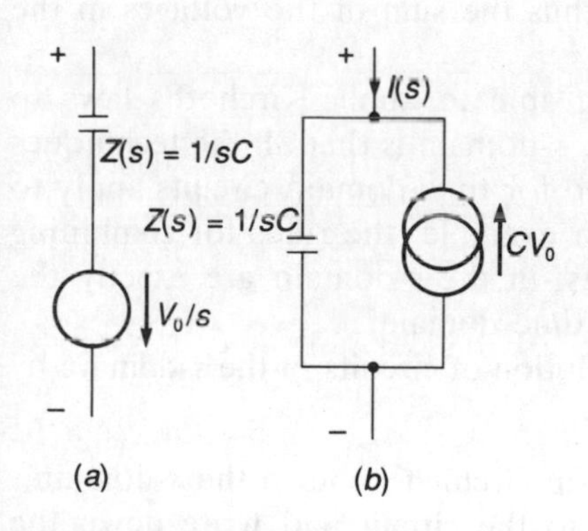

Fig. 11.9 The capacitor element in the s-domain with potential difference V_0 at $t = 0$, (a) series equivalent circuit, (b) parallel equivalent circuit

An alternative equivalent circuit is of a capacitor of impedance $1/sC$ in parallel with an independent current source of $-CV_0$ (Fig. 11.9(b)). Since $V(s)/(1/sC)$ is the current through the capacitor, equation [8] for a parallel circuit gives

$$I(s) = \text{current through } C + \text{current source } (-CV_0) \qquad [13]$$

In general we define *impedance* in the s-domain $Z(s)$ as

$$Z(s) = \frac{V(s)}{I(s)} \qquad [14]$$

The reciprocal of impedance is admittance, therefore *admittance* in the s-domain $Y(s)$ is

$$Y(s) = \frac{I(s)}{V(s)} \qquad [15]$$

A point to notice is the similarity between impedance, or

admittance, expressions in the s-domain and the expressions obtained (see Chapter 3) for sinusoidal signals using complex numbers, i.e. s is similar to $j\omega$. Thus, for example, in the s-domain for a capacitor $Z(s) = 1/sC$ while using complex numbers it is $1/j\omega C$. For this reason s is sometimes called the *complex frequency* and the s-domain the *complex frequency domain*.

Example 7

What is the impedance in the s-domain of a $0.1\,\mu F$ capacitor?

Answer

The impedance is $1/sC$ (equation [10]) and thus

$$Z(s) = \frac{1}{0.1 \times 10^{-6} s}\,\Omega$$

Kirchoff's laws in the s-domain

Kirchoff's laws are applicable to s-domain currents and voltages in exactly the same way as they are used with time-domain currents and voltages. This is because the Laplace transform of the sum of a number of time-domain functions is the sum of the transforms of each function treated independently. Thus, for Kirchoff's current law in the time domain the sum of the currents at a junction is zero. For each current we can obtain the Laplace transform and thus the sum of the transformed currents is also zero. Similarly for Kirchoff's voltage law in the time domain the sum of the voltages around a closed loop is zero and thus the sum of the voltages in the s-domain is also zero.

A consequence of being able to apply Kirchoff's laws to currents and voltages in the s-domain is that all the techniques of circuit analysis developed for time-domain circuits apply to s-domain circuits. Thus, for example, the rules for combining impedances, or admittances, in the s-domain are exactly the same as those used in the time domain.

The procedure for the solution of circuits in the s-domain by means of Kirchoff's laws is:

1. Convert the time-domain circuit to one in the s-domain.
2. Apply Kirchoff's laws to the circuit and write down the equation or equations for the circuit for the elements in the s-domain.
3. Solve the equations to obtain the Laplace transform of the required quantity.
4. Rearrange the equations into a form that can be recognized in the table of Laplace transform pairs.
5. Hence obtain the inverse transformation and so the solution.

Example 8

What is the impedance in the s-domain of a 100 Ω resistor in series with a 4 mH inductor?

Answer

The impedance in the s-domain of the resistor is 100 Ω (equation [6]), while that of the inductor is $sL = 4 \times 10^{-3}s\,\Omega$ (equation [7]). For two impedances in series the total impedance is their sum, hence $100 + 4 \times 10^{-3}s\,\Omega$.

Example 9

What is the admittance in the s-domain of a 100 Ω resistor, a 4 mH inductor and a 0.1 μF capacitor in parallel?

Answer

The admittance in the s-domain of the resistor is $1/R = 0.01$ S (equation [6]), that of the inductor $1/sL = 1/(4 \times 10^{-3}s) = 250/s$ S (equation [7]), and that of the capacitor $sC = 0.1 \times 10^{-6}s$ S (equation [10]). The total admittance is the sum of the three admittances and so

$$0.01 + \frac{250}{s} + 0.1 \times 10^{-6}s = \frac{10^{-7}}{s}(s^2 + 1 \times 10^5 s + 2.5 \times 10^9)\ \text{S}$$

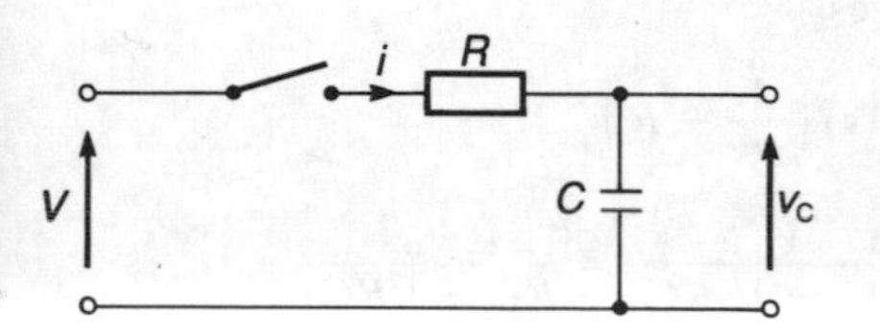

Fig. 11.10 Example 10

Example 10

Derive an expression for the variation with time of the current through and the voltage across the capacitor for the circuit in Fig. 11.10 after the switch is closed and a step voltage V applied to the circuit.

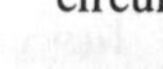

Answer

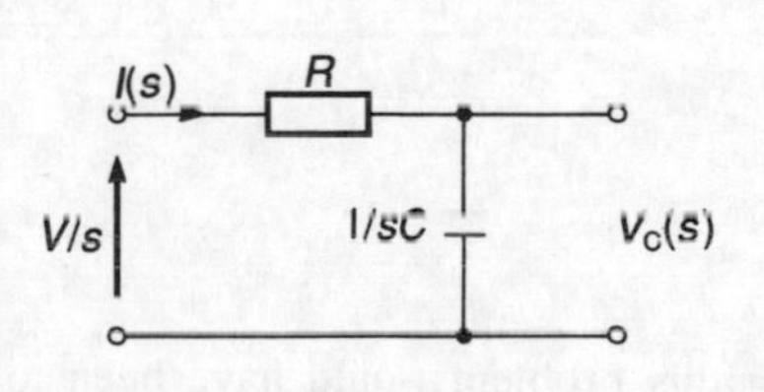

Fig. 11.11 Example 10

Assuming the capacitor is initially uncharged, the circuit in the s-domain is as shown in Fig. 11.11, the step voltage being V/s, the capacitor $1/sC$ and the resistor R. Applying Kirchoff's voltage law to the circuit gives

$$\frac{V}{s} = RI(s) + V_C(s)$$

But $V_C(s) = (1/sC)I(s)$ and so

$$\frac{V}{s} = I(s)\left(R + \frac{1}{sC}\right)$$

$$I(s) = \frac{(V/s)}{R + (1/sC)} = \frac{(V/s)}{R(1 + 1/RCs)} = \frac{V}{R(s + 1/RC)}$$

This is of the form $1/(s + a)$, for which the inverse transform is e^{-at} with $a = 1/RC$. Thus

$$i = \frac{V}{R}e^{-t/RC}$$

The voltage v_C across the capacitor can be found by using

$$v_C = \int_0^t \frac{i}{C}\,dt = \int_0^t \frac{V e^{-t/RC}}{RC}\,dt = \left[\frac{V e^{-t/RC}}{RC(-1/RC)}\right]_0^t = V(1 - e^{-t/RC})$$

Alternatively we can first determine $V_C(s)$. Applying Kirchoff's voltage law to the circuit gives

$$V_C(s) = \frac{V}{s} - RI(s) = \frac{V}{s} - \frac{RV}{R(s + 1/RC)}$$

The inverse transformation of this is

$$v_C = V - V e^{-1/RC} = V(1 - e^{-t/RC})$$

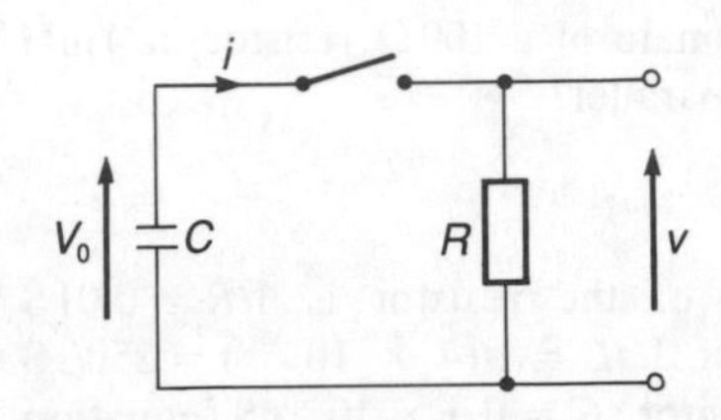

Fig. 11.12 Example 11

Example 11

A charged capacitor, potential difference V_0, is suddenly discharged through a resistor, the circuit being as shown in Fig. 11.12. Determine how the current and voltage vary with time.

Answer

The s-domain equivalent circuit is shown in Fig. 11.13(a). The charged capacitor is represented by a generator of voltage V_0/s in series with an impedance $1/sC$ (equation [12]). Applying Kirchoff's voltage law to the circuit gives

$$\frac{V_0}{s} = \frac{I(s)}{sC} + RI(s) = I(s)\left(\frac{1}{sC} + R\right)$$

$$I(s) = \frac{V_0/s}{R + (1/sC)} = \frac{V_0}{Rs(1 + 1/sRC)} = \frac{V_0}{R(s + 1/RC)}$$

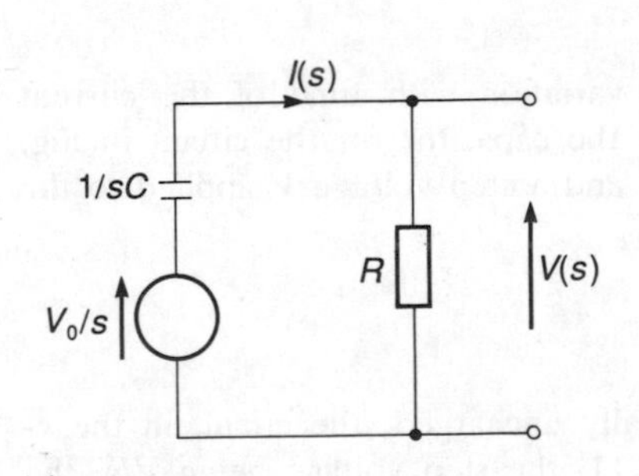

This is of the form $1/(s + a)$ with $a = 1/RC$. Thus the inverse transform is

$$i = \frac{V_0}{R} e^{-t/RC}$$

Since $v = iR$, then

$$v = V_0 e^{-t/RC}$$

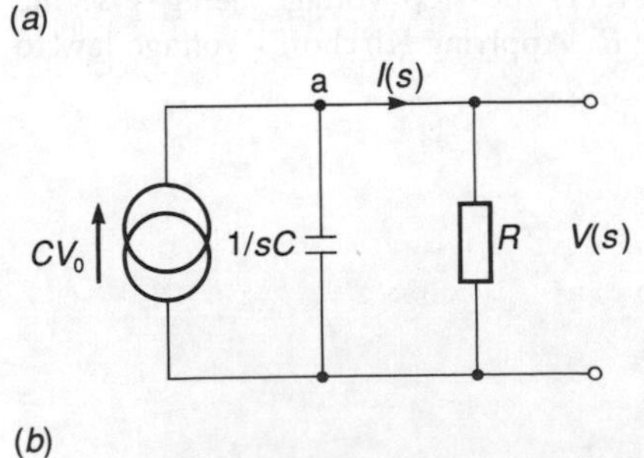

Fig. 11.13 Example 11

An alternative way of tackling this problem would have been to determine the voltage v first by transforming the original circuit into the s-domain by means of the parallel equivalent circuit (using equation [13]). Figure 11.13(b) shows the result, the charged capacitor being replaced by a current source CV_0 and a parallel impedance of $1/sC$. Applying Kirchoff's current law to the circuit gives for node a

$$CV_0 = \frac{V(s)}{(1/sC)} + \frac{V(s)}{R} = V(s)(sC + 1/R)$$

$$V(s) = \frac{CV_0}{sC + 1/R} = \frac{V_0}{s + 1/RC}$$

This is of the form $1/(s + a)$ with $a = 1/RC$. Thus the inverse transform is

$$v = V_0 e^{-t/RC}$$

Example 12

A ramp voltage of $v = kt$ is applied to a circuit consisting of an inductance L in series with a resistance R. If initially at $t = 0$ there is no current, derive an expression for how the current varies with time t.

Answer

In the s-domain a unit ramp voltage of $1t$ is $1/s^2$ (see Table 11.1) and thus a ramp of kt is k/s^2. The inductance has an impedance in the s-domain of sL (equation [7]) and the resistance one of R (equation [6]). Thus applying Kirchoff's voltage law

$$\frac{k}{s^2} = RI(s) + sLI(s)$$

where $I(s)$ is the current through both components. Hence

$$I(s) = \frac{k}{s^2(R + sL)} = \frac{(R/L)(k/R)}{s^2(s + R/L)}$$

This equation is of the form $a/[s^2(s + a)]$ with $a = R/L$, hence using Table 11.1 the inverse transform is

$$i = \frac{k}{R}\left[t - \frac{(1 - e^{-Rt/L})}{(R/L)}\right]$$

$$= \frac{kt}{R} - \frac{kL}{R^2} + \frac{kL}{R^2}e^{-Rt/L}$$

Example 13

A sinusoidal voltage of $V \sin \omega t$ is applied at time $t = 0$ to a circuit consisting of an inductance L in series with a resistance R. Derive an expression indicating how the current varies with time.

Answer

The sinusoidal voltage input in the s-domain is, using Table 11.1, $V\omega/(s^2 + \omega^2)$, the impedance of the inductance is sL (equation [7]) and that of the resistance R (equation [6]). Thus, applying Kirchoff's voltage law to the circuit

$$\frac{V\omega}{s^2 + \omega^2} = RI(s) + sLI(s)$$

where $I(s)$ is the current through both components. Hence

$$I(s) = \frac{V\omega}{(s^2 + \omega^2)(R + sL)} = \frac{V\omega/L}{(s^2 + \omega^2)(s + R/L)}$$

Partial fractions can be used to put the above expression into a form for which Table 11.1 can be used to obtain the inverse transform.

$$\frac{V\omega/L}{(s^2 + \omega^2)(s + R/L)} = \frac{A}{s + R/L} + \frac{Bs + C}{s^2 + \omega^2}$$

$$V\omega/L = As^2 + A\omega^2 + Bs^2 + B(R/L)s + Cs + C(R/L)$$

Equating values of s^2 gives

$$A + B = 0$$

Hence $A = -B$. Equating values of s gives

$$B(R/L) + C = 0$$

Hence $C = -BR/L = AR/L$. Equating values of numbers gives

$$V\omega/L = A\omega^2 + C(R/L)$$

Thus, substituting for C,

$$V\omega/L = A\omega^2 + A(R/L)(R/L)$$

$$A = \frac{V\omega L}{R^2 + \omega^2 L^2}$$

Hence

$$B = \frac{-V\omega L}{R^2 + \omega^2 L^2}$$

$$C = \frac{V\omega R}{R^2 + \omega^2 L^2}$$

Thus

$$I(s) = \frac{\left(\dfrac{V\omega L}{R^2 + \omega^2 L^2}\right)}{s + R/L} - \frac{\left(\dfrac{V\omega L}{R^2 + \omega^2 L^2}\right)s + \left(\dfrac{V\omega R}{R^2 + \omega^2 L^2}\right)}{s^2 + \omega^2}$$

$$= \frac{V\omega}{R^2 + \omega^2 L^2}\left(\frac{L}{s + R/L} - \frac{sL}{s^2 + \omega^2} + \frac{R}{s^2 + \omega^2}\right)$$

Using Table 11.1, the inverse transform is thus

$$i = \frac{V\omega}{R^2 + \omega^2 L^2}\left(L\mathrm{e}^{-Rt/L} - L\cos\omega t + \frac{R}{\omega}\sin\omega t\right)$$

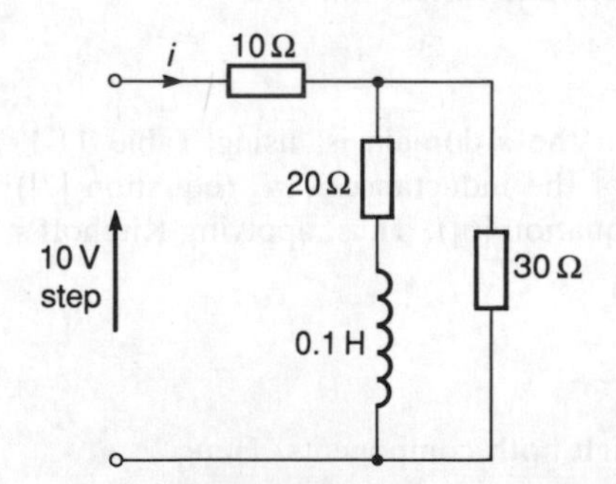

Fig. 11.14 Example 14

Example 14

Derive an expression showing how the current in the circuit in Fig. 11.14 will vary with time when there is a 10 V step input to the circuit.

Answer

In the s-domain, the total circuit impedance $Z(s)$ is, using the conventional equations for summing impedances in series and parallel,

$$Z(s) = 10 + \frac{30(20 + 0.1s)}{30 + 20 + 0.1s} = \frac{300 + 200 + 1s + 600 + 3s}{50 + 0.1s}$$

$$= \frac{1100 + 4s}{50 + 0.1s}$$

In the s-domain the step voltage is $10/s$. Thus the s-domain circuit consists of just such a source connected to the above impedance $Z(s)$, as in Fig. 11.15. Thus

$$I(s) = \frac{10/s}{(1100 + 4s)/(50 + 0.1s)} = \frac{500 + 1s}{s(1100 + 4s)}$$

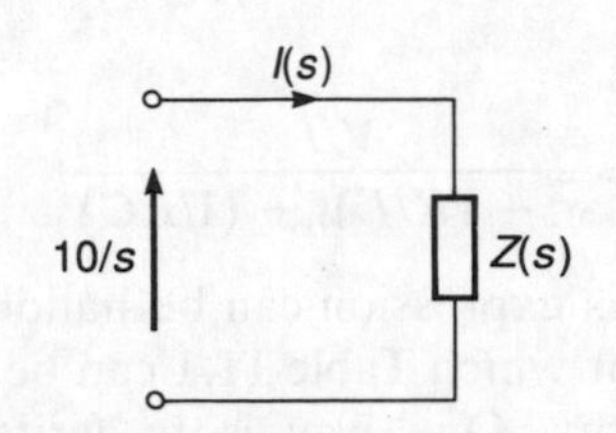

Fig. 11.15 Example 14

$$= \frac{500}{s(1100 + 4s)} + \frac{1}{(1100 + 4s)}$$

$$= \frac{125}{s(s + 275)} + \frac{0.25}{(s + 275)}$$

Table 11.1 indicates that the inverse transform of $1/[s(s + a)]$ is $(1 - e^{-at})/a$ and for $1/(s + a)$ is e^{-at}. Thus the inverse transform of the above is

$$i = \frac{125}{275}(1 - e^{-275t}) + 0.25\,e^{-275t}$$

$$= 0.45 - 0.20\,e^{-275t}$$

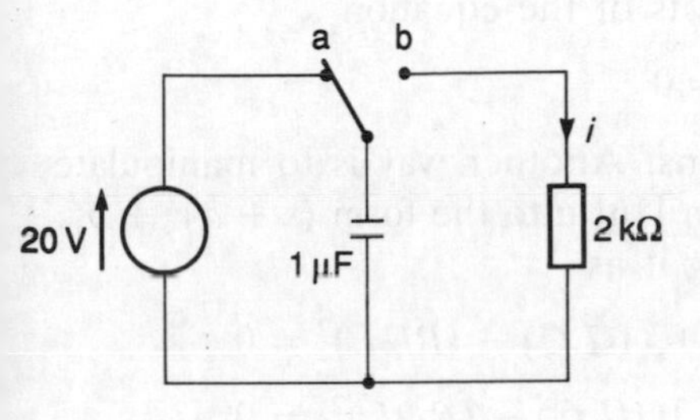

Fig. 11.16 Example 15

Example 15

The switch in the circuit shown in Fig. 11.16 has been in position a for a long time. Derive an equation describing how the current i through the resistor will vary with time when the switch is moved from a to b.

Answer

The capacitor will have become fully charged and so will have a potential difference between its terminals of 20 V. When the switch is moved from a to b then the charged capacitor can be considered to be a voltage generator of 20/s voltage. The discharge circuit is thus as shown in Fig. 11.17. Applying Kirchoff's voltage law to the circuit in the s-domain,

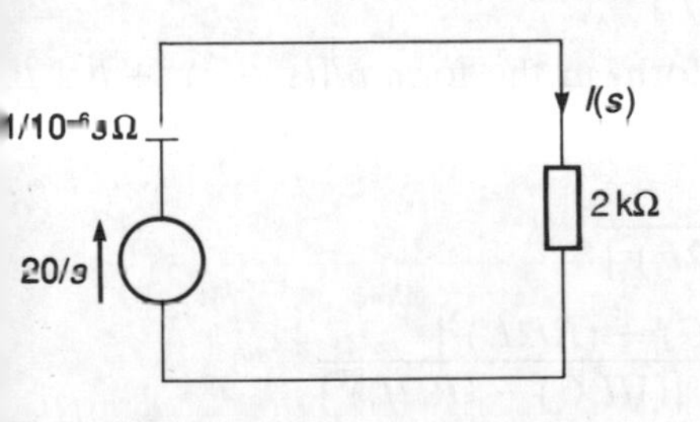

Fig. 11.17 Example 15

$$\frac{20}{s} = I(s)\left(2000 + \frac{1}{10^{-6}s}\right)$$

$$I(s) = \frac{(20/s)}{2000 + 10^6/s} = \frac{20}{2000s + 10^6} = \frac{0.01}{s + 500}$$

This is of the form $1/(s + a)$ with $a = 500$. Table 11.1 gives the inverse of this to be e^{-at}. Hence

$$i = 0.01\,e^{-500t}\text{ A}$$

RLC circuits

Consider a circuit consisting of resistance, inductance and capacitance in series (Fig. 11.18). In the s-domain the series arrangement has an impedance of

$$Z(s) = R + sL + \frac{1}{sC}$$

Hence if there is a step input voltage V at time $t = 0$, the circuit being at zero conditions, then since the step voltage in the s-domain is V/s

$$\frac{V}{s} = I(s)\left(R + sL + \frac{1}{sC}\right) \qquad [16]$$

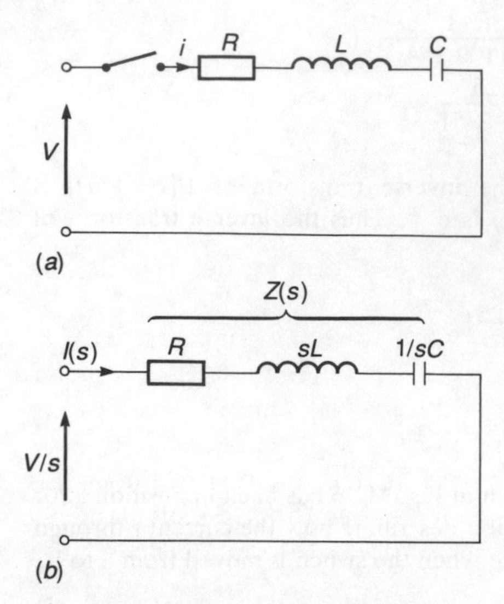

Fig. 11.18 Series *RLC* circuit, (*a*) in the time domain, (*b*) in the *s*-domain

Hence

$$I(s) = \frac{V}{s(R + sL + 1/sC)} = \frac{V/L}{s^2 + (R/L)s + (1/LC)} \quad [17]$$

There are a number of ways this expression can be handled in order to obtain an expression for which Table 11.1 can be used to obtain the inverse transform. One way is to write the equation in the form

$$I(s) = \frac{V/L}{(s - p_1)(s - p_2)} \quad [18]$$

where p_1 and p_2 are the roots of the equation

$$s^2 + (R/L)s + (1/LC) = 0$$

and then use partial fractions. Another way is to manipulate the denominator of equation [16] into the form $(s + a)^2 + b^2$. This can be done by writing it as

$$s^2 + (R/L)s + (R/2L)^2 + (1/LC) - (R/2L)^2 = 0$$

$$(s + R/2L)^2 + [(1/LC) - (R/2L)^2] = 0$$

Thus

$$a = R/2L$$

$$b = \sqrt{[(1/LC) - (R/2L)^2]}$$

Table 11.1 gives for a transform in the form $b/[(s + a)^2 + b^2]$ the inverse of $e^{-at} \sin bt$. Putting equation [16] into this form

$$I(s) = \frac{(V/L)}{\sqrt{[(1/LC) - (R/2L)^2]}}$$

$$\frac{\sqrt{[(1/LC) - (R/2L)^2]}}{(s + R/2L)^2 + [(1/LC) - (R/2L)^2]}$$

results in the inverse transform of

$$i = \frac{(V/L)}{\sqrt{[(1/LC) - (R/2L)^2]}}$$

$$e^{-Rt/2L} \sin \sqrt{[(1/LC) - (R/2L)^2]}t \quad [19]$$

The circuit is *underdamped* when $(R/2L)^2$ is less than $(1/LC)$, *critically damped* when $(R/2L)^2$ equals $(1/LC)$ and *overdamped* when $(R/2L)^2$ is greater than $(1/LC)$ (see Chapter 2).

The above is the situation with a step input to the *RLC* circuit. A similar analysis can be carried out for other forms of input. The same conditions exist for determining whether the motion is underdamped or overdamped. Thus where the input is a sinusoidal voltage of $V \sin \omega t$, then since the Laplace transform of this voltage is $V\omega/(s^2 + \omega^2)$ equation [16] becomes

$$\frac{V\omega}{s^2 + \omega^2} = I(s)\left(R + sL + \frac{1}{sC}\right) \qquad [20]$$

$$I(s) = \frac{V\omega}{(s^2 + \omega^2)(R + sL + 1/sC)}$$

$$= \frac{(V/L)\omega s}{(s^2 + \omega^2)[s^2 + (R/L)s + (1/LC)]}$$

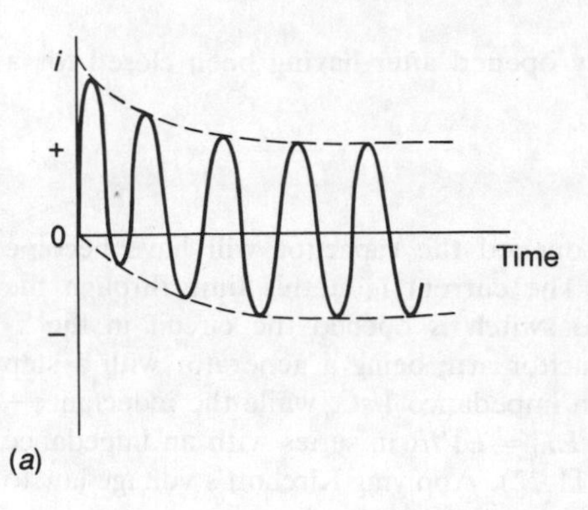

(a)

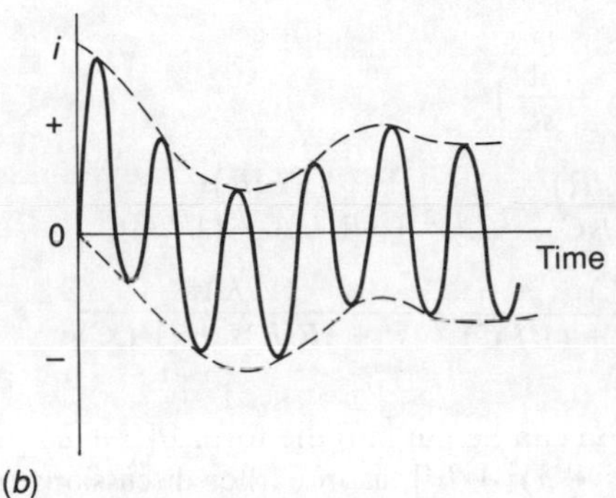

(b)

Fig. 11.19 (*a*) Over-damped, (*b*) under-damped

This can be put into a more suitable form for determining the inverse by the use of partial fractions. Figure 11.19 shows the type of results obtained. With overdamping the current oscillates with the oscillation slowly attaining its steady state. At critical damping it attains the steady state in the shortest time. With underdamping the current oscillates with two frequencies, one due to the source and one due to the circuit elements. Eventually the oscillation due to the circuit elements dies away to leave just the steady state oscillation due to the source.

Example 16

For the circuit shown in Fig. 11.20, determine the equation describing how the current varies with time when the switch is closed if the initial conditions are zero.

Answer

Closing the switch can be considered to apply a step voltage to the circuit. Thus applying Kirchoff's voltage law to the circuit in the s-domain

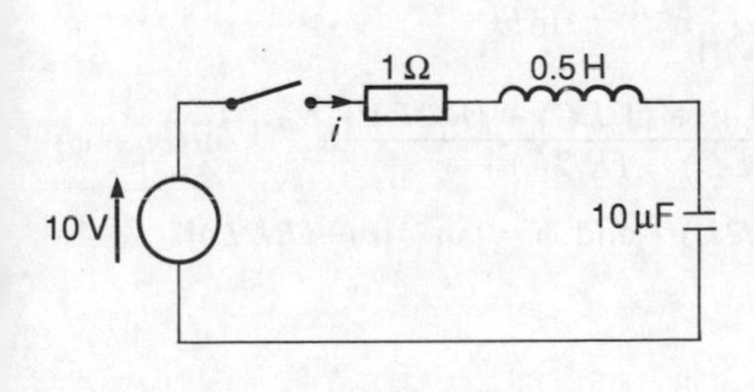

Fig. 11.20 Example 16

$$\frac{10}{s} = I(s)\left(1 + 0.5s + \frac{1}{10^{-5}s}\right)$$

$$I(s) = \frac{1}{0.1s + 0.05s^2 + 10^4} = \frac{20}{s^2 + 2s + 2 \times 10^5}$$

Putting the equation into the form $b/[(s + a)^2 + b^2]$

$$I(s) = \frac{20}{s^2 + 2s + 1 + 2 \times 10^5 - 1} = \frac{20}{(s + 1)^2 + (2 \times 10^5 - 1)}$$

$$= \frac{20}{(2 \times 10^5 - 1)} \frac{1}{(s + 1)^2 + (2 \times 10^5 - 1)}$$

Table 11.1 gives for a transform in the form $b/[(s + a)^2 + b^2]$ the inverse of $e^{-at} \sin bt$. Hence

$$i = \frac{20}{2 \times 10^5 - 1} e^{-1t} \sin (2 \times 10^5 - 1)t$$

$$\approx 10^{-4} e^{-t} \sin 2 \times 10^5 t$$

Example 17

Derive an equation for the current i in the circuit shown in Fig. 11.21

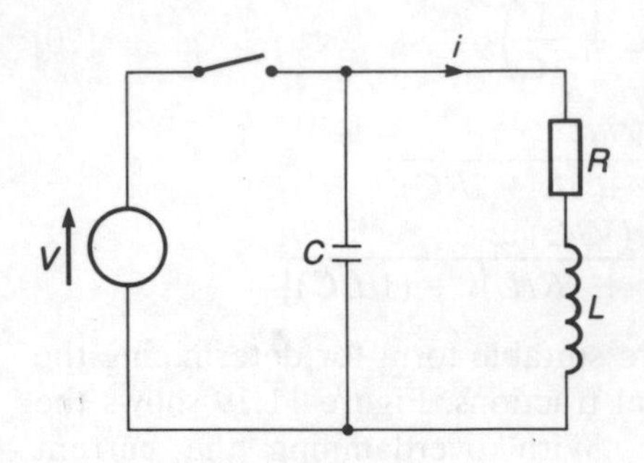

Fig. 11.21 Example 17

when the switch is suddenly opened after having been closed for a long time.

Answer

Prior to the switch being opened the capacitor will have become charged to the voltage V. The current I_0 at this time through the inductor is V/R. When the switch is opened the circuit in the s-domain consists of the capacitor arm being a generator with a step voltage V/s in series with an impedance $1/sC$, while the inductance–resistor arm is a generator $LI_0 = LV/R$ in series with an impedance sL and a resistance R (Fig. 11.22). Applying Kirchoff's voltage law to this s-domain circuit gives

$$\frac{V}{s} + \frac{LV}{R} = I(s)\left(R + sL + \frac{1}{sC}\right)$$

$$I(s) = \frac{(V/s) + (VL/R)}{R + sL + (1/sC)} = \frac{V + (VL/R)s}{L[s^2 + (R/L)s + (1/LC)]}$$

$$= \frac{(V/L)}{s^2 + (R/L)s + (1/LC)} + \frac{(V/R)s}{s^2 + (R/L)s + (1/LC)}$$

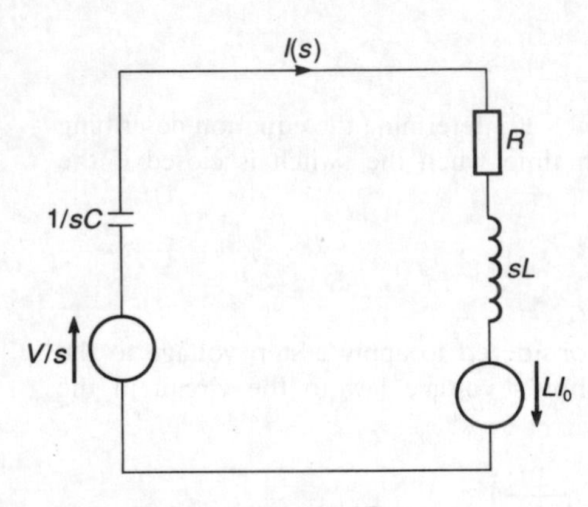

Fig. 11.22 Example 17

The first part of the expression can be put into the form $1/[(s + a)^2 + b^2)$ and the second part $s/[(s + a)^2 + b^2]$, as in earlier discussions of the series RLC circuit.

$$I(s) = \frac{(V/L)}{(s + R/2L)^2 + (1/LC) - (R/2L)^2}$$

$$+ \frac{(V/R)s}{(s + R/2L)^2 + (1/LC) - (R/2L)^2}$$

Hence, using Table 11.1,

$$i = \frac{(V/L)}{\sqrt{[(1/LC) - (R/2L)^2]}} e^{-Rt/2L} \sin bt$$

$$+ \frac{(V/R)\sqrt{[(R/2L)^2 + (1/LC) - (R/2L)^2]}}{\sqrt{[(1/LC) - (R/2L)^2]}} e^{-Rt/L} \sin (bt + \phi)$$

where $b = \sqrt{[(1/LC) - (R/2L)^2]}$ and $\phi = \tan^{-1}[b/-(R/2L)]$.

Problems

1 Determine from first principles the Laplace transformation of the function e^{-at}.

2 Determine, using Table 11.1, the Laplace transforms for:
   (*a*) a step voltage of size 6 V which starts at $t = 0$ s,
   (*b*) a step voltage of size 6 V which starts at $t = 3$ s,
   (*c*) a ramp voltage of 6 V/s which starts at $t = 0$ s,
   (*d*) a ramp voltage of 6 V/s which starts at $t = 3$ s,
   (*e*) an impulse of size 6 V at $t = 0$ s,
   (*f*) an impulse of size 6 V at $t = 3$ s,
   (*g*) a sinusoidal signal of amplitude 6 V and frequency 50 Hz which starts at $t = 0$ s.

3 Determine, using Table 11.1, the Laplace transforms for (*a*) e^{-2t}, (*b*) $5e^{-2t}$, (*c*) $V_0 e^{-t/\tau}$, (*d*) $1 - e^{-2t}$, (*e*) $5(1 - e^{-2t})$, (*f*) $V_0(1 - e^{-t/\tau})$.

4 Determine, using Table 11.1, the inverse of the following Laplace transforms: (*a*) $2/(s + 3)$, (*b*) $2/(3s + 1)$, (*c*) $2/s(s + 3)$, (*d*) $2/s(3s + 1)$.

5 Determine the Laplace transforms of the following voltages which vary with time according to the given equations:
(*a*) $v = 5(1 - e^{-t/50})$,
(*b*) $v = 10 + 5(1 - e^{-t/50})$,
(*c*) $v = 5e^{-t/50}$.

6 Using the final and initial value theorems, what are the final and initial values of the signals which gave the following Laplace transforms?

$$(a)\ \frac{5}{s},\quad (b)\ \frac{5}{s(s+2)}.$$

7 Using partial fractions, determine the signal variation with time which gave the following Laplace transforms:

$$(a)\ \frac{4s-5}{s^2-s-2},\quad (b)\ \frac{6s+8}{s(s+1)(s+2)},\quad (c)\ \frac{1}{s^2+3s+2}.$$

8 What are the impedances in the *s*-domain of (*a*) a 1 kΩ resistance, (*b*) a 0.5 H inductance, (*c*) a 2 μF capacitance?

9 What are the impedances in the *s*-domain of (*a*) a resistance of 100 Ω in series with an inductance of 10 mH, (*b*) a resistance of 1 kΩ in series with a capacitance of 10 μF, (*c*) a resistance of 1 kΩ, an inductance of 10 mH and a capacitance of 10 μF in series, (*d*) a resistance of 100 Ω in parallel with an inductance of 10 mH, (*e*) an inductance of 10 mH in parallel with a capacitance of 10 μF?

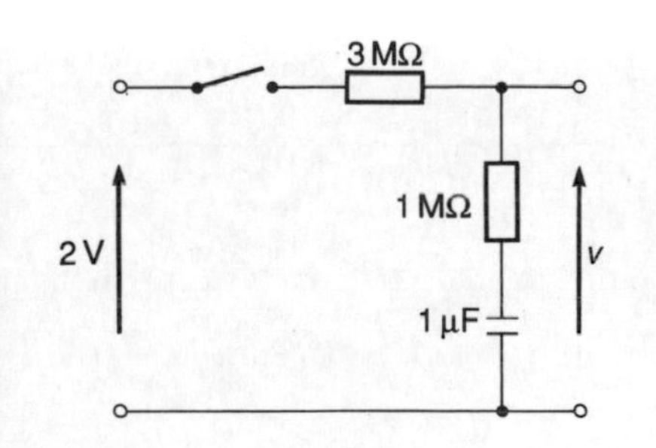

Fig. 11.23 Problem 11

10 Determine how the current varies with time for a circuit consisting of a coil with resistance *R* and inductance *L* when it is suddenly connected to a voltage *V*.

11 Determine how, for the circuit in Fig. 11.23, the potential difference *v* varies with time when the switch is closed. The initial conditions are zero.

12 A circuit consists of a 1 MΩ resistor in series with a 0.1 μF capacitor. Determine how the voltage across the capacitor varies with time when there is a step voltage input of 6 V. The initial conditions are zero.

13 Determine how the current *i* and the voltage *v* will vary with time for the circuit shown in Fig. 11.24 when the switch is closed.

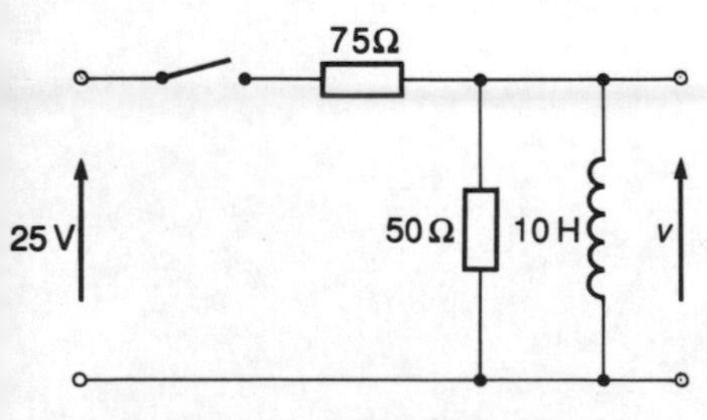

Fig. 11.24 Problem 13

14 A coil of resistance *R* and inductance *L* is in parallel with a capacitance *C*. If the initial conditions are zero, derive a relationship for the current variation with time when a ramp input voltage of $v = kt$ is applied to the combination. The initial conditions are zero.

15 A coil of resistance 1 Ω and inductance 0.5 H is in series with a capacitor of 2 mF. Derive a relationship showing how the current through the circuit will change with time when a voltage of 10 V is suddenly applied to the circuit. The initial conditions are zero.

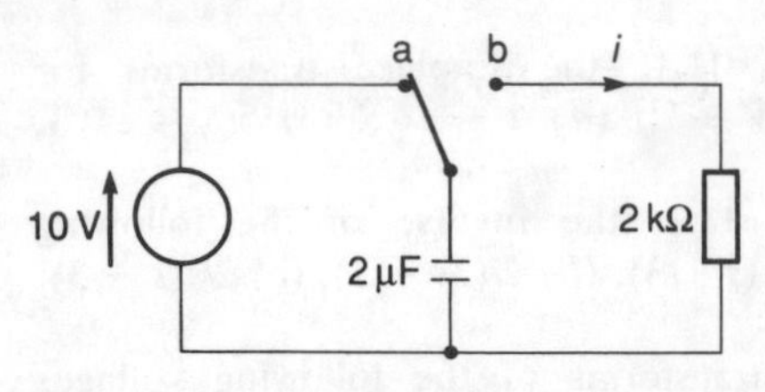

Fig. 11.25 Problem 17

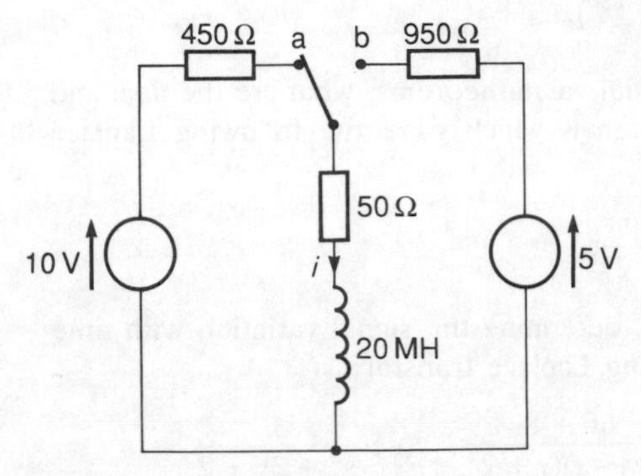

Fig. 11.26 Problem 19

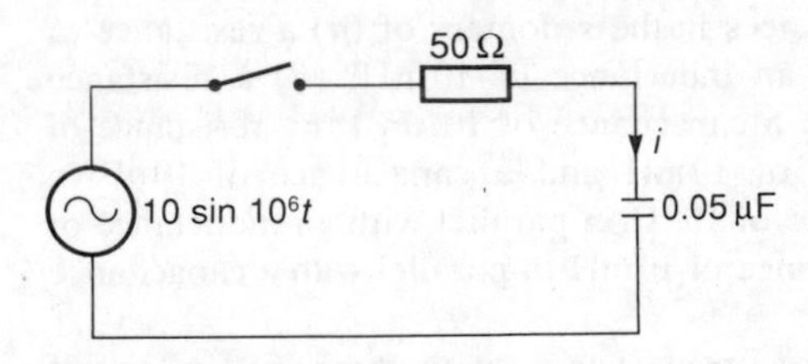

Fig. 11.27 Problem 20

16 A 1.25 kΩ resistor, a 0.25 H inductor and a 1 μF capacitor are connected in series. How does the potential difference across the capacitor vary with time when a step voltage of 30 V is applied to the circuit?

17 The switch in the circuit shown in Fig. 11.25 has been in position a for a long time. Derive an equation describing how the current i through the resistor will vary with time when the switch is moved from a to b.

18 Repeat problem 17 with the 2 kΩ resistor replaced by a 0.5 H inductor.

19 Derive an equation describing how the current i varies with time when the switch is moved from position a to b, after being a long time at a, in the circuit described by Fig. 11.26.

20 Determine how the current varies with time for the circuit shown in Fig. 11.27 when the switch is closed and the sinusoidal voltage $10 \sin 10^6 t$ V is applied. The capacitor is initially uncharged.

21 Determine how the current varies with time for a series RLC circuit when a sinusoidal voltage of $10 \sin 10^5 t$ V is switched on. The resistance is 200 Ω, the inductance 5 mH and the capacitance 0.5 μF. The initial conditions are zero.

12 Complex waveforms

Introduction

In earlier chapters in this book, circuit analysis has been considered for sinusoidal signals. However, many electrical waveforms encountered in electrical and electronic equipment and circuits, e.g. a square waveform, are not sinusoidal. Some might have started out as sinusoidal but become distorted due to non-linear components. However, all repetitive electrical waveforms can be considered to be made up of a combination of sinusoidal waveforms of various frequencies and amplitudes. This is the basis of what is called *Fourier's theorem*. Thus the square waveform can be considered to be made up from a combination of sinusoidal waveforms. This technique of considering a repetitive waveform to be made up of a number of sinusoidal waveforms enables circuit analysis to be carried out.

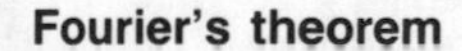

Fourier's theorem

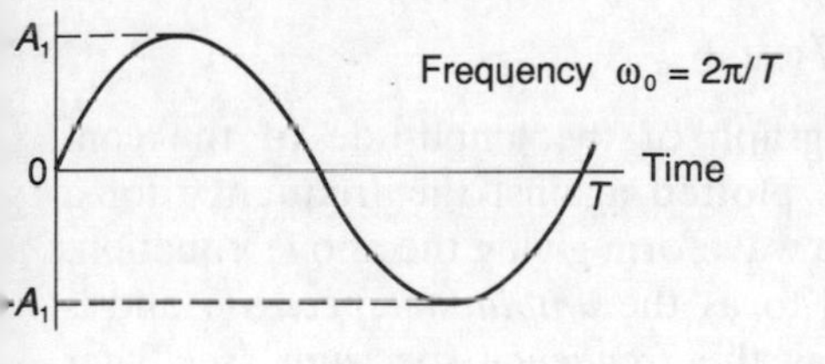

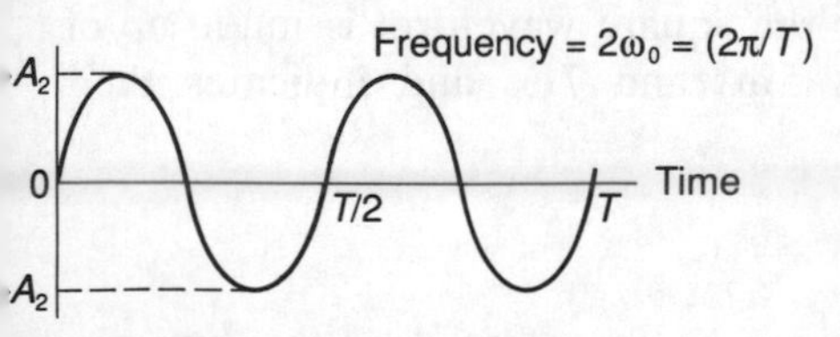

Fig. 12.1 Sinusoidal signals, (*a*) amplitude A_1 frequency ω_0, (*b*) amplitude A_2 frequency $2\,\omega_0$

A periodic signal is one that repeats itself at regular intervals, the time between successive repetitions being called the *periodic time T*. A sinusoidal signal (Fig. 12.1) can be represented by

$$v = A_1 \sin \omega_0 t$$

where A_1 is the amplitude of the signal and ω_0, the angular frequency, is $2\pi/T$. A signal with twice the frequency, and a different amplitude A_2, may be represented by

$$v = A_2 \sin 2\,\omega_0 t$$

A signal with three times the frequency, and a different amplitude A_3, may be represented by

$$v = A_3 \sin 3\,\omega_0 t$$

The above equations all describe sinusoidal signals that have started off with $v = 0$ at $t = 0$. When this is not the case we can represent the signals by

$$v = A_1 \sin(\omega_0 t + \phi_1)$$

$$v = A_2 \sin(2\omega_0 t + \phi_2)$$

$$v = A_3 \sin(3\omega_0 t + \phi_3)$$

where ϕ_1, ϕ_2, and ϕ_3 are the phase angles with respect to $t = 0$.

According to *Fourier's theorem* we can consider any periodic signal to be made up of a combination of sinusoidal waves. Thus we can consider a periodic signal to be presented by

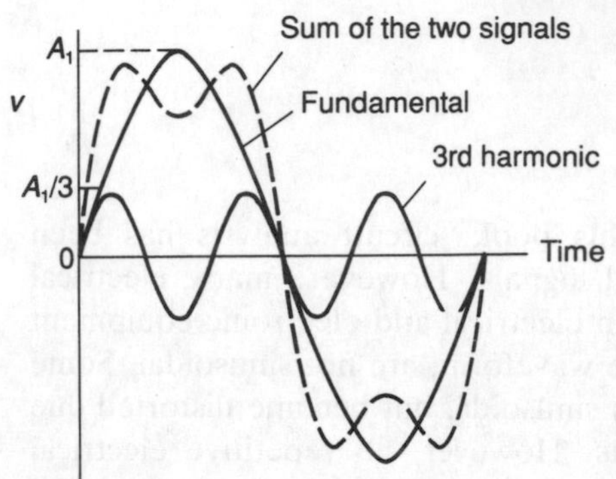

Fig. 12.2 $v = A_1 \sin \omega_0 t + (1/3) A_1 \sin 3\omega_0 t$

$$v = A_1 \sin(\omega_0 t + \phi_1) + A_2 \sin(2\omega_0 t + \phi_2) + A_3 \sin(3\omega_0 t + \phi_3) + \ldots A_n \sin(n\omega_0 t + \phi_n) \quad [1]$$

If the signal includes a d.c. component A_0 then

$$v = A_0 + A_1 \sin(\omega_0 t + \phi_1) + A_2 \sin(2\omega_0 t + \phi_2) + A_3 \sin(3\omega_0 t + \phi_3) + \ldots A_n \sin(n\omega_0 t + \phi_n) \quad [2]$$

where $\omega_0 = 2\pi/T$ and is called the *fundamental frequency* or *first harmonic*. The frequency $2\omega_0$ is called the *second harmonic*, $3\omega_0$ the *third harmonic* and so on to $n\omega_0$ as the nth *harmonic*. A_1, A_2, etc. are the amplitudes of the various harmonics.

Figure 12.2 shows how an almost square waveform can be built up with the fundamental and the third harmonic, the amplitude of the third harmonic being one-third that of the fundamental. For such a wave we can write

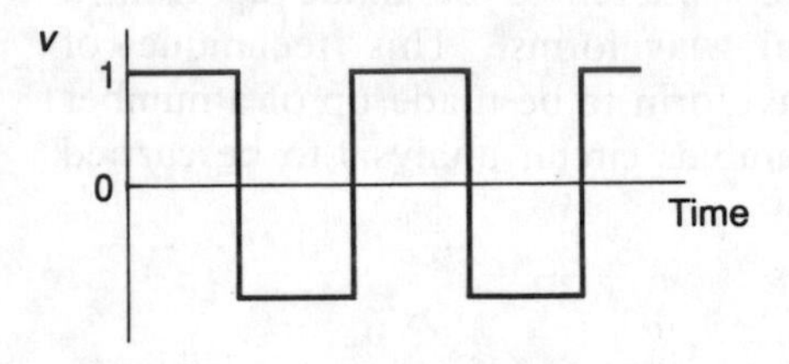

$$v = A_1 \sin \omega_0 t + (1/3)A_1 \sin 3\omega_0 t$$

A better approximation to a square waveform is given by adding higher odd harmonics,

$$v = A_1 \sin \omega_0 t + (1/3)A_1 \sin 3\omega_0 t + (1/5)A_1 \sin 5\omega_0 t + (1/7)A_1 \sin 7\omega_0 t + \ldots$$

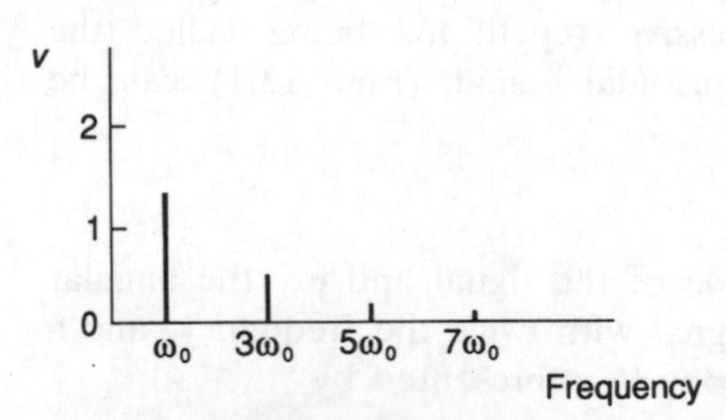

Fig. 12.3 Spectrum for a square wave

Figure 12.3 shows a graph of the amplitude of the constituent sinusoidal waves plotted against the frequency for a square waveform, i.e. the waveform giving the above equation. Such a graph is referred to as the *amplitude spectrum* and is part of what is known as the *frequency spectrum* (see later definition). It shows that the square waveform is made up of frequencies at ω_0, $3\omega_0$, $5\omega_0$ and $7\omega_0$ and indicates their amplitudes.

The Fourier series

Any repetitive waveform can be represented by the sum of a fundamental and a number of harmonics, i.e. their frequencies are integral multiples of some fundamental. Such a sequence of terms is known as a Fourier series. Thus a *Fourier series* can be defined as a sum of sinusoidal waveforms which are harmonically related. Each constituent waveform has its own magnitude and phase shift. The series can be written, in general, for some function of time as

$$f(t) = a_0 + a_1 \cos \omega t + a_2 \cos 2\omega t + \dots a_n \cos n\omega t + \dots b_1 \sin \omega t + b_2 \sin 2\omega t + \dots b_n \sin n\omega t \quad [3]$$

This can be written as

$$f(t) = a_0 + \sum_{n=1}^{n=\infty} (a_n \cos n\omega t + b_n \sin n\omega t) \quad [4]$$

where a_0, a_n and b_n are known as the *Fourier coefficients* and n is the harmonic order. The reason for writing each harmonic as the sum of cosine and sine terms is that the sum of two such terms is just a sinusoidal waveform with some phase angle (Fig. 12.4) and thus the summation is of the sinusoidal waveforms each with their own magnitude and phase. The series is thus sometimes written as

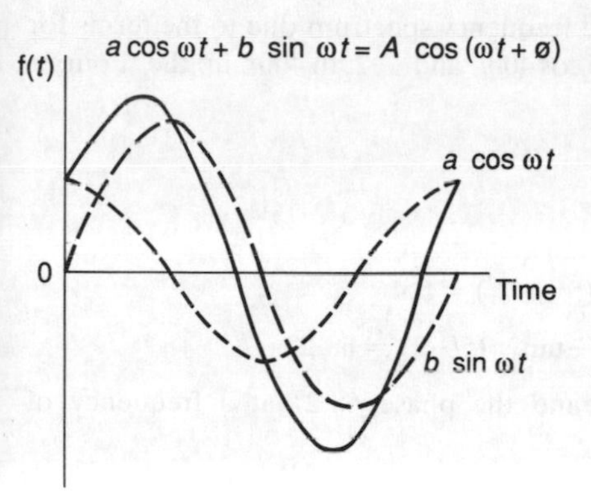

Fig. 12.4 $a \cos \omega t + b \sin \omega t$

$$f(t) = A_0 + \sum_{n=1}^{n=\infty} A_n \cos (n\omega t + \phi_n) \quad [5]$$

The d.c. component in equation [3] is a_0 and this equals the d.c. component A_0 in equation [5]. For the first order we must also have

$$a_1 \cos \omega t + b_1 \sin \omega t = A_1 \cos (\omega t + \phi_1)$$

$$= A_1(\cos \omega t \cos \phi_1 - \sin \omega t \sin \phi_1)$$

Hence we must have $a_1 = A_1 \cos \phi_1$ and $b_1 = -A_1 \sin \phi_1$. Thus

$$a_1^2 + b_1^2 = A_1^2(\cos^2 \phi_1 + \sin^2 \phi_1) = A_1^2$$

$$A_1 = \sqrt{(a_1^2 + b_1^2)}$$

and

$$\frac{b_1}{a_1} = -\tan \phi_1$$

In general,

$$a_n = A_n \cos \phi_n \quad [6]$$

$$b_n = -A_n \sin \phi_n \quad [7]$$

and

$$A_n = \sqrt{(a_n^2 + b_n^2)} \quad [8]$$

$$\phi_n = -\tan^{-1} (b_n/a_n) \quad [9]$$

The collection of all the A_n values as a function of the frequency is called the *amplitude spectrum* and the collection of all the ϕ_n values the *phase spectrum*. The two together constitute the *frequency spectrum*.

Example 1

In terms of the amplitude and phase one of the terms in a Fourier series is $10 \cos(300t + 30°)$. What would be the equivalent terms in the Fourier series written in the form of a pair of sines and cosines?

Answer

Using equations [6] and [7]

$$a_n = A_n \cos \phi_n = 10 \cos 30° = 8.7$$
$$b_n = -A_n \sin \phi_n = -10 \sin 30° = -5.0$$

Hence the equivalent is $8.7 \cos 300t - 5.0 \sin 300t$

Example 2

What are the elements in the frequency spectrum due to the terms for a particular harmonic of $5 \cos 400t$ and $-2 \sin 400t$ in the Fourier series?

Answer

Using equations [8] and [9]

$$A_n = \sqrt{(a_n^2 + b_n^2)} = \sqrt{(5^2 + 2^2)} = 5.4$$
$$\phi_n = -\tan^{-1}(b_n/a_n) = -\tan^{-1}(5/-2) = 68.2°$$

Thus the amplitude is 5.4 and the phase 68.2° at a frequency of 400 rad/s.

The Fourier coefficients

Consider the Fourier series in the sine–cosine form (equation [3])

$$f(t) = a_0 + a_1 \cos \omega t + a_2 \cos 2\,\omega t + \ldots a_n \cos n\,\omega t + \ldots b_1 \sin \omega t + b_2 \sin 2\,\omega t + \ldots b_n \sin n\,\omega t$$

Now consider the effect of integrating both sides of the equation over one period T of the fundamental. Since such an integral for each cosine and sine term is just the area under the graph of that expression for the time T then since each of the sine and cosine terms covers a whole number of cycles in the time T the negative parts of their cycles will cancel out the positive parts and they will all be zero (Fig. 12.5). A consequence of this is that the only term which is not zero is the integral of the a_0 term. Thus, integrating from some arbitrary time t_0 over one period T gives

$$\int_{t_0}^{t_0+T} f(t)\,\mathrm{d}t = a_0 \int_{t_0}^{t_0+T} \mathrm{d}t = a_0 T$$

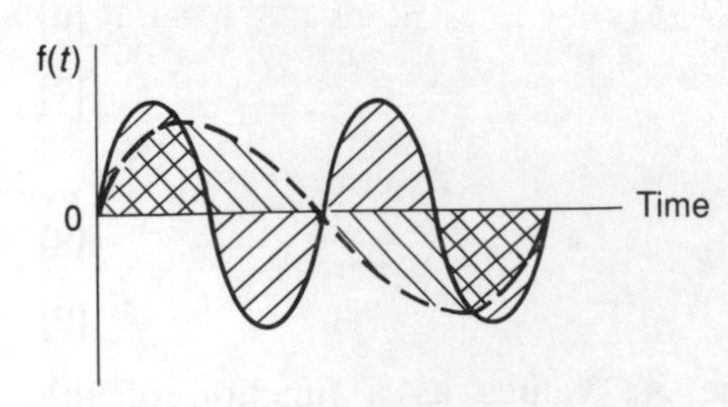

Fig. 12.5 Fundamental and second harmonic

Hence

$$a_0 = \frac{1}{T}\int_{t_0}^{t_0+T} f(t)\,\mathrm{d}t \qquad [10]$$

We can obtain the term a_1 by multiplying the series by $\cos \omega t$ and then integrating over one period T.

$$\begin{aligned} f(t)\cos\omega t &= a_0\cos\omega t + a_1\cos\omega t\cos\omega t + a_2\cos\omega t\cos 2\,\omega t \\ &\quad + \ldots\ b_1\sin\omega t\cos\omega t + b_2\cos\omega t\sin 2\,\omega t \\ &\quad + \ldots \\ &= a_0\cos\omega t + a_1\cos^2\omega t + a_2\cos\omega t\cos 2\,\omega t \\ &\quad + \ldots\ b_1\tfrac{1}{2}\sin 2\,\omega t + b_2\cos\omega t\sin 2\,\omega t \\ &\quad + \ldots \end{aligned}$$

The integration over a period T of all the terms involving $\sin \omega t$ and $\cos \omega t$ will be zero. The integration over one period of $\sin 2\,\omega t$ is also zero. This thus leaves

$$\begin{aligned} \int_{t_0}^{t_0+T} f(t)\cos\omega t\,\mathrm{d}t &= a_1\int_{t_0}^{t_0+T} \cos^2\omega t\,\mathrm{d}t \\ &= a_1\int_{t_0}^{t_0+T} \tfrac{1}{2}(\cos 2\,\omega t + 1)\,\mathrm{d}t \\ &= a_1\int_{t_0}^{t_0+T} \tfrac{1}{2}\cos 2\,\omega t\,\mathrm{d}t + a_1\int_{t_0}^{t_0+T} \tfrac{1}{2}\,\mathrm{d}t \\ &= 0 + \tfrac{1}{2}a_1 T \end{aligned}$$

Thus

$$a_1 = \frac{2}{T}\int_{t_0}^{t_0+T} f(t)\cos\omega t\,\mathrm{d}t \qquad [11]$$

By multiplying the series by $\sin \omega t$ the value of b_1 could have been obtained.

$$b_1 = \frac{2}{T}\int_{t_0}^{t_0+T} f(t)\sin\omega t\,\mathrm{d}t \qquad [12]$$

In general the terms in the series are obtained by multiplying it by $\cos n\omega t$ and $\sin n\omega t$, where $n = 1, 2, 3$, etc. In general

$$a_n = \frac{2}{T}\int_{t_0}^{t_0+T} f(t)\cos n\omega t\,\mathrm{d}t \qquad [13]$$

$$b_n = \frac{2}{T}\int_{t_0}^{t_0+T} f(t)\sin n\omega t\,\mathrm{d}t \qquad [14]$$

Fourier coefficients for rectangular waveforms

Figure 12.6 shows a periodic rectangular waveform consisting of pulses of height V and duration t_1 repeated at regular time

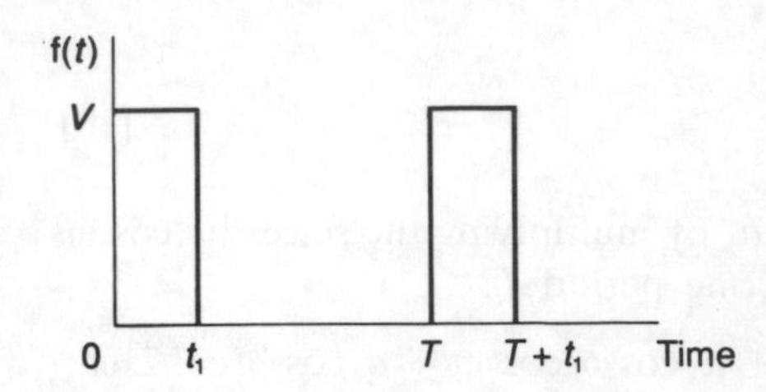

Fig. 12.6 Rectangular waveform

intervals of T. a_0 can be found by means of equation [10].

$$a_0 = \frac{1}{T}\int_{t_0}^{t_0+T} f(t)\,dt$$

If we take t_0 to be 0 then the total area under the graph between 0 and T is Vt_1. This is the value of the integral and so

$$a_0 = \frac{Vt_1}{T} \qquad [15]$$

The values of a_n can be found by means of equation [13]

$$a_n = \frac{2}{T}\int_{t_0}^{t_0+T} f(t)\cos n\omega t\,dt$$

Since $\omega = 2\pi/T$ then $n\omega = 2\pi n/T$. Since $f(t)$ has the value V only up to t_1 and then it is zero up to time T we can write

$$a_n = \frac{2}{T}\int_0^{t_1} V\cos\frac{2\pi nt}{T}\,dt$$

$$= \frac{2}{T}\left[V \times \frac{T}{2\pi n}\sin\frac{2\pi nt}{T}\right]_0^{t_1}$$

$$a_n = \frac{V}{\pi n}\sin\frac{2\pi nt_1}{T} \qquad [16]$$

The values of b_n can be found by means of equation [14].

$$b_n = \frac{2}{T}\int_{t_0}^{t_0+T} f(t)\sin n\omega t\,dt$$

Since $\omega = 2\pi/T$ then $n\omega = 2\pi n/T$. Since $f(t)$ has the value V only up to t_1 and then it is zero up to time T we can write

$$b_n = \frac{2}{T}\int_0^{t_1} V\sin\frac{2\pi nt}{T}\,dt$$

$$= -\frac{2}{T}\left[V \times \frac{T}{2\pi n}\cos\frac{2\pi nt}{T}\right]_0^{t_1}$$

$$b_n = \frac{V}{\pi n}\left(1 - \cos\frac{2\pi nt_1}{T}\right) \qquad [17]$$

The frequency spectrum is given by the use of equations [8] and [9], i.e. the amplitudes by

$$A_n = \sqrt{(a_n^2 + b_n^2)}$$

$$= \sqrt{\left[\left(\frac{V}{\pi n}\sin\frac{2\pi nt_1}{T}\right)^2 + \left(\frac{V}{\pi n}\left\{1 - \cos\frac{2\pi nt_1}{T}\right\}\right)^2\right]}$$

$$= \frac{V}{\pi n}\sqrt{\left[\sin^2\frac{2\pi nt_1}{T} + \cos^2\frac{2\pi nt_1}{T} + 1 - 2\cos\frac{2\pi nt_1}{T}\right]}$$

$$= \frac{V}{\pi n}\sqrt{\left[2 - 2\cos\frac{2\pi n t_1}{T}\right]} \qquad [18]$$

and the phases by

$$\phi_n = -\tan^{-1}\left(\frac{b_n}{a_n}\right) = -\tan^{-1}\left[\frac{\frac{V}{\pi n}\left(1 - \cos\frac{2\pi n t_1}{T}\right)}{\frac{V}{\pi n}\sin\frac{2\pi n t_1}{T}}\right]$$

$$= -\tan^{-1}\left[\frac{1 - \cos 2\pi n t_1/T}{\sin 2\pi n t_1/T}\right] \qquad [19]$$

Example 3

Determine the Fourier series coefficients for a periodic rectangular wave consisting of pulses of size V and duration half the period, as in Fig. 12.7.

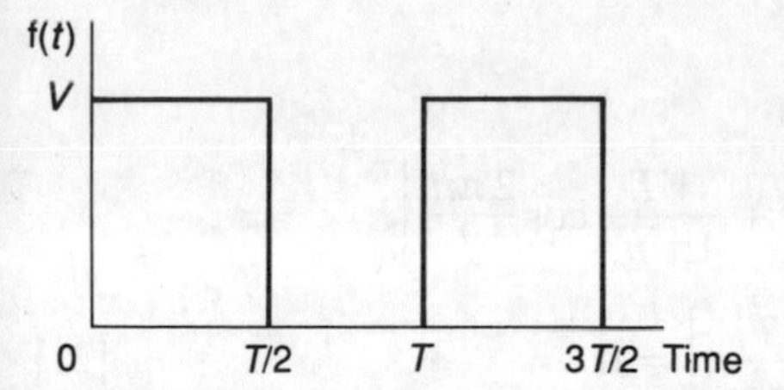

Fig. 12.7 Example 3

Answer

Putting $t_1 = \frac{1}{2}T$ in the equations [15], [16] and [17] gives the coefficients. Thus $a_0 = V/2$ (this is the average value of V over a period),

$$a_1 = \frac{V}{\pi}\sin\pi = 0$$

and

$$a_2 = \frac{V}{2\pi}\sin 2\pi = 0$$

All the a terms are zero.

$$b_1 = \frac{V}{\pi}(1 - \cos\pi) = \frac{2V}{\pi}$$

$$b_2 = \frac{V}{2\pi}(1 - \cos 2\pi) = 0$$

$$b_3 = \frac{V}{3\pi}(1 - \cos 3\pi) = \frac{2V}{3\pi}$$

Thus the series can be written as

$$f(t) = V\left(\frac{1}{2} + \frac{2}{\pi}\sin\omega t + \frac{2}{3\pi}\sin 3\omega t + \ldots\right)$$

Fourier coefficients for a sawtooth waveform

Figure 12.8 shows a sawtooth waveform. a_0 can be found by means of equation [10].

$$a_0 = \frac{1}{T}\int_{t_0}^{t_0+T} f(t)\,dt$$

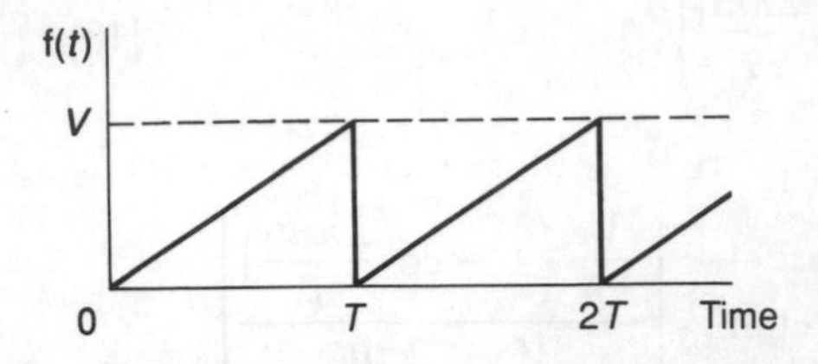

Fig. 12.8 Sawtooth waveform

If we take t_0 to be 0 then the total area under the graph between 0 and T is $\frac{1}{2}VT$. This is the value of the integral and so

$$a_0 = \tfrac{1}{2}V \qquad [20]$$

The values of a_n can be found by means of equation [13]

$$a_n = \frac{2}{T}\int_{t_0}^{t_0+T} f(t)\cos n\omega t\,dt$$

Since $\omega = 2\pi/T$ then $n\omega = 2\pi n/T$. Since $f(t) = (V/T)t$ within a period, then

$$a_n = \frac{2}{T}\int_{t_0}^{t_1} \frac{Vt}{T}\cos\frac{2\pi nt}{T}\,dt$$

Using integration by parts, i.e. the formula

$$\int u\,dv = uv - \int v\,du$$

then

$$a_n = \frac{2}{T}\left[\frac{Vt}{2\pi n}\sin\frac{2\pi nt}{T} + \frac{VT}{4\pi^2 n^2}\cos\frac{2\pi nt}{T}\right]_0^T$$

$$= \frac{2V}{T}\left[\frac{T}{4\pi^2 n^2} - \frac{T}{4\pi^2 n^2}\right] = 0 \qquad [21]$$

The values of b_n can be found by means of equation [14].

$$b_n = \frac{2}{T}\int_{t_0}^{t_0+T} f(t)\sin n\omega t\,dt$$

$$= \frac{2}{T}\int_{t_0}^{t_1} \frac{Vt}{T}\sin\frac{2\pi nt}{T}\,dt$$

Integration by parts gives

$$b_n = \frac{2}{T}\left[-\frac{Vt}{2\pi n}\cos\frac{2\pi nt}{T} + \frac{VT}{4\pi^2 n^2}\sin\frac{2\pi nt}{T}\right]_0^T$$

$$= \frac{2V}{T}\left[-\frac{T}{2\pi n}\right] = -\frac{V}{\pi n} \qquad [22]$$

The frequency spectrum is given by the use of equations [8] and [9], i.e. the amplitudes by

$$A_n = \surd(a_n^2 + b_n^2) = \frac{V}{\pi n} \qquad [23]$$

and the phases by

$$\phi_n = -\tan^{-1}\left(\frac{b_n}{a_n}\right) = \tan^{-1}\infty = 90° \qquad [24]$$

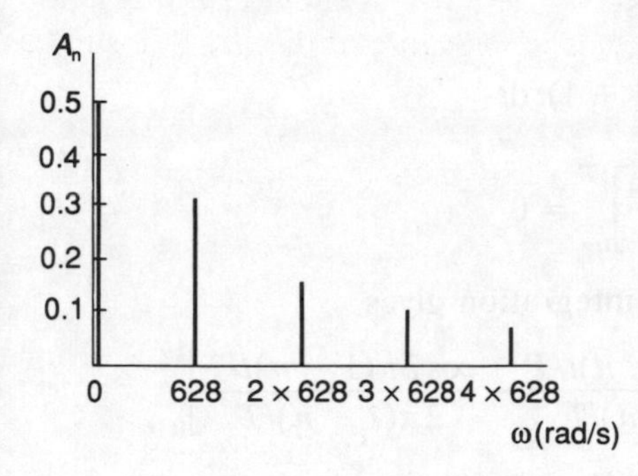

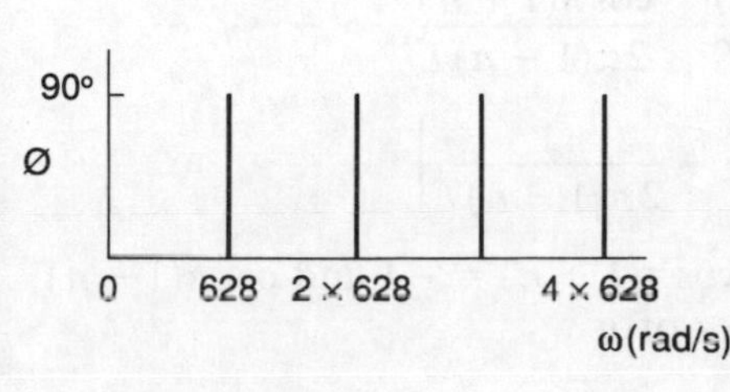

Fig. 12.9 Example 4

Example 4

Sketch the frequency spectrum for a sawtooth waveform which has a period of 10 ms and a maximum amplitude of 1 V.

Answer

Using equation [20] $A_0 = \frac{1}{2}V = 0.5\,\text{V}$. Using equation [23]

$$A_n = \frac{V}{\pi n} = \frac{1}{\pi n}$$

Hence $A_1 = 0.32\,\text{V}$, $A_2 = 0.16\,\text{V}$, $A_3 = 0.11\,\text{V}$, $A_4 = 0.08\,\text{V}$, etc. The phases are given by equation [24] as $\phi = 90°$ for all values of n. Thus the frequency spectrum will be as shown in Fig. 12.9, the fundamental frequency being $2\pi/T = 2\pi/0.010 = 628\,\text{rad/s}$.

Fourier coefficients for rectified sinusoids

Figure 12.10 shows a half-rectified sinusoidal waveform. The angular frequency of the waveform is $2\pi/T$. a_0 can be found by means of equation [10].

$$a_0 = \frac{1}{T}\int_{t_0}^{t_0+T} f(t)\,\mathrm{d}t$$

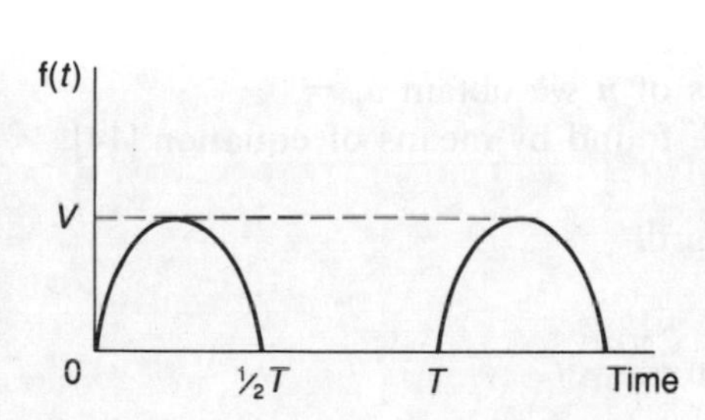

Fig. 12.10 Half-rectified sinusoid

If we take t_0 to be 0 then the total area under the graph between 0 and T is VT/π. This is the value of the integral and so

$$a_0 = \frac{V}{\pi} \qquad [25]$$

The values of a_n can be found by means of equation [13].

$$a_n = \frac{2}{T}\int_{t_0}^{t_0+T} f(t)\cos n\omega t\,\mathrm{d}t$$

Since $\omega = 2\pi/T$ then $n\omega = 2\pi n/T$. Since $f(t) = V\sin(\pi t/\frac{1}{2}T)$ between 0 and $\frac{1}{2}T$ and $f(t) = 0$ between $\frac{1}{2}T$ and T then we need only consider that part between 0 and $\frac{1}{2}T$. Thus

$$a_n = \frac{2}{T}\int_0^{\frac{1}{2}T} V\sin\frac{2\pi t}{T}\cos\frac{2\pi nt}{T}\,\mathrm{d}t$$

Since $2\sin\theta\cos\phi = \sin(\theta+\phi)\sin(\theta-\phi)$, then

$$a_n = \frac{2}{T}\int_0^{\frac{1}{2}T} \frac{V}{2}\left[\sin\frac{2\pi}{T}(1+n)t + \sin\frac{2\pi}{T}(1-n)t\right]\mathrm{d}t$$

When $n = 1$ this becomes

$$a_1 = \frac{2}{T}\int_0^{\frac{1}{2}T} \frac{V}{2}\sin\frac{2\pi}{T}(1+1)t\,\mathrm{d}t$$

$$= \frac{V}{T}\left[\frac{-\cos(4\pi t/T)}{(4\pi/T)}\right]_0^{\frac{1}{2}T} = 0 \qquad [26]$$

For $n = 2, 3, 4$, etc. the integration gives

$$a_n = \frac{V}{T}\left[-\frac{\cos 2\pi(1+n)t/T}{2\pi(1+n)/T} - \frac{\cos 2\pi(1-n)t/T}{2\pi(1-n)/T}\right]_0^{\frac{1}{2}T}$$

$$= \frac{V}{T}\left[-\frac{\cos\pi(1+n)}{2\pi(1+n)T} - \frac{\cos\pi(1-n)}{2\pi(1-n)T} + \frac{1}{2\pi(1+n)T} + \frac{1}{2\pi(1-n)T}\right]$$

For even values of n then $\cos\pi(1+n) = -1$ and $\cos\pi(1-n) = -1$. Thus for even values of n

$$a_n = \frac{V}{T}\left[\frac{2}{2\pi(1+n)T} + \frac{2}{2\pi(1-n)T}\right]$$

$$= \frac{V}{T}\left[\frac{(1+n)T-(1-n)T}{\pi(1+n)(1-n)}\right]$$

$$= \frac{2n}{\pi(1-n^2)} \qquad [27]$$

For odd values of n then $\cos\pi(1+n) = 1$ and $\cos\pi(1-n) = 1$. Thus for even values of n we obtain $a_n = 0$.

The values of b_n can be found by means of equation [14].

$$b_n = \frac{2}{T}\int_{t_0}^{t_0+T} f(t)\sin n\omega t\,\mathrm{d}t$$

$$= \frac{2}{T}\int_0^{\frac{1}{2}T} V\sin\frac{2\pi t}{T}\sin\frac{2\pi nt}{T}\mathrm{d}t$$

$$= \frac{2}{T}\int_0^{\frac{1}{2}T}\frac{V}{2}\left[\cos\frac{2\pi}{T}(1-n)t - \cos\frac{2\pi}{T}(1+n)t\right]\mathrm{d}t$$

For $n = 1$ this becomes

$$b_1 = \frac{2}{T}\int_0^{\frac{1}{2}T}\frac{V}{2}\left[1-\cos\frac{4\pi t}{T}\right]\mathrm{d}t$$

$$= \frac{V}{T}\left[t - \frac{\sin 4\pi t/T}{4\pi/T}\right]_0^{\frac{1}{2}T} = \frac{V}{2} \qquad [28]$$

For $n = 2, 3, 4$, etc. then

$$b_n = \frac{V}{T}\left[\frac{\sin 2\pi(1-n)t/T}{2\pi(1-n)/T} - \frac{\sin 2\pi(1+n)t/T}{2\pi(1+n)/T}\right]_0^{\frac{1}{2}T}$$

Since $\sin \pi(1 - n) = \sin \pi(1 + n) = 0$ then $b_n = 0$ for $n = 2, 3, 4$, etc.

The frequency spectrum is given by the use of equations [8] and [9], thus the amplitudes for even values of n are

$$A_n = \sqrt{(a_n^2 + b_n^2)} = \frac{2V}{\pi(n^2 - 1)} \qquad [29]$$

and the phases for even values of n are

$$\phi_n = -\tan^{-1}\left(\frac{b_n}{a_n}\right) = \tan^{-1} 0 = 180° \qquad [30]$$

Figure 12.11 shows the amplitude spectrum for the half-wave rectified waveform.

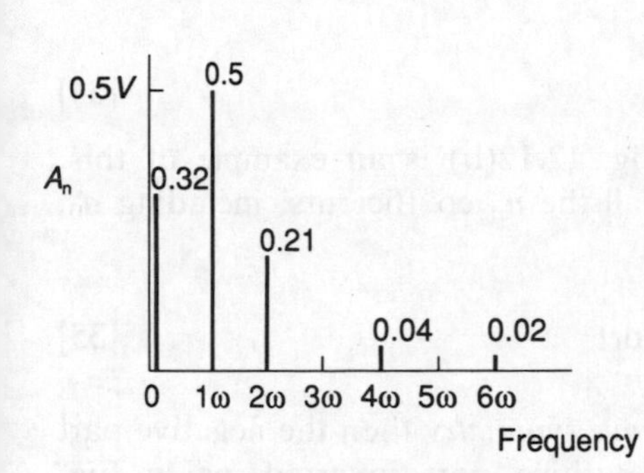

Fig. 12.11 Amplitude spectrum for half-wave rectified waveform

Symmetry

Inspection of a complex waveform can reveal symmetries which considerably simplify the determination of the Fourier coefficients. If the waveform has *even symmetry* then it is symmetrical about the $t = 0$ axis, as the $\cos \omega t$ waveform is in Fig. 12.12(*a*), and

$$f(t) = f(-t) \qquad [31]$$

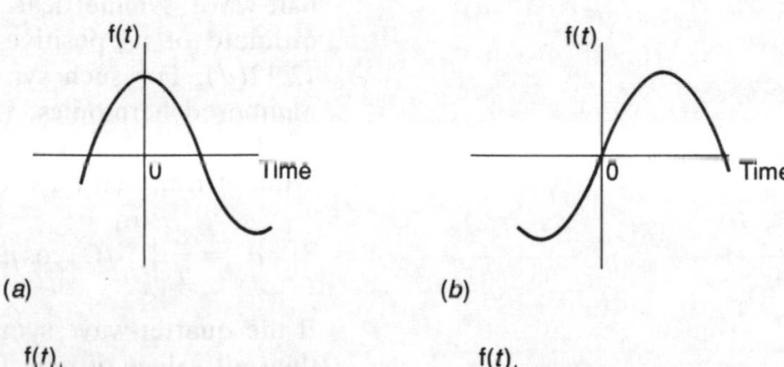

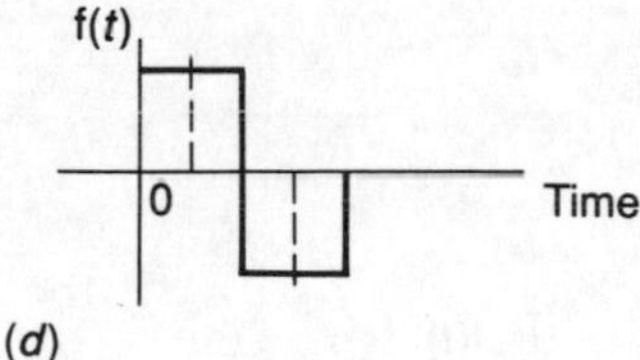

Fig. 12.12 (*a*) Even symmetry, (*b*) odd symmetry, (*c*) half-wave symmetry, (*d*) quarter-wave symmetry

Such symmetry results in all the b_n coefficients being zero, with

$$a_0 = \frac{2}{T}\int_0^{T/2} f(t)\,dt \qquad [32]$$

$$a_n = \frac{4}{T}\int_0^{T/2} f(t)\cos n\omega t\,dt \qquad [33]$$

If the waveform has *odd symmetry* then the symmetry about the $t = 0$ axis is such that

$$f(t) = -f(-t) \quad [34]$$

The $\sin \omega t$ waveform in Fig. 12.12(b) is an example of this. Such symmetry results in all the a_n coefficients, including a_0, being zero and

$$b_n = \frac{4}{T}\int_0^{T/2} f(t) \sin n\omega t \, dt \quad [35]$$

If the waveform has *half-wave symmetry* then the negative part of the cycle is just the positive part inverted, as in Fig. 12.12(*c*). For such symmetry $a_0 = 0$ and there are no even numbered harmonics, i.e. $a_n = b_n = 0$ for even values of n. For odd values of n

$$a_n = \frac{4}{T}\int_0^{T/2} f(t) \cos n\omega t \, dt \quad [36]$$

$$b_n = \frac{4}{T}\int_0^{T/2} f(t) \sin n\omega t \, dt \quad [37]$$

The term *quarter-wave symmetry* is used for a waveform that is half-wave symmetrical and also is symmetrical about mid-ordinate of its positive and negative half cycles, as in Fig. 12.12(*d*). For such symmetry $a_0 = 0$ and there are no even numbered harmonics, i.e. $a_n = b_n = 0$ for even values of n. If the quarter-wave symmetrical waveform is also even-symmetrical then all values of $b_n = 0$ and for odd values of n

$$a_n = \frac{8}{T}\int_0^{T/4} f(t) \cos n\omega t \, dt \quad [38]$$

If the quarter-wave symmetrical waveform is odd-symmetrical then all values of $a_n = 0$ and for odd values of n

$$b_n = \frac{8}{T}\int_0^{T/4} f(t) \sin n\omega t \, dt \quad [39]$$

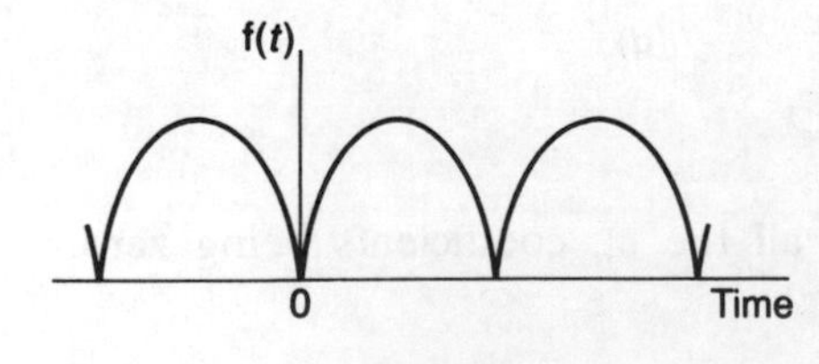

Fig. 12.13 Example 5

Example 5

What symmetries are shown by a full-wave rectified sinusoidal signal (Fig. 12.13)?

Answer

The signal shows just even symmetry, being symmetrical about the $t = 0$ axis.

Example 6

Derive the Fourier coefficients for the triangular waveform shown in Fig. 12.14.

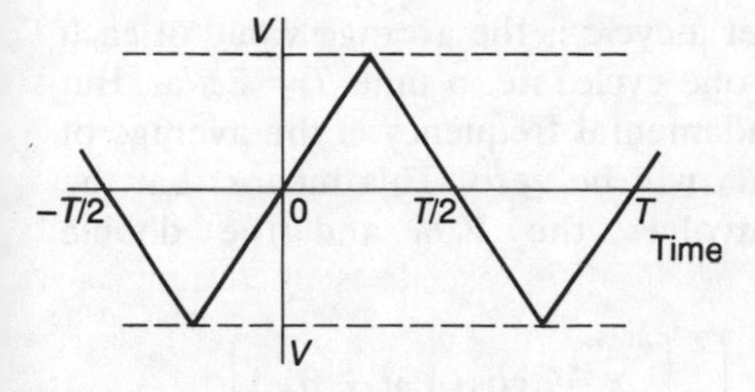

Fig. 12.14 Example 6

Answer

The waveform is odd-symmetrical, half-wave symmetrical and quarter-wave symmetrical. Thus $a_0 = 0$, $a_n = 0$ for all values of n, and $b_n = 0$ for even values of n. Because the waveform is quarter-wave symmetrical then equation [39] can be used for the odd values of n. Hence, since for the time interval 0 to $T/4$ we have $f(t) = Vt/\frac{1}{4}T = 4Vt/T$ and $\omega = 2\pi/T$,

$$b_n = \frac{8}{T}\int_0^{T/4} f(t) \sin n\omega t \, dt = \frac{8}{T}\int_0^{T/4} \frac{4Vt}{T} \sin \frac{2\pi nt}{T} dt$$

Using integration by parts

$$b_n = \frac{32V}{T^2}\left[\frac{\sin 2\pi nt/T}{(2\pi n/T)^2} - \frac{t \cos 2\pi nt/T}{2\pi n/T}\right]_0^{T/4}$$

$$= \frac{8V}{\pi^2 n^2} \sin \frac{n\pi}{2}$$

Power with complex waves

Consider an impedance for which the potential difference across it at some instant is given by (equation [5])

$$v(t) = V_0 + \sum_{n=1}^{n=\infty} V_n \cos(n\omega t + \theta_{vn})$$

where θ_{vn} are the phases of the voltage harmonics. The current through it is given by

$$i(t) = I_0 + \sum_{n=1}^{n=\infty} I_n \cos(n\omega t + \theta_{vn})$$

where θ_{in} are the phases of the current harmonics. The power at an instant is thus

$$p(t) = v(t) \times i(t)$$

$$= \left[V_0 + \sum_{n=1}^{n=\infty} V_n \cos(n\omega t + \theta_{vn})\right] \left[I_0 + \sum_{n=1}^{n=\infty} I_n \cos(n\omega t + \theta_{in})\right]$$

$$= V_0 I_0 + V_0 \sum_{n=1}^{n=\infty} I_n \cos(n\omega t + \theta_{in})$$

$$+ I_0 \sum_{n=1}^{n=\infty} V_n \cos(n\omega t + \theta_{vn})$$

$$+ \left[\sum_{n=1}^{n=\infty} V_n \cos(n\omega t + \theta_{vn})\right] \left[\sum_{n=1}^{n=\infty} I_n \cos(n\omega t + \theta_{in})\right]$$

The average power P over a cycle is the average value of each of the above terms over one cycle, i.e. a time $T = 2\pi/\omega$. But over one cycle of the fundamental frequency ω the average of any term involving $\cos n\omega$ will be zero. This means that the average power only involves the V_0I_0 and the double summation terms. Thus

$$P = \frac{1}{T}\int_0^T V_0I_0\,\mathrm{d}t + \frac{1}{T}\int_0^T \left[\sum_{n=1}^{n=\infty} V_n\cos(n\omega t + \theta_{vn})\right] \left[\sum_{n=1}^{n=\infty} I_n\cos(n\omega t + \theta_{in})\,dt\right]$$

$$= I_0V_0 + \frac{1}{T}\sum_{n=1}^{n=\infty}\int_0^T V_nI_n\cos(n\omega t + \theta_{vn})\cos(n\omega t + \theta_{in})\,dt$$

$$= I_0V_0 + \frac{1}{T}\sum_{n=1}^{n=\infty} \int_0^T \frac{V_nI_n}{2}[\cos(\theta_{vn} - \theta_{in}) - \cos(2n\omega t + \theta_{vn} + \theta_{in})\,\mathrm{d}t$$

$$= I_0V_0 + \frac{1}{T}\sum_{n=1}^{n=\infty} \left[\frac{V_nI_n}{2}\cos(\theta_{vn} - \theta_{in}) - \frac{\sin(2n\omega t + \theta_{vn} + \theta_{in})}{2n\omega}\right]_0^T$$

$$= I_0V_0 + \sum_{n=1}^{n=\infty} \tfrac{1}{2}V_nI_n\cos(\theta_{vn} - \theta_{in}) \qquad [40]$$

The first term represents the power due to the d.c. components of voltage and current and the second term represents the summation of the power due to the sinusoidal components. The second term in fact represents only the power due to sinusoidal components of the same frequency. Currents and voltages of different frequencies do not interact to produce any average power.

Since the root-mean-square value of a current is $I/\sqrt{2}$ and that of a voltage $V/\sqrt{2}$, then in terms of the root-mean-square currents and voltages equation [40] can be written as

$$P = I_0V_0 + I_{1\mathrm{rms}}V_{1\mathrm{rms}}\cos\phi_1 + I_{2\mathrm{rms}}V_{2\mathrm{rms}}\cos\phi_2 + \ldots \qquad [41]$$

where $\phi_n = \theta_{vn} - \theta_{in}$, i.e. the phase difference for a harmonic between the voltage and the current. Then the total power is the sum of the powers due to each individual harmonic.

The *power factor* of an alternating current circuit when there is a complex wave is

$$\text{power factor} = \frac{\text{total power supplied}}{\text{total volt–amperes}} \qquad [42]$$

$$= \frac{\text{total power supplied}}{\text{total r.m.s. voltage} \times \text{total r.m.s. current}}$$

$$= \frac{I_{1rms}V_{1rms}\cos\phi_1 + I_{2rms}V_{2rms}\cos\phi_2 + \ldots}{V_{rms}I_{rms}}$$

where V_{rms} is the root-mean-square value of $v(t)$ and I_{rms} is the root-mean-square value of $i(t)$. Only if the voltage and current are purely sinusoidal, i.e. there is only $n = 1$, is the power factor $\cos\phi$.

The root-mean-square current I_{rms} is the square root of the average value of i^2. Since i^2 is the sum of the squares of the harmonics, then

$$I_{rms} = \sqrt{\left[I_0^2 + \left(\frac{I_1}{\sqrt{2}}\right)^2 + \left(\frac{I_2}{\sqrt{2}}\right)^2 + \ldots\right]}$$

$$= \sqrt{\left[I_0^2 + \frac{I_1^2}{2} + \frac{I_2^2}{2} + \ldots\right]} \qquad [43]$$

Similarly,

$$V_{rms} = \sqrt{\left[V_0^2 + \frac{V_1^2}{2} + \frac{V_2^2}{2} + \ldots\right]} \qquad [44]$$

Example 7

Determine the average power delivered to a 20 Ω resistor by a square wave voltage of

$$v(t) = 30 + 54\cos(\omega t - 45°) + 38\cos(2\omega t - 90°) + 18\cos(3\omega t - 135°)\ \text{V}$$

Answer

Since the component is a resistor the voltage will be in phase with the current and $\cos\phi_n = 1$. Hence, using equation [41]

$$P = I_0V_0 + I_{1rms}V_{1rms}\cos\phi_1 + I_{2rms}V_{2rms}\cos\phi_2 + \ldots$$

$$= \frac{V_0^2}{R} + \frac{V_{1rms}^2}{R} + \frac{V_{2rms}^2}{R} + \ldots$$

$$= \frac{30^2}{20} + \frac{(54/\sqrt{2})^2}{20} + \frac{(38/\sqrt{2})^2}{20} + \frac{(18/\sqrt{2})^2}{20}$$

$$= 45 + 72.9 + 36.1 + 8.1 = 162.1\ \text{W}$$

Example 8

A complex voltage of

$$v = 20\sin\omega t + 10\sin(3\omega t + \pi/4) + 5\sin(5\omega t - \pi/2)\ \text{V}$$

is applied to a circuit and results in a current of

$$i = 0.5\sin(\omega t - \pi/6) + 0.1\sin(3\omega t - \pi/12) + 0.05\sin(5\omega t - 8\pi/12)\ \text{A}$$

What is (*a*) the total power supplied and (*b*) the overall power factor?

Answer

(*a*) The total power is the sum of the power due to each harmonic (equation [4]),

$$P = \tfrac{1}{2} \times 20 \times 0.5 \cos \pi/6 + \tfrac{1}{2} \times 10 \times 0.1 \cos(\pi/4 - (-\pi/12)) + \tfrac{1}{2} \times 5 \times 0.05 \cos(-\pi/2 - (-8\pi/12))$$
$$= 4.33 + 0.25 + 0.11 = 4.69\,\text{W}$$

(*b*) The root-mean-square current is given by equation [43] as

$$I_{\text{rms}} = \sqrt{\left[\frac{I_1^2}{2} + \frac{I_2^2}{2} + \ldots\right]}$$
$$= \sqrt{\left[\frac{0.5^2}{2} + \frac{0.1^2}{2} + \frac{0.05^2}{2}\right]} = 0.36\,\text{A}$$

The root-mean-square voltage is given by equation [44] as

$$V_{\text{rms}} = \sqrt{\left[\frac{V_1^2}{2} + \frac{V_2^2}{2} + \ldots\right]}$$
$$= \sqrt{\left[\frac{20^2}{2} + \frac{10^2}{2} + \frac{5^2}{2}\right]} = 16.2\,\text{V}$$

The overall power factor is given by equation [42] as

$$\text{power factor} = \frac{\text{total power supplied}}{\text{total r.m.s. voltage} \times \text{total r.m.s. current}}$$
$$= \frac{4.69}{16.2 \times 0.36} = 0.80$$

RLC circuits

Consider the application of a complex, single phase, voltage to circuits involving resistance, inductance and capacitance. Consider the voltage to be

$$v(t) = V_1 \cos \omega t + V_2 \cos 2\,\omega t + \ldots \qquad [45]$$

where V_1, V_2, etc. are the maximum values of the voltage.

With a pure *resistance* the current and voltage are in phase and for all values of n we have $R_n = R$. Thus for each harmonic the current is V_n/R and thus

$$i(t) = \frac{V_1}{R} \cos \omega t + \frac{V_2}{R} \cos 2\,\omega t + \ldots \qquad [46]$$

The current and voltage waveforms are thus identical in shape since all the amplitudes are just scaled by the same factor $1/R$ and the frequencies and phases are unchanged.

With a pure *inductance* the reactance $X_{Ln} = n\omega L$ and the current lags the voltage by 90°. Thus the current for each harmonic is V_n/X_{Ln} and so

$$i(t) = \frac{V_1}{\omega L}\cos(\omega t - 90°) + \frac{V_2}{2\,\omega L}\cos(2\,\omega t - 90°) + \ldots \quad [47]$$

The current waveform is thus of a different shape to the voltage waveform, the amplitudes for different values of n being scaled by different factors.

With a pure *capacitance* the reactance $X_{Cn} = 1/n\omega C$ and the current leads the voltage by 90°. The current for each harmonic is V_n/X_{Cn} and so

$$i(t) = \frac{V_1}{1/\omega C}\cos(\omega t + 90°) + \frac{V_2}{1/2\,\omega C}\cos(2\,\omega t + 90°) + \ldots \quad [48]$$

The current waveform is thus of a different shape to the voltage waveform, the amplitudes being scaled by different factors.

With a circuit consisting of a *resistance in series with an inductance* then the impedance

$$Z_n = \sqrt{[R^2 + (n\omega L)^2]}\angle \tan^{-1}(n\omega L/R)$$

The current for each harmonic is V_n/Z_n and thus

$$i(t) = \frac{V_1}{\sqrt{[R^2 + (\omega L)^2]}}\cos(\omega t - \tan^{-1}(\omega L/R)) + \frac{V_2}{\sqrt{[R^2 + (2\,\omega L)^2]}}\cos(2\,\omega t - \tan^{-1}(2\,\omega L/R)) + \ldots \quad [49]$$

The current waveform is thus of a different shape to the voltage waveform, the amplitudes being scaled by different factors and the phases also being changed.

With a circuit consisting of a *resistance in series with a capacitance* then the impedance

$$Z_n = \sqrt{[R^2 + (1/n\omega C)^2]}\angle \tan^{-1}(-1/n\omega CR)$$

The current for each harmonic is V_n/Z_n and so

$$i(t) = \frac{V_1}{\sqrt{[R^2 + (1/\omega C)^2]}}\cos(\omega t + \tan^{-1}(1/\omega CR)) + \frac{V_2}{\sqrt{[R^2 + (1/2\,\omega C)^2]}}\cos(2\,\omega t + \tan^{-1}(1\,\omega CR)) + \ldots \quad [50]$$

The current waveform is thus of a different shape to the voltage waveform, the amplitudes being scaled by different factors and the phases also being changed.

With a *series RLC circuit* then the impedance

$$Z_n = \sqrt{[R^2 + (n\omega L - 1/n\omega C)^2]} \underline{/\tan^{-1}(n\omega L - 1/n\omega C)/R}$$

The current for each harmonic is V_n/Z_n and so

$$i(t) = \frac{V_1}{\sqrt{[R^2 + (\omega L - 1/\omega C)^2]}} \cos(\omega t - \tan^{-1}(\omega L - 1/\omega C)/R) + \frac{V_2}{\sqrt{[R^2 + (2\omega L - 1/2\omega C)^2]}} \cos(2\omega t - \tan^{-1}(2\omega L - 1/2\omega C)/R) \ldots \quad [51]$$

The current waveform is thus of a different shape to the voltage waveform, the amplitudes being scaled by different factors and the phases also being changed.

Example 9

A voltage of

$$v(t) = 100 \sin 200t + 25 \sin(600t + 30°) + 5 \sin(1000t - 90°)\text{ V}$$

is connected across a coil of resistance 10 Ω and inductance 10 mH. What is the current?

Answer

The impedance Z_n is

$$Z_n = \sqrt{[R^2 + (n\omega L)^2]}\underline{/\tan^{-1}(n\omega L/R)}$$
$$= \sqrt{[10^2 + (n \times 200 \times 0.010)^2]}\underline{/\tan^{-1}(n \times 200 \times 0.010/10)}$$

The voltage waveform contains the $n = 1$, $n = 3$ and $n = 5$ harmonics. For $n = 1$ then $Z_1 = 10.2\underline{/11.3°}\,\Omega$. For $n = 3$ then $Z_3 = 11.7\underline{/31.0°}\,\Omega$. For $n = 5$ then $Z_5 = 14.1\underline{/45°}\,\Omega$. The current for each harmonic is V_n/Z_n and thus

$$i(t) = \frac{100}{10.2}\sin(200t - 11.3°) + \frac{25}{11.7}\sin(600t - 1.0°) + \frac{5}{1401}\sin(1000t - 135°)\text{ A}$$
$$= 9.80 \sin(200t - 11.3°) + 2.14 \sin(600t - 1.0°) + 0.35 \sin(1000t - 135°)\text{ A}$$

Example 10

A voltage of

$$v(t) = 100 \cos \omega t + 30 \cos(3\omega t + 60°)\text{ V}$$

is connected to the circuit shown in Fig. 12.15. What will be the current through the 500 Ω resistor if the fundamental frequency is 10^4 rad/s?

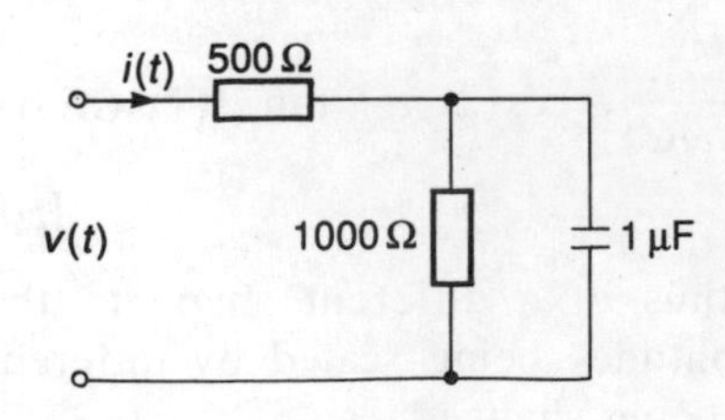

Fig. 12.15 Example 10

Answer

The circuit impedance is

$$Z_n = 500 + \frac{1000 \times (1/jn \times 10^4 \times 10^{-6})}{1000 + (1/jn \times 10^4 \times 10^{-6})}$$

When $n = 1$ then

$$Z_1 = 500 + \frac{10^5/j}{1000 + 100/j} = 500 + \frac{10^5}{100 + j1000}$$

$$= 509.9 - j99.0 = 519.4\angle -11.0°\ \Omega$$

When $n = 3$ then

$$Z_3 = 500 + \frac{0.33 \times 10^5/j}{1000 + 33.3/j} = 500 + \frac{0.33 \times 10^5}{33.3 + j1000}$$

$$= 501.1 - j33.3 = 502.2\angle -3.8°\ \Omega$$

Thus

$$i(t) = \frac{100}{519.4}\cos(\omega t + 11.0°) + \frac{30}{502.2}\cos(3\omega t + 63.8°)$$

$$= 0.193\cos(\omega t + 11.0°) + 0.060\cos(3\omega t + 63.8°)\ \text{A}$$

Selective resonance

With a series *RLC* circuit supplied with a sinusoidal voltage, resonance occurs when $\omega L = 1/\omega C$, the impedance being a minimum at this frequency and just equal to the resistance. If, however, the circuit is supplied with a complex wave having a number of harmonics, resonance can occur not only at the fundamental frequency but each of the harmonic frequencies. When resonance occurs at any frequency then the impedance at that frequency is a minimum. Thus if resonance occurs at some harmonic then that harmonic shows current magnification compared with the other harmonics. The term *selective resonance* is used to describe this effect. The effect of this is to produce a considerable distortion of the circuit current waveform. The high current at the harmonic frequency can result in dangerously high voltages appearing across the inductance or capacitance in the circuit.

The general condition for resonance at a harmonic can be stated as

$$n\omega L = 1/n\omega C \qquad [52]$$

Example 11

A series *RLC* circuit consists of a 10 Ω resistance, 0.1 H inductance and 0.1 μF capacitance and is supplied with a voltage of

$$v(t) = 100\sin\omega t + 20\sin 3\omega t\ \text{V}$$

What is the fundamental frequency if the circuit resonates at the third harmonic frequency, and the current at this resonance condition?

Answer

Using equation [52],

$$\omega^2 = \frac{1}{3^2 \times 0.1 \times 0.1 \times 10^{-6}}$$

Hence the fundamental frequency ω is 3.33 krad/s. At the third harmonic the impedance is just the resistance of 10 Ω. At the fundamental the impedance is

$$Z = \sqrt{[R^2 + (\omega L - 1/\omega C)^2]} \angle \tan^{-1}(\omega L - 1/\omega C)/R$$

$$= \sqrt{[10^2 + (3.33 \times 10^3 \times 0.1 - 1/3.33 \times 10^3 \times 0.1 \times 10^{-6})^2]}$$

$$\angle \tan^{-1}(3.33 \times 10^3 \times 0.1 - 1/3.33 \times 10^3 \times 0.1 \times 10^{-6})/10$$

$$= 2670 \angle -89.8^\circ \ \Omega$$

Thus the current is

$$i(t) = 0.037 \cos(\omega t + 89.8^\circ) + 2.0 \cos 3\omega t \text{ A}$$

Sources of harmonics

While some signals are deliberately produced in forms which have harmonics, e.g. square waves and rectified sinusoidal waves, there are circumstances where a circuit has an input of a sinusoidal waveform and the harmonics are produced as a result of the components in the circuit. Components with non-linear current/voltage characteristics, e.g. semiconductor diodes and thermistors, will produce harmonics. In general the current/voltage characteristic can be represented by the equation

$$i = a + bv + cv^2 + dv^3 + \ldots \qquad [53]$$

where a, b, c, d, etc. are constants, v is the applied voltage and i the current. With some components some of the constants may be zero. Generally the first three terms are sufficient to describe components.

Consider the current waveform resulting from a sinusoidal voltage of $v = A \sin \omega t$ with a non-linear component represented by

$$i = a + bv + cv^2$$

Then

$$i = a + bA \sin \omega t + cA^2 \sin^2 \omega t$$

But $\sin^2\theta = \frac{1}{2}(1 - \cos 2\theta)$, hence

$$i = a + bA \sin \omega t + \tfrac{1}{2}cA^2(1 - \cos 2\omega t)$$

$$= (a + \tfrac{1}{2}cA^2) + bA \sin \omega t - \tfrac{1}{2}cA^2 \cos 2\omega t$$

The current waveform has thus a d.c. component, a fundamental and a second harmonic.

Another example of a non-linear component is an iron-cored transformer. This is because the relationship between the magnetizing current and the magnetic flux set up in the core is not linear.

Example 12

Determine the current waveform produced when a sinusoidal voltage of $10 \sin \omega t$ is applied to a non-linear resistor which has a current–voltage characteristic of $i = 0.1 v^2$.

Answer

The current is

$$i = 0.1 \times 10^2 \sin^2 \omega t$$

Since $\sin^2 \theta = \frac{1}{2}(1 - \cos 2\theta)$,

$$i = 10 \times \tfrac{1}{2}(1 - \cos 2\omega t) = 5 - 5 \cos 2\omega t$$

Problems

1 An element in the frequency spectrum is $20 \cos(100t + 40°)$. What would be the equivalent terms in the Fourier series written in the form of pairs of sines and cosines?

2 What are the amplitude and phase terms in the frequency spectrum due to the terms for a particular harmonic of $10 \cos 100t$ and $20 \sin 100t$ in a Fourier series?

3 Determine the frequency spectrum for the periodic rectangular voltage waveform shown in Fig. 12.16.

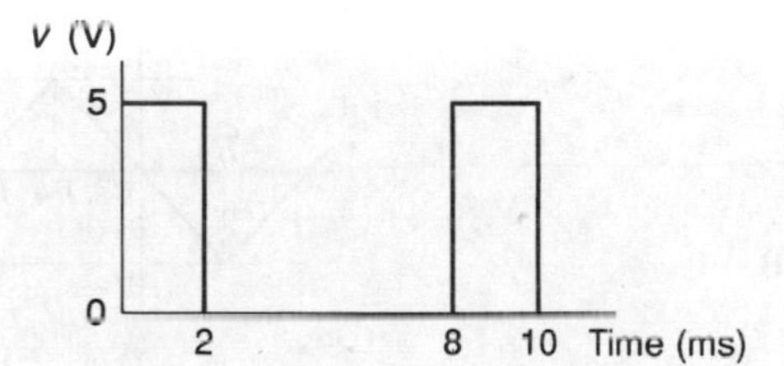

Fig. 12.16 Problem 3

4 Determine the Fourier coefficients for the periodic rectangular waveform shown in Fig. 12.17.

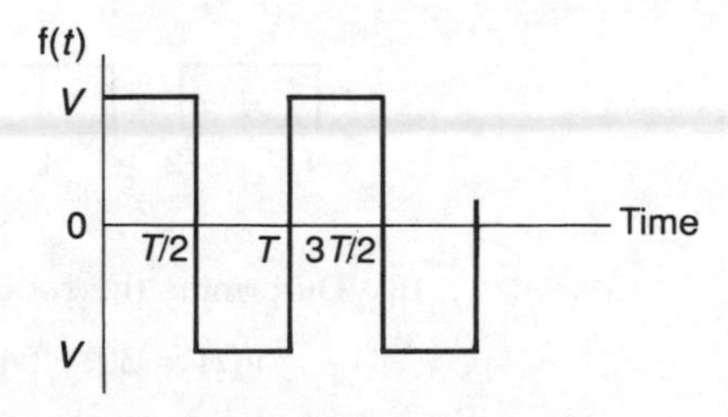

Fig. 12.17 Problem 4

5 Determine the Fourier series for the periodic square waveform shown in Fig. 12.18.

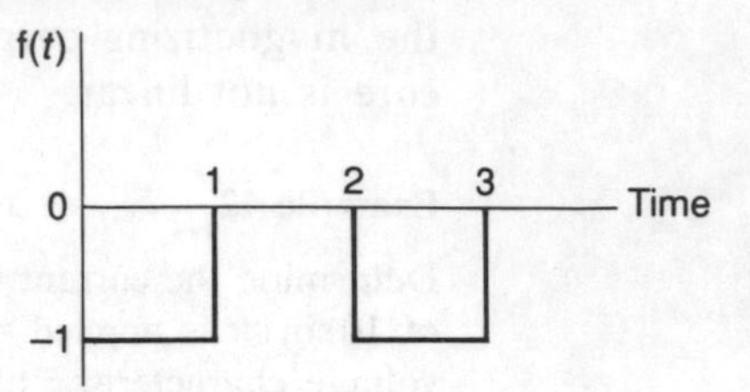

Fig. 12.18 Problem 5

6 Determine the Fourier coefficients and the frequency spectrum for a full-wave rectified sinusoidal waveform.

7 Determine the Fourier coefficients and the frequency spectrum for a triangular waveform (Fig. 12.19).

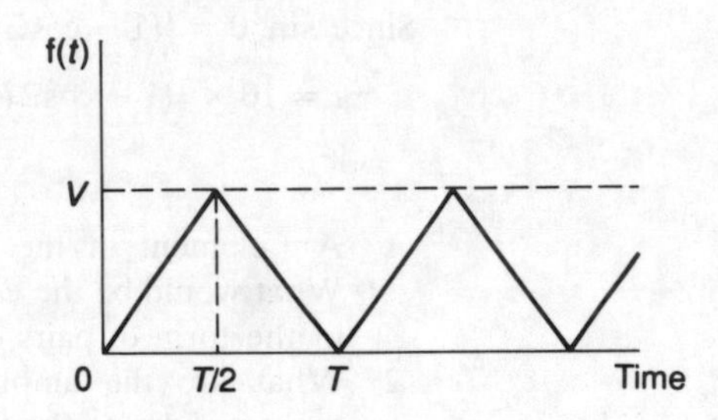

Fig. 12.19 Problem 7

8 What types of symmetry are shown by the waveform in Fig. 12.20?

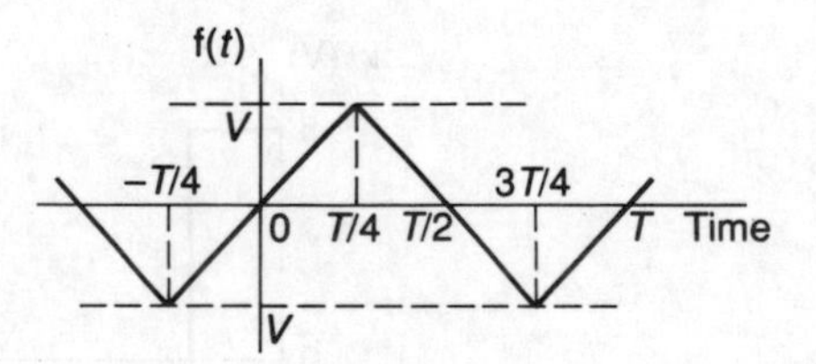

Fig. 12.20 Problem 8

9 Determine the Fourier series for the rectangular waveform shown in Fig. 12.21.

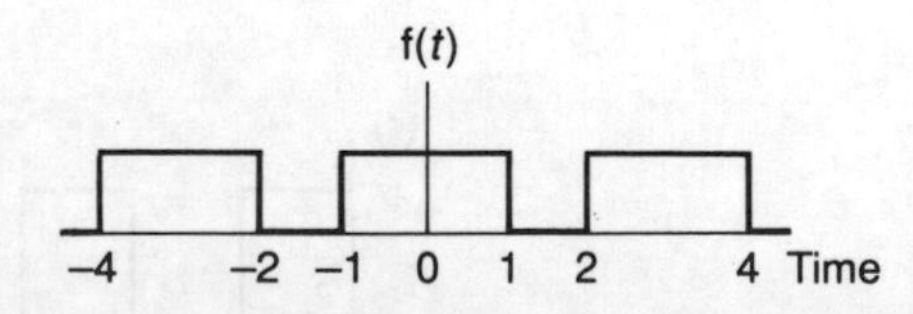

Fig. 12.21 Problem 9

10 Determine the root-mean-square voltage for the complex voltage

$$v(t) = 30 + 54\cos(\omega t - 45°) + 38\cos(2\omega t - 90°) + 11\cos(3\omega t - 135°)\ \text{V}$$

11 Determine the average power delivered to a 50 Ω resistor by a

square wave current of

$$i(t) = 1.4\cos\omega t + 0.4\cos(3\omega t - 30°) + 0.2\cos(5\omega t - 45°)\text{ A}$$

12 A complex voltage of

$$v = 75\sin\omega t + 20\sin(3\omega t - \pi/2) + 15\sin(5\omega t + \pi)\text{ V}$$

is applied to a circuit and results in a current of

$$i = 0.75\sin(\omega t - \pi/2) + 0.20\sin(3\omega t - \pi) + 0.15\sin(\omega t + \pi/2)\text{ A}$$

What is (*a*) the total power supplied and (*b*) the overall power factor?

13 A complex voltage of

$$v = 75\sin\omega t + 20\sin(3\omega t - \pi/2) + 15\sin(5\omega t + \pi)\text{ V}$$

is applied to a circuit and results in a current of

$$i = 0.75\sin(\omega t - \pi/3) + 0.20\sin(3\omega t - \pi/6) + 0.15\sin(\omega t + 2\pi/3)\text{ A}$$

What is (*a*) the total power supplied and (*b*) the overall power factor?

14 A voltage of

$$v(t) = 150\cos\omega t + 15\cos(3\omega t + 60°) + 7.5\cos(5\omega t - 30°)\text{ V}$$

is applied across a coil having a resistance of 10 Ω and an inductance of 1 mH. What will be the current if the frequency is 1 kHz?

15 A voltage of

$$v(t) = 100\cos\omega t + 30\cos(3\omega t + 60°)\text{ V}$$

is applied to a circuit consisting of a 100 Ω resistance in series with a 1 μF capacitor. What will be the current if the frequency is 1000 rad/s?

16 A voltage of

$$v(t) = 240\cos\omega t + 40\cos(3\omega t + 60°) + 30\cos(5\omega t - 30°)\text{ V}$$

is applied to a circuit consisting of a coil having a resistance of 12 Ω and an inductive reactance of 3 Ω. What will be the current?

17 A square wave voltage with a waveform the same as that shown in Fig. 12.2 is applied to a circuit consisting of a resistance *R* in series with an inductance *L*. What is the current?

18 A circuit consisting of a resistance of 1 Ω in series with a capacitor of 1 F is supplied with a half wave rectified sinusoidal voltage of maximum amplitude 1 V and period 2 s. What is the potential difference across the capacitor? Note that this circuit is a low-pass filter.

19 A sawtooth voltage with a maximum amplitude of 120 V is applied across a 10 Ω resistor. What is the average power delivered to the resistor?

20 A voltage of

$$v(t) = 150\sin\omega t + 25\sin(3\omega t - 30°) + 10\sin(5\omega t + 60°)\text{ V}$$

is applied to a circuit consisting of a resistance of 10 Ω in series with an inductance of 50 mH and a capacitor. What will be the capacitance if resonance occurs with the third harmonic and the circuit current when the fundamental frequency is 50 Hz?

21 A series *RLC* circuit consists of a 10 Ω resistance, 10 mH inductance and 0.1 μF capacitance. What is the fundamental frequency if the circuit resonates at the fifth harmonic frequency?

22 Determine the current waveform produced when a sinusoidal voltage of $10 \sin \omega t$ is applied to a non-linear resistor which has a current–voltage characteristic of $i = 2 + 1.2v + 0.5v^2$.

Answers to problems

Chapter 1

1. (*a*) 12 W, (*b*) 1440 J
2. $p = i(10 + 0.5t)$, 316.8 kJ
3. 4 A
4. (*a*) 6.3 Ω, (*b*) 0.82 Ω, (*c*) 0.55 Ω, (*d*) 4 Ω
5. (*a*) 2.7 μF, (*b*) 12 μF
6. (*a*) 8 μC on each, (*b*) 8 V, 4 V, (*c*) 32 μJ, 16 μJ
7. $i = (2/3)t^3$
8. 50 mV
9. (*a*) 12 mH, (*b*) 1.7 mH
10. 12.5 mJ
11. 9 V
12. 10 Ω

Chapter 2

1. (*a*) 6.0 mA, (*b*) 2.0 V, (*c*) 6.0 V, 48 μ*C*
2. (*a*) 20 μA, (*b*) 0, (*c*) 18.1 μA, (*d*) 1.9 V, (*e*) 7.4 μA, (*f*) 12.6 V, (*g*) 0, (*h*) 20 V
3. (*a*) 200 μ*C*, 20 V (*b*) 121 μ*C*, 12.1 V
4. (*a*) −10 V, (*b*) −0.50 mA, (*c*) 10 V, (*d*) 0.18 mA, (*e*) −3.7 V, (*f*) 3.7 V
5. (*a*) 33 mA, (*b*) 55 mA, (*c*) 100 mA
6. $i = 1(1 - e^{-3t})$
7. (*a*) 0.005 s, (*b*) 0.25 A, (*c*) 0.375 A
8. 1.4 ms
9. (*a*) $10 = 16 \times 10^{-6} d^2v_C/dt^2 + 80 \times 10^{-3} dv_C/dt + v_C$, (*b*) $4 = 1 \times 10^{-7} d^2v_C/dt^2 + 1 \times 10^{-4} dv_C/dt + v_C$, (*c*) $0 = 12 \times 10^{-6} d^2v_C/dt^2 + 8 \times 10^{-4} dv_C/dt + v_C$
10. (*a*) Over damped, (*b*) under damped, (*c*) under damped
11. 316 Ω
12. (*a*) 500 rad/s, (*b*) under damped

Chapter 3

1. (*a*) $20\angle 60°$, (*b*) $10\angle 50°$, (*c*) $5\angle 30°$, (*d*) $0.5\angle -60°$
2. (*a*) 1.73 + j1, (*b*) 6.4 + j7.7, (*c*) 2.6 − j1.5,

(*d*) −2.5 + j4.3

3. (*a*) $5.4\angle 68.2°$, (*b*) $5.1\angle 11.3°$, (*c*) $3.6\angle -33.70$, (*d*) $3.6\angle 123.7°$
4. (*a*) 5 + j7, (*b*) 8 + j1, (*c*) −3 + j1, (*d*) −7 + j7
5. (*a*) 2 + j1, (*b*) 2 − j2, (*c*) −3 + j3, (*d*) −7 − j5
6. (*a*) 4 + j8, (*b*) 0 + j29, (*c*) −18 + j1, (*d*) 14 + j2
7. (*a*) 1 − j0.5, (*b*) 0.80 + j0.65, (*c*) 0.24 − j0.90, (*d*) −0.069 + j0.83
8. (*a*) 20 Ω, $20\angle 0°$ Ω, 0.05 S, $0.05\angle 0°$ S, (*b*) j10 Ω, $10\angle 90°$ Ω, −j0.1 S, $0.1\angle -90°$, (*c*) −j50 Ω, $50\angle -90°$, j0.02 S, $0.02\angle 90°$ S
9. $125\angle 0°$ V, $v = 125 \sin 1000t$
10. $1.57\angle 45°$, $i = 1.57 \sin(100t + 45°)$
11. $0.67\angle -47.7°$, $i = 0.67 \sin(5000t - 47.7°)$
12. 50 Ω, 400 Ω
13. (*a*) 1.86 − j2.34 A = $2.99\angle -51.5°$ A, (*b*) $149.5\angle -51.5°$ V, (*c*) $187.8\angle 38.5°$ V
14. $R = 20\,\Omega$, $X_L = 5\,\Omega$.
15. (*a*) $0.108\angle 21.8°$, $i = 0.108 \sin(50t + 21.8°)$ A (*b*) $0.43\angle -71.1°$, $i = 0.43 \sin(500t - 71.1°)$ A, (*c*) $2.55\angle -10.8°$, $i = 2.55 \sin(1000t - 10.8°)$ A
16. (*a*) $0.28\angle 45°$ A, (*b*) $0.20\angle 0°$, (*c*) $0.20\angle 90°$
17. (*a*) $10\angle 0°$ A, (*b*) $70.7\angle -45°$ V, $70.7\angle 45°$ V, (*c*) $5\angle -90°$ A, $11.2\angle 26.6°$ A
18. $R = \sqrt{(L/C)}$
19. $0.146\angle 5.5°$ S
20. (*a*) 2.96 + j7.10 VA, (*b*) 2.96 W, (*c*) 7.10 VAr
21. 1000 W, 750 VAr
22. (*a*) 100.7 W, (*b*) 201.2 VAr
23. (*a*) 15 + j40 or $42.7\angle 69.4°$ Ω, (*b*) $2.81\angle -39.4°$, (*c*) $337.2\angle 69.4°$ or 118.6 + j315.6 VA, (*d*) 118.6 W, (*e*) 315.6 VAr
24. (*a*) 3.78 A, (*b*) 6.44 A, (*c*) 3.84 A, (*d*) 0.90 lagging
25. 324 μF
26. (*a*) 35.4 μF, (*b*) 2.67 A
27. 19.1 kVA, 0.88 lagging
28. 12.9 A, 0.73
29. 143 kW, 0.91
30. 112 kVAr
31. 93.9 μF
32. (*a*) 8 + j6 kVA, (*b*) 4.61 + j3.46 Ω or $5.76\angle 36.9°$ Ω
33. 6.7 μF

Chapter 4

1. 6 Ω
2. $2.0 \sin(1000t + 76°)$ A
3. $5.65\angle -45°$ A

4. $6 + j12\,\Omega$
5. $1 + j2\,\Omega$
6. $2.5\angle 5.5^\circ$
7. $2.9\angle -11^\circ$ A with parallel impedance $10 + j2\,\Omega$
8. $14\angle -76^\circ$ V in series with $2 - j3\,\Omega$
9. (*a*) 0.11 A, (*b*) 1.2 A, (*c*) 1.0 A
10. (*a*) $1.02\angle -84^\circ$ A, (*b*) $2.2\angle -13^\circ$ A
11. (*a*) 0.8 A, (*b*) 0.72 A, (*c*) 0.090 A, (*d*) 3 A
12. (*a*) $8.2\angle -14^\circ$ V, $7 + j1\,\Omega$, (*b*) $5.7\angle 8.1^\circ$ V, $7.98 - j56\,\Omega$, (*c*) $12.8\angle -38.7^\circ$ V, $3.2 + j4\,\Omega$
13. $0.20\angle 78.7^\circ$ A
14. (*a*) 1.3 V, (*b*) 1.1 V, (*c*) 18 V
15. (*a*) $0.89\angle 63.4^\circ$ A, $5 + j10\,\Omega$, (*b*) $0.5\angle 90$ A, $5 - j5\,\Omega$, (*c*) $10\angle 0^\circ$ A, $1.4 + j0.35\,\Omega$
16. $0.89\angle 63.4^\circ$ A, $10 - j20\,\Omega$
17. $8.9\angle 26.6^\circ$ V, $4 + j2\,\Omega$
18. $0.85\angle 25.6^\circ$ A
19. (*a*) 0.17 A, 0.26 A, 0.086 A, (*b*) 0.90 A, 0.34 A, 0.55 A, 0.04 A, 0.38 A, (*c*) 1.0 A, 0.30 A, 0.85 A, 0.55 A, 1.85 A, (*d*) 1.87 A, 1.63 A, 0.84 A, 0.16 A
20. (*a*) $3.5\angle 52.1^\circ$ A, (*b*) $1.1\angle -109.5^\circ$ A, (*c*) $0.35\angle -81.6^\circ$ A, (*d*) $3.23\angle 14^\circ$ A
21. (*a*) 0.17 A, 0.26 A, 0.086 A, (*b*) 0.90 A, 0.34 A, 0.55 A, 0.04 A, 0.38 A, (*c*) 1.0 A, 0.30 A, 0.85 A, 0.55 A, 1.85 A, (*d*) 1.87 A, 1.63 A, 0.84 A, 0.16 A
22. (*a*) $3.5\angle 52.1^\circ$ A, (*b*) $1.1\angle -109.5^\circ$ A, (*c*) $0.35\angle -81.6^\circ$ A, (*d*) $3.23\angle 14^\circ$ A
23. $13.3\,\Omega$
24. (*a*) $10 - j5\,\Omega$, $0.25\angle 0^\circ$ A, 0.63 W, (*b*) $5 + j5\,\Omega$, $0.32\angle -18.4^\circ$ A, 0.51 W, (*c*) $3.75 - j1.25\,\Omega$, $1.05\angle 18$ A, 4.17 W
25. $12 - j10.5\,\Omega$

Chapter 5

1. 4 V
2. 0.35
3. (*a*) $10 + j1800\,\Omega$, (*b*) $20 + j100\,\Omega$, (*c*) $120 - j600\,\Omega$, (*d*) $130 + j1200\,\Omega$, (*e*) $0.083\angle -84^\circ$ A, (*f*) $0.203\angle -72.7^\circ$ A
4. (*a*) $0.098\angle -84^\circ$ A, (*b*) $0.018\angle 74^\circ$ A
5. (*a*) 20, (*b*) 3 A
6. 9.1 A, 41.7 A
7. (*a*) $12.5\,k\Omega$, (*b*) $25 - j25\,k\Omega$
8. 1.4
9. (*a*) 1.8×10^{-3} Wb, (*b*) 400 V
10. 0.21 A, 0.45 A

11. See text
12. 2.48 Ω

Chapter 6

1. 0.63 μF
2. (*a*) 113 Hz, (*b*) 1.25 A rms, (*c*) $88\angle{-90°}$ V rms
3. (*a*) 78.0 Hz, (*b*) $25\angle{0°}$ V rms
4. $(1/\sqrt{LC})\sqrt{[R_2^2 - (L/C)]}/\sqrt{[R_3^2 - (L/C)]}$
5. 506 Hz
6. (*a*) 2251 Hz, (*b*) 177, (*c*) 2257 Hz, 2245 Hz, (*d*) 13 Hz, (*e*) 8 Ω, (*f*) 11.3 Ω
7. 63.7 mH, 99.5 nF, 314 Hz
8. (*a*) 159 nF, 39.8 mH, (*b*) 20 V
9. 4 Ω, 4 μH
10. (*a*) 159 kHz, (*b*) 20, (*c*) 7.95 kHz
11. (*a*) 25.3 nF, (*b*) 79.6 Hz
12. (*a*) 2.0×10^4 rad/s, (*b*) 20 V, (*c*) 1.9×10^4 rad/s, (*d*) 21.4 V
13. (*a*) $0.0625\angle{0°}$ A, (*b*) $0.0123\angle{78.7°}$ A
14. (*a*) 1.0×10^6 rad/s, (*b*) 0.01
15. (*a*) 1.4×10^5 rad/s, (*b*) 70.7
16. (*a*) 10 000 rad/s, (*b*) 5, (*c*) 2000 rad/s, (*d*) 9050 rad/s, 11 050 rad/s
    (Note ω_0 is not in the centre of the band because Q is small.)
17. 2.3 μF, 3.1, 323 Hz
18. (*a*) 167, (*b*) 26.6 kΩ, (*c*) 25.2 kΩ
19. $\omega_0 = \sqrt{[(CR^2 - L)/(LC^2R^2)]}$, $Q = \sqrt{[(CR^2 - L)/L]}$
20. (*a*) 100 V, (*b*) 97.5 V
21. 2.0×10^5 rad/s
22. (*a*) 6.3×10^4 rad/s, (*b*) 1.4×10^4 rad/s
23. 1.5 nF, 34.4 krad/s
24. Single humped
25. (*a*) 0.018, (*b*) 80 krad/s
26. (*a*) 0.05, (*b*) 0.0096

Chapter 7

1. 13.7 Ω, 3.7 Ω
2. (*a*) 3.0 dB, (*b*) 0.35 Np
3. 10 dB
4. 63 mA
5. 20 dB
6. (*a*) 10 dB, (*b*) 0.20 mV
7. (*a*) 161 Ω, (*b*) 12.6 dB
8. 104 Ω, 141 Ω
9. (*a*) 408 Ω, (*b*) 7.5 dB
10. 716 Ω, 1.36 kΩ
11. 733 Ω, 2.97 kΩ
12. 8.4 dB

13. 5.41 dB
14. 5.6 dB
15. See text
16. 159 Hz, 3.01 dB
17. 3.18 kHz, 1.0 kΩ
18. 0.088 μF, 64 mH
19. 22.1 nF, 15.9 mH
20. 88 nF, 16 mH
21. 22.8 dB, 30.6 dB
22. 22.8 dB
23. (*a*) 26.5 Ω, 13.2 Ω, (*b*) 27.7 Ω, 12.7 Ω, (*c*) 3.1 dB

Chapter 8

1. $h_{11} = 14\,\Omega$, $h_{21} = -2/3$, $h_{22} = 1/9\,S$, $h_{12} = 2/3$
2. $z_{11} = 7\,\Omega$, $z_{22} = 8\,\Omega$, $z_{12} = 3\,\Omega$, $z_{21} = 3\,\Omega$
3. (*a*) $\mathbf{V}_2 = 1/4$, $\mathbf{I}_1 = 5/8$, (*b*) $\mathbf{V}_1 = -3/8$, $\mathbf{V}_2 = 23/8$, (*c*) $\mathbf{I}_2 = 1/2$, $\mathbf{I}_1 = 3/2$
4. The sum of the *z* matrices for the two networks.
5. $A = 4$, $B = 20\,\Omega$, $C = 0.02\,S$, $D = 0.35$ and $AD - BC = 1$
6. (*a*) 200 Ω, (*b*) 100 Ω
7. (*a*) $50\angle 20°\,\Omega$, (*b*) $50\angle -40°\,\Omega$
8. (*a*) $A = 1$, $B = 0$, $C = 1/200\,S$, $D = 1$; (*b*) $A = 1 - j2$, $B = 400 - j400\,\Omega$, $C = 1/j100\,S$, $D = 1 - j2$; (*c*) $A = 1 + j5$, $B = j100\,\Omega$, $C = 0.1 + j0.25\,S$, $D = 1 + j5$
9. Symmetrical T with series impedances of 12 Ω and shunt admittance 1/3 S.
10. Series impedances of −j10 Ω and shunt admittance −j0.1 S
11. Series impedance 30 Ω, shunt admittance 1/30 S
12. Series resistances 200 Ω and 100 Ω, shunt resistance 300 Ω
13. $Z_0 = \sqrt{(B/C)}$

Chapter 9

1. (*a*) 157 km, (*b*) 2.5×10^4 km/s
2. (*a*) 2.6×10^5 km/s, (*b*) 0.24 rad/km
3. (*a*) 0.14 rad/km, (*b*) 44.9 km, (*c*) 2.2×10^5 km/s, (*d*) 4.5×10^{-5} s
4. $1101\angle -75.9°\,\Omega$
5. $0.076\angle 61.5°$/km
6. 58 Ω, 5.2×10^{-3} Np/m
7. 0.005 Np/m, 0.31 rad/m
8. $0.475\angle -71.6°$ V
9. (*a*) $7.9\angle -45°$ V, (*b*) $6.3\angle -90°$ V, (*c*) $5.0\angle -135°$ V
10. $0.82\angle -115°$ V, $1.6\angle -95°$ mA
11. (*a*) $1656\angle -5.8°\,\Omega$, (*b*) $8.97\angle -61.4°$ V, $5.42\angle -55.6°$ mA, (*c*) 46.9 km, (*d*) 4.7×10^5 km/s, (*e*) 1.7×10^{-5} s.

12.

Distance (km)	Voltage Magnitude (V)	Phase (rad)	Current Magnitude (MA)	Phase (rad)
0	10.0	0	10.0	$+\pi/4$
3	8.6	$-\pi/2$	8.6	$-\pi/4$
6	7.4	$-\pi$	7.4	$-3\pi/4$
9	6.4	$-3\pi/2$	6.4	$-5\pi/4$
12	5.5	-2π	5.5	$-7\pi/4$

13. $Z_0 = \sqrt{(Z_{oc}Z_{sc})} = 632\angle -5°\,\Omega$
14. (*a*) $516\angle -42.9°\,\Omega$, (*b*) $0.107\angle 46.8° = 0.0732 + j0.0780$, (*c*) 0.0732 Np/km, (*d*) 0.0780 rad/km, (*e*) 80.6 km, (*f*) 8.58×10^4 km/s, (*g*) 1.2×10^{-4} S, (*h*) 26.5 V, 0.0514 A, 12.7 V, 0.0247 A, (*i*) 0.230 W
15. 1.198 H/km
16. (*a*) 0.5, (*b*) 0.083 A, 6.7 V, (*c*) −0.042 A, 3.3 V
17. (*a*) 0, 0, (*b*) −0.33, 0.33, (*c*) 0.20 and −0.20
18. $0.38\angle 24.5°$
19. 787 Ω
20. 89.4 Ω
21. 226 Ω
22. 245 Ω
23. 1.5
24. (*a*) 1.5, (*b*) $136\angle 11.6°\,\Omega$
25. See text
26. See text and Example 17
27. 6, 14.5 + j40 Ω
28. Length 0.14λ, 0.43λ from load
29. (*a*) 8, (*b*) 2190 V, 274 V, 3.65 A, 0.46 A, (*c*) length 213 mm, 32 m from load

Chapter 10

1. 173.2 V
2. 17.3 A
3. (*a*) 11.5 A, (*b*) 115 V
4. (*a*) 20 A, (*b*) 200 V
5. (*a*) 0.80 A, (*b*) 0.80 A, (*c*) 80 V, (*d*) 139 V, (*e*) 0.80 A
6. (*a*) 0.67 A, (*b*) 0.67 A, (*c*) 100 V, (*d*) 100 V, (*e*) 1.15 A
7. (*a*) 0.385 A, (*b*) 46.2 V, (*c*) 0.385 A
8. (*a*) 289 V, (*b*) 2.89 A, (*c*) 2.5 kW
9. (*a*) 415 V, (*b*) 4.15 A, (*c*) 5.2 kW
10. 83%
11. 2.54 A
12. 28.9 A
13. 10 kW, 0.82

14. $37.3\angle -28.9^\circ$ A
15. $21.6\angle -10^\circ$ A, $49.6\angle -129^\circ$ A, $43.5\angle 77^\circ$ A
16. $10\angle 80^\circ$ A, $10\angle 200^\circ$ A
17. (*a*) $j\sqrt{3}$, (*b*) 3, (*c*) 0
18. $5.8\angle 30^\circ$ A, $5.8\angle -30^\circ$ A, 0
19. $2.0\angle 40.6^\circ$, $6.2\angle 36.3^\circ$ A, $2.5\angle 32.6^\circ$ A, $6.0\angle 40.6^\circ$ A
20. $27.3\angle 4.6^\circ$ A, $58.0\angle 43.3^\circ$ A, $19.0\angle 24.9^\circ$ A, $81.9\angle 4.6^\circ$ A
21. $0.94\angle -45^\circ$ A
22. $3460\angle 30^\circ$ A
23. $2222\angle 180^\circ$ A
24. $1925\angle 0^\circ$ V
25. $6893\angle 1.9^\circ$ V
26. Phase $89.3\angle 6.5^\circ$ kV, line $154.7\angle 6.5^\circ$ kV
27. $A = 1$, $B = 2 + j6\,\Omega$, $C = 0$, $D = 1$
28. $A = D = 0.977\angle 0.46^\circ$, $B = 63.2\angle 71.6^\circ\,\Omega$, $C = 7.9 \times 10^{-4}\angle 90^\circ$ S
29. Phase $137\angle 27^\circ$ kV, line $237\angle 27^\circ$ kV, $332\angle 26^\circ$ A, 136 MW, 24.7%
30. Phase $89.2\angle 5.3^\circ$ kV, line $155\angle 5.3^\circ$ kV, $197\angle -28^\circ$ A, 94.9%
31. $A = D = 0.99\angle 0.22^\circ$, $B = 58\angle 75^\circ\,\Omega$, $C = 0.0005\angle 90^\circ$ S
32. A and C unchanged, $B = 18\angle 21^\circ\,\Omega$, D = no significant change
33. Change from 24.7% to 6.7%.

Chapter 11

1. $1/(s + a)$
2. (*a*) $6/s$, (*b*) $(6/s)e^{-3s}$, (*c*) $6/s^2$, (*d*) $(6/s^2)e^{-3s}$, (*e*) 6, (*f*) $6e^{-3s}$, (*g*) $6 \times 100\pi/(s^2 + \{100\pi\}^2)$
3. (*a*) $1/(s + 2)$, (*b*) $5/(s + 2)$, (*c*) $V_0/(s + \{1/\tau\})$, (*d*) $2/s(s + 2)$, (*e*) $10/s(s + 2)$, (*f*) $(V_0/\tau)/s(s + \{1/\tau\})$
4. (*a*) $2e^{-3t}$, (*b*) $(2/3)e^{-t/3}$, (*c*) $(2/3)(1 - e^{-3t})$, (*d*) $2(1 - e^{t/3})$
5. (*a*) $[5 \times (1/50)]/[s(s + \{1/50\})]$, (*b*) $(10/s) + [5 \times (1/50)]/[s(s + \{1/50\})]$, (*c*) $5/[s + (1/50)]$
6. (*a*) 5, 5, (*b*) 5/2, 0
7. (*a*) $e^{2t} + 3e^{-t}$, (*b*) $4 - 2e^{-t} - 2e^{-2t}$, (*c*) $e^{-t} - e^{-2t}$
8. (*a*) 1 kΩ, (*b*) $0.5s\,\Omega$, (*c*) $1/(2 \times 10^{-6}s)\,\Omega$
9. (*a*) $100 + 0.01s\,\Omega$, (*b*) $1000 + 1/(10^{-5}s)\,\Omega$, (*c*) $1000 + 0.01s + 1/(10^{-5}s)\,\Omega$, (*d*) $s/(100 + 0.01s)\,\Omega$, (*e*) $10^{-5}s/(1 + 10^{-7}s^2)$
10. $i = (V/R)(1 - e^{-Rt/L})$
11. $v = 2 - 1.5e^{-t/4}$ V
12. $v = 6(1 - e^{-10t})$ V
13. $i = (1/25)(5 - 2e^{-3t})$ A, $v = 10e^{-3t}$ V
14. $i = kC - kL/R^2 + kt/R + (kL/R^2)e^{-Rt/L}$ A
15. $i = 0.63e^{-t}\sin 31.6t$ A

16. $v = 30 - 40\,e^{-1000t} + 10\,e^{-4000t}$ V
17. $i = 0.005\,e^{-1000t}$ A
18. $i = 0.01 \sin 2000t$ A
19. $i = 5 \times 10^{-3} + 15 \times 10^{-3} e^{-50000t}$ A
20. $i = -0.017\,e^{-400000t} + 0.017 \cos 10^6 t + 0.043 \sin 10^6 t$ A
21. $i = (390 - 0.018t)\,e^{-20000t} + 0.018 \cos 10^5 t - 0.0075 \sin 10^5 t$ A

Chapter 12

1. $15.3 \cos 100t - 12.9 \sin 100t$
2. Amplitude 22.4 and phase $-63.4°$ at $\omega = 100$ rad/s
3. d.c. 1.25 V, 785 rad/s 2.25 V, 2×785 rad/s 1.59 V, 3×785 rad/s 0.75 V; 785 rad/s $-45°$, 2×785 rad/s $-90°$, 3×785 rad/s $-135°$
4. $a_0 = 0$, $a_n = 0$, $b_n = 4V/(\pi n)$ for $n = 1, 3, 5, \ldots$
5. $-(1/2) - (2/\pi) \sin \pi t - (2/3\pi) \sin 3\pi t - (2/5\pi) \sin 5\pi t - \ldots$
6. $a_0 = A_0 = 2V/\pi$, $a_n = 4V/\pi(1 - 4n^2)$, $b_n = 0$, $A_n = 4V/\pi(4n^2 - 1)$, $\phi_n = 180°$
7. $a_0 = A_0 = V/2$, for $n = 1, 3, 5, \ldots$ $a_n = -4V/n^2\pi^2$, $b_n = 0$, for $n = 1, 3, 5, \ldots$ $A_n = 4V/n^2\pi^2$ and $\phi_n = -180°$
8. Odd, half-wave and quarter-wave
9. $0.67 + 0.28 \cos 2\pi t/3 - 0.14 \cos 4\pi t/3 + 0.07 \cos 8\pi t/3$
10. 56.0 V
11. 54 W
12. (*a*) 0, (*b*) 0
13. (*a*) 15.6 W, (*b*) 0.50
14. $12.7 \cos(\omega t - 32°) + 0.7 \cos(3\omega t - 2°) + 0.23 \cos(5\omega t - 102°)$ A
15. $0.0995 \cos(\omega t + 84.3°) + 0.0862 \cos(3\omega t + 133.3°)$ A
16. $19.4 \cos(\omega t - 14.4°) + 2.67 \cos(3\omega t + 23.1°) + 1.56 \cos(5\omega t - 81.3°)$ A
17. $V/2R + 2V/\pi\sqrt{[R^2 + (\omega L)^2]} \sin(\omega t - \theta) + 2V/3\pi\sqrt{[R^2 + (3\omega L)^2]} \sin(3\omega t - \theta) + \ldots$, with $\theta_n = \tan^{-1} n\omega L/R$
18. $0.32 + 0.15 \cos(\pi t - 162°) - 0.03 \cos(2\pi t - 81°) - 0.003 \cos(4\pi t - 85°) - \ldots$ V
19. 480 W
20. 22.5 μF, $1.2 \sin(\omega t + 85°) + 2.5 \sin(3\omega t - 30°) + 0.2 \sin(5\omega t - 19°)$ A
21. 6.3 krad/s
22. $27 + 12 \sin \omega t - 25 \cos 2\omega t$

Index